Springer Proceedings in Mathematics & Statistics

Volume 114

More information about this series at http://www.springer.com/series/10533

Springer Proceedings in Mathematics & Statistics

This book series features volumes composed of select contributions from workshops and conferences in all areas of current research in mathematics and statistics, including OR and optimization. In addition to an overall evaluation of the interest, scientific quality, and timeliness of each proposal at the hands of the publisher, individual contributions are all refereed to the high quality standards of leading journals in the field. Thus, this series provides the research community with well-edited, authoritative reports on developments in the most exciting areas of mathematical and statistical research today.

V.B. Melas • Stefania Mignani
Paola Monari • Luigi Salmaso
Editors

Topics in Statistical Simulation

Research Papers from the 7th International Workshop on Statistical Simulation

Springer

Editors
V.B. Melas
St. Petersburg State University
St. Petersburg, Russia

Paola Monari
University of Bologna
Rimini, Italy

Stefania Mignani
University of Bologna
Rimini, Italy

Luigi Salmaso
University of Padova
Padova, Italy

ISSN 2194-1009 ISSN 2194-1017 (electronic)
ISBN 978-1-4939-2103-4 ISBN 978-1-4939-2104-1 (eBook)
DOI 10.1007/978-1-4939-2104-1
Springer New York Heidelberg Dordrecht London

Library of Congress Control Number: 2014956564

Printed on acid-free paper

Springer is part of Springer Science+Business Media (www.springer.com)

Preface

This volume contains selected and peer-reviewed papers presented during the 7th International Workshop on Simulation, held in Rimini, May 21–25, 2013. This international conference is devoted to statistical techniques in stochastic simulation, data collection and analysis of scientific experiments and studies representing broad areas of interest. Since 1994, all the previous workshops took place in St.Petersburg (Russia). In 2013, for the first time, the conference took place in the Rimini Campus of the University of Bologna, in Rimini (Italy). The 7th International Workshop on Simulation was sponsored by the Unit of Rimini of the Department of Statistical Sciences of the University of Bologna in collaboration with the Department of Management and Engineering of the University of Padova, the Department of Statistical Modelling of Saint Petersburg State University and INFORMS Simulation Society. The scientific program of the meeting included 186 papers presented by a large number of scientists and experts from several countries and scientific institutes. The scientific contributions were related to several topics related, among the others, to the following issues: new methodologies for clinical trials for small population groups, modelling techniques in biostatistics and reliability, optimal fixed and random experimental sizes, experimental designs constructed by computers, simulation-based optimal design, design in computer and simulation experiments, mixture of distributions for longitudinal data, randomization, asymptotic permutation tests in heteroscedastic designs, inference for response-adaptive design, response adaptive randomization, optimal design and simulations in experimental design, queueing systems modelling, stochastic modelling in various applications, random walks and branching processes, ordered random variables and related topics, nonstandard statistical models and their application, special simulation problems, sequential nonparametric methods, Monte Carlo methods for nonlinear kinetics equations, Monte Carlo methods for vector kinetics equations, Monte Carlo methods in optical probing, numerical simulation of random fields with applications, structural change detection and analysis of complex data, simulations and computations for parametric goodness-of-fit tests in reliability and survival analysis, stochastic modelling in clinical trials, design of experiments and computing, developments in design of experiments, spatial minimax and discrimination designs, complexity in statistical modelling, simulation based Bayesian estimation of latent variable models, advances in estimation of complex latent variable models, copula methods and complex dependence, computer intensive methods and simulations for the analysis of longitudinal data, topics in multilevel models, simulation issues for modelling ordinal data, simulation tools and methods in hospital management, design of experiments: algebra, geometry and simulation, computer intensive techniques for time series analysis, some interesting and diverse applications. We wish to thank all the authors, the chair and the discussants of the sessions, the Department of Statistical Sciences of Bologna University, the Department of Management and Engineering of Padova University, the Department of Statistical Modelling of Saint Petersburg State University, the INFORMS Simulation Society and the Italia Statistical Society, which scientifically sponsored the conference, and Comune di Rimini, Provincia di Rimini, Regione Emilia-Romagna, which sponsored the conference. We also wish to thank the Rimini Campus of the University of Bologna, for the hospitality, the Organizing Committee and the Scientific Committee and all the persons who contributed to the organization of the conference, in particular Stefano Bonnini (editorial assistant for this volume) and Mariagiulia Matteucci (for the organizing work made in Rimini).

Rimini, Italy — Paola Monari
St. Petersburg, Russia — V.B. Melas
Padova, Italy — Luigi Salmaso

Contents

Contributors

Larisa Afanasyeva Lomonosov Moscow State University, Moscow, Russia

Patrizia Agati Department of Statistical Sciences "P. Fortunati", University of Bologna, Bologna, Italy

Nina Alexeyeva Saint Petersburg State University, St.Petersburg, Russia

Alexander Andronov Transport and Telecommunication institute, Riga, Latvia

Vladimir V. Anisimov Quintiles, Lothian, UK

Alessandro Barbiero Department of Economics, Management and Quantitative Methods, Università degli Studi di Milano, Milano, Italy

Luigino Barisan Department of Land, Environment, Agriculture and Forestry, University of Padova, Italy

Elena Bashtova Lomonosov Moscow State University, Moscow, Russia

Ya. Belopolskaya St.Petersburg State University for Architecture and Civil Engineering, St.Petersburg, Russia

Harish S. Bhat University of California, Merced, USA

Alexander Bitukov The Hematology Center, Main Military Clinical Hospital named after N.N.Burdenko, Moscow, Russia

Vasco Boatto Department of Land, Environment, Agriculture and Forestry, University of Padova, Italy

Stefano Bonnini Department of Economics and Management, University of Ferrara, Ferrara, Italy

Michèle Boulanger Department of International Business, Rollins College, Winter Park, FL, USA

Adriana Brogini Department of Statistics, University of Padova, Padova, Italy

Aleksandr Burmistrov Institute of Computational Mathematics and Mathematical Geophysics (Siberian Branch of the Russian Academy of Sciences), Novosibirsk, Russia; Novosibirsk State University, Novosibirsk, Russia

Virgile Caron Telecom ParisTech, Paris, France

Eleonora Carrozzo Department of Management and Engineering, University of Padova, Italy

Giorgio Celant Department of Statistics, University of Padova, Padova, Italy

Ekaterina Chimitova Novosibirsk State Technical University, Novosibirsk, Russia

Antonio Ciampi St. Mary's Research Centre, St. Mary's Hospital, Montreal, Quebec, Canada and Department of Epidemiology, Biostatistics, and Occupational Health, McGill University, Montreal, Quebec, Canada

Iulia Cichi Department of Land, Environment, Agriculture and Forestry, University of Padova, Italy

Martin G. Cole St. Mary's Research Centre, St. Mary's Hospital, Montreal, Quebec, Canada Department of Psychiatry, St. Mary's Hospital, Montreal, Quebec, Canada Department of Psychiatry, McGill University, Montreal, Quebec, Canada

Angela Dean University of Southampton, Southampton, UK

Victor Demin Novosibirsk State Technical University, Novosibirsk, Russia

Cees G.H. Diks University of Amsterdam, Center for Nonlinear Dynamics in Economics and Finance (CeNDEF), Roetersstraat 11, NL - 1018 WB Amsterdam, The Netherlands

Fabrizio Durante School of Economics and Management, Free University of Bozen-Bolzano, Bolzano, Italy

Sergej M. Ermakov St.Petersburg State University, Mathematics and Mechanics Faculty, Stary Peterhof, Russia

Sergey M. Ermakov Faculty of Mathematics and Mechanics, Department of Statistical Simulation, Saint Petersburg State University, Saint Petersburg, Russia

Pier Alda Ferrari Department of Economics, Management and Quantitative Methods, Università degli Studi di Milano, Milano, Italy

Silvano Fiorin Department of Statistical Sciences University of Padua, Padova, Italy

Roberto Fontana Politecnico di Torino, Torino, Italy

Enrico Foscolo School of Economics and Management, Free Univeristy of Bozen-Bolzano, Bozen-Bolzano, Italy

Attilio Gardini Department of Statistical Sciences, University of Bologna, Bologna, Italy

Bernard Garel University of Toulouse, INP-ENSEEIHT and Mathematical Institute of Toulouse, Toulouse, France

Albrecht Gebhardt University Klagenfurt, Klagenfurt, Austria

Rosa Arboretti Giancristofaro Department of Land, Environment, Agriculture and Forestry, University of Padova, Italy

Aenne Glass Institute for Biostatistics and Informatics in Medicine and Ageing Research, University Medicine Rostock, Rostock, Germany

Alisa A. Gorbunova Department of Applied Mathematics, Novosibirsk State Technical University, Novosibirsk, Russia

Aurea Grané Universidad Carlos III de Madrid, Getafe, Spain

Celia M.T. Greenwood Lady Davis Institute for Medical Research Centre, Jewish General Hospital, Montreal, Quebec, Canada

Sonja Hahn Department of Psychology, University of Jena, Jena, Germany

Markus Hainy Johannes Kepler University Linz, Linz, Austria

B.P. Harlamov Institute of Problems of Mechanical Engineering of RAS, Saint-Petersburg, Russia

Mark E. Johnson Department of Statistics, University of Central Florida, Orlando, FL, USA

Mark E. Johnson Department of Statistics, University of Central Florida, Orlando, FL, USA

Vanesa Jordá Department of Economics, University of Cantabria, Santander, Spain

Evgeniya Kablukova Institute of Computational Mathematics and Mathematical Geophysics SB RAS, Novosibirsk, Russia

Nina Kargapolova Institute of Computational Mathematics and Mathematical Geophysics SB RAS, prospect Akademika Lavrentjeva, Novosibirsk, Russia

Boris Kargin Institute of Computational Mathematics and Mathematical Geophysics SB RAS, Novosibirsk, Russia
Novosibirsk National Research State University, Novosibirsk, Russia

Frank Konietschke Department of Medical Statistics, University Medical Center Göttingen, Göttingen, Germany

Anna S. Korda Institute of Computational Mathematics and Mathematical Geophysics SB RAS, Novosibirsk State University, Novosibirsk, Russia

Mariya Korotchenko ICM&MG SB RAS, prospect Akademika Lavrentjeva, Novosibirsk, Russia

Dimitris Kugiumtzis Department of Electrical and Computer Engineering, Aristotle University of Thessaloniki, Thessaloniki, Greece

Guenther Kundt Institute for Biostatistics and Informatics in Medicine and Ageing Research, University Medicine Rostock, Rostock, Germany

Ekaterina Kurochkina Institute of Thermophysics SB RAS, Novosibirsk, Russia

Catherine Kyrtsou University of Macedonia, Greece; University of Strasbourg, BETA; University of Paris, Economix, ISC-Paris, Ile-de-France

Erin Leatherman West Virginia University, Morgantown, WV USA

Boris Yu. Lemeshko Department of Applied Mathematics, Novosibirsk State Technical University, Novosibirsk, Russia

Stanislav B. Lemeshko Department of Applied Mathematics, Novosibirsk State Technical University, Novosibirsk, Russia

Jan-Frederik Mai XAIA Investment GmbH, München, Germany

Luigi Malagò Dipartimento di Informatica, Università degli Studi di Milano, Milano, Italy

Marco Marozzi University of Calabria, Cosenza, Italy

Belén Martín-Barragán University of Edinburgh Business School, Edinburgh, UK

Jane McCusker St. Mary's Research Centre, St. Mary's Hospital, Montreal, Quebec, Canada and Department of Epidemiology, Biostatistics, and Occupational Health, McGill University, Montreal, Quebec, Canada

Ilya Medvedev Institute of Computational Mathematics and Mathematical Geophysics SB RAS, Novosibirsk State University, Novosibirsk, Russia

Guillaume Mijoule University of Paris, Nanterre, France

Gennady A. Mikhailov Institute of Computational Mathematics and Mathematical Geophysics SB RAS, Novosibirsk State University, Novosibirsk, Russia

Nathan Minois INSERM Unit 1027, Toulouse, France

Trifon I. Missov Max Planck Institute for Demographic Research, Rostock, Germany & University of Rostock, Rostock, Germany

Karl Moder Institute of Applied Statistics and Computing, University of Natural Resources and Life Sciences, Vienna, Austria

Paola Monari Dipartimento di Scienze Statistiche, Università di Bologna, Bologna, Italy
Department of Statistical Sciences "P. Fortunati", University of Bologna, Bologna, Italy

Werner G. Müller Johannes Kepler University Linz, Altenberger Straße, Linz, Austria

Vasily Ogorodnikov Institute of Computational Mathematics and Mathematical Geophysics SB RAS, prospect Akademika Lavrentjeva, Novosibirsk, Russia and Novosibirsk State University, Pirogova, Novosibirsk, Russia

Angeliki Papana University of Macedonia, Thessaloniki, Greece

Giovanni Pistone de Castro Statistics, Collegio Carlo Alberto, Moncalieri, Italy

Fabio Rapallo Department DISIT, Università del Piemonte Orientale, Alessandria, Italy

Maria Piera Rogantin Department DIMA, Università di Genova, Genova, Italy

Andrey P. Rogozhnikov Department of Applied Mathematics, Novosibirsk State Technical University, Novosibirsk, Russia

Giulia Roli Dipartimento di Scienze Statistiche, Università di Bologna, Bologna, Italy

Oleg Rukavitsyn The Hematology Center, Main Military Clinical Hospital named after N.N.Burdenko, Moscow, Russia

Luigi Salmaso Department of Management and Engineering, University of Padova, Italy

Diana Santalova Tartu University, Tartu, Estonia

Thomas Santner The Ohio State University, Columbus, OH, USA

José María Sarabia Department of Economics, University of Cantabria, Santander, Spain

Nicolas Savy Toulouse Institute of Mathematics, Toulouse, France

Matthias Scherer Technische Universität München, Garching–Hochbrück, Germany

Mariia Semenova Novosibirsk State Technical University, Novosibirsk, Russia

Olga Sereseva Institute of Computational Mathematics and Mathematical Geophysics SB RAS, prospect Akademika Lavrentjeva, Novosibirsk, Russia

Alexander Sipin Vologda State Pedagogical University, Vologda, Russia

Olga Soboleva Institute of Computational Mathematics and Mathematical Geophysics SB RAS, Novosibirsk, Russia

Luisa Stracqualursi Department of Statistical Sciences "P. Fortunati", University of Bologna, Bologna, Italy

Vicky Tagalakis Lady Davis Institute for Medical Research Centre, Jewish General Hospital, Montreal, Quebec, Canada

Tatiana M. Tovstik St.Petersburg State University, St.Petersburg, Russia

Egor Tsvetkov Moscow Institute of Physics and Technology, Dolgoprudny, Russia

Anna Ufliand Saint Petersburg State University, St.Petersburg, Russia

Sergey A. Ukhinov Institute of Computational Mathematics and Mathematical Geophysics SB RAS, Novosibirsk State University, Novosibirsk, Russia

Thomas W. Vastrick Forensic Document Examiner, Apopka, FL, USA

Helena Veiga Instituto Flores de Lemus and Financial Research Center/UNIDE, Universidad Carlos III de Madrid, Getafe, Spain

Helga Wagner Johannes Kepler University Linz, Altenberger Straße, Linz, Austria

Charles C. Watson, Jr. Enki Holdings LLC, Savannah, GA, USA

ChangJiang Xu Sainte-Justine University Hospital Center (CHU) Research Center, Montreal, Quebec, Canada

Igor V. Zolotukhin Russian Academy of Sciences, Institute of Oceanology, St.Petersburg, Russia

Chapter 1
Queueing Systems with Unreliable Servers in a Random Environment

Larisa Afanasyeva and Elena Bashtova

1.1 Introduction

We consider a single-server queueing system with unreliable server operating in a random environment. One would like to point out that the systems with unreliable servers have been intensively investigated for a long time now. Corresponding models are systems with interruptions of the service. This direction of research is represented by a vast collection of literature. The setting of the problems and the solutions can be found in the paper of Krishnamoorty et al. [7].

In this paper one assumes that the breakdowns of the server are connected to a certain external factor. The external environment is a regenerative stochastic process and the breakdowns of the server occur in accordance with the points of regeneration of this process. The similar model was investigated in the pioneering paper [5]. It was assumed that a breakdown can appear only if the server is occupied by a customer. The notion of completion time, which is the generalization of the service time was introduced. This notion made it possible to apply results for a queueing system $M|G|1|\infty$ with a reliable server to investigate a system subjected to interruptions, i.e. with unreliable server.

Key elements of our analysis are the coupling of renewal processes [4] based on the structure of the random environment, construction of auxiliary processes, and relations between their characteristics and characteristics of the basic process. Note that more general models were investigated in [1]. All the statements of this paper are fulfilled for the model under consideration. But here we focus on another problems. Namely, we find the stationary distribution of the virtual waiting time process and prove the limit theorem for this distribution in heavy traffic situation.

L. Afanasyeva • E. Bashtova (✉)
Lomonosov Moscow State University, Moscow, Russia
e-mail: l.g.afanaseva@yandex.ru; bashtovaelena@rambler.ru

V.B. Melas et al. (eds.), *Topics in Statistical Simulation*, Springer Proceedings in Mathematics & Statistics 114, DOI 10.1007/978-1-4939-2104-1_1

The paper is organized as follows. In the next section the queueing system and the random environment are described and basic relations are established. The ergodic condition is also given. In Sect. 1.3 Laplace–Stiltjese transform (LST) for the stationary distribution of the virtual waiting time is obtained and heavy traffic situation is investigated. Section 1.4 is devoted to two examples. The first one concerns the system operating in a Markov random environment. The model can be applied for analysis of systems with a preemptive priority discipline. The second example arose as a mathematical model of unregulated crossroads [2, 6]. The random environment is described by the number of customers in a queueing system with infinite number of servers.

1.2 Model Description: Basic Relations

A single-server queueing system with a Poisson input $A(t)$ with an intensity λ and an unreliable server is considered. In such a system the service of a customer is subjected to interruptions that are caused by a random environment $U(t)$ not depending on $A(t)$. It is assumed that $U(t)$ is a regenerative stochastic process and $\{u_j\}_{j=1}^{\infty}$ is a sequence of its regeneration periods (see, e.g., [4, 10]). Besides, we suppose that

$$u_j = u_j^{(1)} + u_j^{(2)}$$

where $u_j^{(1)}$ and $u_j^{(2)}$ are independent random variables and

$$\mathsf{P}(u_j^{(1)} \leqslant x) = 1 - e^{-\alpha x}, \quad \mathsf{P}(u_j^{(2)} \leqslant x) = G(x).$$

Let us introduce the sequences

$$s_n^{(2)} = \sum_{j=1}^{n} u_j^{(2)}, \quad s_0^{(2)} = 0, \quad s_n^{(1)} = s_{n-1}^{(2)} + u_n^{(1)}, \quad n = 1, 2, \ldots$$

so that

$$0 = s_0^{(2)} < s_1^{(1)} < s_1^{(2)}, \ldots$$

and

$$u_n^{(1)} = s_n^{(1)} - s_{n-1}^{(2)}, \quad u_n^{(2)} = s_n^{(2)} - s_n^{(1)}.$$

The breakdowns of the server occur at time moments $s_n^{(1)}$ and the server is repairing till the moments $s_n^{(2)}, n = 1, 2, \ldots$. We suppose that the service was interrupted by the breakdown of the server is continued after reconstruction from the point at which it was interrupted. Service times $\{\eta_j\}_{j=1}^{\infty}$ are independent identically distributed random variables not depending on $A(t)$ and $U(t)$.

To investigate the model we employ coupling method as it was done for more general model in [1]. Introduce the following notation

$$B(x) = \mathsf{P}(\eta_j \leqslant x),$$

$$b(s) = \int_0^{\infty} e^{-sx} dB(x), \quad g(s) = \int_0^{\infty} e^{-sx} dG(x),$$

$$b = \mathsf{E}\eta_1, \quad a = \frac{1}{\alpha}, \quad g_1 = \mathsf{E}u_1^{(2)}$$

and assume that $b < \infty$ and $g_1 < \infty$.

Let $W(t)$ be the virtual waiting time process and $q(t)$ be the number of customers in the system at a time t.

Theorem 1.1. *Processes $W(t)$ and $q(t)$ are ergodic if and only if traffic coefficient*

$$\rho = \lambda(1 + \alpha g_1)b < 1. \tag{1.1}$$

This result follows from Theorem 3 in [1]. The proof is based on the following representation that will be also applied later on.

We introduce two auxiliary processes $\tilde{W}(t)$ and $\tilde{W}^*(t)$ by the relations

$$\tilde{W}(t) = W\left(t + S_{N_1(t)}^{(2)}\right), \quad \tilde{W}^*(t) = W\left(t + S_{N_2(t)}^{(1)}\right) \tag{1.2}$$

where

$$S_k^{(i)} = \sum_{j=1}^{k} u_j^{(i)}, \quad N_i(t) = \sup\{k : S_k^{(i)} \leqslant t\}, \ i = 1, 2.$$

We see that $\tilde{W}(t)(\tilde{W}^*(t))$ is obtained from $W(t)$ by the deletion of time intervals when the server is restored (is in the working state). To express $W(t)$ by means of $\tilde{W}(t)$ and $\tilde{W}^*(t)$ we define the event

$$C_t = \bigcup_{n=0}^{\infty} \{t \in [s_n^{(2)}, s_{n+1}^{(1)})\} \tag{1.3}$$

and random variables

$$r_t = (t - S_N(t))\chi(C_t), \quad r_t^* = (t - u_{N(t)+1}^{(1)} - S_N(t))\chi(\overline{C}_t)$$

where

$$N(t) = \sup\{k : S_k \leqslant t\}, \quad S_k = S_k^{(1)} + S_k^{(2)}.$$

Then one can easily obtain from (1.2) and (1.3)

$$W(t) = \tilde{W}(r_t + S_{N(t)}^{(1)})\chi(C_t) + \tilde{W}^*(r_t^* + S_{N(t)}^{(2)})\chi(\overline{C}_t). \tag{1.4}$$

Note that the event C_t means that the server is in the working state at a time t. Let

$$V(t) = |\tilde{W}(r_t + S_{N(t)}^{(1)}) - \tilde{W}^*(r_t^* + S_{N(t)}^{(2)})|.$$

It follows from Lemma 1 in [1] that for any fixed $t \geqslant 0$

$$\lim_{y\to\infty} \mathsf{P}(\limsup_{T\to\infty} V(tT < y)) = 1. \tag{1.5}$$

This relation was employed in [1] for the proof of the ergodic theorem as well as for the asymptotic analysis of $W(t)$ and $q(t)$ in heavy traffic situation. Functional limit theorems for these processes were also established. All the statements from [1] are valid for our model but here we focus on the limit distribution

$$\Phi(x) = \lim_{t\to\infty} \mathsf{P}(W(t) \leqslant x).$$

In view of Theorem 1.1 this distribution exists if and only if $\rho < 1$. First we obtain the expression for

$$\varphi(s) = \int_0^\infty e^{-sx} d\Phi(x)$$

and then we investigate the behavior of the function $\Phi(x)$ in the heavy traffic situation ($\rho \uparrow 1$). Our proofs are based on the results for a queueing system $M|G|1|\infty$ with a reliable server (see, e.g., [8]) and relations (1.4) and (1.5).

1.3 Limit Distribution of Virtual Waiting Time

Theorem 1.2. *Let $\rho < 1$. Then the following relation takes place*

$$\varphi(s) = \tilde{\varphi}(s)\left(\frac{1}{1+\alpha g_1} + \frac{\alpha g_1}{1+\alpha g_1} v(s)\right) \tag{1.6}$$

where

$$\tilde{\varphi}(s) = (1-\rho)\left(1-(\lambda+\alpha)\frac{1-\tilde{b}(s)}{s}\right)^{-1}, \tag{1.7}$$

$$\tilde{b}(s) = \frac{\lambda}{\lambda+\alpha}b(s) + \frac{\alpha}{\lambda+\alpha}g(\lambda(1-b(s))), \tag{1.8}$$

$$v(s) = \frac{1-g(\lambda(1-b(s)))}{g_1\lambda(1-b(s))}. \tag{1.9}$$

Proof. Denote by

$$\tilde{\Phi}(x) = \lim_{t\to\infty} \mathsf{P}(\tilde{W}(t) \leqslant x), \quad \tilde{\Phi}^*(x) = \lim_{t\to\infty} \mathsf{P}(\tilde{W}^*(t) \leqslant x),$$

where $\tilde{W}(t)$ and $\tilde{W}^*(t)$ are defined by (1.2). Let $\varphi(s)$ and $\tilde{\varphi}(s)$ be LST of the distribution functions $\Phi(x)$ and $\tilde{\Phi}(x)$, respectively. It follows from (1.4) that

$$\mathsf{E}e^{-sW(t)} = \mathsf{E}e^{-s\tilde{W}(r_t+S^{(1)}_{N(t)})}\chi(C_t) + \mathsf{E}e^{-s\tilde{W}^*(r_t^*+S^{(2)}_{N(t)})}\chi(\overline{C}_t). \tag{1.10}$$

Now employing (1.5) and well-known limit theorems from the renewal theory (see, e.g., [4, 10]) one can take a limit (as $t \to \infty$) in (1.10). It gives the relation

$$\varphi(s) = \frac{1}{1+\alpha g_1}(\tilde{\varphi}(s) + \alpha g_1\tilde{\varphi}^*(s)).$$

To find $\tilde{\varphi}(s)$ we consider an auxiliary queueing system $M|G|1$ with input intensity $\tilde{\lambda} = \lambda + \alpha$ and $\tilde{b}(s)$, defined by (1.8), as the LST of the service time distribution. Since

$$-\tilde{b}'(0) = \frac{\lambda b}{\lambda+\alpha}(1+\alpha g_1)$$

the traffic coefficient of the system $\tilde{\rho} = (\lambda+\alpha)(-\tilde{b}'(0)) = \rho < 1$. Therefore the system is ergodic and LST of the limit distribution of the virtual waiting time in this system is given by (1.7) (see, e.g., [8]).

Since the input flow $A(t)$ of the initial system is a Poisson process and $u_n^{(1)}$ has an exponential distribution with a parameter α the process $\tilde{W}(t)$ is the virtual waiting time process in the auxiliary system $M|G|1$ with input intensity $\lambda+\alpha$. Besides, one can easy verify that LST of the service time distribution is defined by (1.8).

To find $\tilde{\varphi}^*(s)$ let us denote by γ a random variable with the distribution function $g_1^{-1}\int_0^x (1-G(y))dy$ not depending on the sequence $\{\eta_j\}_{j=1}^{\infty}$ as well as on input $A(t)$. Then the LST of the distribution of the total service time of customers arriving during time interval $(0,\gamma)$ is given by the relation

$$v(s) = \mathsf{E}e^{-s\sum\limits_{j=1}^{A(\gamma)}\eta_j} = \frac{1-g(\lambda(1-b(s)))}{g_1\lambda(1-b(s))}.$$

Since $A(t)$ is a Poisson process, then for any sequence $\{t_n\}_{n=1}^{\infty}$, $t_n \underset{n\to\infty}{\to} \infty$ we have

$$\lim_{n\to\infty} \mathsf{P}(\tilde{W}(t_n) \leq x) = \Phi(x).$$

In view of results from the renewal theory it means that

$$\tilde{\varphi}^*(s) = \tilde{\varphi}(s)v(s).$$

To describe the heavy traffic situation we introduce a family of queueing systems $\{S_\varepsilon\}$ with input flow $A_\varepsilon(t)$ with intensity

$$\lambda_\varepsilon = \frac{1-\varepsilon}{b(1+\alpha g_1)}$$

and traffic coefficient $\rho_\varepsilon = 1-\varepsilon \to 1 \quad \text{as } \varepsilon \to 0.$

For a system S_ε we mark by ε stochastic processes and functions introduced previously. So that $W_\varepsilon(t)$ is the virtual waiting time process for S_ε and

$$\Phi_\varepsilon(x) = \lim_{t\to\infty} \mathsf{P}(W_\varepsilon(t) \leq x).$$

Theorem 1.3. *If* $b_2 = \mathsf{E}\eta_i^2 < \infty$, $g_2 = \mathsf{E}(u_n^{(2)})^2 < \infty$, *then for any* $x > 0$

$$\lim_{\varepsilon\to 0}(1-\Phi_\varepsilon(\varepsilon x)) = e^{-\frac{2x}{\sigma^2}}$$

where

$$\sigma^2 = \frac{b_2}{b} + \frac{\alpha g_2}{(\alpha g_1+1)^2}. \tag{1.11}$$

Proof. We apply relations (1.6–1.9) to obtain the result. First we note that

$$v_\varepsilon(\varepsilon s) = \frac{1-g(\lambda_\varepsilon(1-b(\varepsilon s)))}{g_1\lambda_\varepsilon(1-b(\varepsilon s))} \to 1 \quad \text{as} \quad \varepsilon \to 0.$$

It is known (see, e.g., [3]) that for $M|G|1$ system there exists the limit

$$\tilde{\varphi}_{\varepsilon}(\varepsilon s) \to \frac{1}{1 + \frac{\tilde{b}_2}{2\tilde{b}} s}$$

where $\tilde{b}_2 = \tilde{b}''(0)$, $\tilde{b} = -\tilde{b}'(0)$. Therefore it is necessary only to find these constants from the relation (1.8).

1.4 Examples

Example 1. Here we assume that a random environment $U(t)$ is an ergodic Markov chain with the set of states $\{0, 1, 2, \dots\}$. We define the points of regenerations of $U(t)$ as the instants when $U(t)$ gets over the state zero. Then $u_n^{(1)}$ has an exponential distribution with a parameter $\alpha = 1/\mathsf{E}u_n^{(1)}$. Let $\{\pi_j\}_{j=0}^{\infty}$ be a stationary distribution of the process $U(t)$. Taking into account the equality

$$\pi_0 = \frac{\mathsf{E}u_n^{(1)}}{\mathsf{E}u_n^{(1)} + \mathsf{E}u_n^{(2)}} = \frac{1}{1 + \alpha g_1}$$

we see that a traffic coefficient of the system $M|G|1$ operating in a Markov random environment $U(t)$ is of the form

$$\rho = \frac{\lambda b}{\pi_0}.$$

Consider a birth and death Process $U(t)$. Let α_i be an intensity of birth, β_i be an intensity of death in the state i $(i = 0, 1, \dots)$, $\beta_0 = 0$. Then $U(t)$ is ergodic Markov chain if and only if [8]

$$\sum_{j=1}^{\infty} \prod_{i=1}^{j} \frac{\alpha_{i-1}}{\beta_i} < \infty. \tag{1.12}$$

In this case

$$\pi_0 = \left(1 + \sum_{j=1}^{\infty} \prod_{i=1}^{j} \frac{\alpha_{i-1}}{\beta_i}\right)^{-1}$$

so that process $W(t)$ for a system operating in the random environment $U(t)$ is ergodic if and only if

$$\lambda b\left(1+\sum_{j=1}^{\infty}\prod_{i=1}^{j}\frac{\alpha_{i-1}}{\beta_i}\right)<1. \tag{1.13}$$

Consider the case $\alpha_i = \alpha$, $\beta_i = \beta$ so that $U(t)$ is the number of customers at a time t in a system $M|M|1|\infty$. It is well known (see, e.g., [8]) that the LST of the distribution of the busy period is of the form

$$g(s) = 2\beta(s+\alpha+\beta-\sqrt{(s+\alpha+\beta)^2-4\alpha\beta})^{-1}. \tag{1.14}$$

With the help of Theorems 1.2 and 1.3 one can find the LST of the stationary distribution of the virtual waiting time process for a system $M|G|1|\infty$ in a random environment $U(t)$ and analyze its asymptotic behavior in heavy traffic situation. It is evident that we consider, indeed, a system with preemptive priority discipline.

Example 2. Let a random environment $U(t)$ be the number of customers at a time t in a queueing system $M|G|\infty$ with a Poisson input with intensity α and $F(x)$ as a distribution function of service time with finite mean c. The points of regenerations of $U(t)$ are the instants when $U(t)$ gets over the state zero. Then the nth regeneration period u_n is of the form $u_n = u_n^{(1)} + u_n^{(2)}$. Here $u_n^{(1)}$ has an exponential distribution with a parameter α, $u_n^{(2)}$ represents the nth busy period. Besides, $u_n^{(1)}$ and $u_n^{(2)}$ are independent random variables. The LST of distribution of $u_n^{(2)}$ is defined with the help of the relation (see [9])

$$g(s) = \mathsf{E}e^{-su_n^{(2)}} = 1-\frac{s\beta(s)}{\alpha(1-\beta(s))} \tag{1.15}$$

where

$$\beta(s) = \alpha\int_0^{\infty} e^{-sx-\alpha\int_0^x \overline{F}(y)dy}\,\overline{F}(x)dx, \quad \overline{F}(x) = 1-F(x).$$

One can verify that

$$g_1 = -g'(0) = \alpha^{-1}(e^{\alpha c}-1).$$

and ergodicity condition is of the form

$$\rho = \lambda b e^{\alpha c} < 1.$$

If the distribution $F(x)$ has the second moment, we may apply (1.15) to calculate $g''(0)$ and describe the asymptotic behavior of the limit distribution of $W(t)$ with the help of Theorem 1.3. But calculations and formulas are too cumbersome to give them here. Therefore we consider $M|D|\infty$ with a constant service time c. Then

$$g(s) = \frac{s+\alpha}{se^{(s+\alpha)c} + \alpha}.$$

One can easily verify that the normalizing coefficient σ^2 from Theorem 1.3 is given by the equality

$$\sigma^2 = \frac{b_2}{b} + \frac{2}{\alpha}(e^{\alpha c} - 1 - \alpha c).$$

This model can be applied for description of the number of vehicles at the intersection roads. Some results in this direction were obtained in [2].

Acknowledgements The research was partially supported by RFBR grant 13-01-00653.

References

1. Afanasyeva, L.G., Bashtova, E.E.: Coupling method for asymptotic analysis of queues with regenerative input and unreliable server. Queueing Syst. **76**(2), 125–147 (2014)
2. Afanasyeva, L.G., Rudenko, I.V.: $GI|G|\infty$ queueing systems and their applications to the analysis of traffic models. Theory Prob. Appl. **57**(3), 1–26 (2013)
3. Borovkov, A.A.: Stochastic Processes in Queuing Theory. Springer, New York (1976)
4. Cox, D.R.: Renewal Theory. Methuen and Co, London; Wiley, New York (1962)
5. Gaver, D.: A waiting line with interrupted service, including priorities. J. Roy. Stat. Soc. **13**(24), (1962)
6. Gideon, R., Pyke, R.: Markov renewal modelling of Poisson traffic at intersections having separate turn lanes. Semi-Markov Models Appl. 285–310 (1999)
7. Krishnamoorthy, A., Pramod, P.K., Chakravarthy, S.R.: Queues with interruptions: a survey (2012). Doi:10.1007/s 11750-012-02566
8. Saaty, T.L.: Elements of Queueing Theory with Applications. Mc Graw Hill, New York (1961)
9. Stadjie, W.: The busy period of the queueing system $M|G|\infty$. J. Appl. Probab. **3**(22), 697–704 (1985)
10. Thorisson, H.: Coupling, Stationary and Regeneration. Springer, New York (2000)

Chapter 2
Sequential Combining of Expert Information Using Mathematica

Patrizia Agati, Luisa Stracqualursi, and Paola Monari

2.1 Introduction

Knowledge-gaining and decision-making in real-world domains often require reasoning under uncertainty. In such contexts, combining information from several, possibly heterogeneous, sources ('experts,' such as numerical models, information systems, witnesses, stakeholders, consultants) can really enhance the accuracy and precision of the 'final' estimate of the unknown quantity (a risk, a probability, a future random event, ...).

Bayesian paradigm offers a coherent perspective from which to address the problem. It just suggests to regard experts' opinions/outputs as data from an experiment [7]: a likelihood function may be associated with them to revise the prior knowledge. A Joint Calibration Model (JCM) makes the procedure more easier to assess [1, 6]. In such a way, the information combining process just becomes a knowledge updating process.

An issue strictly related to information combining is how to perform an efficient process of sequential consulting. Investigators, indeed, often prefer to consult the experts in successive stages rather than simultaneously: so, they avoid wasting time (and money) by interviewing a number of experts that exceed what they need. At each stage, the investigator can select the 'best' expert to be consulted and choose whether to stop or continue the consulting.

The aim of this work is to rephrase Bayesian combining algorithm in a sequential context and use `Mathematica` to implement suitable selecting and stopping rules. The paper is organized as follows. Section 2.2 gives the notation and suggests a

P. Agati • L. Stracqualursi (✉) • P. Monari
Department of Statistical Sciences "P. Fortunati", University of Bologna, Via Belle Arti 41, 40126 Bologna, Italy
e-mail: patrizia.agati@unibo.it; luisa.stracqualursi@unibo.it; paola.monari@unibo.it

V.B. Melas et al. (eds.), *Topics in Statistical Simulation*, Springer Proceedings in Mathematics & Statistics 114, DOI 10.1007/978-1-4939-2104-1_2

recursive algorithm for information sequential combining, while Sect. 2.3 proposes selecting and stopping criteria for the consulting process. `Mathematica 4.1` [8] was used to develop the code implemented in the `notebook EXPS.nb` and showed in Sect. 2.4. Finally, Sect. 2.5 presents a case-study.

2.2 A Recursive Algorithm for Information Combining

Let's suppose that an investigator A is uncertain about the value of a random quantity θ and decides to consult a number of 'experts' with the aim of gaining knowledge about it. According to [7], experts' answers can be viewed as data from an experiment: a likelihood function may be associated with them and used to revise a prior judgment via Bayes' theorem. In such a way, the information combining process just becomes an information updating process.

This general principle can be applied to the aggregation of any kind of information, ranging from the combination of point estimates to the combination of probability distributions, and Bayes' rule can be implemented:

- In a 'standard' form, to be used when the experts' answers are combined with the prior all at once,
- In a recursive form, to be used when the experts are consulted sequentially and each new answer gets to update the posterior output obtained at the previous stage.

In the following, we rephrase Morris' approach in a sequential context and show both the simultaneous form and the recursive form of Bayes' rule for combining information from different sources.

Let's suppose that A's body of knowledge about θ is represented as a (possibly uninformative) probability density function (in the following, pdf) $\mathrm{h}_0(\cdot)$ on the space of states $\Theta \subseteq \Re$. Due to efficiency reasons, he chooses to consult the experts sequentially: at each stage $k(k = 1, 2, \ldots, K)$, he picks an expert $Q_j(j = 1, 2, \ldots, n)$ from a pool of size n $(n \geqslant K)$. The selected expert $Q^*_{j;k}$ (or, more briefly, Q_k) provides subject A with a pdf $g_k(\cdot)$ on the space of states. Using Bayes' theorem, the posterior pdf of the investigator A at stage k can be written as

$$h_k(\theta \,|g_1, \ldots, g_i, \ldots, g_k) \propto \ell\,(g_1, \ldots, g_i, \ldots, g_k \,|\theta) \cdot h_0(\theta) \tag{2.1}$$

where $\ell(\cdot)$ denotes the likelihood function of θ for the experimental data $\{g_1, \ldots, g_i, \ldots, g_k\}$ and the constant of proportionality is $\left[\int_\Theta \ell\,(g_1, \ldots, g_i, \ldots, g_k \,|\theta) \cdot h_0(\theta)\, d\theta\right]^{-1}$.

What makes the Bayesian approach rather difficult to apply is the assessment of the likelihood function. If the experts' answers are processed simultaneously, the function to be assessed is a joint probability distribution over the whole set of functions $g_i(\cdot)$, for $i = 1, 2, \ldots, k$; on the other hand, if the expert information are processed recursively, the likelihood is assessed at each stage as a conditional probability distribution given the subset of functions $g_i(\cdot)$ already acquired; in both the cases, the assessment requires to account for the different performances as well as the dependences between experts.

Some assumptions allow to express the likelihood function in a form easier to be modeled [7]:

i) Each $g_i(\cdot)$ is parameterized with a location parameter m_i and a shape parameter v_i. For example, $g_i(\cdot)$ denotes the pdf of a Gaussian random variable $N(m_i, v_i)$. Then, Eq. (2.1) becomes

$$h_k\left(\theta \left| m^{(k)}, v^{(k)}\right.\right) \propto \ell\left(m^{(k)} \left| v^{(k)}, \theta\right.\right) \cdot \ell\left(v^{(k)} \left| \theta\right.\right) \cdot h_0(\theta) \tag{2.2}$$

where:

- $m^{(k)}$represents the event "the location parameters supplied by the first k experts will be $m_1, \ldots, m_i, \ldots, m_k$";
- Analogously, $v^{(k)}$ indicates the event "the scale parameters supplied by the first k experts will be $v_1, \ldots, v_i, \ldots, v_k$";
- Both $\ell\left(v^{(k)} \left| \theta\right.\right)$ and $\ell\left(m^{(k)} \left| v^{(k)}, \theta\right.\right)$ are likelihood functions, to be viewed as functions of θ. The former is the likelihood function of θ for the data $v^{(k)}$: it is defined by the probabilities assigned by subject A to the event $v^{(k)}$ for θ varying. The latter, denoted in the following by $\ell_k(\theta)$ for notational convenience, is the conditioned likelihood function of θ for the data $m^{(k)}$, given the event $v^{(k)}$: it represents the joint probability—conditioned upon the event $v^{(k)}$—assigned by subject A to the event $m^{(k)}$, for θ varying;

ii) The probabilities that subject A assigns to the event $v^{(k)}$ do not depend on θ: in symbols, $\ell\left(v^{(k)} \left| \theta\right.\right) = \ell\left(v^{(k)}\right) = c$, where c denotes a constant of proportionality.[1] Using this assumption, Eq. (2.2) takes the form:

$$h_k\left(\theta \left| m^{(k)}, v^{(k)}\right.\right) \propto \ell\left(m^{(k)} \left| v^{(k)}, \theta\right.\right) \cdot h_0(\theta) \tag{2.3}$$

where the constant of proportionality is $\left[\int_\Theta \ell\left(m^{(k)} \left| \mathrm{v}^{(k)}, \theta\right.\right) \cdot h_0(\theta) d\theta\right]^{-1}$;

iii) The conditional probabilities that subject A assigns to the event "the shape parameter given by the *kth* expert will be v_k," given $m^{(k-1)}$ and $v^{(k-1)}$, do not depend on θ: in symbols, $\ell\left(v_k \left| m^{(k-1)}, v^{(k-1)}, \theta\right.\right) = \ell\left(v_k \left| m^{(k-1)}, v^{(k-1)}\right.\right)$. If such an assumption holds, then Eq. (2.3) can be written in a recursive form as

$$h_k\left(\theta \left| m^{(k)}, v^{(k)}\right.\right) \propto \ell\left(m_k \left| m^{(k-1)}, v^{(k)}, \theta\right.\right) \cdot h_{k-1}(\theta) \tag{2.4}$$

[1]Due to the reciprocity of the stochastic independence assumption *ii)* can be also expressed as *invariance to scale* about θ, that is $h\left(\theta \left| v^{(k)}\right.\right) = h(\theta)$: the event $v^{(k)}$ alone gives no information regarding θ.

where:

- $\ell\left(m_k \left| m^{(k-1)}, v^{(k)}, \theta\right.\right)$ is the conditioned likelihood function of θ for the only observation m_k, given the scale parameters of the first k experts (v_k included) and the location parameters of the first $k-1$ experts;
- The constant of proportionality is $\left[\int_\Theta \ell\left(m_k \left| m^{(k-1)}, v^{(k)}, \theta\right.\right) \cdot h_{k-1}(\theta)\right]^{-1}$.

For the purpose of assessing the likelihood $\ell_k(\theta) = \ell\left(m^{(k)} \left| v^{(k)}, \theta\right.\right)$ in Eq. (2.3), Morris introduced the notions of *performance indicator* and *performance function*.

The performance indicator τ_i associated with $g_i(\cdot)$ is defined as the cumulative distribution function $G_i(\cdot \left| m_i, v_k\right.)$ evaluated at the true value θ_0 of θ:

$$\tau_i = \tau_i(m_i, v_i, \theta_0) = \int_{-\infty}^{\theta_0} g_i(\theta \left| m_i, v_i\right.)\, d\theta = G_i(\theta_0 \left| m_i, v_i\right.) \tag{2.5}$$

where $0 \leqslant \tau_i \leqslant 1$. For example, if the observed value is the 0.3-quantile of $g_i(\cdot)$, then $\tau_i = 0.3$.

The performance function, denoted by $\varphi\left(\tau^{(k)} \left| v^{(k)}, \theta\right.\right)$, is defined as a conditional joint density on the event $\tau^{(k)}$ "the performance indicators for the first k experts will be $\tau_1, \ldots, \tau_i, \ldots, \tau_k$," given $v^{(k)}$ and θ.

Given $v^{(k)}$, for any fixed value of θ, a monotonic decreasing relationship exists between corresponding elements τ_i and m_i. So, a change of variable allows to show that:

$$\ell\left(m^{(k)} \left| v^{(k)}, \theta\right.\right) = C_k(\theta) \cdot \prod_{i=1}^{k} g_i(\theta \left| m_i, v_i\right.) \tag{2.6}$$

where:

$$C_k(\theta) = \varphi\left[[G(\theta)]^{(k)} \left| v^{(k)}, \theta\right.\right] = \varphi\left(\tau^{(k)} \left| v^{(k)}, \theta\right.\right) \tag{2.7}$$

Equation (2.6) shows that the likelihood function can be obtained as the product of the pdfs from the experts, adjusted by a *joint calibration function* $C_k(\cdot)$ that models the performance of the experts and their mutual dependence in assessing θ: $C_k(\cdot)$ is nothing but the performance function $\varphi\left(\tau^{(k)} \left| v^{(k)}, \theta\right.\right)$ viewed as a function of θ (for fixed $m^{(k)}$). It expresses the admissibility degrees assigned to each possible θ-value regarded as the realization of the event $\tau^{(k)}$.

By substituting (2.6) into (2.3), the posterior pdf can be written as:

$$h_k\left(\theta \left| m^{(k)}, v^{(k)}\right.\right) \propto C_k(\theta) \cdot \prod_{i=1}^{k} g_i(\theta \left| m_i, v_i\right.) \cdot h_0(\theta) \tag{2.8}$$

which describes the structural form of what we call "Joint Calibration Model"(JCM).

It is worth noting that Eq. (2.6) can be also used to assess the likelihood function in Eq. (2.4), since the relation $\ell\left(m_k\left|m^{(k-1)}v^{(k)},\theta\right.\right) = \ell_k(\theta)/\ell_{k-1}(\theta)$ holds.

Implementing JCM requires that function $C_k(\theta)$ is properly specified. In other words, once the scale parameters are known to subject A, a conditional pdf $\varphi\left(\tau^{(k)}\left|v^{(k)},\theta\right.\right)$ on the k-dimensional performance indicator variate $\tau^{(k)}$should be specified.

This task is less demanding if function $\varphi\left(\tau^{(k)}\left|v^{(k)},\theta\right.\right)$ can be assumed to take the same value whatever the true value of θ be (*equivariance to shift* assumption):

$$\varphi\left(\tau^{(k)}\left|v^{(k)},\theta\right.\right) = \varphi\left(\tau^{(k)}\left|v^{(k)}\right.\right) \tag{2.9}$$

However, it still remains a frustratingly difficult task, especially in the absence of an adequate parametric modelling, which would allow to assess the entire function by means of a relatively small number of parameters.

There exist several suitable choices about a parametric probabilistic model for the k-dimensional performance variate $\tau^{(k)}$. Some preliminary remarks are necessary in order to motivate our choice:

- According to definition (2.5), each element τ_i is a (cumulate) probability;
- When modelling a joint pdf $\varphi\left(\cdot\left|v^{(k)}\right.\right)$ on the variate $\tau^{(k)}$, it needs to take into account that "values [...] near 0 or 1 will ordinarily have smaller standard errors than those around 0.5. [...] A possibility is to suppose some transform of probability, like log-odds, has constant variance" [4];
- Log-odds lie in the range $-\infty$ to $+\infty$: probabilities that are less, equal or greater than 0.5 correspond to negative, zero, or positive log-odds, respectively. Therefore, modelling the performance function in terms of log-odds, instead of probabilities, is advantageous also because the range of log-odds is coherent with a Gaussian distribution, which is attractive for its good analytic properties and the clear interpretation of its parameters.

For these reasons, a reasonable choice is to assume:

$$\tilde{\tau}^{(k)} \sim N_k\left(\tilde{t}^{(k)}, S\right) \tag{2.10}$$

where

- $\tilde{\tau}^{(k)}$ refers to the k-dimensional vector of log-odds $[\tilde{\tau}_i]'_{i=1,\ldots,k}$, with $\tilde{\tau}_i = \ln\left[\tau_i/(1-\tau_i)\right] \in \Re$ for $i = 1, 2, \ldots, k$;
- $\tilde{t}^{(k)}$and S denote the mean vector and the covariance matrix of the k-variate Gaussian distribution, respectively.

The analytical form of function $\varphi\left(\tau^{(k)}\left|v^{(k)}\right.\right)$ can be obtained by using a change of variable from $\tilde{\tau}^{(k)}$ to $\tau^{(k)}$. Denoting by $\psi\left(\cdot\left|v^{(k)}\right.\right)$ the model in (2.10), the well-known change formula yields:

$$\varphi\left(\tau^{(k)}\left|v^{(k)}\right.\right) = \left|J_{\tilde{\tau}^{(k)}\rightarrow\tau^{(k)}}\right| \cdot \psi_{\tilde{\tau}^{(k)}}\left(\tau^{(k)}\left|v^{(k)}\right.\right) \tag{2.11}$$

As the Jacobian of the transformation $\tilde{\tau}^{(k)} \to \tau^{(k)}$ is:

$$J_{\tilde{\tau}^{(k)} \to \tau^{(k)}} = \prod_{i=1}^{k} \frac{1}{\tau_i (1 - \tau_i)} \tag{2.12}$$

the resulting performance function of the variate $\tau^{(k)}$ is:

$$\varphi\left(\tau^{(k)} \middle| v^{(k)}\right) = c \cdot \prod_{i=1}^{k} \frac{1}{\tau_i (1 - \tau_i)} \cdot \exp\left[-\frac{1}{2}\left(\tilde{\tau}^{(k)} - \tilde{t}^{(k)}\right)' \mathbf{S}^{-1} \left(\tilde{\tau}^{(k)} - \tilde{t}^{(k)}\right)\right] \tag{2.13}$$

where c denotes the normalization constant.

Finally, the calibration function $C_k(\theta)$, defined in (2.7), can be obtained as follows. Definition (2.5) implies that:

$$\tilde{\tau}^{(k)} = \left[\tilde{G}(\theta)\right]^{(k)} \tag{2.14}$$

By substituting Eq. (2.14) in (2.13), $C_k(\theta)$ takes the form:

$$\begin{aligned} C_k(\theta) &= \varphi\left([G(\theta)]^{(k)} \middle| v^{(k)}\right) \\ &= c \cdot \prod_{i=1}^{k} \frac{1}{G_i(\theta) \cdot [1 - G_i(\theta)]} \cdot \exp\left\{-\frac{1}{2}\left\{\left[\tilde{G}(\theta)\right]^{(k)} - \tilde{t}^{(k)}\right\}' \right. \\ &\qquad \left. \mathbf{S}^{-1} \left\{\left[\tilde{G}(\theta)\right]^{(k)} - \tilde{t}^{(k)}\right\}\right\} \end{aligned} \tag{2.15}$$

It's worth noting that the calibration function, as expressed by (2.15), is univocally defined by two parameters only: the mean vector and the covariance matrix of the variate $\tilde{\tau}^{(k)}$.

2.3 Selecting and Stopping Rules

In designing and performing the sequential process, the purpose of expert consulting is reducing the uncertainty about the unknown quantity θ. So, it is reasonable to found the selecting and stopping rules on some criterion of informativeness. More precisely, though no single number can convey the amount of information carried by a density function, a synthetic measure of the (*expected*) additional informative value of a not-yet-consulted expert $Q_{j;k}$ is indispensable for selecting the one to be consulted at stage k, especially when the assessment of the calibration parameters, together with the shape parameters provided by the experts, leads to not-coinciding preference orderings. And, analogously, as likelihood functions and

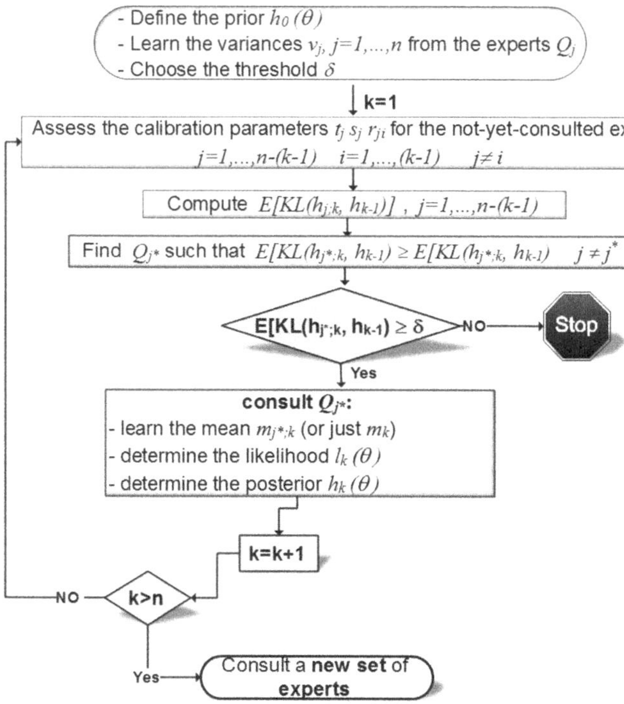

Fig. 2.1 Flow-chart of the sequential procedure with stopping rule based on the expected KL-divergences between contiguous stages

posterior densities can display a wide variety of form, a synthetic measure of the knowledge level achieved about θ is needed for picking out the 'optilmal' stage k^* at which data acquiring can be stopped.

Let's suppose the investigator A is performing the process of revising beliefs in light of new data according to the algorithm described in Sect. 2.2. The prior $h_0(\theta)$ has already been specified; each expert Q_j in the pool of size n has revealed the variance v_j—assumed as uninformative about θ: see assumption *ii)* in Sect. 2.2—of his own density $g_j(\theta)$, and A has already consulted $k-1$ of them, so obtaining the locations of $k-1$ expert densities: A is now at stage k of the process (Figs. 2.1 and 2.2), and must select one among the experts $Q_{j;k}$ not yet consulted $(j = 1, 2, \ldots, n-k+1)$.

For each $Q_{j;k}$, the investigator A assesses—conditionally on v_j, on the basis of the information at his disposal (including all the expert locations m_i revealed up to stage $k-1$)—the parameters of the k-stage calibration function $C_{j;k}(\theta)$: that is, t_j, s_{jj} and the covariances s_{ji} (or the linear correlations r_{ji}) between $Q_{j;k}$ and each expert Q_i already consulted, $i = 1, 2, \ldots, k-1$. At this point of the procedure, no $Q_{j;k}$ has revealed the location value m_j of his own $g_j(\theta)$: the several 'answers' m_j which each expert can virtually give are not all equally informative,

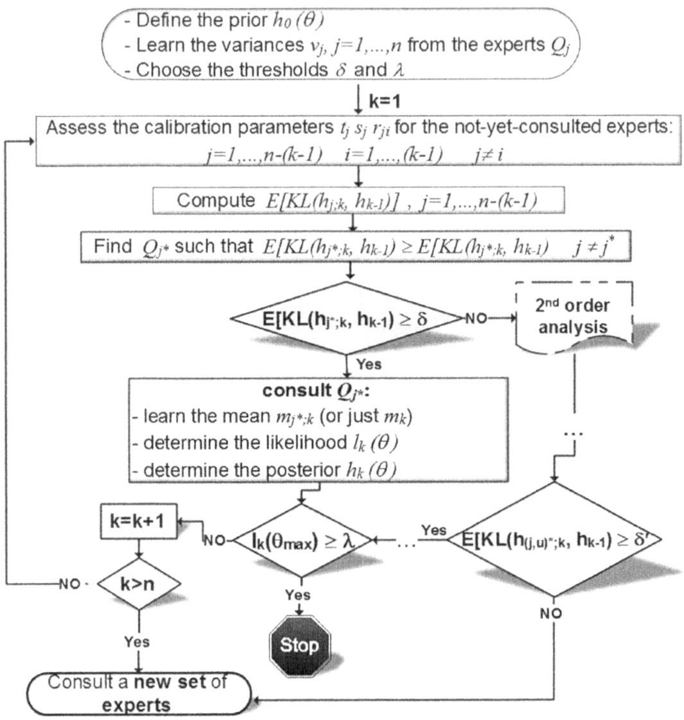

Fig. 2.2 Flow-chart of the sequential procedure with stopping rule based on the observed curvature $I_k(\theta)$ of the log-likelihood valued at $\theta := \theta_{\max}$

so the (informative) value of each expert at stage k—to be measured with regard to A's current knowledge[2] of θ expressed by the posterior density $h_{k-1}(\theta)$ of the previous stage—is an *expected* value, computed by averaging a suitable measure of relevant information about θ in $Q_{j;k}$'s answer over the space M_j of the virtually possible m_j values.

By reasoning in a mere *knowledge* context, which is an *inductive* context, where an expert opinion is more relevant the more it is able to modify the posterior distribution on the unknown quantity, a suitable measure of $Q_{j;k}$'s informative value can be the *expected Kullback–Leibler divergence* of the density $h_{j;k}(\theta)$ with respect to the posterior $h_{k-1}(\theta)$ obtained at the previous stage,

$$\mathrm{E}\left[KL\left(h_{j;k}, h_{k-1}\right)\right] := \int_{M_j} f\left(m_{j;k} | v_{j;k}, m^{(k-1)}, v^{(k-1)}\right) \cdot KL\left(h_{j;k}, h_{k-1}\right) dm_j \tag{2.16}$$

[2]In fact, all the other elements being equal, the more A is uncertain about θ, the more an answer m_j is worthy.

where the KL-divergence [3],

$$KL\left(h_{j;k}, h_{k-1}\right) := \int_{\Theta} h_{j;k}(\theta) \cdot \ln\left[h_{j;k}(\theta) / h_{k-1}(\theta)\right] d\theta \qquad (2.17)$$

measures indirectly the information provided by an answer $m_{j;k}$ in terms of the changes it yields on the density $h_{k-1}(\theta)$. The conditional density $f(\cdot)$ in (2.16) is equal to the reciprocal of the constant of proportionality of Eq. (2.4), regarded as a function of $m_{j;k}$ and normalized; when assumptions *i), ii)*, and *iii)* hold, it can be determined as

$$f\left(m_{j;k}|v_{j;k}, m^{(k-1)}, v^{(k-1)}\right) = f\left(m^{(j;k)}|, v^{(j;k)}\right) / f\left(m^{(k-1)}|v^{(k-1)}\right) \qquad (2.18)$$

where the density $f\left(m^{(j;k)}|, v^{(j;k)}\right)$, and analogously $f\left(m^{(k-1)}|v^{(k-1)}\right)$, is equal, up to the normalization term, to the reciprocal $\int_{\Theta} \ell\left(m^{(k)}|v^{(k)}, \theta\right) \cdot h_0(\theta)\, d\theta$ of the constant of proportionality of Eq. (2.2), regarded as a function of $m^{(k)}$.

The expert $Q^*_{j;k}$ characterized by the greatest expected KL-divergence is, at stage k, the most informative: but is he an expert worth consulting? The answer is yes, if the information he provides is, on average, *enough* different from what A already knows about θ, i.e. if the expected divergence of $h_{j*;k}(\theta)$ with respect to $h_{k-1}(\theta)$ is not less than a preset threshold $\delta (0 \leqslant \delta < \infty)$. About the choice of the threshold δ, a very useful tool is the scheme proposed by McCulloch [5], who suggested to connect any value δ of a KL-divergence to the KL-divergence of a Bernoulli distribution with $p = 0.5$ from a Bernoulli with $p = b(\delta) = \frac{1+\sqrt{1-e^{-2\delta}}}{2}$. Table 2.1 shows a range of correspondences.

So the *selecting rule* can be expressed as follows. *Consult the expert* $Q^*_{j;k}$ *such that*

$$\mathrm{E}\left[KL\left(h_{j*;k}, h_{k-1}\right)\right] \geqslant \mathrm{E}\left[KL\left(h_{j;k}, h_{k-1}\right)\right] \qquad j \neq j^* \qquad (2.19)$$

on condition that

$$\mathrm{E}\left[KL\left(h_{j*;k}, h_{k-1}\right)\right] \geqslant \delta \qquad (2.20)$$

If $Q^*_{j;k}$ *does not satisfy* (2.20), *then proceed to a second order analysis: that is, consult the pair* $\left(Q_{j;k}, Q_{u;k}\right)^*$ *presenting the greatest expected KL-divergence, provided that it is* $\mathrm{E}\left[KL\left(h_{(j,u)*;k}, h_{k-1}\right)\right] \geqslant \delta$.

Table 2.1 Large or small KL-divergences? Relation between δ and $b(\delta)$ values

δ	0	0.0001	0.001	0.005	0.010	0.020	0.050	0.090	0.10	0.14
$b(\delta)$	0.50	0.51	0.52	0.55	0.57	0.60	0.65	0.70	0.71	0.75
δ	0.22	0.34	0.51	0.83	1	2	>3			
$b(\delta)$	0.80	0.85	0.90	0.95	0.96	0.99	≈1.00			

Otherwise contact a new set of experts and perform a new process by using the posterior $h_{k-1}(\theta)$ *as a new prior* $h_0^{'}(\theta)$.

The expert $Q^*_{j;k}$ satisfying both the Eqs. (2.19) and (2.20) becomes just Q_k, the "k-stage expert." By consulting him, A learns the location m_k of the density $g_k(\cdot)$: now, the k-stage calibration function $C_k(\theta)$ is univocally defined, and consequently, the likelihood function $\ell_k(\theta)$ and the posterior density $h_k(\theta)$ too.

It is intuitive, as well as reasonable, that the investigator stops the process only when the knowledge about θ, expressed by the posterior density, is 'inertially stable': i.e., only when additional experts, even if jointly considered, are not able to modify the synthesis distribution appreciably, on the contrary they contribute to its inertiality. So, the *stopping rule* can be defined as follows. *Stop the consulting process at stage* k^* *at which none of the remaining experts satisfies condition (2.20).*

If too many experts are needed for realizing such a stopping condition, it can be weakened by just requiring the knowledge about θ deriving from expert answers to be enough for A's purposes. A measure of the experimental data strength in determining a preference ordering among 'infinitesimally close' values of θ is Fisher's notion of information. The value of the *observed information* $I(\cdot)$ at the maximum of the log-likelihood function

$$I_k(\theta_{\max}) := -\partial^2/\partial\theta^2 \ln \ell_k(\theta_{\max}) \qquad (2.21)$$

is a second-order estimate of the spherical curvature of the function at its maximum: within a second-order approximation, it corresponds to the KL-divergence between two distributions that belong to the same parametric family and differ infinitesimally over the parameter space.

So, an alternative *stopping rule* can be defined as follows. *Stop the consulting at stage* k^* *at which a preset observed curvature* λ *of the log-likelihood valued at* $\theta := \theta_{\max}$ *has been achieved,*

$$I_{k^*}(\theta_{\max}) \geq \lambda \qquad (2.22)$$

In order to decide whether a curvature value $I(\theta_{\max}) = \lambda$ is a large or a small one, a device could be the following. Let's think of a binomial experiment where a number $x = n/2$ of successes is observed in n trials and find x such that $I(\hat{p}_{\text{ML}} = 0.5) = \lambda$, where $\hat{p}_{\text{ML}} = 0.5$ is the maximum likelihood estimate of the binomial parameter p. Table 2.2 shows a range of x values with the corresponding curvature values. The simple relation $x = \lambda/8$ holds: so, for example, if $\lambda = 120$, the width of the curve $\ln \ell_k(\theta)$ near $\theta := \theta_{\max}$ is the same as the curve $\ln \ell(p)$ at $\hat{p}_{\text{ML}} = 0.5$ when $x = 15$ and $n = 30$.

Table 2.2 Large or small curvature values? Relation between x and λ values

x	1	2	5	10	15	20	25	30	40	50
λ	8	16	40	80	120	160	200	240	320	400

2.4 Selecting and Stopping Rules Implemented: A Mathematica Code

The procedure implemented in the notebook file EXPS.nb allows to select the 'best' expert to consult at each stage of a sequential process and decide the 'best' stage at which the process can be stopped. The program computes the expected KL-divergencies between posterior densities at two contiguous stages, for any number of experts: it uses condition (2.20) as stop criterion.

The choice of Mathematica package is due to complexity and accuracy necessary in this recursive procedure [2]. The code begins by importing package Statistics 'NormalDistribution' and opening declarations.

```
<<Statistics`NormalDistribution`;
PDF[NormalDistribution[mu_,sigma_],x_]:=1/(sigma*Sqrt[2 Pi])
                                            Exp[-((x-mu)/sigma)2/2];
CDF[NormalDistribution[mu_,sigma_],x_]:=(Erf[(x-mu)/(Sqrt[2]sigma)]
                                            +1)/2;
```

The program needs the following input quantities:

- '*Initialdata*': it is the matrix which contains the mean m_j, the variance v_j, the calibration parameters t_j and s_j^2 of each expert Q_j $(j = 1, \dots, n)$;
- '*R*': it is the correlation matrix. Its elements are the calibration parameters r_{ji}, denoting the correlation degree between the performances of the experts Q_i and Q_j;
- '*delta*': it is the threshold for the Kullback–Leibler divergence;
- '*v0, m0*': they are the variance and the mean of the prior $h_0(\theta)$;
- '*mmin, mmax*': they are the lower and the upper limits for the unknown θ;
- '*stagemax*': it is the size n of the pool of experts.

The code restores the input matrices: the process starts out.

```
numexpert=Dimensions[Initialdata][[2]];
expert=Table[k,{k,numexpert}];
Matdata=Transpose[Join[{exp,m,v,t,s2},
        Transpose[Join[{expert},Initialdata]]]];
{rin,sin}=Dimensions[Matdata];
 M=Initialdata[[1]];V=Initialdata[[2]];
 T=Initialdata[[3]];S=Initialdata[[4]];
 Cova=Table[Sqrt[S[[i]]*S[[j]]]*R[[i,j]],{i,numexpert},{j,numexpert}];

nextmax=Table[0,{i,stagemax}];
next[n_]:=Table[nextmax[[j]],{j,n-1}];
nextadd[n_,k_]:=Join[next[n],{k}];
mattV[n_]:=Table[V[[nextadd[n,k]]],{k,numexpert}];
mattT[n_]:=Table[T[[nextincr[n,k]]],{k,numexpert}];
mattM[n_]:=Table[M[[nextadd[n,k]]],{k,numexpert}];
mattS[n_]:=Table[S[[nextadd[n,k]]],{k,numexpert}];
mattCova[n_]:=Table[Cova[[nextadd[n,k],nextadd[n,k]]],{k,numexpert}];
Covamin[n_]:=Table[Cova[[next[n],next[n]]]];
nextH=Table[0,{i,stagemax}];
```

```
g[x_,i_]:=PDF[NormalDistribution[0,Sqrt[mattV[n][[k]][[i]]]],x]
G[x_,i_]:=CDF[NormalDistribution[0,Sqrt[mattV[n][[k]][[i]]]],x]
g0[x_]:=PDF[NormalDistribution[m0,Sqrt[v0]],x]

MatrixForm[Matdata]
MatrixForm[R]
```

The code calculates the expected KL-divergences at stage 1 and put them in a vector.

```
Expect1:=(n=1; Print["Stage K ="," " , ];
    Kull=Table[0,{k,numexpert}];
    For[k=1,k<=numexpert,k++,Kull[[k]]=
        NIntegrate[(F1[m]-H1[m]*Log[H1[m]])/K1[m],{m,-5,8}]/
        NIntegrate[H1[m]/K1[m],{m,-5,8}];
        Print[Kull[[k]] ];
        ];
    MatKull=Join[Matdata,{Join[{K},Kull]}];
    max=Position[Kull,Max[Kull]][[1,1]];k=max;H11=H1[M[[max]]];
    nextmax[[1]]=max;nextH[[1]]=H11;
    Return[MatrixForm[MatKull]]
    )
likelihood1[x_,m_]:=Block[{G1x,g0x,B1x,cx,gaux,factx},
            G1x=CDF[NormalDistribution[0,Sqrt[V[[k]]]],x];
            g0x=PDF[NormalDistribution[m0,Sqrt[v0]],x];
            B1x=(G1x+10^-20)/(1-G1x+10^-20)*(1-T[[k]])/T[[k]];
            cx=Log[B1x];
            gaux=Exp[-1/2*(1/S[[k]]*cx\ 2+1/V[[k]]*x\ 2)];
            factx=1/(( G1x+10^-20)(1-G1x+10^-20));
            likeK1=gaux*factx;
            likeH1=likeK1*g0x*Exp[-1/2*m\ 2*1/v0-((x-m0)/v0)*m];
            likeF1=likeH1*Log[likeK1];
            Return[{likeK1,likeH1,likeF1}]
            ]
K1[m_]:=NIntegrate[likelihood1[x,m][[1]],{x,mmin-m,mmax-m}]
H1[m_]:=NIntegrate[likelihood1[x,m][[2]],{x,mmin-m,mmax-m}]
F1[m_]:=NIntegrate[likelihood1[x,m][[3]],{x,mmin-m,mmax-m}]
```

The code computes the expected KL-divergences at a generic stage and put them in a vector.

```
ExpectKull[stage_]:=(n=Rationalize[stage];Print["Stage K ="," " , n];
                    row=Dimensions[MatKull][[1]];
                    If[rin+n-row>1,Print["former stage not perfoms"];
                    Break[]];
                    rownum=Delete[MatKull[[row]],1];
                    max=Position[rownum,Max[rownum]][[1,1]];
                        If[n==2,Hprec=H11,k=max;n=n-1;
                        matrainv=Inverse[mattCova[n][[k]] ];
                        Cov=Inverse[Covamin[n]];
                        Hprec=H[M[[max]],n] ;
                        nextmax[[n]]=max;nextH[[n]]=Hprec;n=n+1
                        ];
                    kullback=Table[0,{k,numexpert}];
                For[k=1,k<=numexpert,k++,
                        If[Abs[Det[mattCova[n][[k]]]]>10^-6,
```

```
                        matrainv=Inverse[mattCova[n][[k]]];
                        Cov=Inverse[Covamin[n]];
                        den=NIntegrate[H[m,n]/K[m,n],{\{}m,-5,8{\}}];
                        num=NIntegrate[F[m,n]*H[m,n]/K[m,n],
                            {\{}m,-5,8{\}}];
                        kullback[[k]]=num/den;
                        ];
                    Print[ kullback[[k]]]
                    ];
                MatKull=Join[MatKull,{Join[{K},kullback]}];
                Return[MatrixForm[MatKull]]
                )
likelihood[x_,m_,stage_]:=(
    vetG=Join[Table[G[x+m-mattM[n][[k]][[i]],i],{i,n-1}],{G[x,n]}];
    vetg=Join[Table[g[x-mattM[n][[k]][[i]],i],{i,n-1}],{g[x,n]}];
    vetB=Table[(vetG[[i]]+10^-20)/(1-vetG[[i]]+10^-20)*
        (1-mattT[n][[k]][[i]])/mattT[n][[k]][[i]],{i,n}];
    vetC=Log[vetB];
    form=-1/2*vetC.matrainv.vetC;
    expo=form-Sum[(m^2/2*1/mattV[n][[k]][[i]]+
         m*(x-mattM[n][[k]][[i]])/mattV[n][[k]][[i]]),{i,n-1}];
    fact=Product[vetg[[i]]/((vetG[[i]]+10^-20)*
        (1-vetG[[i]]+10^-20)),{i,n}];
    likelihoodK=Exp[expo]*fact;
    likelihoodH=likelihoodK*Exp[-1/2*m^2*1/v0-m*(x-m0)/v0]*g0[x];
    vetridC=Delete[vetC,n];
    expogam=form+1/2*vetridC.Cov.vetridC;
    Gam=vetg[[n]]/((vetG[[n]]+10^-20)*(1-vetG[[n]]+10^-20))*
        Exp[expogam];
    Return[{likelihoodK,likelihoodH,Gam}]
    )
H[m_,n_]:=NIntegrate[likelihood[x,m,n][[2]],{x,mmin-m,mmax-m}]
K[m_,n_]:=NIntegrate[likelihood[x,m,n][[1]],{x,mmin-m,mmax-m}]
F[m_,n_]:=(Hm=H[m,n];Fstage=1/Hm*NIntegrate[
        likelihood[x,m,n][[2]]*Log[likelihood[x,m,n][[3]]],{x,mmin-m,
        mmax-m}]-Log[Hm/nextH[[n-1]] ]; Return[Fstage]
        )
Kullmax[stages_] := (Expect1;
 maxim = {Position[Kull, Max[Kull]][[1, 1]], Max[Kull]};
MaxKull = maxim;
 Print["Kmax :", maxim];
 For[n = 2, n <= stages, n++, ExpectKull[n] ;
     maxim={Position[kullback,Max[kullback]][[1, 1]],Max[kullback]};
     Print["delta value: ", delta]; Print["Kmax ", maxim];
 If[maxim[[2]] < delta, Break[], MaxKull = maxim];
 ]; jump = n - 1;
 Return[{MatrixForm[MatKull], jump, MaxKull}]
    )
Kullmax[stagemax]
```

Finally, the output shows:

- The expected KL-divergences of each density $h_{j;k}(\theta)$ from the posterior $h_{k-1}(\theta)$ obtained at the previous stage;

- The label j of the expert that satisfies the selecting criterion;
- A matrix containing the input data, KL-divergences at each stage, the number of experts to consult, and the label of the last expert to consult together with the corresponding expected KL divergence.

2.5 A Case-Study

The behavior of the algorithms and rules proposed in the previous sections, and implemented in `Mathematica`, has been investigated in simulation and experimental studies. Here the results from medical data are synthetically presented to exemplify how the selecting and stopping rules work.

An orthopedist A is uncertain about the long-term failure log-odds θ of a new hip prosthesis. Therefore, he decides to consult a number K of colleagues, sequentially selected from a pool of size $n = 7$. He learns the variance v_j from each orthopedist Q_j ($j = 1,$Ě$, n$) and assesses all the calibration parameters, without modifying them in proceeding from a stage to the successive one: these data are shown in Table 2.3. Subject A has also (subjectively) assessed $m0 = -1$, $v0 = 1$, and set the threshold $\delta = 0.035$: the choice of this value means that, at stage k, the most informative expert $Q^*_{j;k}$ will be consulted only if the expected KL-divergence of $h_{j^*;k}(\theta)$ with respect to $h_{k-1}(\theta)$ will be no less than the KL-divergence of a Bernoulli distribution $B(p)$ with $p = 0.5$ from a Bernoulli distribution with $p = 0.63$; or, in other words, only if stopping the process at stage $k-1$ instead of proceeding to stage k involves, on average, an information loss larger than that one yielded by using a $B(0.63)$ instead of a $B(0.5)$.

Conditions *i), ii),* and *iii)* in Sect. 2.2 are assumed to be satisfied, so that the combining algorithm outlined in Sect. 2.2 can be fairly applied. In fact: *i)* as confirmed by experts, it rests on empirical evidence that the failure logodds θ can be supposed as Gaussian; *ii)* it is reasonable to think the probability the orthopedist A assigns to the event $v^{(k)}$ is the same for all θ values: the surgeons' variances alone give no information able to change the subject A's beliefs about θ; *iii)* it is reasonable as well to assume the conditional probability A assigns to the event "the

Table 2.3 Initial data for the case-study

Q_j	v_j	t_j	s_{jj}	r_{j1}	r_{j2}	r_{j3}	r_{j4}	r_{j5}	r_{j6}	r_{j7}
Q_1	0.90	0.35	3.93	1						
Q_2	0.40	0.60	4.86	+0.20	1					
Q_3	2.00	0.42	1.80	−0.1	−0.60	1				
Q_4	2.25	0.54	1.11	0	+0.30	0	1			
Q_5	1.80	0.50	1.31	0	+0.10	0	0	1		
Q_6	2.92	0.75	3.81	0	+0.20	−0.10	0	0	1	
Q_7	2.35	0.60	4.53	+0.20	0	−0.60	+0.30	+0.10	0	1

expert $Q_{j;k}$ will give the variance v_k," given the values of the shape and location parameters provided by the $k-1$ experts previously consulted, is the same for all θ values. In order to perform an efficient sequential consulting, input data are entered into Mathematica notebook EXPS.nb:

```
Initialdata={{-2, -0.8, -1, -1.5, -1.3, -1.4, -1.35},
             {0.9, 0.4, 2, 2.25, 1.8, 2.92, 2.35},
             {0.35, 0.6, 0.42, 0.54, 0.5, 0.75, 0.6},
             {3.93, 4.86, 1.80, 1.11, 1.31, 3.81, 4.53}};
R={{1, 0.2, -0.1, 0, 0, 0, 0.2},{0.2, 1, -0.6, 0.3, 0.1,
    0.2, 0},{-0.1, -0.6, 1, 0, 0, -0.1, -0.6},{0, 0.3, 0,
    1, 0, 0, 0.3},{0, 0.1, 0, 0, 1, 0, 0.1},{0, 0.2, -0.1,
    0, 0, 1, 0},{0.2, 0, -0.6, 0.3, 0.1, 0, 1}};

delta=0.035;          v0=1; m0=-1;
mmin=-8; mmax=11;       stagemax=7;
```

Mathematica output is the following:

```
Stage k = 1      Stage k = 2        Stage k = 3        Stage k = 4
0.340594         1.36673            1.57449            1.70397
0.519248         0                  0                  0
0.299469         1.65218            0                  0
0.373058         1.37703            1.27123            1.40743
0.394185         1.47585            1.63163            0
0.102336         0.80258            0.42429            0.45441
0.121147         0.90094            1.56959            1.75234
Kmax:{2,0.5192}  delta value:0.035  delta value:0.035  delta value:0.035
                 Kmax:{3,1.65218}   Kmax:{5, 1.63163}  Kmax:{7, 1.75234}

Stage k = 5         Stage k = 6         Stage k = 7
2.14744             2.17423             0
0                   0                   0
0                   0                   0
2.62677             0                   0
0                   0                   0
0.310055            0.0533287           0.034978
0                   0                   0
delta value:0.035   delta value:0.035   delta value:0.035
Kmax:{4,2.62677}    Kmax:{1, 2.17267}   Kmax:{6, 0.034978}
```

The output shows that, at stage $k = 1$, expert Q_2 is selected, due to an expected KL-divergence equal to 0.519248. At stage $k = 2$, the maximum expected KL-divergence corresponds to expert Q_3, characterized by a high negative correlation with $Q_2 (r = -0.60)$ together with a low bias ($t = 0.42 \cong 0.5$). Expert Q_5 is, on average, the most informative at stage $k = 3$, and so on. At stage $k = 7$, the expected KL-divergence for the last expert, Q_6, is 0.034978, that is less than the threshold δ: since Q_6 does not involve a knowledge expected gaining judged as relevant, subject A will not consult him. So, as the output matrix shows too, the sequential consulting stops at stage $k = 6$: the last expert who enters into the process is Q_1, with an expected KL-divergence equal to 0.0533287.

$$\left(\begin{array}{cccccccc}
\text{exp} & 1 & 2 & 3 & 4 & 5 & 6 & 7 \\
m & -2 & -0.8 & -1 & -1.5 & -1.3 & -1.4 & -1.35 \\
v & 0.9 & 0.4 & 2 & 2.25 & 1.8 & 2.92 & 2.35 \\
t & 0.35 & 0.6 & 0.42 & 0.54 & 0.5 & 0.75 & 0.6 \\
s2 & 3.93 & 4.86 & 1.80 & 1.10 & 1.31 & 3.81 & 4.53 \\
K & 0.34059 & 0.51925 & 0.29947 & 0.37306 & 0.39419 & 0.10234 & 0.12115 \\
K & 1.36673 & 0 & 1.65218 & 1.37703 & 1.47585 & 0.80258 & 0.90094 \\
K & 1.57449 & 0 & 0 & 1.27123 & 1.63163 & 0.42429 & 1.56959 \\
K & 1.70397 & 0 & 0 & 1.40743 & 0 & 0.45441 & 1.75234 \\
K & 2.14744 & 0 & 0 & 2.62677 & 0 & 0.310155 & 0 \\
K & 2.17423 & 0 & 0 & 0 & 0 & 0.053329 & 0 \\
K & 0 & 0 & 0 & 0 & 0 & 0.034978 & 0
\end{array}\right), 6, \{1, 2.17267\}$$

Table 2.4 shows the expected KL-divergences for each expert at each stage (the maximum expected KL-divergence is displayed in bold), as well as the location parameters supplied by the selected experts. Posterior distributions from stage 0 (the prior) to stage 6 are shown in Fig. 2.3. The 'final' pdf on the unknown log-odds θ has mean, median, and mode equal to -0.826 and standard deviation equal to 0.202: it can be regarded as the synthesis representation of the expert knowledge about the long-term failure log-odds of the new hip prostheses.

The behavior of the proposed rules and algorithms in the case-study appears to be coherent with the intuition and gives an empirical support about the efficiency of the selecting and stopping criteria.

Table 2.4 Results of the sequential process

	Stage	Stage	Stage	Stage	Stage	Stage
	$k = 1$	$k = 2$	$k = 3$	$k = 4$	$k = 5$	$k = 6$
Q_j	$\mathrm{E}\left[KL\left(h_{j;1}, h_0\right)\right]$	...	...	...	...	$\mathrm{E}\left[KL\left(h_{j;6}, h_5\right)\right]$
Q_1	0.34059	1.36673	1.57449	1.70397	2.14744	**2.17423**
Q_2	**0.51925**	–	–	–	–	–
Q_3	0.29947	**1.65218**	–	–	–	–
Q_4	0.37306	1.37703	1.27123	1.40743	**2.62677**	–
Q_5	0.39419	1.47585	**1.63163**	–	–	–
Q_6	0.10234	0.80258	0.42429	0.45441	0.310155	0.053329
Q_7	0.12115	0.90094	0.56959	**1.75234**	–	–
	↓	↓	↓	↓	↓	↓
$Q^*_{j;k}$	Q_2	Q_3	Q_5	Q_7	Q_4	Q_1
m_k	−0.8	−1	−1.3	−1.35	−1.5	−2

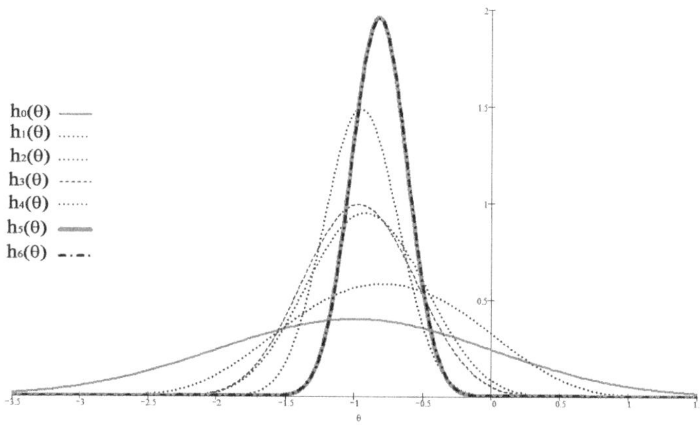

Fig. 2.3 Posterior distributions at stages 0 (i.e., the prior), 1, 2,. . . ,6 of the sequential procedure

References

1. Agati, P., Calò, D.G., Stracqualursi, L.: A joint calibration model for combining predictive distributions. Statistica **2**, 203–212 (2007)
2. Brown, J.R., Harvey, M.E.: Arbitrary precision mathematica functions to evaluate the one-sided one sample K–S cumulative sampling distribution. J. Stat. Softw. **26**(3), 1–55 (2008). http://www.jstatsoft.org/v26/i03
3. Kullback, S.: Information Theory and Statistics. Wiley, New York (1959)
4. Lindley, D.V.: The 1988 Wald Memorial Lectures: the present position in Bayesian statistic. Stat. Sci. **5** (1), 44–89 (1990)
5. McCulloch, R.: Local model influence. J. Am. Stat. Assoc. **84**, 473–478 (1989)
6. Monari, P., Agati, P.: Fiducial inference in combining expert judgements. J. Ital. Stat. Soc. **84**, 81–97 (2001)
7. Morris, P.A.: Combining expert judgments: a Bayesian approach. Manag. Sci. **23**, 679–693 (1977)
8. Wolfram, S.: The `Mathematica` Book. 5th edn. Wolfram Media, Champaign, USA (2003)

Chapter 3
Markov-Modulated Samples and Their Applications

Alexander Andronov

3.1 Problem Setting

The classical sample theory supposes that sample elements are identically distributed and independent (i.i.d.) random variables. Lately a great attention has been granted to dependence in probabilistic structures, for example, dependence between interarrival times of various flows, between service times, etc. Usually it is described by the so-called Markov-modulated processes. They are used widely in environmental, medical, industrial, and sociological researches. We restrict ourselves by a case when elements of the sample are positive random variables. It is convenient to consider them as lifetimes of unreliable elements.

Let us consider sample elements $\{X_i, i = 1, \ldots, n\}$, modulated by a finite continuous-time Markov chain (see [6]). For simplicity we say that the elements operate in the so-called random *environment*. The last is described by an "external" continuous-time ergodic Markov chain $J(t), t \geqslant 0$, with a final state space $E = \{1, 2, \ldots, k\}$. Let $\lambda_{i,j}$ be the transition rate from state i to state j.

Additionally, n binary identical elements are considered. Each component can be in two states: *up*(1) and *down*(0). The elements of system fail one by one, in random order. For a fixed state $i \in E$, all n elements have the same failure rate $\gamma_i(t)$ and are stochastically independent. When the external process changes its state from i to j at some random instant t, all elements, which are alive at time t, continue their life with new failure rate $\gamma_j(t)$. If on interval (t_0, t) the random environment has state $i \in E$, then the residual lifetime $\tau_r - t_0$ (*up*-state) of the rth component, $r = 1, 2, \ldots, n$, has a cumulative distribution function (CDF) with failure rate $\gamma_i(t)$ for time moment t, and the variables $\{\tau_r - t_0, r = 1, 2, \ldots, n\}$ are independent.

A. Andronov (✉)
Transport and Telecommunication institute, 1 Lomonosova street, Riga, LV-1019, Latvia
e-mail: lora@mailbox.riga.lv

V.B. Melas et al. (eds.), *Topics in Statistical Simulation*, Springer Proceedings in Mathematics & Statistics 114, DOI 10.1007/978-1-4939-2104-1_3

We wish to get statistical estimates for the unknown parameters $\beta^{(i)} = (\beta_{1,i}, \beta_{2,i}, \ldots, \beta_{m,i})^T$, $i = 1, \ldots, k$. Note that in the above described process elements of the sample $\{X_i, i = 1, \ldots, n\}$ are no i.i.d. anymore, as it is assumed in the classical sampling theory.

Further we make the following suppositions. Firstly, parameters of the Markov-modulated processes $\{\lambda_{i,j}\}$ are known. Secondly, with respect to hazard rates $\gamma_i(t)$, a parametrical setting takes place: all $\gamma_i(t)$ are known, accurate to m parameters $\beta^{(i)} = (\beta_{1,i}, \beta_{2,i}, \ldots, \beta_{m,i})^T$, so we will write $\gamma_i(t; \beta^{(i)})$. Further, we use the $(m \times k)$-matrix $\beta = (\beta^{(1)}, \beta^{(2)}, \ldots, \beta^{(k)})$ of unknown parameters. Thirdly, with respect to the available sample: sample elements are fixed corresponding to their appearance, so the order statistics $X_{(1)}, X_{(2)}, \ldots, X_{(n)}$ are fixed. Finally, the states of the random environment $J(t)$ are known only for time moments $0, X_{(1)}, X_{(2)}, \ldots, X_{(n)}$.

The maximum likelihood estimates (see [5, 8, 9]) for the unknown parameters β are derived. Results of a simulation study illustrate the elaborated technique. Presented paper continues our previous investigations [1,2].

3.2 Transition Probabilities

In this section we cite a result from the paper of Andronov and Gertsbakh [3]. Define $N(t)$ as the number of elements which are in the *up* state at time moment t. Obviously $P\{N(0) = n\} = 1$. We denote

$$p_{r,i,j}(t_0,t)=P\{N(t)=r,\ J(t)=j\,|\,N(t_0)=r,\ J(t_0) = i\}, r \in \{1, \ldots, n\},\ i, j \in E,$$

$$p_{r,i}(t_0,t)=(p_{r,i,1}(t_0,t), \ldots, p_{r,i,k}(t_0,t))^T,\ P_r(t_0,t)=(p_{r,1}(t_0,t), \ldots, p_{r,k}(t_0,t)),$$

$$\Gamma(t,\beta)=\text{diag}\left(\gamma_1(t,\beta^{(1)}), \ldots, \gamma_1(t,\beta^{(k)})\right),\ \Lambda=\text{diag}\left(-\sum_{i=1}^{k}\lambda_{i,1}, \ldots, -\sum_{i=1}^{k}\lambda_{i,k}\right). \tag{3.1}$$

It has been shown that

$$\dot{P}_r(t_0,t) = -\left(\Lambda + r\Gamma(t,\beta)\right) P_r(t_0,t) + \lambda^T P_r(t_0,t), 0 \leqslant t_0 \leqslant t. \tag{3.2}$$

Below we consider a simple time-homogeneous case when $\gamma_i(t;\beta^{(i)}) = \gamma_i(\beta_i)\ \forall i$ $\Gamma(t,\beta) = \Gamma(\beta) = \text{diag}(\gamma_1(\beta^{(1)}), \ldots, \gamma_k(\beta^{(k)}))$. Therefore,

$$\dot{P}_r(t_0,t) = \left(\lambda^T - (\Lambda + r\Gamma(\beta))\right) P_r(t_0,t),\ 0 \leqslant t_0 \leqslant t.$$

In this case a solution can be represented by matrix exponent (see [4, 7]):

$$P_r(t_0,t) = \exp\left((t - t_0)\left(\lambda^T - (\Lambda + r\Gamma(\beta))\right)\right),\ 0 \leqslant t_0 \leqslant t. \tag{3.3}$$

3.3 Maximum Likelihood Estimates

In the considered case, besides the initial state $j(0)$ of $J(0)$, a sample of size n is given: $(x, j) = \{(x_{j(r)}, j(r)), r = 1, \ldots, n\}$, where $x_{j(r)}$ is the rth order statistic of the sample and $j(r) = J(x_{(r)})$ is a corresponding state of the random environment. Setting $x_{(0)} = 0$ we rewrite the log-likelihood function as

$$\begin{aligned} \mathrm{ll}(\beta; (x, j)) = \sum_{r=0}^{n-1} [\ln p_{n-r, j(r), j(r+1)}(x_{(r)}, x_{(r+1)}; \beta) \\ + \ln(n - r) + \ln \gamma_{j(r+1)}(x_{(r+1)}; \beta^{(j(r+1))})]. \end{aligned} \tag{3.4}$$

Considering gradients with respect to the column vectors $\beta^{(v)}$, $v = 1, \ldots, k$, we get maximum likelihood equations

$$\begin{aligned} \frac{\partial}{\partial \beta^{(v)}} \mathrm{ll}(\beta; (x, j)) = \sum_{r=0}^{n-1} \frac{1}{p_{n-r, j(r), j(r+1)}(x_{(r)}, x_{(r+1)}; \beta)} \\ \times \frac{\partial}{\partial \beta^{(v)}} p_{n-r, j(r), j(r+1)}(x_{(r)}, x_{(r+1)}; \beta) \\ + \sum_{r=0}^{n-1} \frac{1}{\gamma_{j(r+1)}(x_{(r+1)}; \beta^{(j(r+1))})} \\ \times \frac{\partial}{\partial \beta^{(v)}} \gamma_{j(r+1)}(x_{(r+1)}; \beta^{(j(r+1))}) = 0, \ v = 1, \ldots, k. \end{aligned} \tag{3.5}$$

Further, we consider a time-homogeneous case when all rate intensity $\gamma_i(\beta^{(i)})$ have one unknown scalar parameter β_i only, so $\gamma_i(\beta^{(i)}) = \beta_i$, $i = 1, \ldots, k$. We will write $p_{m,i,j}(t - t_0) = p_{m,i,j}(t_0, t)$. Then, the likelihood equations (3.5) have the following form:

$$\begin{aligned} \frac{\partial}{\partial \beta^{(v)}} \mathrm{ll}(\beta; (x, j)) = \sum_{r=0}^{n-1} \frac{1}{p_{n-r, j(r), j(r+1)}(x_{(r+1)} - x_{(r)}; \beta)} \\ \times \frac{\partial}{\partial \beta^{(v)}} p_{n-r, j(r), j(r+1)}(x_{(r+1)} - x_{(r)}; \beta) \\ + \frac{1}{\beta_v} \sum_{r=0}^{n-1} \delta_{v, j(r+1)} = 0, \end{aligned} \tag{3.6}$$

where $\delta_{v,j(r+1)}$ is the Kronecker symbol: $\delta_{v,j(r+1)} = 1$ if $v = j(r + 1)$, and $\delta_{v,j(r+1)} = 0$ otherwise.

Now we must get an expression for the derivative $\frac{\partial}{\partial\beta^{(v)}} p_{n-r,j(r),j(r+1)}(x_{(r+1)} - x_{(r)}; \beta)$. For that we use an expression for a derivative of a matrix exponent (see Lemma of Appendix). Let D_v be a square matrix from zero, where only one non-zero element equals 1 and takes the vth place of a main diagonal. Then, for the homogeneous case when $\Gamma(\beta) = \mathrm{diag}(\gamma_1(\beta), \ldots, \gamma_k(\beta)) = \Gamma = \mathrm{diag}(\beta_1, \ldots, \beta_k)$, according to (3.3), we have for $v = 1, \ldots, k$:

$$\frac{\partial}{\partial\beta_v} P_r(t_0, t) = \frac{\partial}{\partial\beta_v} \exp\{(t - t_0)(\lambda^T - \Lambda - r\Gamma)\} = \sum_{i=1}^{\infty} \frac{1}{i!}(t - t_0)^i$$

$$\sum_{j=0}^{i-1} (\lambda^T - \Lambda - r\Gamma)^j \left(-rD_v \frac{\partial}{\partial\beta_v}\beta_v\right)(\lambda^T - \Lambda - r\Gamma)^{i-1-j} =$$

$$-r\sum_{i=1}^{\infty} \frac{1}{i!}(t - t_0)^i \sum_{j=0}^{i-1} (\lambda^T - \Lambda - r\Gamma)^j D_v(\lambda^T - \Lambda - r\Gamma)^{i-1-j}.$$

Therefore

$$\frac{\partial}{\partial\beta_v} P_r(t_0, t) = -r\sum_{i=1}^{\infty} \frac{1}{i!}(t - t_0)^i \sum_{j=0}^{i-1} \left((\lambda^T - \Lambda - r\Gamma)^j\right)^{\langle v\rangle}$$

$$\left((\lambda^T - \Lambda - r\Gamma)^{i-1-j}\right)_{\langle v\rangle}, \tag{3.7}$$

where $M^{\langle v\rangle}$ and $M_{\langle v\rangle}$ mean γth column and γth row of matrix M.

Now we can use a numerical method for the solution of the likelihood equation (3.6). Note that parameter β_v can be non-trivially estimated if state v has been registered as some $j(r) = v$, $r = 0, 1, \ldots, n$.

3.4 Simulation Study

Below there are the results of a simulation study presented, they are performed for an analysis of the described estimating procedure efficiency. As initial data, data from the paper [3] have been used. Let us describe one. A random environment has three states ($k = 3$, $E = \{1, 2, 3\}$). Transition intensities $\{\lambda_{i,j}\}$ from state i to state j, $(i, j = 1, 2, 3)$ are given by a matrix

$$\lambda = (\lambda_{i,j}) = \begin{pmatrix} 0 & 0.2 & 0.3 \\ 0.1 & 0 & 0.2 \\ 0.4 & 0.2 & 0 \end{pmatrix}. \tag{3.8}$$

Let a number of the considered elements n equals 5. For the environment state $i \in E$, all elements have a constant failure rate $\gamma_i(t) = \beta_i$ and they fail independently. Therefore, a last time till a given element failure (for the same state i of the environment) has the exponential distribution with parameter β_i. These parameters must be estimated. For that purpose a sample is given. It contains a sequence of $n+1$ pairs: $(x, j) = \{(x_{j(r)}, j(r)), r = 0, \ldots, n\}$, where $x_{j(r)}$ is the rth order statistic of n-sample, and $j(r) = J(x_{(r)})$ is an environment state in the instant $x_{(r)}$. The initial pair $(x_{(0)}, j(0))$ equals $(0, 1)$.

All the mentioned sampling data are given and are used in an estimating procedure. Own samples are simulated for the following parameter values: $\beta = (\beta_1\ \beta_2\ \beta_3)^T = (0.1\ 0.2\ 0.3)^T$. It is convenient to present these data as $3 \times (n+1)$ matrix. An example of such matrix for $n = 5$ is the following:

$$Sample = \begin{pmatrix} r & 0 & 1 & 2 & 3 & 4 & 5 \\ x_{(r)} & 0 & 0.624 & 1.502 & 2.009 & 8.711 & 9.429 \\ J(x_{(r)}) & 1 & 1 & 2 & 1 & 3 & 3 \end{pmatrix}.$$

In the simulation process, samples are generated one by one. Various samples are independent. Each sample corresponds to the appointed initial state of the environment: a sample with number $3i + j$ corresponds to the initial state $J(0) = j$; $j = 1, 2, 3$; $i = 0, \ldots$. Further, q such three samples (with the initial states $j = 1, 2, 3$) form a block, containing $3q$ samples. A maximum log-likelihood estimate (MLE) $\tilde{\beta} = (\tilde{\beta}_1, \tilde{\beta}_2, \tilde{\beta}_3)^T$ is calculated for each block.

In broad outline, a procedure is as follows. For each sample, a changing of the environment $J(t)$, $t > 0$, and instants of element failure $x_{(r)}$, $r = 1, \ldots, n$, are simulated. Then, for the sample, a logarithm of likelihood function (3.4) and its gradient (3.5) or (3.6) are recorded. These expressions are used for MLE calculation. As an optimization method, the gradient method has been used.

The gradient method is given by the following parameters: n is a sample size (initial number of system elements); b_0 is an initial value of parameter estimate; d is a step of moving along the gradient; ε is a maximum module of a difference between sequential values of the parameter estimate β, for which a calculation is ended; L is a limit number of a gradient recalculation during moving from an initial point; K is a number of addends, appreciated in an expansion of the matrix exponent (3.7); $3q$ is a number of the samples in the block.

A set of such parameters numerical values $(n, b_0, d, \varepsilon, L, K, q)$ is called an experiment design. Below, the results of simulation study are presented.

In Table 3.1 the corresponding results are presented for the design experiment $n = 5$, $b_0 = (0.08\ 0.22\ 0.328)^T$, $d = 0.015$, $\varepsilon = 0.01$, $L = 20$, $K = 20$, $r = 5$, $q = 5$ and various values of total block number N. In the first column the initial value of the estimate $b_0 = (0.08\ 0.22\ 0.328)^T$ is written. The following columns there are the estimate values given by averaging over N blocks. For a big number of the blocks, the coefficients d and ε have been changed. Namely, for $N = 15, 17, 19, 21$ those values equal 0.002, and for $N = 21$, additionally, $L = 40$.

Table 3.1 Convergence of the estimates for initial value $b_0 = (0.08\ 0.22\ 0.328)^T$

N		1	2	3	4	5	6	7	8
$\tilde{\beta}_1$	0.080	0.076	0.103	0.119	0.111	0.097	0.088	0.098	0.101
$\tilde{\beta}_2$	0.220	0.217	0.208	0.211	0.219	0.219	0.213	0.211	0.213
$\tilde{\beta}_3$	0.328	0.219	0.303	0.333	0.351	0.316	0.280	0.256	0.292
N	9	10	11	12	13	15	17	19	21
$\tilde{\beta}_1$	0.094	0.109	0.103	0.098	0.103	0.100	0.101	0.102	0.102
$\tilde{\beta}_2$	0.209	0.210	0.207	0.209	0.209	0.208	0.208	0.208	0.208
$\tilde{\beta}_3$	0.267	0.298	0.279	0.280	0.298	0.283	0.290	0.295	0.298

An analysis of the Table 3.1 shows that a convergence to true values (0.1 0.2 0.3) takes place but very slow.

In conclusion we would like to remark that considered approach allows improving probabilistic predictions for functioning of various complex technical and economical systems.

Appendix

Lemma 3.1. *If elements of matrix $G(t)$ are differentiable function of t, then*

$$\frac{\partial}{\partial t}G(t)^n = \sum_{i=0}^{n-1} G(t)^i \left[\frac{\partial}{\partial t}G(t)\right] G(t)^{n-1-i},\ n = 1, 2, \ldots,$$

$$\frac{\partial}{\partial t}\exp(G(t)) = \frac{\partial}{\partial t}\sum_{i=0}^{\infty}\frac{1}{i!}G(t)^i = \sum_{i=1}^{\infty}\frac{1}{i!}\sum_{j=0}^{i-1} G(t)^j \left[\frac{\partial}{\partial t}G(t)\right] G(t)^{i-1-j}.$$

Proof. Lemma 3.1 is true for $n = 1$ and 2. If one is true for $n > 1$, then

$$\begin{aligned}\frac{\partial}{\partial t}G(t)^{n+1} &= \left[\frac{\partial}{\partial t}G(t)\right]G(t)^n + G(t)\frac{\partial}{\partial t}G(t)^n\\ &= \left[\frac{\partial}{\partial t}G(t)\right]G(t)^n + G(t)\sum_{i=0}^{n-1}G(t)^i\left[\frac{\partial}{\partial t}G(t)\right]G(t)^{n-1-i}\\ &= \sum_{i=0}^{n}G(t)^i\left[\frac{\partial}{\partial t}G(t)\right]G(t)^{n-i}.\end{aligned}$$

□

References

1. Andronov, A.M.: Parameter statistical estimates of Markov-modulated linear regression. In: Statistical Method of Parameter Estimation and Hypotheses Testing, vol. 24, pp. 163–180. Perm State University, Perm (2012) (in Russian)
2. Andronov, A.M.: Maximum Likelihood Estimates for Markov-additive Processes of Arrivals by Aggregated Data. In: Kollo, T. (ed.) Multivariate Statistics: Theory and Applications. Proceedings of IX Tartu Conference on Multivariate Statistics and XX International Workshop on Matrices and Statistics, pp. 17–33. World Scientific, Singapore (2013)
3. Andronov, A.M., Gertsbakh, I.B.: Signatures in Markov-modulated processes. Stoch. Models **30**, 1–15 (2014)
4. Bellman, R.: Introduction to Matrix Analysis. McGraw-Hill, New York (1969)
5. Kollo, T., von Rosen, D.: Advanced Multivariate Statistics with Matrices. Springer, Dordrecht (2005)
6. Pacheco A., Tang, L.C., Prabhu N.U.: Markov-Modulated Processes & Semiregenerative Phenomena. World Scientific, New Jersey (2009)
7. Pontryagin, L.S.: Ordinary Differential Equations. Nauka, Moscow (2011) (in Russian)
8. Rao, C.R.: Linear Statistical Inference and Its Application. Wiley, New York (1965)
9. Turkington, D.A.: Matrix calculus and zero-one matrices. Statistical and Econometric Applications. Cambridge University Press, Cambridge (2002)

Chapter 4
Simulating Correlated Ordinal and Discrete Variables with Assigned Marginal Distributions

Alessandro Barbiero and Pier Alda Ferrari

4.1 Introduction and Motivation

In many research fields, data sets often include ordinal variables, e.g. measured on a Likert scale, or count variables. This work proposes and illustrates a procedure for simulating samples from ordinal and discrete variables with assigned marginal distributions and association structure, which can be used as a useful computational tool by researchers. In fact, model building, parameter estimation, hypothesis tests, and other statistical tools require verification to assess their validity and reliability, typically via simulated data. Up to now, a few methodologies that address this problem have appeared in the literature. Demirtas [5] proposed a method for generating ordinal data by simulating correlated binary data and transforming them into ordinal data, but the procedure is complex and computationally expensive, since it requires the iterative generation of large samples of binary data. Ruscio and Kaczetow [10] introduced an iterative algorithm for simulating multivariate non-normal data (discrete or continuous), which first constructs a huge artificial population whose components are independent samples from the desired marginal distributions and then reorders them in order to catch the target correlations. The desired samples are drawn from this final population as simple random samples. Recently, Ferrari and Barbiero [1] proposed a method (GenOrd) able to generate correlated point-scale rv (i.e., whose support is of the type $1, 2, \dots, k$) with marginal distributions and Pearson's correlations assigned by the user.

In this work, after briefly recalling the GenOrd method, we show that it is also able to generate discrete variables with any finite/infinite support and/or association structure expressed in terms of Spearman's correlations. The performance of the

A. Barbiero (✉) • P.A. Ferrari
Department of Economics, Management and Quantitative Methods,
Università degli Studi di Milano, Milano, Italy
e-mail: alessandro.barbiero@unimi.it; pieralda.ferrari@unimi.it

V.B. Melas et al. (eds.), *Topics in Statistical Simulation*, Springer Proceedings
in Mathematics & Statistics 114, DOI 10.1007/978-1-4939-2104-1_4

method's extensions is assessed in terms of computational efficiency and precision through two examples of application, which also show the utility and usability of the method even for non-experts.

4.2 Generating Ordinal Data: The GenOrd Procedure

The objective is to simulate from a target m-dimensional rv $\boldsymbol{X}$, with assigned correlation matrix $\mathbf{R}^D$ and marginal cumulative distributions $\boldsymbol{F}_i$, $0 < F_{i1} < F_{i2} < \cdots < F_{il} < \cdots < F_{i(k_i-1)} < 1$, $i = 1, \ldots, m$, where $F_{il} = P(X_i \leqslant x_{il})$, being $(1, 2, \ldots, k_i)$ the support of the ith component X_i of $\boldsymbol{X}$. The method starts from an m-dimensional variable $\boldsymbol{Z} \sim N(\mathbf{0}, \mathbf{R}^C = \mathbf{R}^D)$. For each component Z_i of $\boldsymbol{Z}$, the $k_i - 1$ probabilities in $\boldsymbol{F}_i$ are chosen and define the corresponding normal quantiles $r_{i1} < r_{i2} < \cdots < r_{il} < \cdots < r_{i(k_i-1)}$. The variables Z_i are converted into ordinal variables with support $1, 2, \ldots, k_i$ as follows:

$$\begin{aligned} \text{if } Z_i &< r_{i1} \rightarrow X_i = 1 \\ \text{if } r_{i1} \leqslant Z_i &< r_{i2} \rightarrow X_i = 2 \\ &\cdots \\ \text{if } r_{i(k_i-1)} \leqslant Z_i &\rightarrow X_i = k_i. \end{aligned} \tag{4.1}$$

Let $\boldsymbol{X}^{(1)} = (X_1, \ldots, X_m)$. Although $\boldsymbol{X}^{(1)}$ has the desired marginal distributions, unfortunately $\mathbf{R}^{D(1)} \neq \mathbf{R}^D$. This is a known issue, see, for example, [4, 7]. An iterative algorithm is adopted in GenOrd in order to recover the "right" $\mathbf{R}^C$ able to reproduce the target $\mathbf{R}^D$. The final continuous correlation matrix $\mathbf{R}^C$ is then used to generate any $n \times m$ ordinal matrix with target ordinal correlation matrix $\mathbf{R}^D$ and with the desired marginals. The generation of samples is then carried out through the inverse transform technique: given a set of marginal distributions and a feasible correlation matrix $\mathbf{R}^D$, a random sample of chosen size n is drawn from $\boldsymbol{Z} \sim N(\mathbf{0}, \mathbf{R}^C)$ and the ordinal data are obtained according to the discretization process (4.1).

4.3 Extensions of GenOrd Procedure

Extensions of the GenOrd procedure can be carried out with respect to the correlation measure and the support (non-point-scale support or infinite support).

4.3.1 Extension to Any Finite/Infinite-Support Discrete Distribution

The extension of the GenOrd procedure to the case of any finite-support discrete distribution is straightforward. To this aim, the discretization process in (4.1) has to be modified by simply substituting to the integers $1, 2, \ldots, k_i$ the ordered values of the finite support $(x_{i1}, \ldots, x_{ik_i})$.

On the contrary, the extension to the case of unbounded discrete variables is not straightforward, because the algorithm computing the correlation matrix of the multivariate ordinal/discrete rv implicitly requires a finite number of cut-points. The support of the rv has to be somehow truncated. For example, for positive rv, such as the Poisson or the geometric, this can be carried out by right-truncating the support to a proper value, say $k_{\max} = F^{-1}(1 - \epsilon)$, with ϵ as small as possible. With the (approximate) marginal distributions obtained once the support has been truncated, the procedure follows the same steps described in Sect. 4.2 in order to compute the proper $\mathbf{R}^C$. To generate the desired discrete data, a sample is drawn from the multivariate standard normal rv with correlation matrix $\mathbf{R}^C$ and is then discretized directly recalling the inverse cdf of the target rv. This way, the marginal distribution of each of the unbounded-support discrete rv is ensured, and a (possibly small) approximation error is introduced only in terms of pairwise correlations.

4.3.2 Extension to Spearman's Correlation

The extension of the GenOrd procedure to the case of association among variables expressed in terms of Spearman's correlation coefficient requires more caution. It is well known that for a bivariate sample (x_i, y_i), $i = 1, \ldots, n$, Spearman correlation is defined as $\rho_S = \text{cor}(\text{rank}(\boldsymbol{x}), \text{rank}(\boldsymbol{y}))$, with $\boldsymbol{x} = (x_1, \ldots, x_n)'$, and similarly for $\boldsymbol{y}$. For (continuous) rv X_1 and X_2 with cdf F_1 and F_2, Spearman correlation is defined as $\rho_S(X_1, X_2) = \text{cor}(F_1(X_1), F_2(X_2))$. Attention is needed with discrete variables, characterized by a step-wise cumulative distribution function and whose observed values may present ties. Conventionally, in ρ_S, rank of equal sample values is the arithmetic mean of what their ranks would otherwise be. Then, for consistency, for discrete rv, Spearman correlation should be defined as $\rho_S(X_1, X_2) = \text{cor}(F_1^*(X_1), F_2^*(X_2))$ with $F_{il}^* = (F_{i,l} + F_{i,l-1})/2$, $i = 1, 2$, $l = 1, \ldots, k_i$; if $l = 1$, then $F_{i1}^* = F_{i1}/2$. The generalization of the simulation technique is then straightforward, since for the bivariate normal distribution, the following relationship between Pearson and Spearman's correlations holds: $\rho_S = \frac{6}{\pi} \arcsin(\rho/2)$ [9].

4.3.3 Performance

Generally speaking, the features a simulation technique should met are basically "generality", "accuracy", and "computational efficiency". With generality, we mean the capability of covering as many feasible scenarios as possible. With accuracy we mean the capability of producing samples coming from rv respecting some assigned "target", here, marginal distributions and pairwise correlations assigned by the user. For this case, although in the literature there are not standard measures, however, one can compute the Monte Carlo mean of each pairwise sample correlation over all the simulated samples, and compare it with the target one. Analogously, one can compute the Monte Carlo distribution of each variable and compare it with the assigned one through some goodness-of-fit test. With computational efficiency, we mean the capability of simulating samples in a short time. Focusing on the first feature, the only limitations GenOrd is practically bounded to are those related to the minimum and maximum admissible correlations for those marginal distributions, which any simulation technique has to take into account. As to its accuracy, GenOrd has been shown to produce accurate results when dealing with point-scale variables [6]. With regard to computational efficiency, GenOrd has been assessed also as computationally convenient [6]. In the next section, the accuracy and efficiency of its extended version will be assessed and its usefulness will be claimed through a new application to an inferential problem.

4.4 Examples of Application

In the following two subsections, GenOrd is further empirically explored with regard to its extensions described in Sect. 4.3, namely discrete variables with infinite support and correlation expressed via Spearman's ρ_S.

4.4.1 Example 1: Simulation of Correlated Geometric Variables

Here we show how GenOrd is able to generate from correlated geometric distributions with assigned parameters and correlation coefficient. For the two geometric rv X_1 and X_2, we consider all the possible combinations arising from the values $0.3, 0.5, 0.7$ for the parameters p_1 and p_2, combined with the values $-0.4, -0.2, 0.2, 0.4, 0.6, 0.8$ for the correlation coefficient ρ. In applying GenOrd, we set the value of the "threshold" parameter ϵ at 0.0001. We generate 50,000 samples of size $n = 100$ under each scenario.

In order to assess the accurateness of the procedure in terms of correlation, we focused on the sample correlation coefficient r and in its somehow bias-corrected version $r' = r[1 + (1 - r^2)/(2n)]$. In Table 4.1, we reported the MC mean of

Table 4.1 Simulation results: MC expected value of sample Pearson's correlation r and r_c

	$p_1 = 0.3$		$p_1 = 0.3$		$p_1 = 0.3$		$p_1 = 0.5$		$p_1 = 0.5$		$p_1 = 0.7$	
ρ	$p_2 = 0.3$		$p_2 = 0.5$		$p_2 = 0.7$		$p_2 = 0.5$		$p_2 = 0.7$		$p_2 = 0.7$	
	r	r'	r	r'	r	r'	r	r'	r	r'	r	r'
−0.4	−0.4085	−0.4102	−0.4095	−0.4112	−0.4110	−0.4127	−0.4107	−0.4124	NA	NA	NA	NA
−0.2	−0.2027	−0.2036	−0.2027	−0.2037	−0.2028	−0.2038	−0.2029	−0.2039	−0.2030	−0.2040	−0.2035	−0.2044
0.2	0.2003	0.2012	0.2000	0.2009	0.1999	0.2008	0.1998	0.2007	0.1995	0.2005	0.1990	0.1999
0.4	0.3994	0.4010	0.3989	0.4005	0.3973	0.3989	0.3983	0.3999	0.3969	0.3985	0.3959	0.3975
0.6	0.5981	0.6000	0.5974	0.5992	0.5956	0.5974	0.5968	0.5987	0.5953	0.5972	0.5939	0.5957
0.8	0.7979	0.7993	0.7972	0.7986	0.7951	0.7965	0.7965	0.7979	0.7948	0.7962	0.7938	0.7952

(NA: correlation not feasible with the assigned marginals)

these two estimates. These values are both quite close to the target values of ρ; moreover, for positive ρ, r' often sensibly reduces the sample bias in absolute terms. In order to assess the accurateness of the procedure in terms of marginal distribution, a goodness-of-fit test can be applied in order to verify if the two empirical distributions actually come from the two target geometric distributions. To this aim, we employ the test suggested by [8] for discrete variables and discussed for the geometric distribution by [3], who empirically checked its good performance in terms of actual significance level. Table 4.2 reports the (percentage) significance levels, $\hat{\alpha}_1$ and $\hat{\alpha}_2$ of such test performed at a nominal level $\alpha = 5\%$ for the two marginal distribution under each scenario; they all are very close to the nominal level. Simulating 50,000 bivariate samples required just a few seconds independently of the scenario examined.

4.4.2 *Example 2: Performance of Confidence Intervals for Spearman's Correlation*

In inferential problems involving ordinal or continuous data, it is usually necessary to determine the sampling distribution of Spearman's sample correlation coefficient r_S for hypotheses testing or to construct confidence intervals for ρ_S. If observations come from a bivariate normal rv, an approximate distribution of the sample correlation coefficient r_S can be adopted, and an approximate $(1-\alpha)$ confidence interval (CI), which is claimed to work well even for small sample size and high values of ρ_S, is [2]

$$(\rho_S^L, \rho_S^U) = \left(\frac{e^{2L}-1}{e^{2L}+1}, \frac{e^{2U}-1}{e^{2U}+1}\right) \tag{4.2}$$

where

$$L = 0.5\left[\log(1+r_S) - \log(1-r_S)\right] - z_{\alpha/2}\sqrt{(1+r_S{}^2/2)/(n-3)} \tag{4.3}$$

$$U = 0.5\left[\log(1+r_S) - \log(1-r_S)\right] + z_{\alpha/2}\sqrt{(1+r_S{}^2/2)/(n-3)} \tag{4.4}$$

$z_{\alpha/2}$ being the value of the standard normal rv Z such that $P(Z > z_{\alpha/2} = \alpha/2)$. The CI in (4.2) can be used also for any strictly monotonic transformation of bivariate normal random variables, because of the invariance property of Spearman's correlation. The problem is of course more complex when ρ_S concerns non-normal variables that are not derived by such monotonic transformations, especially discrete/ordinal variables. There is no evidence as to whether the performance of the CI in (4.2) remains satisfactory in these cases. For example, the discretization process in (4.1) from a bivariate normal variable, being a non-strictly monotonic function, distorts the resulting Spearman correlation coefficient of final discrete

Table 4.2 Simulation results: significance level of the goodness-of-fit test for the geometric marginals (NA: correlation not feasible with the assigned marginals)

	$p_1 = 0.3$		$p_1 = 0.3$		$p_1 = 0.3$		$p_1 = 0.5$		$p_1 = 0.5$		$p_1 = 0.7$	
ρ	$p_2 = 0.3$		$p_2 = 0.5$		$p_2 = 0.7$		$p_2 = 0.5$		$p_2 = 0.7$		$p_2 = 0.7$	
	$\hat{\alpha}_1$	$\hat{\alpha}_2$	$\hat{\alpha}_1$	$\hat{\alpha}_2$	$\hat{\alpha}_1$	$\hat{\alpha}_2$	$\hat{\alpha}_1$	$\hat{\alpha}_2$	$\hat{\alpha}_1$	$\hat{\alpha}_2$	$\hat{\alpha}_1$	$\hat{\alpha}_2$
−0.4	4.918	4.938	4.914	4.992	4.876	4.834	5.014	4.912	NA	NA	NA	NA
−0.2	5.004	4.886	4.982	4.958	5.018	4.974	4.938	4.916	4.994	4.928	4.858	4.890
0.2	4.996	5.046	4.988	4.900	4.920	4.922	4.956	4.926	4.932	4.886	4.964	4.946
0.4	4.930	4.972	4.946	5.074	4.968	5.024	4.910	5.072	4.920	5.036	4.970	5.010
0.6	4.906	4.956	4.894	5.050	4.830	4.932	4.978	5.084	4.906	4.926	4.824	4.960
0.8	4.844	4.880	4.870	4.972	4.834	4.864	4.846	4.956	4.922	4.832	4.828	4.962

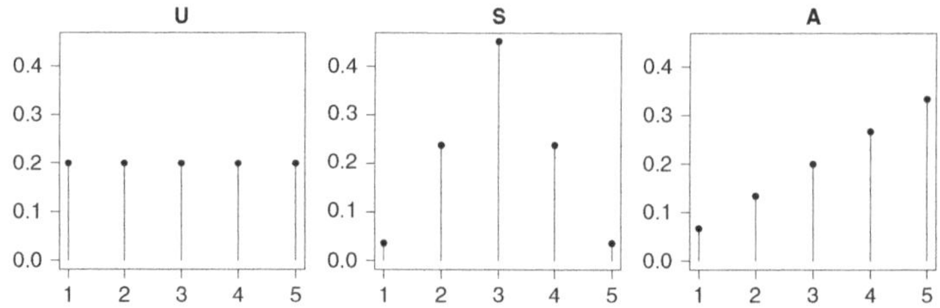

Fig. 4.1 Marginal distributions considered in the simulation study: U=uniform, S=symmetrical, A=asymmetrical (k=5)

variables. It is then interesting to empirically investigate the performance of the CI (4.2) by focusing on the actual coverage and average length. Our simulation procedure offers the possibility of finding the actual coverage probability of the CI for discrete variables under different experimental conditions.

For this purpose, we consider a pair of ordinal variables, with k categories, and $\rho_S = \{0.2, 0.4, 0.6, 0.8\}$. Specifically, we consider the following three "types" of marginal distributions: discrete uniform (U, with constant mass $1/k$ for each value of the support), unimodal symmetrical (S, resembling the continuous normal distribution), and asymmetrical (A, with mass $p(i) = ip(1), i = 1, \ldots, k$), with $k = 3, 5, 7$ (see also Fig. 4.1). By combining the possible marginal distributions and the values of ρ_S, a number of scenarios are obtained. Under each of these scenarios and following our procedure, we generate a matrix of bivariate ordinal data with size $n = 20, 50, 100, 200, 500$, compute the sample correlation coefficient r_S, and then construct the 95 % CI for ρ_S by (4.2). We iterate these steps 20,000 times; at the end of the simulation plan, we compute the Monte Carlo distribution of the sample correlation coefficient r_S, the coverage of the CIs and their average width. The values of the actual (MC) coverage rate for each scenario, displayed in Fig. 4.2, indicate an effect of ρ_S, n, and the marginal distribution. In particular, the marginal distribution seems to play an important role. Although discrete uniform distributions keep the coverage probability quite close to the nominal level (the actual coverage rate is always between 0.933 and 0.967), unimodal symmetrical and asymmetrical distributions apparently distort the coverage probability, often with a negative "bias", especially when the number of categories is low ($k = 3$). The effect of ρ_S is quite important too: high values of ρ_S, combined with symmetrical unimodal distribution with few categories, strongly reduce the coverage probability of CIs. Note that for high values of correlation ($\rho_S = 0.8$) and for small sample size ($n = 20$) some difficulty may arise in the construction of the CI: in fact, r_S is likely to take value 1 in some samples, and then formulas (4.3, 4.4) are clearly no longer applicable. This issue may dramatically decrease the actual coverage rate of CIs, as can be seen looking at the two last plots of the panel of Fig. 4.2. The sample size n seems to have a relevant role only for symmetrical distributions: the coverage

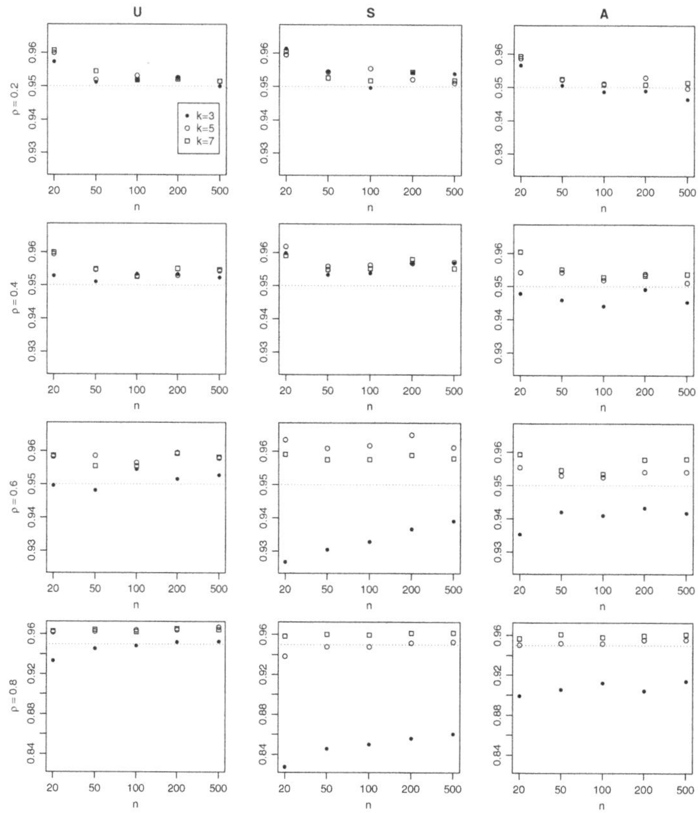

Fig. 4.2 Coverage probability of the CI for Spearman's correlation coefficient under the simulated scenarios

probability gets close to the nominal one as n increases. Note that however even very large values of n (namely, 500) do not ensure a convergence of the actual coverage probability to the nominal value 0.95; this means that relaxing the hypothesis of bivariate normality can significantly distort the actual coverage probability even for large samples. Thus, attention should be paid when building CIs for Spearman correlation based on non-normal samples (ordinal or discrete data), even when the size is large. Departures from the (multi)normality assumption can lead to a severe decrease of the actual coverage probability of such (approximate) CIs when the cardinality of the support is very low; or, vice versa, to an increase if the support comprises several values, and the distribution is symmetrical but unimodal. On the contrary, uniform marginal distributions seem able to keep the coverage probability close to the nominal one.

References

1. Barbiero, A., Ferrari, P.A.: GenOrd: Simulation of ordinal and discrete variables with given correlation matrix and marginal distributions. R package version 1.2.0. http://CRAN.R-project.org/package=GenOrd (2014)
2. Bonett, D.G., Wright, T.A.: Sample size requirements for estimating Pearson, Kendall and Spearman correlations. Psychometrika **1**, 23–28 (2000)
3. Bracquemond, C., Crétois, E., Gaudoin, O.: A comparative study of goodness-of-fit tests for the geometric distribution and application to discrete time reliability, Technical Report, http://www-ljk.imag.fr/SMS/ftp/BraCreGau02.pdf (2002)
4. Cario, M.C., Nelson, B.L.: Modeling and generating random vectors with arbitrary marginal distributions and correlation matrix. Technical report, Department of Industrial Engineering and Management Sciences, Northwestern University, Evanston (1997)
5. Demirtas, H.: A method for multivariate ordinal data generation given marginal distributions and correlations. J. Stat. Comput. Simul. **76**(11), 1017–1025 (2006)
6. Ferrari, P.A., Barbiero, A.: Simulating ordinal data. Multivariate Behav. Res. **47**(4), 566–589 (2012)
7. Guilford, J.P.: Fundamental Statistics in Psychology and Education. McGraw-Hill, New York (1965)
8. Kocherlakota, S., Kocherlakota, K.: Goodness-of-fit tests for discrete distributions. Commun. Stat. Theory Methods **15**, 815–829 (1986)
9. Pearson, K.: Mathematical contributions to the theory of evolution: XVI. On further methods of determining correlation. In: Draper's Research Memoirs. Biometric Series, vol. 4. Cambridge University Press, Cambridge (1907)
10. Ruscio, J., Kaczetow, W.: Simulating multivariate nonnormal data using an iterative algorithm. Multivariate Behav. Res. **3**, 355–381 (2008)

Chapter 5
Probabilistic Counterparts for Strongly Coupled Parabolic Systems

Ya. Belopolskaya

5.1 Motivation

Since the fundamental work by Amann [1] a number of people were interested in the study of strongly coupled parabolic systems, that is systems of parabolic equations with nondiagonal principal part. Systems of this type arise as models for various phenomena in biology, chemistry, hydrodynamics and other fields. Let us mention, for example, the Keller–Segel model [7] of chemotaxis which is a macroscopic model presented via the system of parabolic equations

$$\begin{cases} u_t = \operatorname{div}[\varkappa(u,v)\nabla u - \chi(u,v)\nabla v] + \gamma(u,v), & u(0,x) = u_0(x), \\ v_t = \operatorname{div}[\alpha(u,v)\nabla v + \beta(u,v)\nabla u] + g(u,v), & v(0,x) = v_0(x). \end{cases} \tag{5.1}$$

Here ∇u denotes the gradient of u, u_t is the time derivative, v is the density of the chemical substance, u is the population density and $\beta \geqslant 0, \gamma \geqslant 0$ are the production and decay rates of the chemical, respectively. The function $\chi(u,v)$ is the chemotactic sensitivity which generally takes the form $\chi(u,v) = u\chi(v)$.

Another example is a mathematical model of cell growth, where one can observe cases when cells closely approach and come into contact with each other. This phenomenon is called a contact inhibition of growth between cells. The model describing contact inhibition between normal and abnormal cells was studied in [2]. This model is given by

$$\begin{cases} u_t = \operatorname{div}[u\nabla(u+v)] + (1-u-v)u & t \geqslant 0, x \in R^d, \quad u(0,x) = u_0(x) \\ v_t = \operatorname{div}[v\nabla(u+v)] + \gamma(1-\alpha(u+v))v, & t \geqslant 0, x \in R^d, \quad v(0,x) = v_0(x), \end{cases} \tag{5.2}$$

Ya. Belopolskaya (✉)
St. Petersburg State University for Architecture and Civil Engineering, St. Petersburg, Russia
e-mail: yana@yb1569.spb.edu

V.B. Melas et al. (eds.), *Topics in Statistical Simulation*, Springer Proceedings in Mathematics & Statistics 114, DOI 10.1007/978-1-4939-2104-1_5

where functions $u(t,x)$, $v(t,x)$ represent the densities of normal and abnormal cells while γ and α are positive constants.

Systems of parabolic equations of the type (5.1) and (5.2) give the macroscopic description of the investigated phenomenon. To describe its microscopic picture one needs to construct the probabilistic representation of a solution to the corresponding problem. But as far as we know there is no probabilistic representations of the Cauchy problem solution to systems of this type. Actually there is a number of papers where nondiagonal systems of parabolic equations were studied from a probabilistic point of view. Let us mention papers [3–6, 11] where various types of solutions of the Cauchy problem, namely classical, weak (distributional) and viscosity ones were studied via construction of the correspondent probabilistic representations. In other words in [3–6] there were derived stochastic equations for diffusion processes and their multiplicative operator functionals (MOF) that allow to construct the corresponding probabilistic representation of the solutions to original problems. Eventually one can generalize the model and consider for the underlying stochastic process both a diffusion process and a Markov chain as well as the corresponding MOF. Nevertheless it gives a possibility to consider systems of parabolic equations specified to have mere diagonal principal parts.

Here we derive the correspondent representations for a fully coupled parabolic system (5.2) in terms of a solution to a special system of stochastic equations. Our approach is crucially based on the Kunita theory of stochastic flows [8–10].

5.2 Classical and Weak Solutions of Nondiagonal Parabolic Systems and Their Stochastic Counterparts

Let $\mathscr{D} = \mathscr{D}(R^d)$ be the set of all C^∞-functions with compact supports equipped with the Schwartz topology, $\mathscr{D}'$ be its dual space and Z be the set of all integers. Elements of $\mathscr{S}'$ are called Schwartz distributions. Given $k \in Z$ we denote by $\mathscr{H}^k$ the Sobolev space of real valued functions u, defined on R^d such that u and its generalized derivatives up to the kth order belong to $L^2(R^d)$. The completion $\mathscr{H}^k$ of $\mathscr{D}$ by the norm $\|u\|_k = \left(\sum_{|\alpha|\leqslant k} \int_{R^d} \|\nabla^\alpha u(x)\|^2 dx\right)^{\frac{1}{2}}$ is a Hilbert space. We denote by $\langle u, h\rangle$ pairing between $\mathscr{D}$ and $\mathscr{D}'$. Here $\|\nabla^k u\| = \sum_{|\alpha|\leqslant k} |\frac{\partial^\alpha u}{\partial x_1^{\alpha^1} \ldots \partial x_d^{\alpha_d}}|$ with $\sum_{j=1}^d \alpha_j = k$. We use notations $y\cdot x = \sum_{i=1}^d y_i x_i$, $\langle h, u\rangle = \int_{R^d} h(x)u(x)dx$ and $\langle h, u\rangle = \int_{R^d} h(x)\cdot u(x)dx$ for $L^2(R^d; R^d)$. A distribution u is said to belong to $\mathscr{H}^k_{loc}$ if $hu \in \mathscr{H}^k$ for any $h \in \mathscr{D}$. For any $k \in Z$ and positive $l \in Z$ such that $l \geq |k| + [d/2] + 1$ one can deduce that there exist positive constants C, C_1 such that inequalities

$$\|fu\|_k \leqslant C\|f\|_{k,\infty}\|u\|_k \leqslant C_1\|f\|_l\|u\|_k$$

hold for any $f \in \mathscr{H}^l$, $u \in \mathscr{H}^k$.

Let $\mathscr{H}^k_T = \{u \in L^2([0,T]\times R^d; R) : \nabla^k u \in L^2([0,T]\times R^d; R^d)\}$.

Definition 5.1. A pair of functions $(u, v) \in \mathscr{H}_T^1 \times \in \mathscr{H}_T^1$ is called a weak (distributional) solution of (5.2), provided $\nabla u, \nabla v \in L^2_{\mathrm{loc}}([0,T] \times R^d; R^d)$ and for arbitrary test functions $h(t), g(t) \in \mathscr{H}^1([0,T]; \mathscr{D})$ the following integral identities hold

$$-\int_0^\infty \langle u(\theta), h_\theta(\theta)\rangle d\theta - \langle u(0), h(0)\rangle + \int_0^\infty \langle u(\theta)[\nabla u(\theta) + \nabla v(\theta)], \nabla h(\theta)\rangle\, d\theta$$
$$= \int_0^\infty \langle u(\theta), (1 - u(\theta) - v(\theta)), h(\theta)\rangle d\theta, \tag{5.3}$$

$$-\int_0^\infty \langle v(\theta), g_\theta(\theta)\rangle d\theta - \langle v(0), h(0)\rangle + \int_0^\infty \langle v(\theta)[\nabla v(\theta) + \nabla u(\theta)], \nabla g(\theta)\rangle\, d\theta$$
$$+ \int_0^\infty \langle v(\theta)\gamma(1 - u(\theta) - \kappa v(\theta)), g(\theta)\rangle\, d\theta. \tag{5.4}$$

Denote by $u^1 = u$ and $u^2 = v$. To construct a probabilistic representation of a weak solution $u^q \in \mathscr{H}_T^k$, $q = 1, 2$ of (5.2), we rewrite (5.3), (5.4) as follows

$$\int_0^\infty \left\langle u^1, \left[h_\theta + (u^1 + u^2)\Delta h + (u^1 + u^2)\frac{\nabla u^1}{u^1} \cdot \nabla h + (1 - u^1 - u^2)h\right]\right\rangle d\theta$$
$$= \langle u^1(0), h(0)\rangle, \tag{5.5}$$

$$\int_0^\infty \left\langle u^2, \left[g_\theta + (u^1 + u^2)\Delta g + (u^1 + u^2)\frac{\nabla u^2}{u^2} \cdot \nabla g + \gamma(1 - u^1 - \varkappa u^2)g\right]\right\rangle d\theta$$
$$= \langle u^2(0), h(0)\rangle, \tag{5.6}$$

where Δ is the Laplace operator. The above integral identities prompt that, given functions $u^1, u^2 \in \mathscr{H}_T^2$, one can consider the Cauchy problem for parabolic equations

$$h_\theta + \left(1 + \frac{u^2}{u^1}\right) \nabla u^1 \cdot \nabla h + [u^1 + u^2]\Delta h + (1 - u^1 - u^2)h = 0, \tag{5.7}$$

$$g_\theta + \left(1 + \frac{u^1}{u^2}\right) \nabla u^2 \cdot \nabla g + [u^1 + u^2]g_{xx} + \gamma(1 - u^1 - \varkappa u^2)g = 0. \tag{5.8}$$

Set

$$m^1_{u^1,u^2}(x) = (u^1(x) + u^2(x))\frac{\nabla u^1(x)}{u^1(x)}, \quad m^2_{u^1,u^2}(x) = (u^1(x) + u^2(x))\frac{\nabla u^2(x)}{u^2(x)}, \tag{5.9}$$

$$\frac{1}{2}M^2_{u^1,u^2}(x) = u^1(x)+u^2(x), \quad f_1 = 1-u^1-u^2, \quad f_2 = \gamma(1-u^1-\varkappa u^2). \tag{5.10}$$

Given a probability space $(\Omega, \mathscr{F}, P)$ and a standard Wiener process $w(t) \in R^d$, we consider stochastic equations for $0 \leqslant t \leqslant \theta \leqslant T < \infty$,

$$d\xi^q(\theta) = -m^q_{u^1(\theta),u^2(\theta)}(\xi^q(\theta))d\theta - M_{u^1(\theta),u^2(\theta)}(\xi^q(\theta))dw(\theta), \tag{5.11}$$

where $q = 1, 2$. Let $\xi^q_{t,x}(\theta)$ denote a solution to (5.11) such that $\xi^q_{t,x}(t) = x$.

Assume that u, v are classical solutions of (5.3), (5.4), $u_0, v_0 \in C^{2+\epsilon}$, $\epsilon > 0$, and $u_0 \geqslant \varpi > 0$, $v_0 \geqslant \varpi > 0$. Then, by the SDE theory results we know that there exists a unique solution $\xi^q_{t,x}(\theta) \in R^d$ of (5.11) and by the Feynman–Kac formula one can check that the functions

$$h(t,x) = E\left[\exp\left\{\int_t^T f_1(u(\theta, \xi^1_{t,x}(\theta)), v(\theta, \xi^1_{t,x}(\theta)))d\theta\right\} h_0(\xi^1_{t,x}(T))\right],$$

$$g(t,x) = E\left[\exp\left\{\int_t^T f_2(u(\theta, \xi^2_{t,x}(\theta)), v(\theta, \xi^2_{t,x}(\theta)))d\theta\right\} g_0(\xi^2_{t,x}(T))\right],$$

define the classical solution of the Cauchy problem for parabolic equations (5.7), (5.8) with the Cauchy data $h(T, x) = h_0(x), \quad g(T, x) = g_0(x)$.

To find a link between the above processes $\xi^q(t), q = 1, 2$, and weak solutions of (5.2) we need some additional results from the stochastic flow theory [8–10].

Under the above assumptions set $\varphi^q_{s,t}(x) = \xi^q_{s,x}(t)$, $q = 1, 2$, and note that $\varphi^q_{s,t} : R^d \to R^d$ is a C^2-diffeomorphism of R^d called the stochastic flow. Let $[\varphi^q_{s,t}]^{-1} = \psi^q_{t,s}$ be the inverse maps of the stochastic flows $\varphi^q_{s,t}$. We denote by $\hat{\xi}^q(\theta) = \xi^q(t-\theta)$ the stochastic process with the stochastic flow $\psi^q_{t,\theta}$.

Given a distribution u_0 we define a distribution valued processes

$$T^q(t) = \exp\left\{\int_0^t f_q \circ \varphi^q_{0,\theta} d\theta\right\} u^q_0.$$

Next, we consider a composition $T^q(t) \circ \psi^q_{t,0}$, where

$$T^q(t) \circ \psi^q_{t,0}(x) = \exp\left\{\int_0^t f_q(\psi^q_{\theta,0}(x))d\theta\right\} u^q_0(\psi^q_{t,0}(x)),$$

and its generalized expectation

$$U^q(t) = E\left[T^q(t) \circ \psi^q_{t,0}\right].$$

Finally, by the results from [9, 10] we know that $U^q(t)$ is an evolution family of bounded operators acting in $\mathscr{H}^k_T$. Now we are ready to state our main results.

Theorem 5.1. *Assume that* $u_0, v_0 \in \mathscr{H}^1 \cap C^2$ *and there exists a unique weak solution* (u, v) *of the Cauchy problem (5.2) such that* $u(t), v(t) \in \mathscr{H}_T^1 \cap C^2(R^d)$. *Let the processes* $\xi^1(t), \xi^2(t)$ *satisfy (5.11), while* $\hat{\xi}^1(t)$, $\hat{\xi}^2(t)$ *are the correspondent reversal processes. Then, for any test functions* h, g *functionsc*

$$U^q(t,x) = E_{t,x}\left[\exp\left\{\int_0^t f_q(\hat{\xi}^q(\theta))d\theta\right\} u_0^q(\hat{\xi}^q(t))\right], \quad q = 1,2, \tag{5.12}$$

satisfy integral identities

$$\int_t^T \left\langle U^1(\theta), \left[h_\theta(\theta) + [u^1(\theta) + u^2(\theta)]\Delta h(\theta) + [u^1(\theta) + u^2(\theta)]\frac{\nabla u^1}{u^1} \cdot \nabla h\right]\right\rangle d\theta$$
$$= \langle U^1(T), h(T)\rangle - \langle U^1(t), h(t)\rangle + \int_t^T \langle U^1(\theta), (1 - u^1(\theta) - u^2(\theta))h(\theta)\rangle d\theta, \tag{5.13}$$

$$\int_t^T \left\langle U^2(\theta), \left[g_\theta(\theta) + [u^1(\theta) + u^2(\theta)]\Delta g(\theta) + [u^1(\theta) + u^2(\theta)]\frac{\nabla u^2}{u^2}\right]\right\rangle d\theta$$
$$= \left\langle U^2(T), g(T)\rangle - \langle U^2(t), g(t)\rangle + \int_t^T \langle U^2(\theta), \gamma(1 - u^1(\theta) - \varkappa u^2(\theta))g(\theta)\right\rangle d\theta. \tag{5.14}$$

Theorem 5.2. *Assume that Theorem 5.1 assumptions hold and distributions* $u_0^q, q = 1,2,$ *belong to* $\mathscr{H}^k$. *Then, the pair of functions* $u^q = E[U^q(t) \circ \psi_{t,0}^q] \in \mathscr{H}_T^k$ *gives the unique distributional solution of (5.2).*

We prove these statements in the next section.

Corollary 5.1. *Under Theorem 5.1 assumptions the functions* $U^1(t,x), U^2(t,x)$ *given by (5.12) are twice differentiable in space variable and*

$$U^1(t,x) = u^1(t,x), U^2(t,x) = u^2(t,x). \tag{5.15}$$

Hence, relations $\xi^q(t) = x, \quad q = 1,2,$

$$d\xi^q(\theta) = -m^q_{u^1(\theta),u^2(\theta)}(\xi^q(\theta))d\theta - M_{u^1(\theta),u^2(\theta)}(\xi^q(\theta))dw(\theta), \tag{5.16}$$

$$u^1(t,x) = E_{t,x}\left[\exp\left\{\int_0^t f_1(\hat{\xi}^1(\theta))d\theta\right\} u_0^1(\hat{\xi}_1(t))\right], \tag{5.17}$$

$$u^2(t,x) = E_{t,x}\left[\exp\left\{\int_0^t f_2(\hat{\xi}^2(\theta))d\theta\right\} u_0^2(\hat{\xi}^2(t))\right] \tag{5.18}$$

make a closed system.

Since both $(U^1(t,x), U^2(t,x))$ and $(u^1(t,x), u^2(t,x))$ satisfy the same integral identity, the corollary statement results from the assumed uniqueness of a distributional solution of (5.2).

5.3 Stochastic Flow Theory

To prove Theorem 5.1 we consider the Jacobian $\hat{J}^q_{\theta,t}(\omega)) = \det\nabla\hat{\xi}^q_{t,y}(\theta)$ of the map $\psi^q_{t,\theta}$ and note that $\hat{J}^q_{\theta,t}(\omega) > 0$ and $\hat{J}^q_{t,t}(\omega)) = 1$. To simplify notations we omit indices u^1, u^2 and use Stratonovich form of (5.11)

$$d\xi^q_{t,x}(\theta) = -\tilde{m}^q(\xi^q_{t,x}(\theta))d\theta - M(\xi^q_{t,x}(\theta)) \circ dw(\theta), \tag{5.19}$$

where $\tilde{m}^q(x) = m^q(x) - \nabla M(x)M(x)$ and $M(\xi^q(\theta)) \circ dw(\theta) = M(\xi^q(\theta))dw(\theta) + \nabla M(\xi^q(\theta))M(\xi^q(\theta))d\theta$. One can check that $\hat{\xi}^q(t) = \psi^q_{t,\theta}(y)$, $q = 1,2$ satisfy SDE

$$d\psi^q_{t,\theta}(y) = [\nabla\varphi^q_{\theta,t}]^{-1}(\psi^q_{t,\theta})\tilde{m}^q(y)d\theta + \nabla\varphi^q_{t,\theta}(\psi^q_{t,\theta})^{-1}M(y) \circ dw(\theta), \tag{5.20}$$

where $[\nabla\varphi^q_{t,\theta}(y)]^{-1}$ is the Jacobian matrix inverse to the Jacobian matrix $\nabla\varphi^q_{t,\theta}(x)$ of the map $\varphi^q_{t,\theta}(x)$. To be more precise, we deduce from the Kunita theorem (see [8], Theorem 4.2.2) the following result.

Theorem 5.3. *Let $\varphi^q_{t,\theta}(x)$ satisfy (5.11) with $m^q, M \in C^k$, where $(k \geqslant 3)$. Then, the inverse flow $[\varphi^q_{\theta,t}]^{-1} = \psi^q_{t,\theta}$ satisfies (5.20).*

Proof. To verify the assertion we consider a stochastic equation that governs the Jacobian matrix $\hat{\mathbf{J}}^q(\theta) = [\nabla\varphi^q_{\theta,t}]^{-1}$

$$d\hat{\mathbf{J}}^q(\theta) = \nabla[\tilde{m}^q(\varphi^q_{t,\theta}(x))]\hat{\mathbf{J}}^q(\theta)d\theta + \nabla[M(\varphi^q_{t,\theta}(x))]\hat{\mathbf{J}}^q(\theta) \circ dw(\theta), \quad \hat{\mathbf{J}}^q(\theta) = I. \tag{5.21}$$

Consider the process

$$G^q(x,\theta) = \int_t^\theta [\nabla\varphi^q_{t,\tau}]^{-1}(x)\tilde{m}^q(\varphi^q_{t,\tau}(x))d\tau + \int_t^\theta [\varphi^q_{t,\tau}]^{-1}_x(x)M(\varphi^q_{t,\tau}(x)) \circ dw(\tau),$$

and evaluate $\varphi^q_{\theta,t}(\psi^q_{t,\theta}(y))$, where $\psi^q_{t,\theta}(y)$ is a random process with stochastic differential $d\psi^q_{t,\theta}(y) = dG^q(\psi^q_{t,\theta}(y),\theta)$. Set $\varphi^q_{t,\theta}(x) = \varphi^q(x,\theta)$. By the Ito–Wentzel formula we have

$$\varphi^q_{t,\theta}(\psi^q_{\theta,t}(y)) = y + \int_t^\theta d^S\varphi^q(\psi^q_{\tau,t}(y),\tau) + \int_t^\theta \nabla\varphi^q(\psi^q_{\tau,t}(y),\tau)\circ d\psi^q_{\tau,t}(y)$$
$$= y - \int_t^\theta \tilde{m}^q(\varphi^q_{t,\tau}(\psi^q_{\tau,t}(y)))d\tau - \int_t^\theta M(\varphi^q_{t,\tau,}(\psi^q_{\tau,t}(y)))\circ dw(\tau)$$
$$+\int_t^\theta \nabla\varphi^q(\psi^q_{\tau,t}(y),\tau)[\nabla\varphi^q(\psi^q_{\tau,t}(y),\tau)]^{-1}\tilde{m}^q(\varphi^q_{t,\tau}(\psi^q_{\tau,t}(y)))d\tau$$
$$+\int_t^\theta \nabla\varphi^q(\psi^q_{\tau,t}(y),\tau)[\nabla\varphi^q(\psi^q_{t,\tau}(y),\tau)]^{-1}M(\varphi^q_{t,\tau}(\psi^q_{\tau,t}(y)))\circ dw(\tau)$$
$$= y.$$

Hence , $\varphi^q_{t,\theta}(\psi^q_{\theta,t}(y)) = y$.

Given a distribution $u^q \in \mathscr{D}'$ we define its composition with a stochastic flow $\psi^q_{\theta,t}(\omega)$ as a random variable valued in $\mathscr{D}'$. Given a function $h \in \mathscr{D}$ the product $h\circ\varphi^q_{t,\theta}(\omega)J^q_{t,\theta}(\omega)$ belongs to $\mathscr{D}$, where $J^q_{t,\theta}$ is the Jacobian of $\varphi^q_{t,\theta}$. Set

$$T^q_{t,\theta}h(\omega) = \langle u^q, h\circ\varphi^q_{t,\theta}(\omega)J^q_{t,\theta}(\omega)\rangle, \quad h\in\mathscr{D}. \tag{5.22}$$

One can easily check that in this way a linear functional over $\mathscr{D}$ is defined. We denote it by $u^q\circ\psi^q_{\theta,t}$. Provided $u^q = u^q(x)dx$ where $u^q(x)$is a continuous function, $u^q\circ\psi^q_{\theta,t}$ is just a composition of u^q with $\psi^q_{\theta,t}$ due to the following integral by parts formula

$$\int_{R^d} u^q(\psi^q_{\theta,t}(x,\omega))h(x)dx = \int_{R^d} u^q(y)h(\varphi^q_{t,\theta}(y,\omega))J^q_{t,\theta}(y,\omega)dy, \quad h\in\mathscr{D}. \tag{5.23}$$

Let $\langle[\mathscr{L}^q]^*u^q, h\rangle = \langle u^q, \mathscr{L}^q h\rangle\rangle$. Applying the generalized Ito formula derived by Kunita [9], we deduce the following result.

Lemma 5.1. *Let $u^q(t)\in\mathscr{H}^1_T$ be a nonrandom continuous function and $\varphi^q_{t,\theta}$, $\psi^q_{\theta,t}$ be the above defined stochastic flows generated by solutions to (5.11). Then we have*

$$u^q(t)\circ\psi^q_{\theta,t} = u^q(\theta) + \int_t^\theta d_\tau u^q(\tau)\circ\psi^q_{\tau,t} + \int_t^\theta [L^q_0]^*[u^q(\tau)\circ\psi^q_{\tau,s}]d\tau$$
$$+\int_t^\theta \nabla u(\tau)\circ\psi_{\tau,t}]M\,dw, \tag{5.24}$$

where $\mathscr{L}^q_0 f = m^q\cdot\nabla f + \frac{1}{2}M^2\Delta f$ and $[L^q_0]^*$ is defined in the distribution sense.

By Lemma 5.1 we deduce the following assertion.

Let $u^q \in \mathscr{H}_T^k$. Then one can prove (see [10]) that $\langle Z_{t,0}^q, h\rangle = E[\langle u^q(t) \circ \psi_{t,0}^q, h\rangle]$ exists for any $h \in \mathscr{H}_T^{-k}$, defines a continuous linear functional on $\mathscr{H}^{-k}$ and hence can be considered as an element $Z_{t,0}^q = E[u^q(t) \circ \psi_{t,0}]$ from $\mathscr{H}_T^k$ which is called the generalized expectation.

Let

$$\eta_0^q(t) = \exp\left(\int_0^t f_q \circ \psi_{\theta,t}^q d\theta\right) \tag{5.25}$$

and $\varkappa_0^q(t) = \eta_0^q(t) u^q(t)$. By the generalized Ito formula we can verify that $U^q(t) = E[\varkappa_0^q(t) \circ \psi_{t,0}^q]$ is the unique generalized solution to the Cauchy problem

$$\frac{dU^q}{dt} = [L^q]^* U^q(t), \quad u^q(0) = u_0^q, \tag{5.26}$$

where $[L^q]^*$ is the operator defined in a distributional sense and dual to

$$L^q = \frac{1}{2} M_{u^1,u^2}^2(x)\Delta + m_{u^1,u^2}^q(x)\nabla + f_q(x).$$

This concludes the proof of Theorem 5.1.

Coming back to (5.2) recall that we have assumed that $u^1 = u, u^2 = v$ are unique distributional solutions to (5.2) as well and hence $U^q(t,x) \equiv u^q(t,x)$ that yields the statement of Theorem 5.2.

Let us mention some final remarks. In this paper we have constructed a probabilistic representation of a weak solution to the problem (5.2) which belongs to $\mathscr{H}_T^1 \cap C^2(R^d)$. Actually, Theorem 5.1 states that if we have a unique weak solution (u, v) of the problem (5.2) from this class, the functions $u = u^1, v = v^1$ admit the probabilistic representations of the form

$$u^1(t,y) = E_{t,y}[\eta_0^1(t) u^1(0) \circ \psi_{t,0}^1(y)], \quad u^2(t,y) = E_{t,y}[\eta_0^2(t) u^2(0) \circ \psi_{t,0}^2(y)]. \tag{5.27}$$

Notice that relations (5.11), (5.20), (5.25) and (5.27) make a closed system and our next problem to be discussed somewhere else is to prove that under suitable conditions this system has a unique solution $(\xi^q(t), \eta_0^q(t), u^q(t,x))$, $q = 1, 2$. Finally we have to check that setting $u = u^1, v = u^2$ we get a distributional solution of the problem (5.2) as well.

Acknowledgements Financial support of grant RFBR 12-01-00427-a and the Minobrnauki project N 2074 are gratefully acknowledged.

References

1. Amann, H.: Dynamic theory of quasilinear parabolic systems - III. Global existence. Math. Z. **202**, 219–250 (1989)
2. Bertsch, M., Hilhorst, D., Izuhara,H. , Mimura, M.: A nonlinear parabolic-hyperbolic system for contact inhibition of cell-growth. Differ. Equ. Appl. **4**(1), 137–157 (2012)
3. Belopolskaya, Ya., Dalecky, Yu.: Investigation of the Cauchy problem for systems of quasilinear equations via Markov processes. Izv. VUZ Matematika. N **12**, 6–17 (1978)
4. Belopolskaya, Ya., Dalecky, Yu.L.: Stochastic Equations and Differential Geometry. Kluwer, Boston (1990)
5. Belopolskaya, Ya., Woyczynski,W.: Generalized solution of the Cauchy problem for systems of nonlinear parabolic equations and diffusion processes. Stoch. Dyn. **11**(1), 1–31 (2012)
6. Belopolskaya, Ya., Woyczynski, W.: Probabilistic approach to viscosity solutions of the Cauchy problem for systems of fully nonlinear parabolic equations. J. Math. Sci. **188**, 655–672 (2013)
7. Corrias, L., Perthame, B., Zaag,H.: A model motivated by angiogenesis. Milan J. Math. **72** 1–29 (2004)
8. Kunita, H.: Stochastic Flows and Stochastic Differential Equations. Cambridge University Press, Cambridge (1990)
9. Kunita, H.: Stochastic Flows Acting on Schwartz Distributions. J. Theor. Pobab. **7**(2), 247–278 (1994)
10. Kunita, H.: Generalized solutions of stochastic partial differential equations. J. Theor. Pobab. **7**(2), 279–308 (1994)
11. Pardoux, E., Pradeilles, F., Rao, Z.: Probabilistic interpretation of semilinear parabolic partial differential equations. Annales de l'I.H.P, Sec. B **33**(4), 467–490 (1997)

Chapter 6
Algorithms for Linear Stochastic Delay Differential Equations

Harish S. Bhat

6.1 Introduction

We consider the stochastic delay differential equation (SDDE)

$$d\mathbf{X}_t = A\mathbf{X}_t dt + B\mathbf{X}_{t-\tau} dt + C d\mathbf{W}_t. \tag{6.1}$$

Here A, B, and C are $N \times N$ constant coefficient matrices, the time delay $\tau > 0$ is constant, $\mathbf{W}_t$ is N-dimensional Brownian motion, and the unknown $\mathbf{X}_t$ is an $\mathbb{R}^N$-valued stochastic process.

System (6.1) models phenomena in neuroscience [7] and mechanics [4, 10], among several other fields. For each $t \geqslant 0$, let $p(\mathbf{x}, t)$ denote the probability density function of $\mathbf{X}_t$. In many scientific contexts, the quantities of interest are functionals of p—for example, the mean and variance of the solution of (6.1). In these contexts, the sample paths $\mathbf{X}_t$ of (6.1) are of interest only to the extent that they help to compute p or functionals of p.

Let $\dagger$ denote matrix transpose. If we remove the time delay term, say by setting $B = 0$, then we can solve for $p(\mathbf{x}, t)$ directly via the partial differential equation

$$\frac{\partial p}{\partial t} + \operatorname{trace}(A) p + (A\mathbf{x})^\dagger \nabla p = \frac{1}{2} C C^\dagger \nabla^2 p \tag{6.2}$$

This equation is known as either the Fokker–Planck or Kolmogorov equation associated with the stochastic differential equation $d\mathbf{X}_t = A\mathbf{X}_t dt + C d\mathbf{W}_t$.

The Fokker–Planck equation associated with the time-delayed equation (6.1) suffers from a closure problem [6]. This problem prevents the application of

H.S. Bhat (✉)
University of California, Merced, CA, USA
e-mail: hbhat@ucmerced.edu

V.B. Melas et al. (eds.), *Topics in Statistical Simulation*, Springer Proceedings in Mathematics & Statistics 114, DOI 10.1007/978-1-4939-2104-1_6

deterministic methods from numerical analysis to solve for the density function $p(\mathbf{x}, t)$. As a result, Monte Carlo methods are commonly employed; in this framework, one simulates a sufficiently large number of sample paths of (6.1) in order to estimate the density function or functionals thereof.

In this note, we develop a new algorithm to directly solve for the density function of (6.1). By first discretizing (6.1) in time, we bypass the closure issues of Fokker–Planck approaches. The resulting scheme involves no sampling, and is thus capable of computing the density function of (6.1) faster than Monte Carlo methods, for the same desired level of accuracy.

6.2 Algorithm

Starting from (6.1), we apply the Euler–Maruyama time-discretization. Specifically, let ℓ denote a positive integer, and set $h = \tau/\ell$. Let $\mathbf{Y}_n$ denote the numerical approximation to $\mathbf{X}_{nh}$. Then, by definition, $\mathbf{Y}_{n-\ell}$ is the numerical approximation to $\mathbf{X}_{nh-\tau}$. Set I equal to the $N \times N$ identity matrix, and let $\{\mathbf{Z}_n\}_{n\geqslant 1}$ denote an i.i.d. sequence of $\mathcal{N}(\mathbf{0}, I)$ random variables—here $\mathcal{N}(\boldsymbol{\mu}, \Sigma)$ denotes the multivariate Gaussian with mean vector $\boldsymbol{\mu}$ and covariance matrix Σ. Then the Euler–Maruyama discretization of (6.1) is

$$\mathbf{Y}_{n+1} = (I + Ah)\mathbf{Y}_n + Bh\mathbf{Y}_{n-\ell} + Ch^{1/2}\mathbf{Z}_{n+1}. \tag{6.3}$$

Thus far we have not mentioned initial conditions. For the original differential equation (6.1), the initial conditions consist of the segment $\{\mathbf{X}_t \mid -\tau \leqslant t \leqslant 0\}$. Discretizing this segment yields $\mathcal{I} = \{\mathbf{Y}_n \mid -\ell \leqslant n \leqslant 0\}$, where $\mathbf{Y}_n = \mathbf{X}_{nh}$. In what follows, we assume that $\mathcal{I}$ is given and that each $\mathbf{Y}_n \in \mathbb{R}^N$ is a constant (deterministic) vector.

With (6.3) together with the initial segment $\mathcal{I}$, we can certainly generate sample paths $\{\mathbf{Y}_n\}_{n\geqslant 1}$. Note that this involves sampling the random variables $\{\mathbf{Z}_n\}_{n\geqslant 1}$. See [2] for numerical analysis of this approach.

Let us now give a convenient representation of the solution of (6.3):

Theorem 6.1. *For each $n \geqslant -\ell$, there exist $N \times N$ matrices $\{\alpha^n_m\}^0_{m=-\ell}$ and $\{\beta^n_r\}^n_{r=1}$ such that*

$$\mathbf{Y}_n = \sum_{m=-\ell}^{0} \alpha^n_m \mathbf{Y}_m + \sum_{r=1}^{n} \beta^n_r \mathbf{Z}_r. \tag{6.4}$$

Proof. When $-\ell \leqslant n \leqslant 0$, the statement is true by definition: in this range, the β^n_r matrices are all zero, $\alpha^n_n = I$, and $\alpha^n_m = 0$ for $m \neq n$.

The rest of the proof is by induction. For the $n = 1$ case, we note that (6.3) implies

$$\mathbf{Y}_1 = (I + Ah)\mathbf{Y}_0 + Bh\mathbf{Y}_{-\ell} + Ch^{1/2}\mathbf{Z}_1.$$

Hence $\alpha_0^1 = I + Ah$, $\alpha_{-\ell}^1 = Bh$, and $\alpha_m^n = 0$ for $-\ell < m < 0$. Setting $\beta_1^1 = Ch^{1/2}$, we see that (6.4) holds for $n = 1$.

Next assume that (6.4) holds for $1 \leqslant n \leqslant n'$. For $-\ell \leqslant m \leqslant 0$, set

$$\alpha_m^{n'+1} = (I + Ah)\alpha_m^{n'} + Bh\alpha_m^{n'-\ell}. \tag{6.5}$$

For $1 \leqslant r \leqslant n' + 1$, set

$$\beta_r^{n'+1} = \begin{cases} (I + Ah)\beta_r^{n'} + Bh\beta_r^{n'-\ell} & 1 \leqslant r \leqslant n' - \ell \\ (I + Ah)\beta_r^{n'} & n' - \ell + 1 \leqslant r \leqslant n' \\ Ch^{1/2} & r = n' + 1. \end{cases} \tag{6.6}$$

A calculation now shows that $\mathbf{Y}_{n'+1}$ defined by (6.4), (6.5), and (6.6) satisfies the $n = n' + 1$ case of (6.3). □

The system (6.5) and (6.6) is an algorithm for determining the solution of the discretized equation (6.3). This algorithm does not involve sampling any random variables. There are several points we wish to make about this algorithm:

1. The α equation (6.5) is decoupled from the β equation (6.6). The equations can be stepped forward in time independently of one another.
2. The dynamics of (6.5) and (6.6) are independent of the initial conditions $\mathscr{I}$. Once we have computed α and β, we can evaluate the solution (6.4) for any choice of initial conditions.
3. Once we have the solution in the form (6.4), it is simple to determine the distribution of $\mathbf{Y}_n$. Each $\beta_r^n \mathbf{Z}_r$ has a $\mathscr{N}(\mathbf{0}, \beta_r^n(\beta_r^n)^\dagger)$ distribution. Using the independence of each $\mathbf{Z}_r$ and the fact that the initial vectors $\{\mathbf{Y}_m\}_{m=-\ell}^0$ are constant, we have

$$\mathbf{Y}_n \sim \mathscr{N}\left(\sum_{m=-\ell}^{0} \alpha_m^n \mathbf{Y}_m, \sum_{r=1}^{n} \beta_r^n (\beta_r^n)^\dagger\right). \tag{6.7}$$

   The upshot is that the α and β coefficients describe, respectively, the mean and the variance/covariance of the computed solution.
4. The α equation (6.5) can be derived in a much more direct fashion. Let us first take the expected value of both sides of (6.1) to derive the deterministic DDE (delay differential equation):

$$\frac{d}{dt}E[\mathbf{X}_t] = AE[\mathbf{X}_t] + BE[\mathbf{X}_{t-\tau}].$$

   Applying the standard Euler discretization to this equation yields (6.5). Numerous prior works have studied Euler discretizations of a deterministic DDE. Therefore, for the empirical convergence tests described below, we consider problems where $E[\mathbf{X}_t]$ is zero and focus our attention on (6.6).

5. Several methods exist to approximate SDDE by Markov chains [1, 8, 9]. Such methods necessarily involve creating a number of discrete states to approximate the continuous state space of (6.1); often the number of such states scales with ℓ, the discrete delay. While the Markov chain method of [1] is accurate and fast for delayed random walks where ℓ is small and fixed, the number of states scales like 4^ℓ. Hence the method breaks down when τ is large; in this case, in order for the time step $h = \tau/\ell$ to be acceptable, we must choose a large value of ℓ. Algorithm (6.5) and (6.6) does not discretize the state space of (6.1), and it is much less sensitive to the magnitude of the time delay τ than Markov chain methods.

6.3 Implementation and Tests

We have implemented algorithm (6.5) and (6.6) in R, an open-source framework for statistical computing. The implementation simplifies considerably in the case of a scalar equation, i.e., when $N = 1$. We therefore separate our discussion into scalar and vector cases.

6.3.1 *Scalar Case* ($N = 1$)

When $N = 1$, the coefficients A, B, and C in (6.1) and the coefficients $\{\alpha_m^n\}$ and $\{\beta_r^n\}$ in (6.4) are all scalars. Then $\alpha^n = (\alpha_{-\ell}^n, \ldots, \alpha_0^n)$ and $\beta^n = (\beta_1^n, \ldots, \beta_n^n)$ are vectors, of respective dimension $\ell + 1$ and n. With this notation, (6.5) and (6.6) can be written in matrix–vector form as

$$\alpha^{n+1} = (1 + Ah)\alpha^n + Bh\alpha^{n-\ell} \tag{6.8}$$

$$\beta^{n+1} = \begin{bmatrix} (1 + Ah)\beta^n \\ 0 \end{bmatrix} + \begin{bmatrix} Bh\beta^{n-\ell} \\ \mathbf{0} \end{bmatrix} + Ch^{1/2}\mathbf{e}_{n+1}. \tag{6.9}$$

Here $\mathbf{0}$ is the zero vector in $\mathbb{R}^{\ell+1}$, and $\mathbf{e}_{n+1} = (0, \ldots, 0, 1) \in \mathbb{R}^{n+1}$.

As explained above, algorithm (6.8) and (6.9) yields the exact probability density function of the stochastic delay *difference* equation (6.3). To explore the practical benefits of this fact, we compare our algorithm against the following Monte Carlo procedure: (i) fix a value of the time step $h = \tau/\ell$, (ii) sample the random variables $\{Z_n\}_{n\geq 1}$ and step forward in time using (6.3), (iii) stop when we obtain a sample of Y_n at a fixed final time $T > 0$. Running this procedure many times, we obtain a corpus of samples of Y_n at time T. In what follows, we will compare the variance of these samples against the variance computed using (6.9).

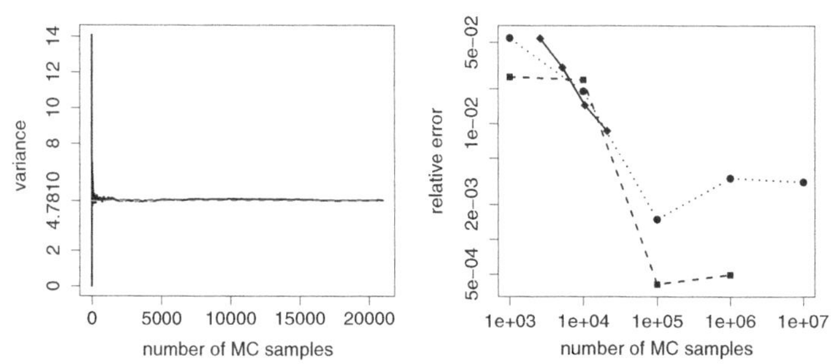

Fig. 6.1 On the *left*, we fix $h = 10^{-4}$, and plot in *solid black* the variance of the first N Monte Carlo (MC) samples of (6.3) as a function of N, together with the variance computed using (6.9) in *dashed red*. Convergence to the reference variance is not monotonic, seen more clearly on the right log–log plot. Here we show results from runs with $h = 10^{-2}$ (*circles, dotted*), $h = 10^{-3}$ (*squares, dashed*), and $h = 10^{-4}$ (*diamonds, solid*). Each point is the relative error between the computed MC variance and the reference variance. In general, a very large number of MC samples may be necessary to achieve the accuracy of (6.9)

For concreteness, let us fix the parameters $\tau = 1$, $T = 5$, $A = -0.2$, $B = 0.3$, and $C = 1.0$. Recall that the time step h is determined by $h = \tau/\ell$ where ℓ is a fixed positive integer. In the left half of Fig. 6.1, we set $\ell = 10^4$ (so that $h = 10^{-4}$) and plot in solid black the variance of the first N Monte Carlo samples as a function of N. The total number of samples computed here is $N = 21{,}000$. We also plot in dashed red the variance computed using (6.9), which to four decimals is 4.7810.

In the right half of Fig. 6.1, we show three sets of numerical tests. Each point here is the relative error between the computed Monte Carlo variance and the reference variance computed using (6.9), plotted on a log–log scale. In circles (dotted line), we have data for $h = 10^{-2}$. In squares (dashed line), we have data for $h = 10^{-3}$. In diamonds (solid line), we have data for $h = 10^{-4}$. The main point that we take from this plot is that the convergence of the Monte Carlo method to the solution computed using (6.9) is likely to be slow and non-monotonic. This implies that algorithm (6.8) and (6.9) can be used to significantly speed up simulations of linear SDDE. Algorithm (6.8) and (6.9) computes a solution with an accuracy that can only be approached by Monte Carlo methods with an extremely large number of samples.

In terms of convergence results, what we are most interested in is the $h \to 0$ convergence of the algorithm (6.5) and (6.6) or its scalar variant (6.8) and (6.9), without regard to any Monte Carlo scheme. In the left half of Fig. 6.2, we plot the variance computed using (6.9) as a function of h, the time step. The horizontal axis has been scaled logarithmically. The convergence shown is consistent with first-order convergence, i.e., an error that scales likes h. This comes as no surprise; the Euler–Maruyama method used to derive (6.3) exhibits first-order weak convergence. To state this more formally, let $\mathscr{C}_P^k(\mathbb{R}^N)$ denote the space of k times continuously differentiable real-valued functions on $\mathbb{R}^N$, such that the

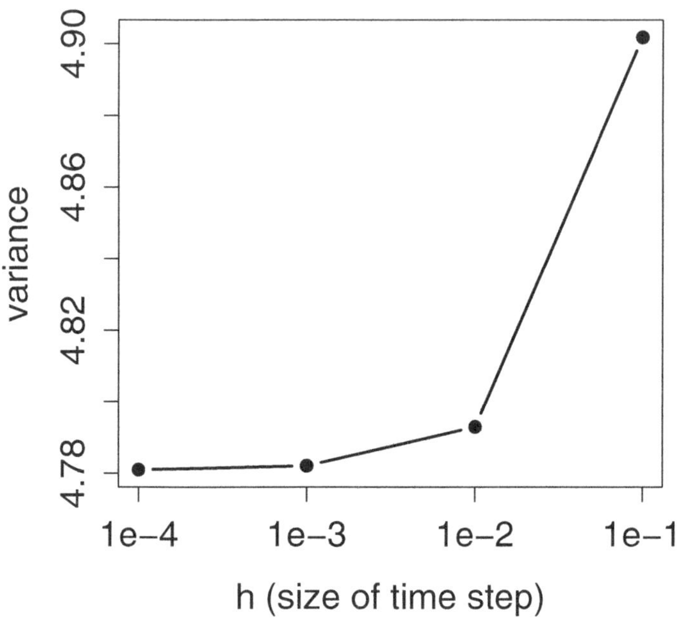

Fig. 6.2 We compute the variance using (6.9) at $T = 5$ using $h = 10^{-j}$ for $j = 1, 2, 3, 4$. The variance appears to converge as h decreases, and the rate is consistent with first-order convergence. Note that the horizontal axis is logarithmically scaled

functions and their derivatives have polynomial growth [5]. Then it is known [3] that there exist $0 < H < 1$ and C (independent of h) such that for all $0 < h < H$ and all $g \in \mathscr{C}_P^{2(\gamma+1)}(\mathbb{R}^N)$,

$$|E(g(\mathbf{X}_T)) - E(g(\mathbf{Y}_{T/h}))| \leqslant Ch. \tag{6.10}$$

In future work, we aim to build on this result to prove convergence of (6.5) and (6.6).

6.3.2 *Vector Case* ($N > 1$)

Now we return to the fully vectorial algorithm (6.5) and (6.6). Let $N = 2$ and define

$$A = \begin{bmatrix} -0.8 & -1.25 \\ 1 & 0 \end{bmatrix}, \quad B = \begin{bmatrix} -0.05 & -0.21 \\ 0.19 & -0.36 \end{bmatrix}, \quad C = \begin{bmatrix} 0.014 & 0.028 \\ 0.042 & 0.014 \end{bmatrix}. \tag{6.11}$$

We fix $\tau = 1$ and set the initial conditions $\mathbf{X}_t = [1, 0]$ for $-\tau \leqslant t \leqslant 0$. We then seek the solution $\mathbf{X}_t$ of (6.1).

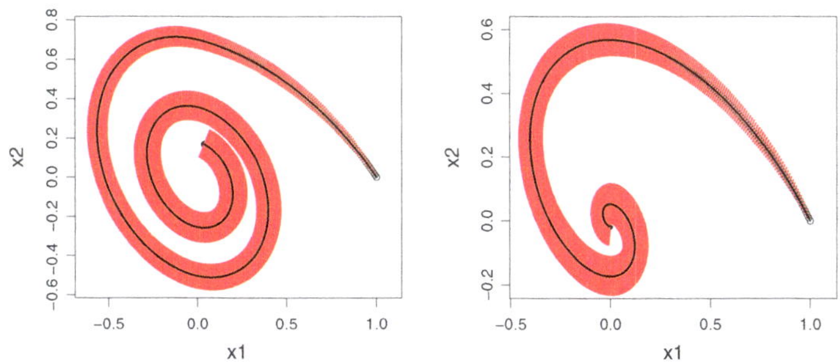

Fig. 6.3 We illustrate vector-valued solutions of (6.1) using algorithm (6.5) and (6.6) with $\tau = 1$, $h = 10^{-2}$, and initial conditions fixed at [1, 0]. For both plots, the *black line* gives the evolution of the mean vector $E[\mathbf{X}_t]$; at each point in time, the *red band* has total width equal to twice the spectral norm of the variance–covariance matrix $\mathrm{Var}[\mathbf{X}_t]$. For the plot on the *left*, all three matrices A, B, and C are nonzero and given by (6.11). For the plot on the *right*, we retain the A and C matrices, but shut off the time delay by setting $B = 0$. These plots demonstrate the utility of algorithm (6.5) and (6.6)

Using algorithm (6.5) and (6.6), we compute the α and β matrices up to $T = 10$ using a time step of $h = 10^{-2}$. We then use (6.7) to compute the mean vector and variance–covariance matrix of the solution at each point in time from $t = 0$ to $t = T$. In the left half of Fig. 6.3, we plot using a black solid line the time evolution of the mean vector. At each point in time at which the solution is computed, we also plot a red line segment whose total length is twice the spectral norm of the variance–covariance matrix at that time. These segments are intended to help visualize the uncertainty in the mean solution, and they are plotted orthogonally to the tangent vectors of the black line.

We see from (6.11) that if B and C were instead equal to zero, the dynamics of (6.1) would be governed by A. The resulting linear system has a globally attracting spiral-type equilibrium at [0, 0]. This stable spiral dynamic can be seen in the left plot of Fig. 6.3. The width of the red band is due entirely to the C matrix in (6.11). If we solve (6.1) with the noise matrix shut off (i.e., $C = 0$) and A and B as in (6.11), the solution would be given by the black line.

To analyze the effect of the time-delay term governed by B, we solve the system again using algorithm (6.5) and (6.6), but this time with $B = 0$. The solution in this case is plotted in the right half of Fig. 6.3. Though the attracting fixed point at [0, 0] remains, the dynamics are noticeably different. In this case, we can see that the time delay term acts to slow the system's approach to equilibrium.

Note that producing both plots in Fig. 6.3 requires less than 15 min on a single core of a desktop computer with a 2.0 GHz Intel Xeon chip. To produce plots of a similar quality using Monte Carlo simulations of (6.1) would require much more computational effort.

Earlier we remarked that the scalar case was simpler than the vector case. In the scalar case, we compute *all* "1 × 1" matrices β_r^{n+1} at once. Thus in the scalar algorithm (6.8) and (6.9), the only loop variable is n, discrete time. In the vector case, we must loop over both n and r, since each β^n is now a collection of n different $N \times N$ matrices. This last fact further complicates matters: $\beta^{n-\ell}$ contains a different number of matrices than β^n. At the moment, we use the list data structure in R to store all these objects. In ongoing work, we seek large performance gains by reimplementing the algorithm using more efficient data structures in C++.

Conclusion

In this paper, we have derived, implemented, and tested a new algorithm for the numerical simulation of linear N-dimensional SDDE of the form (6.1). The algorithm does not involve sampling any random variables, nor does it compute sample paths. Instead, the algorithm computes matrices that yield the full probability density function of the solution. Overall, the results indicate that the new algorithm produces accurate solutions much more efficiently than existing Monte Carlo approaches. Specific features of the algorithm include (i) the ability to generate solutions for many different initial conditions after running the algorithm only once, and (ii) the decoupling of the mean and the variance portions of the algorithm. Future work shall involve establishing the convergence and stability of the algorithm, and applying the algorithm to realistic modeling problems.

Acknowledgements This work was partially supported by a grant from UC Merced's Graduate Research Council.

References

1. Bhat, H.S., Kumar, N.: Spectral solution of delayed random walks. Phys. Rev. E **86**(4), 045701 (2012)
2. Buckwar, E.: Introduction to the numerical analysis of stochastic delay differential equations. J. Comput. Appl. Math. **125**, 297–307 (2000)
3. Buckwar, E., Kuske, R., Mohammed, S.E., Shardlow, T.: Weak convergence of the Euler scheme for stochastic differential delay equations. LMS J. Comput. Math. **11**, 60–99 (2008)
4. Crawford III, J.H., Verriest, E.I., Lieuwen, T.C.: Exact statistics for linear time delayed oscillators subjected to Gaussian excitation. J. Sound Vib. **332**(22), 5929–5938 (2013)
5. Kloeden, P.E., Platen, E.: Numerical Solution of Stochastic Differential Equations. Applications of Mathematics: Stochastic Modelling and Applied Probability. Springer, Berlin (1992)
6. Longtin, A.: Stochastic delay-differential equations. In: Atay, F.M. (ed.) Complex Time-Delay Systems, pp. 177–195. Springer, Berlin (2009)
7. Milton, J.G., Townsend, J.L., King, M.A., Ohira, T.: Balancing with positive feedback: the case for discontinuous control. Phil. Trans. Roy. Soc. A **367**(1891), 1181–1193 (2009)

8. Ohira, T., Milton, J.G.: Delayed random walks: Investigating the interplay between delay and noise. In: Gilsinn, D.E., Kalmár-Nagy, T., Balachandran, B. (eds.) Delay Differential Equations, pp. 305–335. Springer, New York (2009)
9. Sun, J.Q.: Finite dimensional Markov process approximation for stochastic time-delayed dynamical systems. Commun. Nonlinear Sci. Numer. Simulat. **14**(5), 1822–1829 (2009)
10. Sun, J.Q., Song, B.: Solutions of the FPK equation for time-delayed dynamical systems with the continuous time approximation method. Probab. Eng. Mech. **27**(1), 69–74 (2012)

Chapter 7
Combined Tests for Comparing Mutabilities of Two Populations

Stefano Bonnini

7.1 Introduction

Mutability is the aptitude of a qualitative variable to assume different categories [3]. With numerical variables, dispersion and heterogeneity of values, that is variability, may be measured by means of range, interquartile range, variance, standard deviation, coefficient of variation, mean absolute deviation, and several other indexes. With categorical data, in particular with nominal variables, the concept of mutability takes the place of that of variability. Mutability may be measured by other indexes mainly based on the observed frequencies: index of Gini [3], entropy of Shannon [6], family of indexes proposed by Rényi [5], and many others.

An index of mutability must satisfy the following properties:

- It takes value 0 if the same category is observed on all the statistical units (degenerate distribution);
- It takes the maximum value if all the categories are observed with the same frequencies (uniform distribution).

In general, the closer to uniform the distribution, the larger the mutability, and the larger the differences in frequencies across categories, the smaller the mutability. In several real problems, in presence of categorical data, the interest is focused on the inferential problem of comparing mutabilities of two or more populations, similarly to the comparison of variabilities for numerical variables, which is often faced with the test on variances.

For this problem, a permutation test has been proposed by Arboretti Giancristofaro et al. [1]. This test is based on the computation of an index of mutability for both

S. Bonnini (✉)
Department of Economics and Management, University of Ferrara, Via Voltapaletto 11, 44121 Ferrara, Italy
e-mail: stefano.bonnini@unife.it; bnnsfn@unife.it

V.B. Melas et al. (eds.), *Topics in Statistical Simulation*, Springer Proceedings in Mathematics & Statistics 114, DOI 10.1007/978-1-4939-2104-1_7

the samples and on the difference of the such sampling indexes as test statistic. After a preliminary transformation of data, according to the rule of the Pareto diagram, the permutation test follows a procedure similar to that of the test for stochastic dominance (see [4]). The good power behavior under the null and the alternative hypotheses is proved through a Monte Carlo simulation study. In [2] an alternative nonparametric solution, based on a bootstrap resampling strategy, is studied and compared with the permutation method. Even if, for both the permutation and the bootstrap solution, the power behaviors of the tests based on different indexes are similar, some differences of the rejection rates, highlighted by simulation studies, justify the attempt of looking for an index free test, that is a test based on a statistic which is not function of just one specific index of mutability.

In the present paper, for overcoming the cited drawback, a new permutation test, based on the combination of different tests for mutability, is proposed. In Sect. 7.2 the testing procedure is presented. In Sect. 7.3 the results of a simulation study, for comparing the power behavior of the proposed test with other tests based on specific indexes, are shown and discussed. Section 7.4 is dedicated to the application of the test to a real case study. Section 7.5 contains some final remarks.

7.2 Two-Sample Permutation Test for Mutability

Let us consider two populations and the categorical random variable X whose support is given by the set of K categories $\{A_1, \ldots, A_K\}$. Let us denote the proportion or the probability related to the kth category for the jth population with θ_{jk}, with $j = 1, 2$ and $k = 1, \ldots, K$. In other words, by denoting the categorical random variable under study for the jth population with X_j, it follows that

$$\theta_{jk} = Pr\{X_j = A_k\}. \tag{7.1}$$

The vectors $\boldsymbol{\theta_1} = [\theta_{11}, \ldots, \theta_{1K}]'$ and $\boldsymbol{\theta_2} = [\theta_{21}, \ldots, \theta_{2K}]'$ are unknown parameters of the respective populations and we are interested to compare $\mathrm{mut}(\boldsymbol{\theta_1})$ and $\mathrm{mut}(\boldsymbol{\theta_2})$, where $\mathrm{mut}(\boldsymbol{\theta_j})$ denotes the mutability of the jth population. For example, without loss of generality, let us consider the following two-sample one sided testing problem:

$$H_0 : \mathrm{mut}(\boldsymbol{\theta_1}) = \mathrm{mut}(\boldsymbol{\theta_2}) \tag{7.2}$$

against

$$H_1 : \mathrm{mut}(\boldsymbol{\theta_1}) > \mathrm{mut}(\boldsymbol{\theta_2}). \tag{7.3}$$

According to what we said above, the mutability of a population is related to the degree of "concentration" of the proportions or probabilities among the categories in the population. As a matter of fact, a greater concentration implies less mutability and less concentration implies greater mutability. Degenerate distribution corresponds to maximum concentration and uniform distribution corresponds to minimum concentration. Accordingly, the comparison of mutabilities can be defined by using the cumulative sums of the ordered parameters $\theta_{j(1)}, \ldots, \theta_{j(K)}$, with $j = 1, 2$, where $\theta_{j(k_1)} \geqslant \theta_{j(k_2)}$ if and only if $k_1 \leqslant k_2$. Hence the problem can be formally defined as follows:

$$H_0 : \sum_{k=1}^{s} \theta_{1(k)} = \sum_{k=1}^{s} \theta_{2(k)} \ \forall s \in \{1, \ldots, K-1\} \tag{7.4}$$

against

$$H_1 : \sum_{k=1}^{s} \theta_{1(k)} \leqslant \sum_{k=1}^{s} \theta_{2(k)} \text{ and } \exists s \in \{1, \ldots, K-1\} \text{ such that } \sum_{k=1}^{s} \theta_{1(k)} < \sum_{k=1}^{s} \theta_{2(k)}. \tag{7.5}$$

The cumulative sums in (7.4) and (7.5) do not include the case $s = K$ because trivially $\sum_{k=1}^{K} \theta_{1(k)} = \sum_{k=1}^{K} \theta_{2(k)} = 1$ is always true. In the presence of maximum mutability in the jth population we have $\sum_{k=1}^{s} \theta_{j(k)} = s/K \ \forall s \in \{1, \ldots, K-1\}$. Instead in the presence of minimum mutability $\sum_{k=1}^{s} \theta_{j(k)} = 1 \ \forall s \in \{1, \ldots, K-1\}$.

Let us consider the class of indexes of mutability defined in the parameter space, such that, when (7.2) and (7.4) are true, the index takes the same value in the two populations and when (7.3) and (7.5) are true the index takes a greater value in the first population. Let us denote with $\nu_j = \nu(\boldsymbol{\theta}_j)$ one of these indexes, computed for the jth population. Among these measures of mutability we mention the following:

- Index of Gini: $\nu_{G,j} = 1 - \sum_{k=1}^{K} \theta_{jk}^2$;
- Index of Shannon: $\nu_{S,j} = -\sum_{k=1}^{K} \theta_{jk} \log \theta_{jk}$;
- Index of Rényi of order 3: $\nu_{R_3,j} = -\frac{1}{2} \log \sum_{k=1}^{K} \theta_{jk}^3$;
- Index of Rényi of order ∞: $\nu_{R_\infty,j} = -\log \sup(\theta_{j1}, \ldots, \theta_{jK})$.

Each of these indexes reaches its maximum value when the distribution in the jth population is uniform: for the index of Gini we have $\max[\nu_G(\boldsymbol{\theta}_j)] = (K-1)/K$; for the other indexes $\max[\nu_S(\boldsymbol{\theta}_j)] = \max[\nu_{R_3}(\boldsymbol{\theta}_j)] = \max[\nu_{R_\infty}(\boldsymbol{\theta}_j)] = \log K$. In case of degenerate distribution each of these indexes is equal to zero. Let us note that such indexes are order invariant, that is

$$\nu(\theta_{j1}, \ldots, \theta_{jK}) = \nu(\theta_{j(1)}, \ldots, \theta_{j(K)}). \tag{7.6}$$

According to (7.4), under H_0 the cumulative sums of the ordered parameters for the two populations are equal. Similarly, from (7.5) follows that under H_1 the cumulative sums of the ordered parameters of the second population are greater than or equal to those of the first population. In the latter case we speak of dominance in mutability. Hence a two-sample directional test on mutability can be considered as a test on stochastic dominance for the variables transformed according to the Pareto diagram rule by considering the ordered parameters. Formally, for each population, the following transformation should be considered:

$$Y_j = \varphi_j(X_j) \text{ with } \varphi_j(A_k) = r \text{ iff } \theta_{jk} = \theta_{j(r)}. \tag{7.7}$$

The transformed variables Y_1 and Y_2 are ordered categorical and $Pr\{Y_j = r\} = \theta_{j(r)}$, $j = 1,2$ and $r = 1,\ldots,K$. The testing problem under study can equivalently be defined as $H_0 : Y_1 =^d Y_2$ against $H_1 : Y_1 >^d Y_2$, where $=^d$ denotes equality in distribution and $>^d$ means stochastic dominance.

Thus a suitable permutation testing procedure is the following:

1. Compute the contingency table of the observed dataset and, for each sample, calculate the ordered absolute frequencies $f_{j(r)}$ with $j = 1,2$ and $r = 1,\ldots,K$;
2. Consider some indexes of mutability and, for each index ν, compute the observed value of the test statistic based on the difference of the sampling indexes: $T_{\nu;0} = \hat{\nu}_1 - \hat{\nu}_2$, where $\hat{\nu}_j = \nu(\hat{\theta}_{j(1)},\ldots,\hat{\theta}_{j(K)})$, with $\hat{\theta}_{j(r)} = f_{j(r)}/n_j$ and $n_j = \sum_r f_{j(r)}$;
3. Perform B independent permutations of the dataset, transformed according to (7.7), by randomly reassigning the statistical units to the two samples;
4. For each permutation of the transformed dataset, compute the corresponding contingency table with frequencies $\tilde{f}_{b;j(r)}$; each frequency corresponds to the number of times the new transformed variable takes value r in the jth sample after the bth permutation, thus $\tilde{f}_{b;1(r)} + \tilde{f}_{b;2(r)} = f_{1(r)} + f_{2(r)}$ and $\sum_r \tilde{f}_{b;j(r)} = n_j$ with $b = 1,\ldots,B$;
5. For each permutation of the transformed dataset and for each ν index, compute the corresponding permutation value of the test statistic $T_{\nu;b} = \tilde{\nu}_{b;1} - \tilde{\nu}_{b;2}$, where $\tilde{\nu}_{b;j} = \nu(\tilde{\theta}_{b;j(1)},\ldots,\tilde{\theta}_{b;j(K)})$, with $\tilde{\theta}_{b;j(r)} = \tilde{f}_{b;j(r)}/n_j$;
6. For each ν index compute the p-value according to the permutation distribution of T_ν.

The permutation p-value is computed as follows:

$$\lambda_\nu = \frac{\sum_{b=1}^{B} I_{[T_{\nu;0},\infty)}(T_{\nu;b}) + 0.5}{B+1}, \tag{7.8}$$

where $I_{[a,b)}(t)$ denotes the indicator function of the interval $[a,b)$, which takes value 1 if $t \in [a,b)$ and value 0 otherwise.

It is worth noting that the described permutation solution is data driven because the transformation of the dataset according to the Pareto diagram rule depends on data themselves. If the transformation considered the true order of the θ_{jk} parameters, under H_0 exchangeability would be exact. Since the parameters' values are unknown, their order must be estimated with the sample observed frequencies and exchangeability is only approximate. See [1] for a deep discussion.

Another aspect of the described procedure is related to the test statistic. According to the chosen ν index, the test statistic is different and the decision of rejecting or not the null hypothesis could change according to which is the index chosen for the procedure. For this reason, in the present paper, a different statistic based on the combination of different ν-dependent tests is proposed.

To this purpose, let us define the significance level function (SLF) of a permutation test based on the T test statistic as:

$$L_T(t) = \frac{\sum_{b=1}^{B} I_{[t,\infty)}(T_b) + 0.5}{B+1}. \tag{7.9}$$

Hence the p-value of the test based on the index of mutability ν is equal to $L_\nu(T_{\nu;0})$. For each ν index, after step (5) of the described procedure, compute $L_\nu(T_{\nu;b})$, $b = 1, \ldots, B$. Then consider the matrix with B rows (corresponding to the permutations) and a number of columns equal to the number of tests (indexes), for representing the multivariate distribution of the test statistic whose marginal components are the test statistics based on specific indexes. In other words, by using, for example, the four test statistics defined above, the bth row of the $B \times 4$ matrix is $[T_{\nu G;b}, T_{\nu S;b}, T_{\nu R_3;b}, T_{\nu R_\infty;b}]$, with $b = 1, \ldots, B$. The computation of the SLF for each column of the matrix provides the $B \times 4$ matrix whose bth row is $[L_{\nu G}(T_{\nu G;b}), L_{\nu S}(T_{\nu S;b}), L_{\nu R_3}(T_{\nu R_3;b}), L_{\nu R_\infty}(T_{\nu R_\infty;b})] = [l_{G;b}, l_{S;b}, l_{R_3;b}, l_{R_\infty;b}]$.

By choosing a combining function $\psi(\bullet)$ satisfying some reasonable, intuitive and easy to justify properties (see [4]), it is possible to derive a test statistic suitable for solving the problem. The bth permutation value of the test statistic is

$$T_{\psi;b} = \psi[l_{G;b}, l_{S;b}, l_{R_3;b}, l_{R_\infty;b}], \tag{7.10}$$

and the permutation p-value is

$$\lambda_\psi = \frac{\sum_{b=1}^{B} I_{[T_{\psi;0},\infty)}(T_{\psi;b}) + 0.5}{B+1}, \tag{7.11}$$

where $T_{\psi;0} = \psi[l_{G;b}, l_{S;b}, l_{R_3;b}, l_{R_\infty;b}]$.

Many non-increasing functions may be used for combining the tests. Assuming that q different tests must be combined, some of the most used functions are:

- Fisher combining function: $\psi_F = -2\sum_{i=1}^{q} \log l_{i;b}$;
- Tippett combining function: $\psi_T = -\max[1 - l_{i;b}]$;
- Liptak combining function: $\psi_L = -\sum_{i=1}^{q} \Phi^{-1}[1 - l_{i;b}]$,

where $\Phi(\bullet)$ denotes the standard normal cumulative distribution function.

For the two-sided test where $H_1 : \text{mut}(\boldsymbol{\theta}_1) \neq \text{mut}(\boldsymbol{\theta}_2)$ the procedure may be easily adapted by using $T_{\nu;0} = |\hat{\nu}_1 - \hat{\nu}_2|$ and $T_{\nu;b} = |\tilde{\nu}_{1;b} - \tilde{\nu}_{2;b}|$.

7.3 Monte Carlo Simulation Study

Let us consider the two-sample one-sided test defined in the previous section. In the present section a Monte Carlo simulation study for comparing the power behavior of some tests is described. The tests taken into account are the ones based on the indexes of Gini (T_G), Shannon (T_S), Rényi of order 3 (T_{R_3}) and ∞ (T_{R_∞}), and the solutions proposed in the present paper, based on the combination of the four cited tests, through the application of the rule of Fisher (T_F), Liptak (T_L) and Tippett (T_T).

For a given simulation setting, for the jth population, data are generated by a continuous uniform distribution:

$$Z_j \sim U(0, 1) \tag{7.12}$$

and transformed according to the following rule:

$$Y_j = \text{int}(K \cdot Z_j^{\xi_j}) + 1 \tag{7.13}$$

where int(x) denotes the integer part of x and ξ_j is a parameter taking values in $[1, \infty)$ decreasingly related to mutability. When $\xi_j = 1$, in the jth population the mutability is maximum. The larger ξ_j the lower the mutability.

Each simulation setting is defined by the number of categories K, the parameter values ξ_1 and ξ_2 and the sample sizes n_1 and n_2. A number $B = 1{,}000$ of permutations is considered for estimating the permutation distribution and 1,000 datasets are generated for each setting for estimating the power through the rejection rates. Two significance levels are taken into account: $\alpha = 0.05$ and $\alpha = 0.10$.

Table 7.1 shows that, when the null hypothesis of equality in mutability is true, in presence of small sample sizes, all the rejection rates do not exceed the nominal alpha levels. The only exception is represented by the T_{R_3} test which, for large ξ values (low mutabilities) and small number of categories K, tends to be anticonservative. For large sample sizes the anticonservative behavior of the test based on the Rényi's index of order 3, especially for low mutabilities and small K, tends to accentuate and sometimes it extends to other tests. However the rejection rates are in general very near the significance levels and, in particular for the combined tests, we can speak of good approximation of the testing procedures.

In Table 7.2 the estimated powers of the tests, under the alternative hypothesis of greater mutability for the first population, are reported. It is evident that the power is increasing function of ξ_1 and ξ_2 (hence it grows as mutabilities decrease). Furthermore, as expected, the larger the sample sizes and the difference of mutabilities ($\xi_2 - \xi_1$) the greater the power. When the sample sizes are not equal the rejection rates are slightly lower. In presence of unbalanced samples, when the sample size of the first sample (which comes from the population with greater mutability) is larger, the estimated power is greater.

Table 7.1 Rejection rates of the two-sample tests on mutability ($B = 1{,}000$ permutations and $CMC = 1{,}000$ generated datasets) under H_0

Setting						Rejection rates						
						Index dependent tests				Combined tests		
n_1	n_2	K	ξ_1	ξ_2	$\xi_2 - \xi_1$	T_G	T_S	T_{R_3}	T_{R_∞}	T_F	T_L	T_T
						$\alpha = 0.05$						
10	10	4	1.0	1.0	–	0.008	0.008	0.009	0.026	0.009	0.009	0.008
			1.5	1.5	–	0.018	0.019	0.023	0.031	0.022	0.022	0.018
			2.0	2.0	–	0.040	0.038	0.049	0.046	0.043	0.043	0.039
			2.5	2.5	–	0.045	0.042	0.050	0.043	0.047	0.048	0.041
						$\alpha = 0.10$						
			1.0	1.0	–	0.040	0.046	0.052	0.062	0.044	0.044	0.044
			1.5	1.5	–	0.049	0.049	0.063	0.079	0.058	0.057	0.056
			2.0	2.0	–	0.102	0.101	0.108	0.095	0.103	0.103	0.099
			2.5	2.5	–	0.100	0.100	0.108	0.097	0.099	0.099	0.095
						$\alpha = 0.05$						
		6	1.0	1.0	–	0.006	0.013	0.021	0.042	0.013	0.014	.019
			1.5	1.5	–	0.026	0.029	0.039	0.043	0.029	0.029	0.029
			2.0	2.0	–	0.029	0.030	0.037	0.037	0.034	0.035	0.032
			2.5	2.5	–	0.041	0.040	0.047	0.041	0.042	0.042	0.037
						$\alpha = 0.10$						
			1.0	1.0	–	0.037	0.043	0.066	0.072	0.054	0.055	0.063
			1.5	1.5	–	0.068	0.072	0.094	0.088	0.085	0.084	0.081
			2.0	2.0	–	0.071	0.078	0.090	0.069	0.077	0.079	0.071
			2.5	2.5	–	0.082	0.081	0.097	0.079	0.090	0.091	0.085
						$\alpha = 0.05$						
50	50	4	1.0	1.0	–	0.009	0.011	0.009	0.027	0.011	0.011	0.017
			1.5	1.5	–	0.044	0.038	0.054	0.055	0.050	0.050	0.044
			2.0	2.0	–	0.054	0.048	0.056	0.047	0.052	0.053	0.051
			2.5	2.5	–	0.055	0.046	0.060	0.049	0.052	0.052	0.051
						$\alpha = 0.10$						
			1.0	1.0	–	0.040	0.038	0.042	0.064	0.046	0.047	0.047
			1.5	1.5	–	0.097	0.091	0.108	0.101	0.101	0.102	0.096
			2.0	2.0	–	0.103	0.102	0.106	0.094	0.103	0.104	0.099
			2.5	2.5	–	0.112	0.107	0.118	0.110	0.112	0.113	0.107
						$\alpha = 0.05$						
		6	1.0	1.0	–	0.009	0.013	0.011	0.027	0.013	0.013	0.016
			1.5	1.5	–	0.038	0.037	0.044	0.050	0.044	0.044	0.045
			2.0	2.0	–	0.046	0.046	0.049	0.044	0.047	0.047	0.047
			2.5	2.5	–	0.056	0.052	0.060	0.055	0.056	0.056	0.054
						$\alpha = 0.10$						
			1.0	1.0	–	0.031	0.035	0.037	0.058	0.039	0.039	0.039
			1.5	1.5	–	0.096	0.086	0.109	0.100	0.100	0.101	0.098

(continued)

Table 7.1 (continued)

Setting						Rejection rates						
						Index dependent tests				Combined tests		
n_1	n_2	K	ξ_1	ξ_2	$\xi_2 - \xi_1$	T_G	T_S	T_{R_3}	T_{R_∞}	T_F	T_L	T_T
			2.0	2.0	–	0.099	0.092	0.108	0.089	0.101	0.101	0.096
			2.5	2.5	–	0.107	.106	0.110	0.102	0.107	0.108	0.105
						$\alpha = 0.05$						
30	70	6	1.0	1.0	–	0.008	0.007	0.014	0.020	0.007	0.009	.007
			1.5	1.5	–	0.038	0.038	0.051	0.053	0.044	0.045	0.045
			2.0	2.0	–	0.046	0.040	0.048	0.049	0.048	0.048	0.044
			2.5	2.5	–	0.055	0.051	0.061	0.053	0.056	0.055	0.053
						$\alpha = 0.10$						
			1.0	1.0	–	0.035	0.035	0.036	0.056	0.037	0.038	0.045
			1.5	1.5	–	0.089	0.087	0.095	0.101	0.093	0.094	0.091
			2.0	2.0	–	0.097	0.083	0.107	0.103	0.097	0.098	0.096
			2.5	2.5	–	0.103	0.109	0.100	0.098	0.101	0.101	0.097
70	30	6	1.0	1.0	–	0.014	0.016	0.022	0.034	0.023	0.024	0.017
			1.5	1.5	–	0.044	0.042	0.049	0.042	0.049	0.049	0.043
			2.0	2.0	–	0.054	0.050	0.057	0.046	0.051	0.052	0.050
			2.5	2.5	–	0.051	0.051	0.050	0.043	0.051	0.051	0.049
						$\alpha = 0.10$						
			1.0	1.0	–	0.058	0.058	0.069	0.078	0.063	0.063	0.059
			1.5	1.5	–	0.081	0.083	0.089	0.082	0.080	0.080	0.076
			2.0	2.0	–	0.110	0.098	0.117	0.102	0.109	0.110	0.102
			2.5	2.5	–	0.101	0.096	0.107	0.091	0.101	0.101	0.100

When the mutability is high and when the difference of mutabilities is low, the test based on Rényi's index of order ∞ (T_{R_∞}) seems to be the most powerful among the compared solutions. In the other cases under H_1, the rejection rates of T_{R_3} are in general the largest but we cannot ignore that the rejection rates of this test tend to exceed the nominal alpha levels under H_0. For this reason other tests, which respect the α levels under the null hypothesis, are preferable to T_{R_3}. Among the index dependent tests, T_G (based on the index of Gini) seems to be the most powerful. But in general the combined tests, in particular T_F and T_L (with very similar performance), whose rejection rates under H_0 are very near the significance levels, present powers greater than T_G.

Table 7.2 Rejection rates of the two sample tests on mutability ($B = 1{,}000$ permutations and $CMC = 1{,}000$ generated datasets) under H_1

Setting						Rejection rates						
						Index dependent tests				Combined tests		
n_1	n_2	K	ξ_1	ξ_2	$\xi_2 - \xi_1$	T_G	T_S	T_{R_3}	T_{R_∞}	T_F	T_L	T_T
						$\alpha = 0.05$						
10	10	4	1.0	1.5	0.5	0.018	0.017	0.020	0.028	0.020	0.020	0.015
			1.5	2.0	0.5	0.060	0.056	0.061	0.074	0.061	0.061	0.049
			1.5	2.5	1.0	0.127	0.125	0.138	0.140	0.136	0.136	0.117
			1.5	3.0	1.5	0.193	0.178	0.212	0.202	0.208	0.208	0.169
						$\alpha = 0.10$						
			1.0	1.5	0.5	0.066	0.066	0.071	0.081	0.066	0.066	0.062
			1.5	2.0	0.5	0.132	0.125	0.141	0.136	0.135	0.133	0.132
			1.5	2.5	1.0	0.229	0.224	0.237	0.240	0.231	0.232	0.221
			1.5	3.0	1.5	0.349	0.323	0.359	0.324	0.348	0.348	0.340
						$\alpha = 0.05$						
		6	1.0	1.5	0.5	0.034	0.045	0.055	0.072	0.051	0.051	0.054
			1.5	2.0	0.5	0.063	0.076	0.093	0.070	0.083	0.083	0.077
			1.5	2.5	1.0	0.118	0.130	0.151	0.103	0.129	0.130	0.122
			1.5	3.0	1.5	0.216	0.221	0.250	0.195	0.230	0.231	0.215
						$\alpha = 0.10$						
			1.0	1.5	0.5	0.089	0.104	0.119	0.115	0.113	0.114	0.115
			1.5	2.0	0.5	0.143	0.149	0.170	0.128	0.158	0.161	0.146
			1.5	2.5	1.0	0.235	0.248	0.266	0.213	0.253	0.254	0.225
			1.5	3.0	1.5	0.346	0.344	0.383	0.317	0.367	0.367	0.346
						$\alpha = 0.05$						
50	50	4	1.0	1.5	0.5	0.189	0.179	0.195	0.219	0.207	0.207	0.203
			1.5	2.0	0.5	0.252	0.250	0.258	0.252	0.258	0.258	0.249
			1.5	2.5	1.0	0.586	0.548	0.594	0.561	0.594	0.594	0.579
			1.5	3.0	1.5	0.758	0.734	0.764	0.752	0.760	0.762	0.753
						$\alpha = 0.10$						
			1.0	1.5	0.5	0.305	0.298	0.326	0.343	0.330	0.331	0.316
			1.5	2.0	0.5	0.399	0.378	0.404	0.372	0.403	0.442	0.388
			1.5	2.5	1.0	0.721	0.699	0.726	0.709	0.723	0.724	0.710
			1.5	3.0	1.5	0.860	0.842	0.862	0.860	0.859	0.859	0.861
						$\alpha = 0.05$						
		6	1.0	1.5	0.5	0.170	0.139	0.188	0.209	0.184	0.185	0.179
			1.5	2.0	0.5	0.294	0.250	0.312	0.294	0.301	0.300	0.288
			1.5	2.5	1.0	0.607	0.566	0.612	0.575	0.610	0.609	0.594
			1.5	3.0	1.5	0.826	0.786	0.832	0.815	0.828	0.827	0.814
						$\alpha = 0.10$						
			1.0	1.5	0.5	0.301	0.263	0.345	0.355	0.335	0.336	0.308
			1.5	2.0	0.5	0.428	0.399	0.445	0.414	0.436	0.439	0.420
			1.5	2.5	1.0	0.751	0.721	0.760	0.723	0.753	0.755	0.733
			1.5	3.0	1.5	0.900	0.880	0.904	0.892	0.903	0.903	0.895

(continued)

Table 7.2 (continued)

Setting						Rejection rates						
						Index dependent tests				Combined tests		
n_1	n_2	K	ξ_1	ξ_2	$\xi_2 - \xi_1$	T_G	T_S	T_{R_3}	T_{R_∞}	T_F	T_L	T_T
						$\alpha = 0.05$						
30	70	6	1.0	1.5	0.5	0.119	0.109	0.134	0.142	0.134	0.133	0.130
			1.5	2.0	0.5	0.242	0.221	0.260	0.239	0.250	0.254	0.230
			1.5	2.5	1.0	0.511	0.489	0.530	0.521	0.519	0.520	0.519
			1.5	3.0	1.5	0.769	0.728	0.778	0.740	0.776	0.777	0.758
						$\alpha = 0.10$						
			1.0	1.5	0.5	0.240	0.235	0.255	0.235	0.255	0.255	0.233
			1.5	2.0	0.5	0.385	0.355	0.397	0.380	0.390	0.390	0.373
			1.5	2.5	1.0	0.675	0.633	0.683	0.650	0.682	0.682	0.664
			1.5	3.0	1.5	0.865	0.830	0.868	0.857	0.867	0.866	0.862
70	30	6	1.0	1.5	0.5	0.189	0.153	0.208	0.223	0.205	0.206	0.189
			1.5	2.0	0.5	0.289	0.243	0.312	0.282	0.300	0.300	0.281
			1.5	2.5	1.0	0.525	0.480	0.543	0.515	0.532	0.534	0.514
			1.5	3.0	1.5	0.780	0.732	0.788	0.753	0.783	0.784	0.767
						$\alpha = 0.10$						
			1.0	1.5	0.5	0.315	0.276	0.345	0.362	0.340	0.343	0.330
			1.5	2.0	0.5	0.426	0.371	0.440	0.423	0.434	0.432	0.419
			1.5	2.5	1.0	0.680	0.608	0.686	0.664	0.682	0.683	0.667
			1.5	3.0	1.5	0.866	0.842	0.871	0.868	0.869	0.870	0.865

7.4 Case Study

In 2013 a survey on student's living and study conditions was performed by University of Ferrara (Italy). One of the questions of the interview, performed on a sample of 747 students, was related to the frequency of the sporting practice. Females are expected to play sports less often than males. As a matter of fact most of the females declared that they never do it. Here we wish to investigate the homogeneity of the answers. We are interested in testing whether the category of females who never play sports is very representative of the behavior of women respect to sports, while for males the answers are much more diversified because of a greater heterogeneity of behaviors. Formally we wish to test

$$H_0 : \text{mut(males)} = \text{mut(females)} \tag{7.14}$$

against

$$H_1 : \text{mut(males)} > \text{mut(females)}. \tag{7.15}$$

Table 7.3 Frequency of sporting practice of students by gender: contingency table

	Frequency					
Gender	Never	Few times a year	Few times a month	Once/twice a week	3–5 times a week	6–7 times a week
Males	103	1	15	97	113	19
Females	170	7	18	113	76	15

Source: 2013 Survey on student's living and study conditions at University of Ferrara (Italy)

Table 7.4 Frequency of sporting practice of students by gender: table of cumulative ordered relative frequencies

	Rank					
Gender	1	2	3	4	5	6
Males	0.325	0.621	0.899	0.954	0.997	1.000
Females	0.426	0.709	0.900	0.945	0.982	1.000

Table 7.5 Normalized indexes of mutability

	Indexes			
Gender	Gini	Shannon	Rényi-3	Rényi-∞
Males	0.869	0.777	0.698	0.628
Females	0.838	0.765	0.623	0.476

The contingency table shows that, as expected, most of the females never play sports, while most of the males do it from 3 to 5 times a week (see Table 7.3).

According to the cumulative ordered relative frequencies, the curve of females is not always over that of males, that is the former does not dominate the latter, thus, from a descriptive point of view, it is not evident whether mutability of males is greater (see Table 7.4).

By computing the indexes of mutability for both the samples we obtain greater values in the sample of males (see Table 7.5). To shed light on the problem let us perform the combined tests for mutability comparisons with $B = 10{,}000$ permutations.

At the significance level $\alpha = 0.05$ the null hypothesis of equality in mutability must be rejected in favor of the alternative hypothesis of greater mutability for males's answers. As a matter of fact all the p-values (0.015 for T_F, 0.020 for T_L and 0.005 for T_T) are less than α.

7.5 Conclusions

In the study of the power behavior, the combined tests for mutability comparisons proposed in the present paper show a good approximation under the null hypothesis. The power is increasing function of the sample sizes and of the difference between the mutabilities, and ceteris paribus decreasing functions of the mutabilities. Among the non anticonservative procedures, the combined tests on mutability are not only well approximated but, especially when using the combinations of Fisher and Liptak, the most powerful under H_1 also. The application of the combined test to real data of a survey on living and study conditions of University students, to prove the greater mutability of the behavior of males respect to females regards to sporting habit, shows the usefulness of the proposed methodology.

References

1. Arboretti Giancristofaro, R., Bonnini, S., Pesarin, F.: A permutation approach for testing heterogeneity in two-sample problems. Stat. Comput. **19**, 209–216 (2009)
2. Bonnini, S.: Testing for heterogeneity with categorical data: permutation solution vs. bootstrap method. Commun. Stat.: Theory Methods **43**(4), 906–917 (2014)
3. Gini, C.: Variability and mutability, contribution to the study of statistical distributions and relations. Studi Economico-Giuridici della R. Universita di Cagliari (1912)
4. Pesarin, F., Salmaso, L.: Permutation Tests for Complex Data. Theory, Applications and Software. Wiley, Chichester (2010)
5. Rényi, A.: Calculus des probabilitès. Dunod, Paris (2009)
6. Shannon, C.E.: A mathematical theory of communication. Syst. Technol. J. **27**, 379–423 (1948)

Chapter 8
Development of an Extensive Forensic Handwriting Database: Statistical Components

Michèle Boulanger, Mark E. Johnson, and Thomas W. Vastrick

8.1 Introduction

The discipline of Forensics Sciences is at a crucial juncture where critical research programs are necessary to forestall undue criticism and to strengthen the credibility of certain areas, such as handwriting analysis. The National Research Council (NRC) commissioned a comprehensive report (*Strengthening Forensic Science in the United States: A Path Forward, [8]*) that outlined the various sub-disciplines, evaluated them on the basis of their scientific underpinnings and identified important research directions. Handwriting analysis received rather faint praise:

> The scientific basis for handwriting comparisons needs to be strengthened. Recent studies [7] have increased our understanding of the individuality and consistency of handwriting and computer studies [11] suggest that there may be a scientific basis for handwriting comparison, at least in the absence of intentional obfuscation or forgery. Although there has been only limited research to quantify the reliability and replicability of the practices used by trained document examiners, the committee agrees that *t*here may be some value in handwriting analysis.

Kam et al. [7], cited above, enlisted over 100 document examiners and a control group of comparably educated individuals who were not trained in document examination. Although the document examiners performed better than this control

---

M. Boulanger (✉)
Department of International Business, Rollins College, Winter Park, FL, USA
e-mail: mboulanger@rollins.com; mboulanger@rollins.edu

M.E. Johnson
Department of Statistics, University of Central Florida, Orlando, FL, USA
e-mail: mejohnso@mail.ucf.edu

T.W. Vastrick
Forensic Document Examiner, Apopka, FL, USA
e-mail: vastrick@yahoo.com

V.B. Melas et al. (eds.), *Topics in Statistical Simulation*, Springer Proceedings in Mathematics & Statistics 114, DOI 10.1007/978-1-4939-2104-1_8

Table 8.1 Scientific interpretation and reporting of results [3]

Examiner finding	Elaboration
Identification	A definite conclusion that the questioned writing matches another sample
Strong probability	Evidence is persuasive, yet some critical quality is missing
Probable	Points strongly towards identification
Indications same person created both samples	There are a few significant features
No conclusion	There are limiting factors (e.g., disguise) or lack of comparable writing
Indications	Same weight as indications with a weak opinion
Probably did not	Evidence is quite strong
Strong probability did not	Virtual certainty
Elimination	Highest degree of confidence

group (thankfully!), they had a 6.5 % error rate in document identification. In another study addressing forensic examiner expertise, [9] noted a 3.4 % error rate related to signature identification. Situations in which experts disagree on the source or identification of documents could generate reasonable doubts in the minds of jurors.

The computer study referenced in the NRC report is by Srihari et al. [11]. This paper, as the title indicates, makes a rigorous case for the individuality of handwriting, which is a fundamental premise of forensic document examination. Srihari's work is pioneering in applying pattern recognition tools in the area and extracting software defined features from handwriting specimens [10]. This study included 1,568 such specimens whose originators ranged the gamut across the US population, covering gender, age, and ethnicity. In contrast, our study will eventually include 5,000 specimens and consider approximately 900 handwriting attributes which collectively can address the handwriting individuality issue.

In response to the calls for additional forensics research, the National Institute of Justice (NIJ) funded a large-scale study at the University of Central Florida involving a statistically appropriate sampling of the overall US population or sub-groups with the following objectives (taken directly from the funded proposal):

1. Develop statistically valid proportions of characteristics of handwriting and hand printing based on specimen samples throughout the United States.
2. Provide practitioners of forensic document examination with statistical basis for reliability and measurement validity to accurately state their conclusions.
3. Provide courts with the reliable data needed to understand the underlying scientific basis for the examinations and the conclusions.

Ultimately, the results of this NIJ study will be incorporated into trial testimony by forensic document examiner experts [4]. Table 8.1 provides the current wording of document experts as standardized by [3].

A primary objective of the study is to strengthen the underpinnings of the statements of Table 8.1, and thus tightening the elaborations in column 2 of the table.

Before embarking on the collection of handwriting specimens, the collaborators dealt with some fundamental sampling issues specific to this investigation. First, there is no sampling frame available for the study. The target population has been defined as adults 18 years old or older who are capable of providing writing samples in English (with some exclusions to be described). Attempts were made to obtain samples from a constituency that is at least representative of the target population regarding demographics and other factors known to influence handwriting. This was accomplished by developing stratification variables corresponding to demographics and handwriting factors. Quotas were then set to guide the collection of samples (i.e., to avoid an over-abundance of specimens, for example, from college age students who are readily located and coerced into participating). A substantial number of forensic document examiners volunteered to collect these specimens with the guidance of our protocols. Thus, the overall characteristics of the providers of the written specimens are in accordance roughly with the proportions of characteristics found in the target population. The determination of the surrogate population for obtaining writing specimens is described in Sect. 8.2, including the lengths taken to obtain samples according to our constraints.

A second fundamental issue is the reliability and replicability of the examiners themselves in performing their review of written specimens. To assess this effort first required a determination and delineation of the potential characteristics that would be considered by each examiner. In particular, it was decided that multiple characteristics would be considered for each letter in both cursive and printing styles as well as numbers and special symbols ("?" , "," , ";" , and so forth). As described in Sect. 8.3, this led to over 2,500 possible features for consideration. With a large candidate set of features for consideration, we then developed an attribute agreement analysis study, which placed a heavy burden on three expert examiners, but led to an evaluation of features that survive a test of agreement on presence/absence of features between and within examiners. Ultimately, about 900 feature attributes had perfect agreement by examiners across and within examiners for multiple written specimens. This attribute agreement analysis study is described in Sect. 8.4. One key and original result of the attribute agreement analysis is the elimination of characteristics that did not reach an agreement by examiners across multiple specimens (which again would cause problems with court testimony).

With this substantial groundwork in place, the continued collection of specimens and their detailed evaluation continues. Once this effort is completed, the authors plan to publish the findings in a follow-up paper with an enlarged focus on statistical results. A major contribution of this paper is a description of the attribute agreement analysis that is essential to establishing the viability of examiner reliability with respect to specific handwriting characteristics. The database of examiner determinations of characteristics should provide a rich source of information that examiners can draw upon to quantify more precisely their opinions regarding ownership of specific "questioned" documents. A full discussion of how the final database will be used in practice will be provided in future publications.

8.2 Target Population Definition and Sampling of Writers

The first key statistical component of the study related to the identification of what constitutes a representative sample of the US population and of appropriate (and measurable) sub-groupings of the population. Our target sample size is 5,000 handwriting specimens.

To obtain a probabilistically valid sample, one needs to define the target population and to develop a sampling frame for that population from which individual units or group of units are drawn according to pre-defined set of probabilities. Our target population is the USA population excluding children but including foreigners traveling or living in the USA. Given the forensic nature of our project, i.e. the use of science and data to evaluate and assess facts pertinent to handwritten specimens such as contracts, licenses, or wills in a court of law, we defined our target population, as follows:

i. Adults (at least 18 years of age)
ii. Residing or traveling in the USA (including foreign tourists)
iii. Able to provide writing samples in English (though not necessarily able to speak English)

We excluded from our target population people with physiological constraints and types of infirmities that would prevent them for being able to write the specimen we developed. These exclusions are documented more precisely in the next section.

Having identified our target population, we quickly realized that we were facing two major hurdles: (1) the development of a sampling frame for that population, and (2) the ability to reach a person once selected from the sampling frame and to obtain two handwritten copies of a standard letter from that person (one script and one cursive). It takes about 20 min to write both versions of the letter and the chances that we obtained these handwritten specimen without direct contact with the person selected are extremely low due to the effort required by the person selected, the lack of interest or reward, the suspicions encountered with giving a handwritten sample, and many other factors. Moreover, we must validate ownership of all handwritten specimen before entering them in our database, and this requires that we actually witness who writes a particular specimen. Thus, a direct contact between a data collector and the person providing us with a handwritten specimen is necessary to establish a viable chain of custody. All these reasons render the feasibility of a large probabilistic sample unrealistic.

Thus our approach to data collection changed from a probabilistic sampling process to the development of a data collection process that will lead to a large sample of "writers" deemed representative of the target population. The approach we followed was based on a study done to evaluate the performance of the national telecommunications network before the breakout of the monopoly service provider, AT&T [1, 2, 5]. There, as in our situation with handwriting, it was not possible to construct a sampling frame of all the potential telecommunication paths in the USA and a multi-level sampling approach based on identification of strata and clusters was developed to lead to a quasi-representative sample. We used a similar planning approach that consisted of the following seven steps to guide the data collection process specific to our study:

a. Research factors influencing handwriting
b. Define stratification variables based on
    i. Key factors influencing handwriting
    ii. Key variables describing target population
c. Define strata for selected stratification variables
d. Estimate proportions of strata in target population
e. Define a data collection process to obtain a sample that will be "deemed" a representation of the target population (i.e., meet the quota specifications)
f. Provide guidelines to data collectors on the data collection process
g. Audit data collection process for adherence to data collection plan and for quality control

8.2.1 Factors Influencing Handwriting

Huber and Headrick [6] provide a list of factors known to potentially influence handwriting (Table 8.2). Each factor was reviewed and a decision made by the collaborators regarding how to handle it in our sampling process: (1) accept all writers with any values of that factor without any recording of these values, (2) accept all writers with any values of that factor but record the value of that factor for each writer in the sample, or (3) reject writers with some values for that factor from the sample. Table 8.2 provides the disposition we made for each factor identified in [6].

8.2.2 Definition of Stratification Variables, Strata, and Proportion Allocation

Our next step was to identify the stratification variables we needed in order to obtain as representative as possible a sample of 5,000 writers. The rationale for our choice of stratification variables was to include factors known or suspected to influence writing as defined in Table 8.2 and to provide coverage for other characteristics of the USA population. Table 8.3 provides our selection of stratification variables, strata, and proportion allocation in the USA. We adopted race (White, Black, Hispanic, Asian) as one of our coverage factors and regions within the USA (NE, NW, MW, SE, SW) as the other.

Table 8.2 Factors influencing handwriting

Factors influencing handwriting [6]			Our sampling process	
Section in H&H	Reference in H&H	Factors in H&H	How handled in our sample ?	Information requested
External factors (E)				
B.37	A	Writing systems: national, cultural, and occupational	People who are in the USA (including foreigners traveling), able to write in English (not necessarily speaking English	Location of third grade schooling
B.37	B	Physiological constraints:		
	B1	Foot & mouth	Do not accept in sample	
	B2	Use of artificial aids (protheses)	Accept in sample	Do you have any physical imparities or injuries?
	B3	Deafness and/or sightlessness	Do not include blind in sample, accept deaf if communications are possible	Do you have any physical imparities or injuries?
B.37	C	Genetic factors: sex	Familial relationship and multiple birth are of no interest in this study. Ignore	Record gender
B.37	D	Physical (Normal)		
	D1	Maturity, practise, and development	Accept only people 18 years or older	Record age
	D2	Handedness	No control of which hand should write in the case of ambidexterity	Record hand doing the writing Do not record grasp

B.37	E	Physical (abnormal state of health)		
	E1	Handwriting as diagnostic tool	Not relevant to our study—ignore	
	E2	Illness organically related	Accept in sample	Do you have any physical imparities or injuries?
B.37	F	Medications	No control, no asking	No information asked
B.37	G	Infirmity		
	G1	Senility	Do not accept in sample	
	G2	Guided hands	Do not accept in sample	
B.37	H	Mental state of writer (emotional stress, nervousness, instability)	Accept in sample	No information requested or noted
B.37	I	Instability	Accept in sample	Do you have any physical imparities or injuries?
Internal factors (I)				
B.38	A	Imitation	Not relevant to our study. Ignore	
B.38	B	Circumstantial	Control environment	Provide pen and paper, Provide comfortable position for the person to write with adequate support level.
B.38	C	Temporal state of the writer		
	C1	Alcohol	Accept in sample	
	C2	Hallucinogens and hard drugs	Accept in sample	
	C3	Hypnosis	Accept in sample	
	C4	Fatigue and physical stress	Accept in sample	Flip-flop printing and cursive writing—record order
B.38	D	Literacy and education		Record information on education level

Table 8.3 Factors used for stratification in our sampling process

Reference in H&H	Stratification variable	Strata definition	Strata proportion in USA (confirmed except writing system) (%)
A	Writing systems	Location of third grade schooling in USA	80.0
		Location of third grade schooling NOT in USA	20.0
B	Gender	Male	49.0
		Female	51.0
C	Age	18–30	33.0
		> 30–50	36.0
		> 50	41.0
D	Handedness	R	90.0
		L	10.0
C	Temporal state	Night (after 8pm)	
		Day (before 8 pm)	
D	Education	HS or less	49.0
		> HS	51.0
N/A	Race	W	63.7
		B	12.6
		H	16.3
		A	4.8
N/A	US region (where samples are taken)	North West	
		North East	
		Middle West	
		South West	
		South East	

8.2.3 Data Collection Process, Guidelines, and Audit

Having identified the strata to be represented in our sample, the next step was to provide guidelines to the data collectors to ensure their compliance with the stratification plan developed thus far. We established the following process to achieve representativeness of the sample:

i. Fix minimal quota specification for 80 % of the sample for each stratification variable (Table 8.4). This was viewed as important by the collectors in order to provide them with some flexibility in meeting the stratification quotas and allowing them to introduce some level of randomness to cover for unforeseen factors that could introduce a bias in the process.
ii. Pre-select type of locations to ensure quotas for representativeness are met and randomness is achieved as much as possible at an affordable cost Within each region, select:

- 20 %: Universities (young adults, education at and beyond high school, foreigner)
- 20 % Churches (mature adults). Select White churches, Black churches, Asian temples and Hispanic churches
- 20 % Night entertainment locations (after 8 pm)
- 20 % Restaurants and Fast food (education less than high school)
- 20 % Survey or discretion

iii. Achieve coverage by letting the surveyor select places within the types and guidelines mandated by the study
iv. Give latitude to surveyor to obtain samples
v. Collect information for potential correction during analysis

The results of these efforts led to the values given in Table 8.4.

8.2.4 *Auditing the Collection Process*

Finally, to ensure compliance with the data collection plan, we regularly audited the collection of the sample specimens and provided guidance to the collectors to adjust for deviations from the quota ranges. This process is illustrated in Table 8.5.

This data collection process is expected ultimately to provide us with 5,000 handwritten specimens, all validated by the data collectors themselves in terms of ownership and in terms of documenting any unusual or useful criteria that may be of use during the analysis process. The auditing process done by the collaborators of this paper is also ensuring the validity of the overall plan and providing confidence in the quality of the specimens whose characteristics are to be entered in the database.

8.3 Scope of Document Examiner Review

A standardized letter was used as the basis of the specimens to be provided by the participants. Our version of the letter is a slight modification of the letter given by Srihari et al. [11], with the exception being the addition of the middle name Raj to the addressee (Fig. 8.1). All upper and lowercases of each letter are found in the letter and lower case letters are found at the beginning, middle and end of words. For each letter, our forensic document examiner (Vastrick) defined several specific features for each letter (both cursive and printed), number and symbol and illustrated them within an ACCESS database. All examiners in our study used this template in order to determine presence/absence of each specified feature attribute. The development of the attribute feature list was a major undertaking but was necessary in order to establish a subset of attributes that examiners would find unambiguous to determine.

Table 8.4 Final data collection plan

Reference in H&H	Stratification variable	Strata definition	Strata proportion in USA (confirmed except writing system) (%)	Minimal quota specification (80 % per factor) (%)
A	Writing Systems	Location of third grade schooling in USA	80.0	>70.0
		Location of third grade schooling NOT in USA	20.0	>10.0
B	Gender	Male	49.0	>40.0
		Female	51.0	>40.0
C	Age	18–30	33.0	>20.0
		>30–50	36.0	>30.0
		>50	41.0	>30.0
D	Handedness	R	90.0	>75.0
		L	11.0	>5.0
C	Temporal state	Night (after 8pm)		>20.0
		Day (before 8 pm)		>60.0
D	Education	HS or less	49.0	>30.0
		>HS	51.0	>50.0
N/A	Race	W	63.7	>55.0
		B	12.6	>10.0
		H	16.3	>11.0
		A	4.8	>4.0
N/A	US region (where samples are taken)	North West		>15.0
		North East		>15.0
		Middle West		>15.0
		South West		>15.0
		South East		>15.0
N/A	Location	College and universities		>20.0
		Religious places		>20.0
		Social and non social gathering		>40.0

Each specimen was generated by a participant using a common pen type (BIC medium point) and 20-pound lined paper. Each participant copied the letter with both cursive and printed styles and per Institutional Research Board protocols that could allow the termination of participation at any time for any reason by the specimen provider. The participation rate was found to be improved when the examiner solicited help on the basis of a "university research project" rather than a project to assess potential criminal behavior.

Table 8.5 Results from auditing process for correction of on-going target population of writers

Stratification variable	Strata definition	Minimal quota specification (80 % per factor) (%)	Expectations for pilot convenience pilot (size 335 on May 3/26)		What we have in pilot	Notes on pilot
Writing systems	Location of third grade schooling in US	>70.0	>235	For a total of 335	285	41 do not know
	Location of third grade schooling NOT in US	>10.0	>34		9	
Gender	Male	>40.0	>134	For a total of 335	85	1 misclassified as w
	Female	>40.0	>134		249	
Age	18–30	>20.0	>67	For a total of 335	300	1 without age
	>30–50	>30.0	>101		73	
	>50	>30.0	>101		58	
Handedness	R	>75.0	>251	For a total of 335	300	
	L	>5.0	>17		335	
Temporal state	Night (after 8pm)	>20.0	>67	For a total of 335		
	Day (before 8 pm)	>60.0	>201			

(continued)

Table 8.5 (continued)

Stratification variable	Strata definition	Minimal quota specification (80 % per factor) (%)	Expectations for pilot convenience pilot (size 335 on May 3/26)		What we have in pilot	Notes on pilot
Education	HS or less	>30.0	>101	For a total of 335	24	
	>HS	>50.0	>168		311	
Race	W	>55.0	>184	For a total of 335	261	Other = 11
	B	>10.0	>34		17	No information = 3
	H	>11.0	>34		32	
	A	>4.0	>13		11	
US region (where samples are taken)	North West	>15.0	0			
	North East	>15.0	0			
	Middle West	>15.0	205?			
	South West	>15.0	0			
	South East	>15.0	130?			
Location	College and universities	>20.0	>67	For a total of 335		
	Religious places	>20.0	>67			
	Social and non social gathering	>40.0	>134			

From: Jim Elder
829 Loop Street, Apt. 300
Allentown, New York 14707

To: Dr. Bob Raj Grant
602 Queensberry Parkway
Omar, West Virginia 25638

We were referred to you by Xena Cohen at the University Medical Center. This is regarding my friend, Kate Zack.

It all started about six months ago while attending the "Rubeq" Jazz Concert. Organizing such an event is no picnic, and as President of the Alumni Association, a co-sponsor of the event. Kate was overworked. But she enjoyed her job and did what was required of her with great zeal and enthusiasm.

However, the extra hours affected her health; halfway through the show she passed out. We rushed her to the hospital, and several questions, x-rays and blood tests later, were told it was just exhaustion.

Kate's been in very bad health since. Could you kindly take a look at the results and give us your opinion?

Thank you!

Jim

Fig. 8.1 Typed version of specimen letter

The contents of the database are illustrated with uppercase cursive $\mathscr{M}$. Figures 8.2 and 8.3 provide a few of the characteristics given in the database with the corresponding description. Other letters are treated similarly and will be available in the final report to the National Institute of Justice.

8.4 Attribute Agreement Analysis

As noted in the previous section, for purposes of our study, the examiners determine presence or absence of numerous characteristics for cursive and printed letters, numbers and various symbols. Since only presence/absence of each feature is

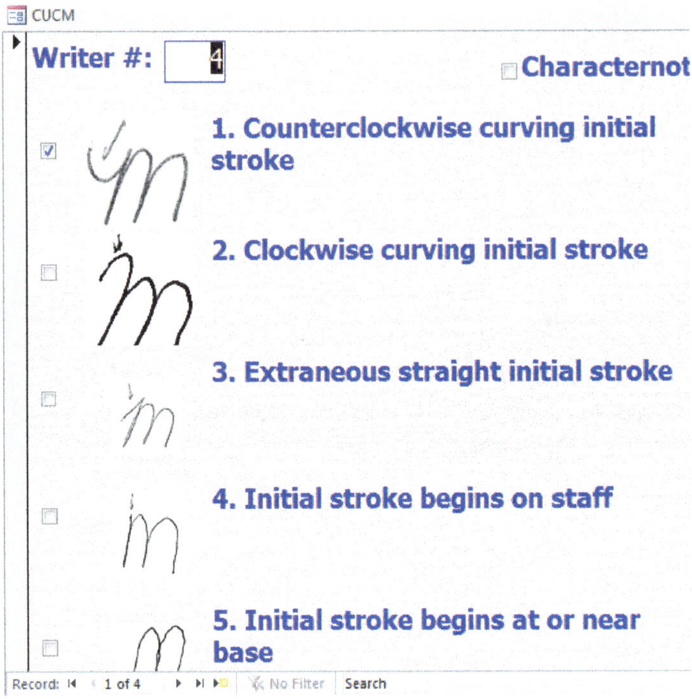

Fig. 8.2 Initial five characteristics of the capital letter "*M*"

reported, particular interest is to be paid to those characteristics that are agreed upon across several specimens (possibly some with and some without the specific feature). An attribute agreement study was designed for which three professional forensic document examiners reviewed several writing specimens. In this section, this study is described and the conclusions presented.

The original design of the attribute agreement study was to provide five distinct specimens to three examiners, and following a lag period, re-provide two of them again to each examiner. Although the process for checking presence/absence of characteristics is somewhat automated, completing a review for all 2,500+ features takes on the order of 8 h per examiner per specimen. Owing to the location of the three examiners in different regions of the country and some time constraints, the actual specimens by examiners were performed, as follows in Table 8.6:

As can be seen from Table 8.6, seven rather than five specimens were circulated, and the same two specimens were not considered twice by each examiner. Nevertheless, in spite of imperfect balance, there are numerous instances of multiple examiners reviewing the same specimens, so considerable data was collected in this exercise. Further, the specific reviews for cursive and printing specimens were slightly different, but such a discrepancy does not detract from the numerous concurrent comparisons made. Examiner R was the only individual who looked at

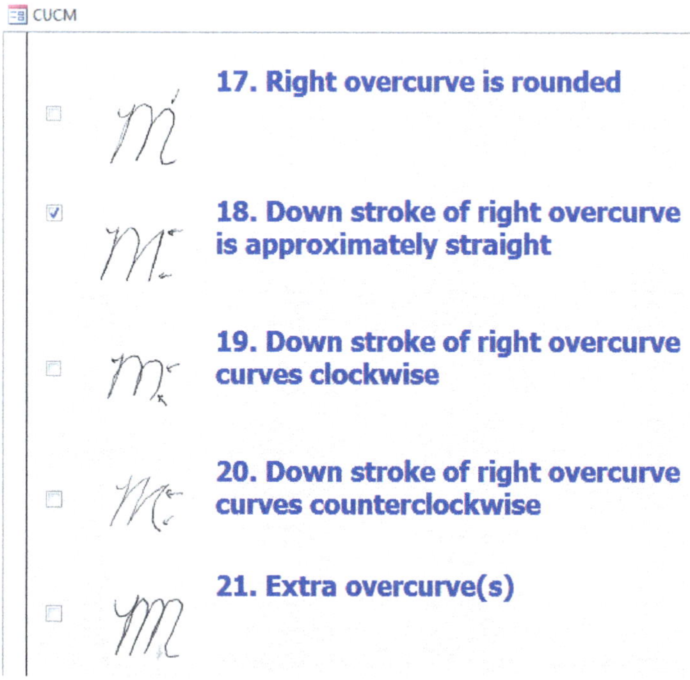

Fig. 8.3 Final five characteristics of the capital letter "ℳ"

Table 8.6 Actual implementation of the attribute agreement study

Cursive letters (both upper and lower case):		
Examiner T:	4, 7, 111, 201, 222	+ 4, 7
Examiner E:	4, 111, 201, 222	+ 4, 111, 222
Examiner R:	4, 7, 95, 111, 201, 222	+ 7
Printed letters (both upper and lower case):		
Examiner T:	4, 7, 111, 201, 222,	+ 4, **222**
Examiner E:	4, **7**, 111, 201, 222	+ 4, 111, 222
Examiner R:	4, 7, 95, 111, 201, 222	+ 7

specimen #95, which provides no information on the agreement across examiners. However, its inclusion did facilitate checking the algorithm for determining which features showed agreement across all reviews (both among examiners and within when an examiner looked at a specimen once again at a later date).

The data collected from the attribute agreement study was tabulated in a spread sheet with 2,500 columns (one per characteristic) and 21 rows for cursive and 22 rows for printed (each row corresponding to one specimen and one examiner review). To determine if the examiners agreed within each specimen (all true or all false), a SAS/JMP analysis was used. The distribution of each column by specimen was summarized and then it was determined if the number of realizations was one (corresponding to either all true or all false) for each specimen. If this were the case

Table 8.7 Characteristics for cursive m, M with complete agreement by the examiners

m	CLCM_1. Internal or terminal letter connected to previous lower case letter
m	CLCM_10. Middle leg is enclosed loop
m	CLCM_17. Internal m not connected to both previous and subsequent letter
m	CLCM_18. Initial and internal letter connected to subsequent letter
m	CLCM_3. Left peak is clearly pointed
m	CLCM_4. Left peak is enclosed loop
m	CLCM_7. Left leg is enclosed loop
$\mathscr{M}$	CUCM_1. Counterclockwise curving initial stroke
$\mathscr{M}$	CUCM_3. Extraneous straight initial stroke
$\mathscr{M}$	CUCM_4. Initial stroke begins on staff
$\mathscr{M}$	CUCM_6. Upward stroke to first overcurve is retrace (open or closed)
$\mathscr{M}$	CUCM_7. Upward stroke to first overcurve is clearly counterclockwise curve (no angular point)
$\mathscr{M}$	CUCM_10. Left overcurve is clearly taller than right overcurve
$\mathscr{M}$	CUCM_23. Not connected to subsequent letter

for each specimen, then the implication was that the examiners agreed across each specimen. Thus, it is a fairly straightforward manned to determine all characteristics that survive this scrutiny a single disagreement eliminates the characteristic from further consideration.

Recall from the previous section the illustration with the cursive letter *m*. From an initial set of 19 characteristics, there were seven characteristics for which the examiners agreed across all specimens considered (Table 8.7). For uppercase cursive $\mathscr{M}$ there were 23 characteristics considered of which seven made the final cut.

A subset of the results for the number of agreed characteristics by case for cursive writing style is given in Table 8.8. The number of surviving characteristics varied considerably by letter and case. The identification and specification of handwriting characteristics represents a tour de force by our forensic document examiner (Vastrick).

8.5 Future Directions

This paper has provided a description of the development of the handwriting features data base and the corresponding attribute agreement analysis. These are essential first steps to establishing the validity of the database and lay the foundation for subsequent quantitative analyses on future documents. These unambiguously identifiable features can be used to discriminate among documents containing these items.

Table 8.8 Summary results for cursive handwriting

	Uppercase cursive		Lowercase cursive	
Letter	# characteristics	# agree	# characteristics	# agree
A	59	11	12	0
B	34	12	13	1
C	16	2	12	5
D	24	13	31	12
E	28	12	8	4
F	30	11	27	11
G	33	21	30	7
H	41	30	16	7
I	29	7	21	7
J	19	11	29	19
K	38	3	14	5
L	26	16	11	6
M	21	7	19	7
N	22	9	13	7
O	17	10	22	9
P	28	7	19	4
Q	12	3	26	11
R	19	7	14	3
S	19	5	15	4
T	42	16	20	11
U	24	7	22	6
V	19	6	24	9
W	35	15	37	11
X	27	16	17	6
Y	19	9	24	5
Z	27	14	22	4

Acknowledgements This work was supported by a National Institute of Justice grant through the National Center for Forensic Science at the University of Central Florida.

References

1. ASTM-E105-04: Standard Practice for Probability Sampling of Materials. ASTM International, West Conshohocken (2004)
2. ASTM-E141-91: Standard Practice for Acceptance of Evidence on the Results of Probability Sampling. ASTM International, West Conshohocken (1991)
3. ASTM-E1658: Standard Terminology for Expressing Conclusions of Forensic Document Examiners. ASTM International, West Conshohocken (2008)
4. ASTM-E2290: Standard Guide for Examination of Handwritten Items. ASTM International, West Conshohocken (2007)

5. Boulanger Carey, M., Chen, H., Descloux, A., Ingle, J., Park, K.: 1982–83 end office connections study: analog voice and voiceband data transmission performance characterization of the public switched network. AT&T Bell Labs. Tech. J. **9**(63), 2059–2119 (1999)
6. Huber, R., Headrick, A.: Handwriting Identification: Facts and Fundamentals. CRC Press, Boca Raton (1999)
7. Kam, M., Fielding, G., Conn, R.: Writer identification by professional document examiners. J. Forensic Sci. **42**(5), 778–786 (1997)
8. NRC: Strengthening Forensic Science in the United States: A Path Forward. National Research Council, Washington, DC, committee on Identifying the Needs of the Forensic Sciences Community and Committee on Applied and Theoretical Statistics (2009)
9. Sita, J., Found, B., Rogers, D.: Forensic handwriting examiners' expertise for signature comparison. J. Forensic Sci. **47**, 1117–1124 (2002)
10. Srihari, S.N., Leedham, G.: A survey of computer methods in forensic document examination. In: Proceedings of the 11th International Graphonomics Society Conference, Scottsdale, AZ, pp. 278–281 (2003)
11. Srihari, S.N., Cha, S.H., Arora, H., Lee, S.: Individuality of handwriting. J. Forensic Sci. **47**, 1–17 (2002)

Chapter 9
Bayes Factors and Maximum Entropy Distribution with Application to Bayesian Tests

Adriana Brogini and Giorgio Celant

9.1 Introduction

The Bayes factor [3–5] has taken a renewed interest in the Bayesian statistics being an instrument less binding than the posterior probability distribution, on which is based the Bayesian response to inferential problems as the parametric hypothesis testing.

Although the Bayes factor, in general, depends on the a priori information of the experimenter, it eliminates some of its influence on the likelihood, measuring the evidence in favour of the hypothesis of interest due to the sampling observations.

It should also be noted as the context of the hypothesis testing is not compatible with an entirely uninformative a priori Bayesian approach, since the same formulation of the problem assumes the subdivision of the parametric space into at least two subsets, involving a latent information on the chosen statistical model, in particular toward the hypothesis that has to be tested.

For the given reasons, in this paper we propose the analysis and the solution of some parametric two-sided tests, combining the Bayes factor's logic with the maximum entropy method, that allows to obtain the less informative a priori probability distribution, taking into account the amount of initial information available to the experimenter.

A. Brogini (✉) • G. Celant
Department of Statistics, University of Padova, Padova, Italy
e-mail: adriana.brogini@unipd.it; giorgio.celant@unipd.it

V.B. Melas et al. (eds.), *Topics in Statistical Simulation*, Springer Proceedings in Mathematics & Statistics 114, DOI 10.1007/978-1-4939-2104-1_9

9.2 Bayes Factor and Testing of Hypothesis

9.2.1 Definitions

Let $\mathcal{X} \times \Theta \subseteq \mathbb{R}^n \times \mathbb{R}$; let $(\mathcal{X} \times \Theta, B_{\mathcal{X}} \bigotimes \mathcal{B}_\Theta, \{P(x,\theta) : (x,\theta) \in \mathcal{X} \times \Theta\})$ be a Bayesian experiment in which: $(\mathcal{X}, \mathcal{B}_{\mathcal{X}}, \{P_\theta : \theta \in \Theta\})$ is a parametric statistical model and $\{\Theta, \mathcal{B}_\Theta, \Pi\}$ is a probability space.

Let us suppose that $P_\theta \ll \mu$ and $\Pi \ll \mu$, where μ is either the Lebesgue measure or the counter measure; $\frac{dP_\theta}{d\mu} = f(x/\theta)$ and $\frac{d\Pi}{d\mu} = \pi(\theta)$ will be, respectively, the Radon–Nikodyn derivative of P_θ and Π respect to μ.

Let $[\Theta_0, \Theta_1]$ be a partition of the parametric space, Θ in which $H_0 : \theta \in \Theta_0$ is the subset of the hypotheses of interest, possibly reduced to a point and $H_1 : \theta \in \Theta_1$ is the subset of the alternative hypotheses. We will suppose to explain the a priori distribution Π on these two events only,

$$\Pi_0 = \text{Prob}(\theta \in \Theta_0) = \int_{\Theta_0} \pi(\theta)\, d\mu(\theta)$$

and

$$\Pi_1 = \text{Prob}(\theta \in \Theta_1) = \int_{\Theta_1} \pi(\theta)\, d\mu(\theta).$$

The problem is in the updating of the a priori information through Bayes theorem, using the likelihood, i.e. the sampling information.

Definition 1. We define $B^\Pi(x)$, Bayes factor in favour of H_0, the amount:

$$B^\Pi(x) = \frac{\Pi(\theta \in \Theta_0/x)}{\Pi(\theta \in \Theta_1/x)} = \frac{\Pi_0}{\Pi_1}$$

where $\Pi(\theta \in \Theta_i/x)$, $i = 0,1$ is the posterior probability computed with the priori Π and Π_i, $i = 0,1$, is the a priori probability of the parameter to belong to the subsets Θ_0 and Θ_1.

It is evident that $B^\Pi(x)$ depends both on likelihood than from a priori information, even if the latter is partially evaded, as we will see in formula (9.2). To highlight this dependence, in the case of our interest, i.e. for $\Theta_0 = \{\theta_0\}$, it is convenient to write the a priori distribution as follows:

$$\Pi(\theta) = \begin{cases} \Pi_{0 g_0}(\theta) & \text{if } \theta \in \Theta_0 \\ \Pi_{1 g_1}(\theta) & \text{if } \theta \in \Theta_1 \end{cases} \tag{9.1}$$

where $g_i(\theta) = \frac{\Pi(\theta)}{\Pi_i(\theta)} \chi_{\Theta_i}(\theta)$ (χ = indicator function). The $g_i(\theta)$, $i = 0,1$ represents proper conditional densities, that describe how the "mass" of a priori probability extends on the two considered subsets; thus, the Bayes factor has the form:

$$B^{\Pi}(x) = \frac{f(x/\theta_0)\,\Pi_0}{\int_{\Theta_1} f(x/\theta)\,g_1(\theta)\,d\mu(\theta)\,(1-\Pi_0)} \Big/ \frac{\Pi_0}{(1-\Pi_0)} \tag{9.2}$$

In the case that both hypotheses are punctual, the amount $B^{\Pi}(x)$ coincides with the classical likelihood rate. In fact from (9.2), setting $\Theta_1 = \{\theta_1\}$, we obtain:

$$B^{\Pi}(x) = \frac{f(x/\theta_0)}{f(x/\theta_1)} \tag{9.3}$$

Formula (9.2) highlights how the Bayes factor, for at least a punctual hypothesis, cannot be defined if the a priori probability distribution is absolutely continuous respect to the Lebesgue measure or more general whenever it is $\Pi_0 = 0$ or $\Pi_1 = 0$. In this case, the use of a priori distributions that are spread, vague or Jeffreys a priori, absolutely continuous respect to the Lebesgue measure, the use of which is consolidated in absence of a well formulated subjective a priori, is not possible.

The following example emphasizes the inadequacy of the use of the spread distribution.

Example 1. Let $X \sim N(\theta, 1)$. We want to test the following hypothesis:

$$\begin{cases} H_0 : \theta = 0 \\ H_1 : \theta \neq 0 \end{cases}$$

The use of a non-informative law $\Pi(\theta) = 1$, where $\theta \neq 0$ leads to

$$\Pi(\theta = 0/x) \equiv \frac{e^{-\frac{x^2}{2}}}{e^{-\frac{x^2}{2}} + \int_{-\infty}^{+\infty} e^{-\frac{(x-\theta)^2}{2}}\,d\theta} = \frac{1}{1+\sqrt{2\pi}\,e^{\frac{x^2}{2}}}$$

if $\Pi_0 = 1/2$. Consequently the posterior probability of H_0 is increased by, $\frac{1}{(1+\sqrt{2\pi})} = 0.285$ giving little advantage to the null hypothesis even in the most favourable case. It can be stated that, in general, the use of spread probability distributions leads to biased results in favour of the alternative hypothesis.

Below, we try to solve the problem of the description of our a priori "ignorance" related to a system of hypothesis, through some a priori that don't have the inconveniences just described.

Of relevant importance it will be the entropy functional, that is defined as follows:

$$H(g(\theta)) = \begin{cases} -\sum_{i=1}^{n} g(\theta_i)\ln g(\theta_i) & \text{(discrete case)} \\ -\int_{\Theta} g(\theta)\ln\left(\frac{g(\theta)}{\tilde{g}(\theta)}\right) d\mu(\theta) & \text{(continuous case)} \end{cases}$$

where $\tilde{g}(\theta)$ is a suitable function (see Sect. 9.3).

9.3 A Priori Distribution of Maximum Entropy

The concept of entropy (untidiness, uncertainty) introduced in thermodynamics was quantified within theory of information, since the fundamental works of [7, 8], as measure of the degree of untidiness or the lack of information associated with the statistical description of a system.

The maximization of the entropy was proposed by Jaynes (1968) as a method to assign a priori probability distributions, considering the entropy as a suitable measure of the lack of information or analogously, of the amount of uncertainty represented by a probability distribution.

The method is based on the idea that we must assign as probability distribution the one which is consistent with the observable evidence, i.e. with the amount of initial information controllable and translatable in a series of constraints, leaving to the rest of variability, maximum freedom. It should be specified, as Jaynes defines "controllable an information about a size of a system, if for each allocation of probability is possible to determine unambiguously if there is agreement or not with the information itself" [6] Jaynes (1981) and that "any type of information you have it provides the most honest description of our knowledge" Jaynes (1967) [6].

If no information is available, and the amount of interest can assume only a finite number of values, the solution of maximum entropy reduces to that proposed by the postulate of Bayes–Laplace (principle of indifference) and the proposed probability distribution is the discrete uniform distribution.

If the amount of interest is continuous, one needs to choose a prior dominant measure ($\tilde{g}$), that allows the invariance of the entropy functional and that corresponds to the situation which is totally uninformative. The solution of maximum entropy will depend from this choice. For example, when a structure of group for the model is available, you choose as $\tilde{g}$ the right measure of Haar, defined for this group.

In the particular case of a parameter of position, case that will be examined in Sect. 9.4, in which there is translation invariance, the measure results that of Lebesgue, that coincides with the invariant location measure a priori of Jeffreys.

9.4 An Application to Hypothesis Testing: The Problem of Punctual Hypothesis

The punctual hypotheses may be considered unrealistic [2], however we observe that:

(a) The hypotheses of the form, $H_0 : \theta \in [\theta_0 - \epsilon, \theta_0 + \epsilon]$, in general may be approximated by punctual hypotheses, $H_0 : \theta = \theta_0$, without any change of the a posteriori probabilities, when the likelihood is constant around θ_0 [1].

(b) In many practical cases the hypotheses are punctual. Just think to an experiment in which the variables are treated. In this case it is essential to check the extent of treatment.

It is worth to assess the following problem of hypotheses testing:

$$\begin{cases} H_0 : \theta = \theta_0 \\ H_1 : \ \theta \in [a, b] \setminus \{\theta_0\}, \ a < b \ (a, b) \in \mathbb{R}^2 \end{cases}$$

Hence: $\Theta = \Theta_0 \bigcup \Theta_1$ and $\Theta_0 = \{\theta_0\}$ and $\Theta_1 = [a, b] \setminus \{\theta_0\}$ Let be:

$$\epsilon = b - a, \quad M_1 = \int_{\Theta_1} \theta g_1(\theta)\, d\mu(\theta), \quad M_2 = \int_{\Theta_1} [\theta - M_1]^2 g_1(\theta)\, d\mu(\theta)$$

where $\mu_1$is any fixed point in the interval. $[a, b] \setminus \{\theta_0\}$.

The information contained into the system of hypothesis above can be translated in the following constraints:

1. $\Pi_0 + \Pi_1 = 1$
2. $\theta \in \Theta_1$, conditionally to H_1.

If we want to use the Bayes factor to solve the hypotheses testing, it is necessary to set the values: Π_i, $i = 0$, 1and $g_1(\theta)$.

This clarification will occur through the method of maximum entropy.

On the basis of this method, it is easy to see that $\Pi_1 = 1/2$, since we know that (Sect. 9.3) the solution is uniform and discrete on $\{\Theta_0, \Theta_1\}$.

To obtain $g_1(\theta)$, through the method of maximum entropy, it is necessary to modify the constraint 2. Note that the only deduction we can draw from the relation $\theta \in \Theta_1$is that:

$$|M_i - \mu_i| \leqslant \epsilon, \quad i = 1, 2.$$

In fact for $i = 1$ we obtain the Cauchy condition; for $i = 2$, we know θ less than an error ϵ, thus the variance cannot exceed ϵ. Hence, the constraint will become:

$$|M_2 - \mu_2| \leqslant \epsilon, \quad \mu_2 = 0.$$

This position allows us to write the Cauchy condition and those related to the variance in the following more compact form:

$$\sum_{j=1}^{2} \left(\frac{M_j - \mu_j}{\epsilon} \right)^2 \leqslant 2. \tag{9.4}$$

The $g_1(\theta)$ is determined by the following problem of optimum

$$\begin{cases} \max\limits_{g_1(\theta)} H\left(g_1\left(\theta\right)\right) \\ \Pi_1 = \frac{1}{2} \\ \sum_{j=1}^{2}\left(\frac{M_j-\mu_j}{\epsilon}\right)^2 \leqslant 2 \end{cases}$$

which is equivalent for convexity of (9.4) to:

$$\begin{cases} \max_{g_1(\theta)} H\left(g_1\left(\theta\right)\right) \\ \Pi_1 = \frac{1}{2} \\ \sum_{j=1}^{2}\left(\frac{M_j-\mu_j}{\epsilon}\right)^2 - 2 = 0 \end{cases}$$

In the following paragraph we show the computations in detail and the solution of the problem of optimum in the case when the constraint (9.4) is replaced with only the Cauchy condition. The general case with the constraint (9.4) doesn't lead to an explicit solution, so it is omitted.

9.5 Determination of the Distribution of Maximum Entropy

The determination of the function $g_1(\theta) : [a, b] \rightarrow \mathbb{R}^+$ which satisfies the constrained optimum problem, constraint of Cauchy, is similar to the determination of the constrained optimum of a real function.

The following theorem holds:

Theorem 1 (Kolmogorov, Fomin). *Let be X, a normed space, A an open of X, $x_0 \in A$, $g : A \rightarrow \mathbb{R}$, Fréchet differentiable in x_0, if x_0 is a point of maximum or minimum for g, then $dg(x_0)(h) = 0$, for whatever h.*

Given that:

$$f(\theta) = \frac{g(\theta)}{\int_a^b g(\theta)\, d\theta}; \qquad M = \int_a^b \theta f(\theta)\, d\theta \tag{9.5}$$

and μ_1 a fixed value in $[a, b]$.

The described problem is equivalent to:

$$\begin{bmatrix} \max_f H(f) = \max\limits_f \left[-\int_a^b f(\theta) \ln f(\theta)\, d\theta\right] \\ \int_a^b f(\theta)\, d\theta - 1 = 0 \\ \frac{1}{\epsilon^2}\left(M_1 - \mu_1\right)^2 - 1 = 0 \end{bmatrix} \tag{9.6}$$

indicating with $\mathcal{I}_0$ an interval around zero, and supposing that $f(\theta)$ is limited in $\mathcal{I}_0 \setminus \{0\}$ When the Fréchet differential exists, the Gateaux differential also exists and they coincide.

We proceed with the computation of the "directional derivative" of the Lagrangian: $\mathcal{L}[f(\theta) + dh(\theta)]$ where

$$\mathcal{L}(f) = f(\theta) + \lambda_0 \left[\int_a^b f(\theta)\, d\theta - 1 \right] + \lambda_1 \left[\frac{1}{\epsilon^2} \left(\int_a^b \theta f(\theta)\, d\theta - \mu_1 \right)^2 - 1 \right] \tag{9.7}$$

We derive term by term the sum

$$\begin{aligned} &\frac{\partial}{\partial \alpha} \left\{ - \int_a^b (f(\theta) + \alpha h(\theta)) \ln (f(\theta) + \alpha h(\theta))\, d\theta \right\} \\ &\quad = - \int_a^b \frac{\partial}{\partial \alpha} [(f(\theta) + \alpha h(\theta)) \ln (f(\theta) + \alpha h(\theta))]\, d\theta \\ &\quad = - \int_a^b [h(\theta) \ln f(\theta) + \alpha h(\theta) + h(\theta)]\, d\theta \\ &\quad = - \int_a^b h(\theta) [\ln (f(\theta) + \alpha h(\theta)) + 1]\, d\theta \end{aligned}$$

hence for $\alpha = 0$ we have

$$- \int_a^b h(\theta) [\ln f(\theta) + 1]\, d\theta$$

$$\frac{\partial}{\partial \alpha} \left\{ \lambda_0 \left[\int_a^b (f(\theta) + \alpha h(\theta))\, d\theta - 1 \right] \right\}$$

$$= \lambda_0 \left[\int_a^b \frac{\partial}{\partial \alpha} (f(\theta) + \alpha h(\theta))\, d\theta \right] = \lambda_0 \int_a^b h(\theta)\, d\theta$$

$$\frac{\partial}{\partial \alpha} \left\{ \lambda_1 \left[\frac{1}{\epsilon^2} \left(\int_a^b \theta (f(\theta) + \alpha h(\theta))\, d\theta - \mu_1 \right)^2 - 1 \right] \right\}$$

$$= \lambda_1 \left\{ \frac{1}{\epsilon^2} \left[\frac{\partial}{\partial \alpha} \left(\int_a^b \theta \left(f(\theta) + \alpha h(\theta) \right) d\theta - \mu_1 \right)^2 - 1 \right] \right\}$$

$$= \frac{\lambda_1}{\epsilon^2} \left\{ 2 \left[\int_a^b \theta \left(f(\theta) + \alpha h(\theta) \right) d\theta - \mu_1 \right] \left[\int_a^b y h(y)\, dy \right] \right\}$$

and for $\alpha = 0$ we have:

$$\frac{2\lambda_1}{\epsilon^2} \left(\int_a^b \theta f(\theta)\, d\theta - \mu_1 \right) \cdot \left(\int_a^b y h(y)\, dy \right)$$

hence, we obtain

$$d\mathcal{L}_{(f)}(h) = \int_a^b h(\theta) \left\{ -\ln f(\theta) - 1 + \lambda_0 + \lambda_1 \left(-2 \frac{\theta \mu_1}{\epsilon^2} \right) + \lambda_1 \int_a^b 2 \frac{\theta y}{\epsilon^2} f(y)\, dy \right\} d\theta. \tag{9.8}$$

In order to obtain the extreme points, we have to solve $d\mathcal{L}_{(f)}(h) = 0$, whatever the direction h for this, it is sufficient to set the term in curly brackets of (9.8) equal to zero, i.e.

$$\ln f(\theta) = \lambda_0 - 1 + \lambda_1 \left(-2 \frac{\theta \mu_1}{\epsilon^2} \right) + \lambda_1 \int_a^b 2 \frac{\theta y}{\epsilon^2} f(y)\, dy. \tag{9.9}$$

Formula (9.9), since $f(\theta)$ is unknown, is a nonlinear integral equation. Since $f(\theta) \in L_1$ and $\left(2 \frac{\theta y}{\epsilon^2} \right)$ is a monomial, we can state that their product is a function of L_1 and we can determine a polynomial solution by setting

$$\ln f(\theta) = a_0 + a_1 \theta. \tag{9.10}$$

We replace this expression in both members of the integral equation (9.9) and, applying the principle of identity between polynomials, we proceed to the determination of the coefficients a_0 and a_1, using the constraints

$$\int_a^b f(\theta)\, d\theta = 1 \Rightarrow e^{a_0} = \frac{a_1}{e^{a_1} - 1}. \tag{9.11}$$

For the sake of simplicity and without loss of generality of the results, we provide the following transformation $\theta = \frac{t-a}{b-a}$; $a, b \in \mathbb{R}$, $a < b$

Computation of a_1:

$$\left[\frac{1}{\epsilon^2}\left(\int_0^1 \theta e^{a_0+a_1\theta} d\theta - \mu_1\right)^2 - 1\right] = 0.$$

where

$$\int_0^1 \theta e^{a_0} e^{a_1\theta} d\theta = e^{a_0}\left[\frac{1}{a_1}e^{a_1} - \frac{1}{a_1^2}e^{a_1} + \frac{1}{a_1^2}\right]$$

and we obtain:

$$\frac{1}{\epsilon^2}e^{a_0}\left[\frac{a_1 e^{a_1} - e^{a_1} + 1 - a_1^2\mu_1}{a_1^2}\right]^2 = 1;$$

$$\frac{1}{\epsilon^2}\frac{a_1}{e^{a_1}-1}\left[\frac{a_1 e^{a_1} - e^{a_1} + 1 - a_1^2\mu_1}{a_1^2}\right]^2 = 1$$

set $e^{a_1} = z$, we obtain the following equation:

$$(\mu_1)^2 (\ln z)^4 - \left[2z\mu_1 + \epsilon^2 (z-1)\right](\ln z)^3 + \left[z^2 - 2\mu_1 + 2z\mu_1\right](\ln z)^2 + \left(2z - 2z^2\right)(\ln z) + 1 + z^2 - 2z = 0. \tag{9.12}$$

Equation (9.12) is equivalent to the following:

$$z = \left[2 - 2(\ln z) - \mu_1 (\ln z)^2 + \epsilon^2 (\ln z)^3 + 2\theta (\ln z)^3 + \epsilon (\ln z)^2\right] \times \sqrt{4 - 4(\ln z) - 4\theta(\ln z) + \epsilon^2 (\ln z)^2 + 4\theta (\ln z)^2}/2\left[1 - 2(\ln z) + (\ln z)^2\right] \tag{9.13}$$

9.6 Computation of the a Posterior and of the Bayes Factor

We explicit the computations and the system of hypothesis described before, with the assumptions and the same notation. Indicating with $\pi(\theta_0/x)$the a posterior density under H_0 and $L(x/H_0)$, the likelihood under H_0, we have:

$$\Pi(\theta_0/x) = \frac{L(x/\theta_0)\,\Pi_0}{\int_0^1 L(x/\theta)\,\pi(\theta)\,d\theta} \tag{9.14}$$

$$= \frac{L(x/\theta_0)\,\Pi_0}{L(x/\theta_0)\,\Pi_0 + (1-\Pi_0)\int_{(0,1]} L(x/\theta)\,g_1(\theta)\,d\theta}$$

In the same way we obtain the Bayes factor that is:

$$B^{\pi}(x) = \frac{L(x/\theta_0)\,\Pi_0}{\left(\int_{(0,1]} L(x/\theta)\,g_1(\theta)\,d\theta\right)(1-\Pi_0)} \cdot \frac{1-\Pi_0}{\Pi_0}$$

$$= \frac{\frac{1}{\sqrt{2\pi}} e^{-\frac{1}{2(x-\theta_0)^2}}}{\frac{1}{\sqrt{2\pi}} \int_0^1 e^{-\frac{1}{2(x-\theta)^2}} e^{a_0+a_1\theta} d\theta}$$

$$= \frac{1}{e^{a_0+1/2(a_1+x)^2[\Phi(0)-\Phi(1)]}},$$

where Φ is the cumulative distribution function of a $N\left[(a_1+x),1\right]$.

9.7 Final Observations

Formula (9.13) of the paragraph 5, which explicits the value of z shows that z is a function of μ_1 and ϵ. In order to obtain a real example, we consider the limits of integration imposed by the transformation $\theta = \frac{t-a}{b-a}$, i.e. $\theta \in (0,1]$. In this case:

$$\epsilon = b - a = 1;\ \mu_1 = \frac{1}{2}$$

and

$$H_0 : \mu = 0$$

$$H_1 : \mu \in (0,\ 1]$$

Using the computer program "MATEMATICA", we have the following two solutions: $z = 1$ and $z \simeq 8,776$. The first solution $z = 1$ is not acceptable because it leads to an undetermined constant a_0. In fact $e^{a_1} = 1$, i.e. $a_1 = 0$ and $e^{a_0} = \frac{a_1}{e^{a_1}-1} = \frac{0}{0}$.

The second solution is compatible and leads to the "approximated" determination of the constants: $a_0 \simeq -1.275$ and $a_1 \simeq 2.172$.

Note that the undetermination, which follows to the trivial solution, $z = 1$ is a confirmation that the uniform distribution, which reflects a more vague information, is not acceptable as solution of the testing problem, as we had already identified in the introduction.

To conclude we observe that at the objection that the constraint (4) doesn't reflect exactly the information described by the system of hypotheses is possible to answer back that, if we are not able to translate the information in a more precise analytic form for the application of the method of maximum entropy, it is necessary to consider, as source of uncertainty (and then to increase the value of the entropy), the information not expressible by the inequality constraints.

References

1. Berger, J.O.: Statistical Decision Theory and Bayesian Analysis. Springer, New York (1985)
2. Casella, G., Berger, R.: Reconciling Bayesian and frequentist evidence in the one-sided testing problem. JASA **82**, 106–111 (1987)
3. Good, I.J.: Probability and the Weighing of Evidence. Griffin, London (1950)
4. Good, I.J.: The Estimation of Probability: An Essay on Modern Bayesian Methods. M.I.T. Press, Cambridge (1965)
5. Jeffreys, H.: Scientific Inference. M.A., D.Sc., F.R.S. Cambridge University Press, Cambridge (1957)
6. Rosa, R.: Massima entropia: E.T. Jaynes e dintorni. Statistica **Anno XLV**(2), 181–208 (1985)
7. Shannon, E.C.: A mathematical theory of communication. Bell Syst. Tech. J. **27**, 379–423 (1948)
8. Weyner, N.: Cybernetics. Wiley, New York (1948)

Chapter 10
Monte Carlo Algorithm for Simulation of the Vehicular Traffic Flow Within the Kinetic Model with Velocity Dependent Thresholds

Aleksandr Burmistrov and Mariya Korotchenko

10.1 Acceleration Oriented Kinetic Model

We develop our algorithms in the frame of the kinetic VTF model suggested in [5]. A distinctive feature of this model consists in introducing of the acceleration variable into the set of the phase coordinates along with the velocity coordinate of the car. Such a modification of the phase space allowed to describe not only a partly constrained traffic but also a higher car density regimes.

According to this model, in the case of homogeneous traffic flow, the probability density $f(a, v, t)$ for a single car with acceleration a and velocity v solves the integro-differential equation of Boltzmann type:

$$\left(\frac{\partial}{\partial t} + a\frac{\partial}{\partial v}\right) f(a, v, t) = \int\limits_{\bar{a},\bar{v},a'} \left[\Sigma(a' \to a|v, \bar{a}, \bar{v}) f(a', v, t) \right. \tag{10.1}$$

$$\left. - \Sigma(a \to a'|v, \bar{a}, \bar{v}) f(a, v, t)\right] f(\bar{a}, \bar{v}, t)\, \mathrm{d}\bar{a}\, \mathrm{d}\bar{v}\, \mathrm{d}a',$$

Here we distinguish two types of vehicles: the *leader* with the kinematic state $(\bar{v}, \bar{a})$ which interacts with the *follower* (the current car situated directly behind the leader).

A. Burmistrov
Institute of Computational Mathematics and Mathematical Geophysics (Siberian Branch of the Russian Academy of Sciences), Prospect Akademika Lavrentjeva, 6, Novosibirsk 630090, Russia

Novosibirsk State University, Pirogova st., 2, Novosibirsk 630090, Russia
e-mail: burm@osmf.sscc.ru

M. Korotchenko (✉)
ICM&MG SB RAS, Prospect Akademika Lavrentjeva, 6, Novosibirsk 630090, Russia
e-mail: kmaria@osmf.sscc.ru; rmary@ngs.ru

V.B. Melas et al. (eds.), *Topics in Statistical Simulation*, Springer Proceedings in Mathematics & Statistics 114, DOI 10.1007/978-1-4939-2104-1_10

As the car acceleration a is added to the phase coordinates, there are only acceleration jumps (no velocity jumps, as in other kinetic models) produced by the pairwise interactions in the system, which are expressed by the Boltzmann-like interaction integral (the right side of Eq. (10.1)).

The function $\Sigma(\cdot)$ in the latter integral is a *weighted interaction rate function*. It determines the type of interaction in the system and is a compound of the *interaction rate* $Q(\cdot)$, the *acceleration change probability density* $\sigma(\cdot)$ and the *distance correlation function* $D(\cdot)$:

$$\Sigma(a' \to a | v, \bar{a}, \bar{v}, \mathbf{m}_f)$$
$$= \int_{h_{\min}}^{\infty} \sigma(a' \to a | h, v, \bar{a}, \bar{v}) \cdot Q(h, a', v, \bar{a}, \bar{v}) \cdot D(h | a', v, \mathbf{m}_f) \, dh.$$

Here $h_{\min}$ is the minimal distance between two cars (a mean length of a car). The interaction rate $Q(\cdot)$ depends on a current microscopic state of the interacting car pair and the distance h between them.

The function $\sigma(\cdot)$ defines the probability of changing the acceleration of the follower from a' to a when the interaction between the cars with the states (v, a') and $(\bar{v}, \bar{a})$ takes place at distance h.

The function $D(\cdot)$ is a conditioned probability density of the distance h. It does not depend on the leader's state $(\bar{v}, \bar{a})$, because for the follower it is difficult to evaluate this state even qualitatively. But the distance behavior depends on the whole traffic flow, which is determined by the probability density $f(\cdot)$ and the car density $\mathscr{K}$. The driver cannot observe $f(\cdot)$ itself, but some of its moments (mean velocity, scattering, etc.). Therefore $D(\cdot)$ depends on the vector of moments $\mathbf{m}_f$, in which we also include the car density $\mathscr{K}$.

We would like to underline that, as the leader does not change its acceleration after the interaction takes place, the function $\Sigma(\cdot)$ is not symmetric, unlike in gas dynamics.

We supplement Eq. (10.1) with the initial distribution $f(a, v, 0) = f_0(a, v)$ and boundary conditions, which ensure that there are no cars with negative velocities and there is a maximum velocity of the VTF, which cannot be exceeded.

Further we define the functions $Q(\cdot)$, $\sigma(\cdot)$ and $D(\cdot)$.

10.2 Integral Equation of the Second Kind

In our previous work [1] we succeeded to construct the basic integral equation of the second kind $F = \mathbf{K}F + F_0$. Its solution F is a *distribution density of the interactions* in the N-particle system of vehicles. It is closely connected with the solution $f(a, v, t)$ to Eq. (10.1), and the kernel $\mathbf{K}$ describes the evolution of the many-particle system. The integral equation enables us to use well-developed techniques of the Monte Carlo simulation, including the majorant frequency principle [2], for estimating the functionals of solution to Eq. (10.1), as well as to perform parametric analysis [1].

Let us denote vectors $A = (a_1, \ldots, a_N)$, $V = (v_1, \ldots, v_N)$ for the given ensemble of N cars. In the phase space with coordinates $(Z, t) \equiv (\pi = (i, j), A, V, t)$ the distribution density $F(Z, t)$ of the interactions in the N-particle system solves the following integral equation:

$$F(Z,t) = \delta(t) P_0(A, V)\delta(\pi_0) + \int_0^t \int F(Z', t') K(Z', t' \to Z, t)\, \mathrm{d}Z'\, \mathrm{d}t'. \tag{10.2}$$

Here $\delta(\cdot)$ is the Dirac delta function, $P_0(\cdot)$ is initial distribution, $\mathrm{d}Z = \mathrm{d}A\, \mathrm{d}V\, \mathrm{d}\mu(\pi)$, and integration with respect to μ means the summation over all possible ordered pairs $\pi = (i, j)$. The kernel $K(Z', t' \to Z, t)$ is a product of transitional densities:

$$K(Z', t' \to Z, t)$$
$$= K_t(t' \to t | A', V') K_V(V' \to V | A', t - t') K_\pi(\pi) K_a(a_i' \to a_i | \pi, V).$$

10.2.1 Markov Chain Simulation

The transition in the *Markov chain*, which is related to the integral Eq. (10.2), consists of several elementary transitions in the following order:

1. the instant t of the next interaction in the system is chosen according to the exponential transition density ($\Theta(\cdot)$ is the Heaviside step function)

$$K_t(t' \to t | A', V') = \Theta(t - t') \nu(A', V' + A'(t - t')) \mathrm{e}^{\left\{ -\int_{t'}^{t} \nu(A', V' + A'(\tau - t'))\, \mathrm{d}\tau \right\}},$$
$$\text{with } \nu(A, V) = \frac{1}{N-1} \sum_{i \neq j} \int \Sigma(a_i \to a_i'' | v_i, a_j, v_j)\, \mathrm{d}a_i'' = \sum_{\pi} \frac{\nu_{(i,j)}}{N-1};$$

2. the velocities of all cars are calculated at time t according to the transition density $K_V(V' \to V | A', t - t') = \delta(V - V' - A'(t - t'))$;
3. the pair number (i, j) is chosen by the probabilities $K_\pi(i, j) = \frac{1}{N-1} \cdot \frac{\nu_{(i,j)}}{\nu(A', V)}$;
4. the new acceleration of the car with the number i is changed according to the transition density $K_a(a_i' \to a_i | \pi, V) = \Sigma(a_i' \to a_i | v_i, a_j, v_j)/\nu_{(i,j)}$.

10.2.2 Monte Carlo Estimation of Functionals

The following functionals of the one-particle distribution function $f(\cdot)$ are of our interest:

$$I_{\mathbf{h}}(T) = \int\int \mathbf{h}(a, v) f(a, v, T)\,\mathrm{d}v\,\mathrm{d}a = (\mathbf{h}, f).$$

By analogy with [3] we can prove that

$$I_{\mathbf{h}}(T) = \int\int_{0}^{T} \mathbf{h}_N(A, V + A(T - t'))$$
$$\times \exp\left\{-\int_{t'}^{T} \nu(A, V + A(\tau - t'))\,\mathrm{d}\tau\right\} F(Z, t')\,\mathrm{d}Z\,\mathrm{d}t',$$

where $\mathbf{h}_N(A, V) = \frac{1}{N}\sum_{i=1}^{N} \mathbf{h}(a_i, v_i)$. As a result we have $I_{\mathbf{h}}(T) = (\tilde{\mathbf{h}}_N, F)$.

For numerical estimation of $I_{\mathbf{h}}(T)$ we can use the collision or absorption estimator, which are functionals of the Markov chain trajectory.

For estimating the velocity and acceleration distribution we choose functions $\mathbf{h}(a, v)$ equal to indicators of some partitioning of the corresponding (velocity or acceleration) interval.

Since the interaction rate is not constant in the profiles we used for numerical experiments, we make use of the majorant frequency principle (see [2]) in our simulations.

10.3 Velocity Dependent Thresholds

We consider an interaction model with dependence on the distance between cars, and study the velocity and acceleration distributions with respect to the car density $\mathcal{K}$.

For the spatially homogeneous case we use two interaction profiles based on the velocity dependent thresholds, which were introduced in [4, 6]. In such profiles an interaction occurs only if the distance between interacting vehicles is equal to one of the threshold distances, which depend on the velocity of the follower. On each of these thresholds for the follower an individual acceleration change occurs.

10.3.1 Interaction with Single Threshold

We take the first example of interaction profile from [6]. In a given interacting car pair the follower with velocity v interacts (i.e., changes its acceleration state) if the distance h to the leading car is equal to threshold distance $H(v)$. In this case the interaction rate is the following:

$$Q(h, a', v, \bar{a}, \bar{v}) = |S(\bar{v}, v, a')| \cdot \delta(h - H(v)),$$

with $H(v) = [\alpha \cdot v + h_{\min}]$ and $S(\bar{v}, v, a') = \left[\bar{v} - v - \frac{dH}{dv}(v) \cdot a'\right]$. We consider threshold parameter α to be constant, though in the general case each driver in the flow has its own α_i and $H_i(v)$. Note that low values of α correspond to a more aggressive driving manner.

It is necessary to define the probability density of the follower's acceleration $\sigma(\cdot)$ only on the threshold $h = H(v)$, depending on the fact, whether the distance increases ($S(\bar{v}, v, a') > 0$) or decreases ($S(\cdot) \leqslant 0$). It is given by the formula:

$$\sigma(a' \to a|H(v), v, \bar{a}, \bar{v}) = \Theta(S(\bar{v}, v, a')) \cdot \delta(a - a^+) + \Theta(-S(\bar{v}, v, a')) \cdot \delta(a - a^-).$$

Here the acceleration a^+ strongly depends on the actual velocity v of the car. It increases at very low velocities and it decreases at higher velocities, having maximum at some given velocity v_m and vanishing near the maximum velocity w.

$$a^+ = \Theta(v - v_m) \cdot a_{\max} \frac{w - v}{w - v_m} + \Theta(v_m - v) \cdot \left[a_0 + \frac{a_{\max} - a_0}{v_m} v\right].$$

In order to prevent accidents we choose the value of deacceleration a^- equal to the *total braking value* [6]. It means that the follower with the current velocity v should stop in the distance $\alpha \cdot v + \bar{h}$, where $\bar{h}$ is the distance that the leader with current velocity $\bar{v}$ covers with the maximum braking value $\bar{a} = a_{\min} < 0$):

$$a^- = \frac{-v^2}{2(\alpha \cdot v + \bar{h})}, \quad \bar{h} = \frac{\bar{v}^2}{2|a_{\min}|}. \tag{10.3}$$

Distance measurements in traffic flows are often approximated by the gamma densities. We use the Gaussian density for convenience with the following parameters:

- the mean distance, i.e. the mean value of $\tilde{D}(\cdot)$, is equal to $1/\mathscr{K}$;
- the scattering of $\tilde{D}(\cdot)$ here is proportional to the mean velocity of all cars V.

The spatial correlation is given by the following distance probability density $D(h|a', v, \mathbf{m}_f)$, which depends on a driver [6]:

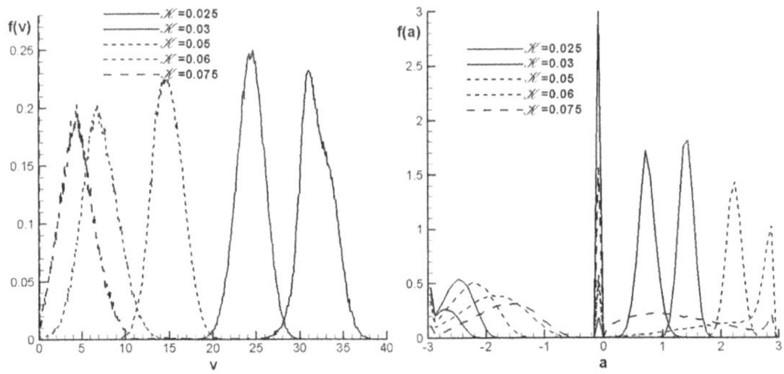

Fig. 10.1 Numerical estimates of the velocity (*left*) and acceleration (*right*) distributions for one threshold

$$D(H(v)|a' < 0, \mathbf{m}_f) = \frac{\tilde{D}(H(v)|\mathbf{m}_f)}{\int\limits_{h<H(v)} \tilde{D}(h|\mathbf{m}_f)\, dh}, \quad D(H(v)|a' \geq 0, \mathbf{m}_f) = \frac{\tilde{D}(H(v)|\mathbf{m}_f)}{\int\limits_{h>H(v)} \tilde{D}(h|\mathbf{m}_f)\, dh}.$$

Numerical estimates of the velocity and acceleration distributions are presented in Fig. 10.1.

The simulation results for the velocity distribution is in a good agreement with measured ones presented in [6]. But the acceleration distribution shows significant deviations. To improve the interaction profile, a second, more distant threshold with relative velocity dependence of the acceleration change is suggested in [4].

10.3.2 Interaction with Two Thresholds

In this subsection we are going to use the interaction profile suggested in [4]. For this profile, the way how the drivers accelerate or deaccelerate depends on the relative velocity between the leader and the follower. This behavior of the drivers makes the traffic flow more homogeneous.

The interaction rate for two thresholds is

$$Q(h, a', v, \bar{a}, \bar{v}) = |S_1(\bar{v}, v, a')| \cdot \delta(h - H_1(v)) + |S_2(\bar{v}, v, a')| \cdot \delta(h - H_2(v)),$$

with $H_1(v) = \alpha_1 \cdot v + h_{\min} < H_2(v) = \alpha_2\sqrt{v + \beta} + \gamma$, here α_1, α_2, β, γ are constants (see [4] for more details concerning the forms of these thresholds).

For simplicity, we use a linear velocity dependence for approximation of the first threshold H_1 as for this quantity the measured data are widely scattered. We use a square root function for approximation of the larger threshold H_2, taking into consideration the driver's behavior shown in the measured data.

It is necessary to define the probability density of the follower's acceleration $\sigma(\cdot)$ only on the thresholds $h = H_i(v)$, $i = 1, 2$.

In the deacceleration case on the first threshold H_1 the value a' is changed into the total braking value a^- according to (10.3). In the acceleration case on H_1, the value a' changes to $a \equiv 0$, i.e. the driver hesitates with accelerating and waits until the distance is equal to H_2:

$$\sigma(a' \to a|H_1(v), v, \bar{a}, \bar{v}) = \Theta(S_1(\bar{v}, v, a')) \cdot \delta(a) + \Theta(-S_1(\bar{v}, v, a')) \cdot \delta(a - a^-).$$

On the second threshold H_2, there are two deacceleration cases. If $\bar{v} < v$, the new deacceleration value a is calculated by a car following approach, which is proportional to the relative velocity $(\bar{v} - v)$ and inverse proportional to the distance between the cars $h = H_2(v)$. For $\bar{v} \geqslant v$, because of the lack of information, this deacceleration value is assumed to be uniformly distributed. In the case of increasing distances on H_2, there are two acceleration cases and the analogous approach is used:

$$\sigma(a' \to a|H_2(v), v, \bar{a}, \bar{v}) = \operatorname{sgn}(S_2) \cdot \Bigg[\Theta(S_2 \cdot (\bar{v} - v)) \cdot \delta\left(a - \min\left\{a_*, \varepsilon \frac{\bar{v} - v}{H_2(v)}\right\}\right)$$
$$+ \Theta(S_2 \cdot (v - \bar{v})) \cdot \mathbb{U}_{(0, a_*)}\Bigg].$$

Here a_* and ε are model parameters, which are supposed to be constant.

The distance correlation function $D(\cdot)$ is constructed in the same way as in the single threshold interaction model. The basic idea is an assignment of the current acceleration value a' of a car to the threshold H_i, on which occurs a change to the new value a [4]:

$$D\left(H_1(v)\,|v, \mathbf{m}_f, a' < -a_* \text{ or } (a' = 0 \text{ if } v = 0)\right) = \frac{\tilde{D}(H_1(v)|\mathbf{m}_f)}{\int_{h<H_1(v)} \tilde{D}(h|\mathbf{m}_f)\, \mathrm{d}h},$$

$$D\left(H_2(v)\,|v, \mathbf{m}_f, a' > 0 \text{ or } (a' = 0 \text{ if } v = w)\right) = \frac{\tilde{D}(H_2(v)|\mathbf{m}_f)}{\int_{h>H_2(v)} \tilde{D}(h|\mathbf{m}_f)\, \mathrm{d}h},$$

$$D\left(H_i(v)\,|v, \mathbf{m}_f, a' \in [-a_*, 0) \text{ or } (a' = 0 \text{ if } v \neq 0, w)\right) = \frac{\tilde{D}(H_i(v)|\mathbf{m}_f)}{\int_{H_1(v)}^{H_2(v)} \tilde{D}(h|\mathbf{m}_f)\, \mathrm{d}h},$$
$$i = 1, 2.$$

Here $\tilde{D}(\cdot)$ is the Gaussian distribution from previous section.

Numerical estimates of the velocity and acceleration distributions are presented in Fig. 10.2.

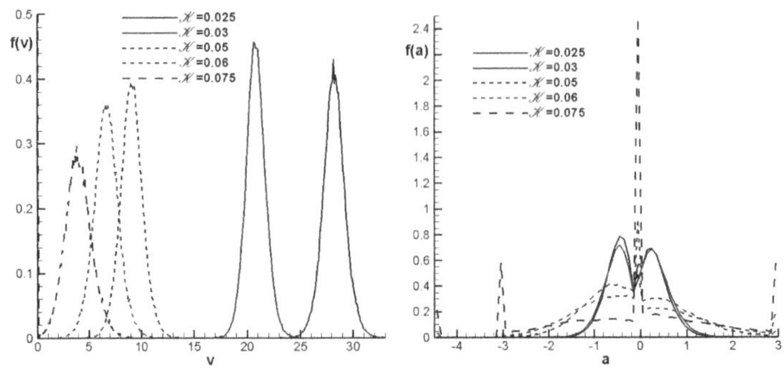

Fig. 10.2 Numerical estimates of the velocity (*left*) and acceleration (*right*) distributions for two thresholds

Conclusion

This work is a continuation of the research and a development of simulation methods started by the authors earlier. In particular, the approach suggested by the authors in [1] is applied for more realistic interaction profiles. Numerical results show practical suitability and efficiency of transition to the integral equation of the second kind and a Markov chain simulation in the VTF problems.

Possible directions for improving the model include consideration of various aspects such as:

- a mixture of both driver behaviors and vehicle classes;
- multi-lane traffic with possibility of overtaking;
- cluster formation on the road;
- spatial inhomogeneities.

Acknowledgements The authors acknowledge the kind hospitality of the Unit of Rimini of the Department of Statistical Sciences of Bologna University and organizers of the Seventh International Workshop on Simulation IWS-2013.

The work is partly supported by SB RAS (Interdisciplinary Integration Project No. 47), the Program "Leading Scientific Schools" (grant SS-5111.2014.1), and the Russian Foundation for Basic Research (grants 11-01-00252, 12-01-00034, 12-01-31134, 13-01-00746, 14-01-00340, 14-01-31451).

References

1. Burmistrov, A.V., Korotchenko, M.A.: Application of statistical methods for the study of kinetic model of traffic flow with separated accelerations. Russ. J. Numer. Anal. Math. Model. **26**, 275–293 (2011)
2. Ivanov, M.S., Rogasinsky, S.V.: The Direct Statistical Simulation Method in Dilute Gas Dynamics. Publ. Comput. Center, Sib. Branch, USSR Acad. Sci., Novosibirsk (1988, in Russian)
3. Mikhailov, G.A., Rogasinsky, S.V.: Weighted Monte Carlo methods for approximate solution of the nonlinear Boltzmann equation. Sib. Math. J. **43**, 496–503 (2002)
4. Waldeer, K.T.: Vergleich der Ergebnisse eines beschleunigungsorientierten, Boltzmannartigen Verkehrsflußmodells mit Messungen. In: Proceedings of 19th Dresden Conference on Traffic and Transportation Science, pp. 84.1–84.16 (2003, in German)
5. Waldeer, K.T.: A vehicular traffic flow model based on a stochastic acceleration process. Transp. Theory Stat. Phys. **33**, 7–30 (2004)
6. Waldeer, K.T.: Numerical investigation of a mesoscopic vehicular traffic flow model based on a stochastic acceleration process. Transp. Theory Stat. Phys. **33**, 31–46 (2004)

Chapter 11
Importance Sampling for Multi-Constraints Rare Event Probability

Virgile Caron

11.1 Introduction and Context

In this paper, we consider efficient estimation of the probability of large deviations of a multivariate sum of independent, identically distributed, light-tailed, and non-lattice random vectors.

Consider $\mathbf{X}_1^n := (\mathbf{X}_1, \ldots, \mathbf{X}_n)$ n i.i.d. random vectors with known common density $p_{\mathbf{X}}$ on $\mathbb{R}^d$, $d \geqslant 1$, copies of $\mathbf{X} := \left(\mathbf{X}^{(1)}, \ldots, \mathbf{X}^{(d)}\right)$. The superscript (j) pertains to the coordinate of a vector and the subscript i pertains to replications. Consider also u a measurable function defined from $\mathbb{R}^d$ to $\mathbb{R}^s$. Define $\mathbf{U} := u(\mathbf{X})$ with density $p_{\mathbf{U}}$ and

$$\mathbf{U}_{1,n} := \sum_{i=1}^{n} \mathbf{U}_i .$$

We intend to estimate for large but fixed n

$$P_n := P\left(\mathbf{U}_{1,n} \in nA\right) \tag{11.1}$$

where A is a non-empty measurable set of $\mathbb{R}^s$ such as $E[u(\mathbf{X})] \notin A$. In [3], the authors consider in detail the case where $d = s = 1$, $A := A_n = (a_n, \infty)$ and a_n is a convergent sequence.

The basic estimate of P_n is defined as follows: generate L i.i.d. samples $X_1^n(l)$ with underlying density $p_{\mathbf{X}}$ and define

V. Caron (✉)
Telecom ParisTech, 37-39 rue Dareau, 75014 Paris, France
e-mail: virgile.caron@telecom-paristech.fr; virgile.caron@upmc.fr

V.B. Melas et al. (eds.), *Topics in Statistical Simulation*, Springer Proceedings in Mathematics & Statistics 114, DOI 10.1007/978-1-4939-2104-1_11

$$\widetilde{P_n} := \frac{1}{L} \sum_{l=1}^{L} \mathbb{1}_{\mathscr{E}_n} \left(X_1^n(l) \right)$$

where

$$\mathscr{E}_n := \left\{ (x_1, \ldots, x_n) \in \left(\mathbb{R}^d \right)^n : (u(x_1) + \cdots + u(x_n)) \in nA \right\}. \tag{11.2}$$

The Importance Sampling estimator of P_n with sampling density g on $\left(\mathbb{R}^d\right)^n$ is

$$\widehat{P_n} := \frac{1}{L} \sum_{l=1}^{L} \hat{P}_n(l) \mathbb{1}_{\mathscr{E}_n} \left(Y_1^n(l) \right) \tag{11.3}$$

where $\hat{P}_n(l)$ is called "importance factor" and can be written

$$\hat{P}_n(l) := \frac{\prod_{i=1}^{n} p_{\mathbf{X}}(Y_i(l))}{g\left(Y_1^n(l) \right)} \tag{11.4}$$

where the L samples $Y_1^n(l) := (Y_1(l), \ldots, Y_n(l))$ are i.i.d. with common density g; the coordinates of $Y_1^n(l)$ however need not be i.i.d. It is known that the optimal choice for g is the density of $\mathbf{X}_1^n := (\mathbf{X}_1, \ldots, \mathbf{X}_n)$ conditioned upon $\left(\mathbf{X}_1^n \in \mathscr{E}_n \right)$, leading to a zero variance estimator. We refer to [5] for the background of this section.

The state-independent IS scheme for rare event estimation (see [6] or [12]), rests on two basic ingredients: the sampling distribution is fitted to the so-called dominating point (which is the point where the quantity to be estimated is mostly captured; see [11]) of the set to be measured; independent and identically distributed replications under this sampling distribution are performed. More recently, a state-dependent algorithm leading to a strongly efficient estimator is provided by [2] when $d = s$, $u(x) = x$ and A has a smooth boundary and a unique dominating point. Indeed, adaptive tilting defines a sampling density for the $i-$th r.v. in the run which depends both on the target event $(\mathbf{U}_{1,n} \in nA)$ and on the current state of the path up to step $i-1$. Jointly with an ad hoc stopping rule controlling the excursion of the current state of the path, this algorithm provides an estimate of P_n with a coefficient of variation independent upon n. This result shows that nearly optimal estimators can be obtained without approximating the conditional density.

The main issue of the method described above is to find dominating point. However, when the dimension of the set A increases, finding a dominating point can be very tricky or even impossible. A solution will be to divide the set under consideration into smaller subset and, for each one of this subset, find a dominating point. Doing so makes the implementation of an IS scheme harder and harder as the dimension increases.

Our proposal is somehow different since it is based on a sharp approximation result of the conditional density of long runs. The approximation holds for any point conditioning of the form $(\mathbf{U}_{1,n} = nv)$. Then sampling v in A according to the distribution of $\mathbf{U}_{1,n}$ conditioned upon $(\mathbf{U}_{1,n} \in nA)$ produces the estimator. By its very definition this procedure does not make use of any dominating point, since it randomly explores the set A. Indeed, our proposal hints on two choices: first do not make use of the notion of dominating point and explore all the target set instead (no part of the set A is neglected); secondly, do not use i.i.d. replications, but merely sample long runs of variables under a proxy of the optimal sampling scheme.

We will propose an IS sampling density which approximates this conditional density very sharply on its first components $y_1, \ldots, y_k$ where $k = k_n$ is very large, namely $k/n \to 1$. However, but in the Gaussian case, k should satisfy $(n-k) \to \infty$ by the very construction of the approximation. The IS density on $\left(\mathbb{R}^d\right)^n$ is obtained multiplying this proxy by a product of a much simpler state-independent IS scheme following [13].

The paper is organized as follows. Section 11.2 is devoted to notations and hypothesis. In Sect. 11.3, we expose the approximation scheme for the conditional density of $\mathbf{X}_1^k$ under $(\mathbf{U}_{1,n} = nv)$. Our IS scheme is introduced in Sect. 11.4. Simulated results are presented in Sect. 11.5 which enlighten the gain of the present approach over state-dependent Importance Sampling schemes.

We rely on [7] where the basic approximation (and proofs) used in the present paper can be found. The real case is studied in [4] and applications for IS estimators can be found in [3].

11.2 Notations and Hypotheses

We consider approximations of the density of the vector $\mathbf{X}_1^k$ on $\left(\mathbb{R}^d\right)^k$, when the conditioning event writes (11.1) and $k := k_n$ is such that

$$0 \leqslant \limsup_{n\to\infty} \frac{k}{n} \leqslant 1 \tag{K1}$$

$$\lim_{n\to\infty} (n-k) = +\infty. \tag{K2}$$

Therefore we may consider the asymptotic behavior of the density of the random walk on long runs.

Throughout the paper the value of a density $p_{\mathbf{Z}}$ of some continuous random vector $\mathbf{Z}$ at point z may be written $p_{\mathbf{Z}}(z)$ or $p\left(\mathbf{Z} = z\right)$, which may prove more convenient according to the context.

Let p_{nv} (and distribution P_{nv}) denote the density of $\mathbf{X}_1^k$ under the local condition $(\mathbf{U}_{1,n} = nv)$

$$p_{nv}\left(\mathbf{X}_1^k = Y_1^k\right) := p(\mathbf{X}_1^k = Y_1^k \,|\, \mathbf{U}_{1,n} = nv) \tag{11.5}$$

where Y_1^k belongs to $\left(\mathbb{R}^d\right)^k$ and v belongs to A.

We will also consider the density p_{nA} (and distribution P_{nA}) of $\mathbf{X}_1^k$ conditioned upon $(\mathbf{U}_{1,n} \in nA)$

$$p_{nA}\left(\mathbf{X}_1^k = Y_1^k\right) := p(\mathbf{X}_1^k = Y_1^k \,|\, \mathbf{U}_{1,n} \in nA). \tag{11.6}$$

The approximating density of p_{nv} is denoted g_{nv}; the corresponding approximation of p_{nA} is denoted g_{nA}. Explicit formulas for those densities are presented in the next section.

11.3 Multivariate Random Walk Under a Local Conditioning Event

Let ε_n be a positive sequence such as

$$\lim_{n\to\infty} \varepsilon_n^2(n-k) = \infty \tag{E1}$$

$$\lim_{n\to\infty} \varepsilon_n(\log n)^2 = 0 \tag{E2}$$

It will be shown that $\varepsilon_n\,(\log n)^2$ is the rate of accuracy of the approximating scheme.

We assume that $\mathbf{U} := u\,(\mathbf{X})$ has a density $p_{\mathbf{U}}$ (with p.m. $P_{\mathbf{U}}$) absolutely continuous with respect to Lebesgue measure on $\mathbb{R}^s$. Furthermore, we assume that u is such that the characteristic function of $\mathbf{U}$ belongs to L^r for some $r \geqslant 1$.

Denote $\underline{0}$ is the vector of $\mathbb{R}^s$ with all coordinates equal to 0 and $V(\underline{0})$ a neighborhood of $\underline{0}$.

We assume that $\mathbf{U}$ satisfy the Cramer condition, meaning

$$\Phi_{\mathbf{U}}(t) := E[\exp < t, \mathbf{U} >] < \infty, \;\; t \in V(\underline{0}) \subset \mathbb{R}^s.$$

and define

$$m(t) := {}^t\nabla \log(\Phi_{\mathbf{U}}(t)), \;\; t \in V(\underline{0}) \subset \mathbb{R}^s$$

and

$$\varkappa(t) := {}^t\nabla\nabla \log(\Phi_{\mathbf{U}}(t)), \;\; t \in V(\underline{0}) \subset \mathbb{R}^s.$$

as the mean and the covariance matrix of the tilted density defined by

$$\pi_u^{\alpha}(x) := \frac{\exp < t, u(x) >}{\Phi_{\mathbf{U}}(t)} p_{\mathbf{X}}(x). \tag{11.7}$$

where t is the only solution of $m(t) = \alpha$ for α in the convex hull of $P_{\mathbf{U}}$. Conditions on $\Phi_{\mathbf{U}}(t)$ which ensure existence and uniqueness of t are referred to steepness properties (see [1], p153 ff, for all properties of moment generating function used in this paper).

We now state the general form of the approximating density. Let $v \in A$ and denote

$$g_0(y_1|y_0) := \pi_u^v(y_1) \tag{11.8}$$

with an arbitrary y_0 and π_u^v defined in (11.7).

For $1 \leqslant i \leqslant k-1$, we recursively define $g(y_{i+1}|y_1^i)$. Set $t_i \in \mathbb{R}^s$ to be the unique solution to the equation

$$m(t_i) = m_{i,n} := \frac{n}{n-i}\left(v - \frac{u_{1,i}}{n}\right) \tag{11.9}$$

where $u_{1,i} = u(y_1) + \cdots + u(y_i)$.

Denote

$$\varkappa_{(i,n)}^{j,l} := \frac{d^2}{dt^{(j)}dt^{(l)}}\left(\log E_{\pi_{\mathbf{U}}^{m_{i,n}}} \exp < t, \mathbf{U} >\right)(\underline{0})$$

and

$$\varkappa_{(i,n)}^{j,l,m} := \frac{d^3}{dt^{(j)}dt^{(l)}dt^{(m)}}\left(\log E_{\pi_{\mathbf{U}}^{m_{i,n}}} \exp < t, \mathbf{U} >\right)(\underline{0}).$$

for j, l and m in $\{1, \ldots, s\}$. In the sequel, $\varkappa_{(i,n)}$ will denote the matrix with elements $\left(\varkappa_{(i,n)}^{j,l}\right)_{1\leqslant j,l\leqslant s}$.

Denote

$$g(y_{i+1}|y_1^i) := C_i \mathfrak{n}_s\left(u(y_{i+1}); \beta\alpha + v, \beta\right) p_{\mathbf{X}}(y_{i+1}) \tag{11.10}$$

where C_i is a normalizing factor, $\mathfrak{n}_s\left(u(y_{i+1}); \beta\alpha + v, \beta\right)$ is the normal density at $u(y_{i+1})$ with mean $\beta\alpha + v$ and covariance matrix β. α and β are defined by

$$\alpha := \left(t_i + \frac{\varkappa_{(i,n)}^{-2}\gamma}{2(n-i-1)}\right)$$

and

$$\beta := \varkappa_{(i,n)}(n-i-1)$$

and γ defined by

$$\gamma := \left(\sum_{j=1}^{s} \varkappa_{(i,n)}^{j,j,p} \right)_{1 \leqslant p \leqslant s}.$$

Then

$$g_{nv}(y_1^k) := g_0(y_1|y_0) \prod_{i=1}^{k-1} g(y_{i+1}|y_1^i) \tag{11.11}$$

Theorem 1. *Assume (E1), (E2), (K1) and (K2).*

- *Let* Y_1^k *be a sample from density* p_{nv}. *Then*

$$p\left(\mathbf{X}_1^k = Y_1^k | \mathbf{U}_{1,n} = nv\right) = g_{nv}(Y_1^k)(1 + o_{P_{nv}}(1 + \varepsilon_n (\log n)^2)) \tag{11.12}$$

- *Let* Y_1^k *be a sample from density* g_{nv}. *Then*

$$p\left(\mathbf{X}_1^k = Y_1^k | \mathbf{U}_{1,n} = nv\right) = g_{nv}(Y_1^k)(1 + o_{G_{nv}}(1 + \varepsilon_n (\log n)^2)) \tag{11.13}$$

Remark 11.1. The approximation of the density of $\mathbf{X}_1^k$ is not performed on the sequence of entire spaces $\left(\mathbb{R}^d\right)^k$ but merely on a sequence of subsets of $\left(\mathbb{R}^d\right)^k$ which contains the trajectories of the conditioned random walk with probability going to 1 as n tends to infinity. The approximation is performed on *typical paths*. For the sake of applications in Importance Sampling, (11.13) is exactly what we need. Nevertheless, as proved in [7], the extension of our results from typical paths to the whole space $\left(\mathbb{R}^d\right)^k$ holds: convergence of the relative error on large sets imply that the total variation distance between the conditioned measure and its approximation goes to 0 on the entire space.

Remark 11.2. The rule which defines the value of k for a given accuracy of the approximation is stated in Sect. 5 of [7].

Remark 11.3. When the $\mathbf{X}_i$'s are i.i.d. multivariate Gaussian with diagonal covariance matrix and $u(x) = x$, the results of the approximation theorem are true for $k = n - 1$ without the error term. Indeed, it holds $p(\mathbf{X}_1^{n-1} = x_1^{n-1} \big| \mathbf{U}_{1,n} = nv) = g_{nv}\left(x_1^{n-1}\right)$ for all x_1^{n-1} in $\left(\mathbb{R}^d\right)^{n-1}$.

As stated above the optimal choice for the sampling density is p_{nA}. It holds

$$p_{nA}(x_1^k) = \int_A p_{nv}\left(\mathbf{X}_1^k = x_1^k\right) p(\mathbf{U}_{1,n}/n = v | \mathbf{U}_{1,n} \in nA) dv \tag{11.14}$$

so that, in contrast with [2] or [6], we do not consider the dominating point approach but merely realize a sharp approximation of the integrand at any point of A and consider the dominating contribution of all those distributions in the evaluation of the conditional density p_{nA}.

11.4 Adaptive IS Estimator for Rare Event Probability

The IS scheme produces samples $Y := (Y_1, \ldots, Y_k)$ distributed under g_{nA}, which is a continuous mixture of densities g_{nv} as in (11.11) with $p\left(\mathbf{U}_{1,n}/n = v|\mathbf{U}_{1,n} \in nA\right)$.

Simulation of samples $\mathbf{U}_{1,n}/n$ under this density can be performed through Metropolis–Hastings algorithm, since

$$r(v, v') := \frac{p(\mathbf{U}_{1,n}/n = v|\, \mathbf{U}_{1,n} \in nA)}{p(\mathbf{U}_{1,n}/n = v'|\, \mathbf{U}_{1,n} \in nA)}$$

turns out to be independent upon $P\left(\mathbf{U}_{1,n} \in nA\right)$. The proposal distribution of the algorithm should be supported by A.

The density g_{nA} is extended from $\left(\mathbb{R}^d\right)^k$ onto $\left(\mathbb{R}^d\right)^n$ completing the $n - k$ remaining coordinates with i.i.d. copies of r.v's $Y_{k+1}, \ldots, Y_n$ with common tilted density

$$g_{nA}\left(y_{k+1}^n \middle|\, y_1^k\right) := \prod_{i=k+1}^{n} \pi_u^{m_k}(y_i) \tag{11.15}$$

with $m_k := m(t_k) = \frac{n}{n-k}\left(v - \frac{u_{1,k}}{n}\right)$ and

$$u_{1,k} = \sum_{i=1}^{k} u(y_i).$$

The last $n - k$ r.v's $\mathbf{Y}_i$'s are therefore drawn according to the state independent i.i.d. scheme in phase with Sadowsky and Bucklew [13].

We now define our IS estimator of P_n. Let $Y_1^n(l) := Y_1(l), \ldots, Y_n(l)$ be generated under g_{nA}. Let

$$\widehat{P_n}(l) := \frac{\prod_{i=1}^{n} p_{\mathbf{X}}(Y_i(l))}{g_{nA}(Y_1^n(l))} \mathbb{1}_{\mathcal{E}_n}\left(Y_1^n(l)\right) \tag{11.16}$$

and define

$$\widehat{P_n} := \frac{1}{L}\sum_{l=1}^{L}\widehat{P_n}(l). \tag{11.17}$$

in accordance with (11.3).

Remark 11.4. In the real case and for $A = (a,\infty)$, the authors of [3] show that under certain regularity conditions the resulting relative error of the estimator is proportional to $\sqrt{n-k_n}$ and drops by a factor $\sqrt{n-k_n}/\sqrt{n}$ with respect to the state independent IS scheme. In [8], the authors propose a slight modification in the extension of g_{nA} which allows to prove the strong efficiency of the estimator (11.17) using arguments from both [2] and [3].

11.5 When the Dimension Becomes Very High

This section compares the performance of the present approach with respect to the standard tilted one using i.i.d. replications under (11.7) on an extension of a well-known example developed in [9] and in [10]. Let $B := (\mathscr{E}_{100})^d$ which is the d-Cartesian product of $\mathscr{E}_{100}$ defined by

$$\mathscr{E}_{100} := \left\{ x_1^{100} : \frac{|x_1 + \cdots + x_{100}|}{100} > 0.28 \right\}.$$

We want to estimate $P_{100} = P[B]$ and explore the gain in relative accuracy when the dimension of the measured set increases. Consider 100 r.v.'s X_i 's i.i.d. random vectors in $\mathbb{R}^d$ with common i.i.d. $N(0.05,1)$ distribution. Our interest is to show that in this simple asymmetric case our proposal provides a good estimate, while the standard IS scheme ignores a part of the event B. The standard i.i.d. IS scheme introduces the dominating point $a =^t (0.28,\ldots,0.28)$ and the family of i.i.d. tilted r.v's with common $N(a,1)$ distribution. It can be seen that a large part of B is never visited through the procedure, inducing a bias in the estimation. Indeed, the *rogue path curse* (see [9]) produces an overwhelming loss in accuracy, imposing a very large increase in runtime to get reasonable results. Under the present proposal the distribution of the Importance Factor concentrates around P_{100} avoiding *rogue path*.

This example is not as artificial as it may seem; indeed, it leads to a 2^d dominating points situation which is quite often met in real life. Exploring at random the set of interest avoids any search for dominating points. Drawing L i.i.d. points $v_1,\ldots,v_L$ according to the distribution of $\mathbf{U}_{1,100}/100$ conditionally upon B we evaluate P_{100} with $k = 99$; note that in the Gaussian case Theorem 1 provides an exact description of the conditional density of X_1^k for all k between 1 and n. The following figure shows the gain in relative accuracy w.r.t. the state independent IS scheme according to the growth of d. The value of P_{100} is 10^{-2d} (Fig. 11.1).

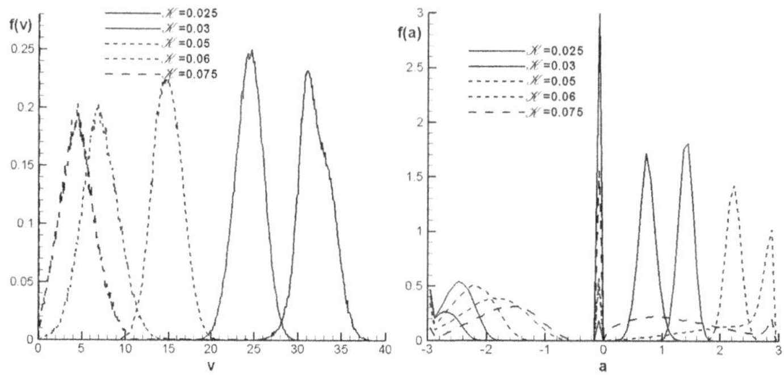

Fig. 11.1 Relative Accuracy of the adaptive estimate (*dotted line*) w.r.t. i.i.d. tilted one (*solid line*) as a function of the dimension d for $L = 1{,}000$

Conclusion

In this paper, we explore a new way to estimate multi-constraints large deviation probability. In future work, the author will investigate the theoretical behavior of the relative error of our proposed estimator.

References

1. Barndorff-Nielsen, O.E.: Information and Exponential Families in Statistical Theory. Wiley, New York (1978)
2. Blanchet, J.H., Glynn, P.W., Leder, K.: Efficient simulation of light-tailed sums: an old-folk song sung to a faster new tune.... In: L'Ecuyer, P., Owen, A.B. (eds.) Monte Carlo and Quasi-Monte Carlo Methods, pp. 227–248. Springer, Berlin (2009)
3. Broniatowski, M., Caron, V.: Towards zero variance estimators for rare event probabilities. ACM TOMACS: Special Issue on Monte Carlo Methods in Statistics **23**(1) (2013) [article 7]
4. Broniatowski, M., Caron, V.: Long runs under a conditional limit distribution. Ann. Appl. Probab (2014, to appear)
5. Bucklew, J.A.: Introduction to Rare Event Simulation. Springer Series in Statistics. Springer, New York (2004)
6. Bucklew, J.A., Ney, P., Sadowsky, J.S.: Monte Carlo simulation and large deviations theory for uniformly recurrent markov chains. J. Appl. Probab. **27**(11), 49–61 (1990)
7. Caron, V.: Approximation of a multivariate conditional density (2013). arxiv:1401.3256
8. Caron, V., Guyader, A., Munoz, Z.M., Tuffin, B.: Some recent results in rare event estimation. In: ESAIM Proceedings (2013, to appear)
9. Dupuis, P., Wang, H.: Importance sampling, large deviations, and differential games. Stoch. Stoch. Rep. **76**, 481–508 (2004)
10. Glasserman, P., Wang, Y.: Counterexamples in importance sampling for large deviations probabilities. Ann. Appl. Probab. **7**(3), 731–746 (1997)
11. Ney, P.: Dominating points and the asymptotics of large deviations for random walk on $\mathbb{R}^d$. Ann. Probab. **11**(1), 158–167 (1983)

12. Sadowsky, J.S.: On Monte-Carlo estimation of large deviations probabilities. Ann. Appl. Probab. **9**(2), 493–503 (1996)
13. Sadowsky, J.S., Bucklew, J.A.: On large deviations theory and asymptotically efficient Monte Carlo estimation. IEEE Trans. Inform. Theory **36**(3), 579–588 (1990)

Chapter 12
Generating and Comparing Multivariate Ordinal Variables by Means of Permutation Tests

Eleonora Carrozzo, Alessandro Barbiero, Luigi Salmaso, and Pier Alda Ferrari

12.1 Introduction

In many applicative problems, it is usually necessary to compare two or more correlation matrices. Nevertheless only few efforts have been done to find a solution, most of them assume multivariate normality. Let us consider now the particular case in which one is interested to define whether the variables under analysis are uncorrelated or not. Formalizing, let us suppose to have a multivariate $n \times m$ sample with sample correlation matrix $\hat{\mathbf{R}}$ we want to test if data come from an m-dimensional random variable with correlation matrix $\mathbf{R}_0 = \mathbf{I}_m$, where $\mathbf{I}_m$ is the identity matrix. Thus the null hypothesis is $H_0 : \mathbf{R} = \mathbf{R}_0$ against the general alternative.

In 1970, Jennrich proposed a test which, assuming multivariate normality, rejects the null hypothesis for large values of the statistic, $T_{\text{Jen}} = \frac{1}{2}tr\left(\mathbf{W}^2\right) - dg'\left(\mathbf{W}\right)\mathbf{T}^{-1}dg\left(\mathbf{W}\right)$ where $\mathbf{W} = \sqrt{n} \cdot \mathbf{R}_0^{-1}\left(\hat{\mathbf{R}} - \mathbf{R}_0\right)$ and $[\mathbf{T}]_{ij} = \delta_{ij} + \rho_{ij,0} \cdot \rho_0^{ij}$ $\rho_0^{ij} = \left[\mathbf{R}_0^{-1}\right]_{ij}$ and δ_{ij} is the Kronecker's delta. T_{Jen} has an asymptotic Chi-squared distribution with $m\,(m-1)/2$ degrees of freedom under H_0 [7]. However, Jennrich's test is a large sample test and can lead to poor performance for small samples. In 1985, Larntz and Perlman proposed a statistic which determines, under multivariate normality assumption, a test with reasonable small sample properties and with power comparable to that of Jennrich's test for large samples [8]. The test

E. Carrozzo • L. Salmaso (✉)
Department of Management and Engineering, University of Padova, Padua, Italy
e-mail: carrozzo@gest.unipd.it; luigi.salmaso@unipd.it

A. Barbiero • P.A. Ferrari
Department of Economics, Management and Quantitative Methods, Università degli Studi di Milano, Milan, Italy
e-mail: alessandro.barbiero@unimi.it; pieralda.ferrari@unimi.it

V.B. Melas et al. (eds.), *Topics in Statistical Simulation*, Springer Proceedings in Mathematics & Statistics 114, DOI 10.1007/978-1-4939-2104-1_12

statistic is $T_{LP} = \sqrt{n-3} \cdot d$ where $d = \max_{1 \leqslant i<j \leqslant m} |z_{ij} - \xi_{ij}|$ and where z_{ij} is the Fisher transform of $\hat{\rho}_{ij}$ and ξ_{ij} is the Fisher transform of $\rho_{ij,0}$. By Sidak's theorem, the test which rejects the null hypothesis if $T_{LP} > b_\alpha$, where $b_\alpha > 0$ is chosen such that $[\varphi(b_\alpha) - \varphi(-b_\alpha)]^{m(m-1)/2} = 1 - \alpha$, is a (possibly conservative) α-level test of H_0.

In this paper we propose a nonparametric approach based on permutation test and nonparametric combination methodology (NPC). After an overview of permutation inference and the NPC methodology, we discuss the nonparametric procedure and we show a simulation-based comparative study among the three abovementioned procedures. In particular, we deal with simulations from multivariate ordinal random variables, in order to show the performance of the procedures when the assumption of normality is not satisfied. In this regard we consider a new proposal for generating samples from multivariate ordinal data whose details and properties are described in the fourth section.

12.2 Nonparametric Combination

In this section, we introduce the method of the nonparametric combination (NPC) of a finite number of dependent permutation tests as a useful tool to solve complex problems when several variables are involved or many different aspects are of interest. Consider an m-dimensional problem, with $m \geqslant 2$. With NPC method the global null hypothesis can be broken down into m sub-hypotheses, each appropriate for each aspect of interest and it is true if all of the sub-hypotheses are true. More formally the null hypothesis consists of the intersection of m partial sub-hypotheses: $\bigcap_{j=1}^{m} H_{0j}$. Similarly the alternative hypothesis can be written as the union of m sub-hypotheses: $\bigcup_{j=1}^{m} H_{1j}$, so the global null hypothesis is false if at least one of the sub-alternatives is true. When partial tests are stochastically independent, the combination of them into a global test is not difficult (see [3] for a review), but in most situations this independence is not a plausible assumption. In fact, partial tests are typically dependent since they are function of the same dataset. When distributional assumptions can be made (e.g., that the data are multivariate normal) or asymptotic results hold, then this dependence may be estimated from the data, leading to methods such as Hotelling's T^2 statistic [5], multivariate ANOVA and regression [13], and omnibus chi-square tests (e.g., [4]). Alternatively, a latent variable may be estimated with factor analysis or item-response theory models [6, ch. 9]. If their scales are comparable, responses can be combined (e.g., in a summative index), either directly or using ranks [10, 11]. Under suitable conditions, each of the above techniques may be used to test complex hypotheses. In particular, the NPC methodology allows the experimenter to combine tests which involve variables with different scales or levels of measurement (e.g., continuous and nominal), multiple tests on different aspects of the same variable (e.g., mean and variance), and even tests in which the number of the variables is greater than the number of the units (for details, see [9]). All these partial p-values, after a suitable adjustment for multiplicity, may be assessed for evidence on the sub-hypotheses. Since the combination of tests is done nonparametrically, without the need to explicitly model the dependence among tests, no further assumptions are needed other than those required by the partial permutation tests themselves. If the partial tests are exact and unbiased, also the combined global test is exact and unbiased [9]. Once an appropriate partial test has been chosen for each sub-hypothesis, we need to select the function with which to combine p-values. Let λ_j be the p-value related to the j-th partial hypothesis, consider some practical examples of combining function: (a) Fisher omnibus combining function based on the statistic $\psi_F = -2\sum_j \log(\lambda_j)$; (b) Liptak combining function based on the statistic $\psi_L = \sum_j \Phi^{-1}(1-\lambda_j)$, where Φ is the standard normal CDF; (c) Tippett combination function based on the statistic $\psi_T = \max_{1\leqslant j\leqslant K}(1-\lambda_j)$. Another combination function is the truncated combination function defined as $\psi_{\text{trunc}} = \prod_j \lambda_j^{I(\lambda_j<\tau)}$ where τ is the truncation point (usually equal to) α. Truncated forms of combinations were mostly introduced to deal with multiplicity issues in genomewide association scans and microarray studies characterized by a huge amount of true null hypotheses and small amount of false hypotheses.

The NPC method can be carried out using the following algorithm [9]:

1. Calculate the vector $T^0 = \left(T_1^0,\ldots,T_j^0,\ldots,T_m^0\right)'$ of observed test statistics corresponding to m partial tests.
2. Repeat the following B times:
   (a) randomly permute the group (e.g., "treated" and "control") labels without replacement;
   (b) calculate the vector $T_b^* = \left(T_{1b}^*,\ldots,T_{jb}^*,\ldots,T_{mb}^*\right)'$ of values of the m test statistics in permutation, $b \in \{1,\ldots,B\}$.
3. Presuming that the partial test statistics are expected to be large in the alternative, let $\hat{L}_j(t) = B^{-1}\sum_{b=1}^{B} I\left(T_{jb}^* \geqslant t\right)$ be the estimated significance level for any test statistic value $t \in R^1$ corresponding to partial test j. Calculate the vector of estimated significance levels for observed data: $\hat{\lambda} = \left(\hat{\lambda}_1,\ldots,\hat{\lambda}_j,\ldots,\hat{\lambda}_m\right)'$, where $\hat{\lambda}_j = \hat{L}_j\left(T_j^0\right)$. Then do the same for each permutation b, calculating $\hat{L}_b^* = \left(\hat{L}_{1b}^*,\ldots,\hat{L}_{jb}^*,\ldots\hat{L}_{mb}^*\right)'$, where $\hat{L}_{jb}^* = \hat{L}_j\left(T_{jb}^*\right)$
4. Use a suitable function ψ to combine the vector of m estimated significance levels into a global test statistic $T''^0 = \psi\left(\hat{\lambda}\right)$. Calculate the analogous statistic $T_b''^* = \psi\left(\hat{L}_b^*\right)$ for each permutation b.

5. Estimate the combined significance level (p-value) of the global test as

$$\hat{\lambda}''_{\psi} = B^{-1} \sum_{b=1}^{B} I\left(T''^{*}_{b} \geq T''^{0}\right)$$

In practice, permutation significance levels can be estimated to an arbitrary degree of accuracy by randomly sampling of a large number of (e.g., $B = 10{,}000$) permutations from the permutation sample space.

12.3 Permutation Testing Procedure for Correlation Matrices

This section has the aim to describe the procedure based on a permutation approach to test the equality of a general correlation matrix with the identity matrix (i.e., with the situation of no correlation among variables).

Let's start from a simple situation where we have a bivariate random variable (X, Y) with correlation matrix $\mathbf{R}_{XY}$, and suppose to have a random sample of size n from this variable. For the sake of simplicity let us consider to test the null hypothesis $H_0 : \mathbf{R}_{XY} = \mathbf{I}_2$ where $\mathbf{R}_2$ is the 2×2 identity matrix. Note that, in this case, the null hypothesis can be written as $H_0 : \rho_{XY} = 0$, where ρ_{XY} is the correlation coefficient (e.g., Pearson's correlation coefficient) between X and Y. A permutation procedure based on the sample correlation coefficient to test this type of hypothesis entails the following steps:

1. Compute a sample correlation coefficient $\hat{\rho}$ from the original paired data (x_i, y_i), where $i = 1, \ldots, n$.
2. Compute a random permutation of one of the two vectors of the observations, obtaining $(x_i, y_{i'})$ where $i = 1, \ldots, n$.
3. Compute the sample correlation coefficient r on the permuted data.
4. Repeat steps 2–3 for B times.
5. The p-value of the permutation test is obtained as the proportion of correlation coefficients r^* computed in step 3 that are greater than r_{obs} computed on the original data. Note that if we consider the two-sided alternative then we need to compute $\dfrac{\#(|r^*| > |r_{\text{obs}}|)}{B}$.

Let's generalize the problem in the case where we have $m > 2$ variables. Hence we wish to test $H_0 : \mathbf{R} = \mathbf{I}_m$ against the general alternative $H_1 : \{\mathbf{R} \neq \mathbf{I}_m\} = \{\exists\, \rho_{ij} \neq 0,\ i \neq j\},\ i, j = 1, \ldots, m$ where $\mathbf{R}$ is the true $m \times m$ correlation matrix and $\mathbf{I}_m$ is the identity matrix of order m. Now we can test this complex hypothesis by breaking down the null hypothesis in $m(m-1)/2$ sub-hypothesis of the type $H_0 : \rho_{ij} = 0$ against the alternative $H_1 : \rho_{ij} \neq 0,\ i < j$. Note that all these sub-hypotheses can be tested by a permutation approach using the above algorithm. Finally we can test the global null hypothesis by applying the NPC to the $m(m-1)/2$ partial tests (see e.g., [9]).

Let us suppose to have a trivariate random variable (X, Y, Z) with a general correlation matrix $\mathbf{R}_{XYZ}$ and want to test if $H_0 : \mathbf{R}_{XYZ} = \mathbf{I}_3$ against a two-sided alternative. Note that in this case $m = 3$. Thus we have to consider the following $\frac{3(3-1)}{2} = 3$ sub-hypotheses: $H_{0(XY)} : \rho_{12} = 0$, $H_{0(XZ)} : \rho_{13} = 0$ and $H_{0(YZ)} : \rho_{23} = 0$. After testing separately all this sub-hypotheses we obtain the three related p-values $\hat{\lambda} = \left(\hat{\lambda}_{XY}, \hat{\lambda}_{XZ}, \hat{\lambda}_{YZ}\right)'$ on the observed data and for each permutation b, $\hat{L}^*_b = \left(\hat{L}^*_{(XY)b}, \hat{L}^*_{(XZ)b}, \hat{L}^*_{(YZ)b}\right)'$. Note that all partial tests must be based on the same permutations. Combining the three p-values of the observed data and of each permutation, we obtain the global test statistic $T''^0 = \psi\left(\hat{\lambda}\right)$ and $T''^*_b = \psi\left(\hat{L}^*_b\right)$ and it is possible to compute the combined p-value $\hat{\lambda}''_\psi$ and of course, if this is less than the significance level α we reject the global null hypothesis.

12.4 A Comparative Simulation Study

In the present section we wish to evaluate and compare the performance of the permutation procedure for testing correlation matrices described in the previous section, with that of some competitors in the literature. To this aim, a simulation study has been carried out where we considered the case of multivariate ordinal variable with the aim to show the performance of the procedures when the assumption of normality does not hold. We consider several situations to investigate the effect of distribution shape (uniform, symmetrical or asymmetrical), number of categories (5, 7) of the variables, sample size (100, 50, 20), and correlation matrix (differing from the identity matrix for one or more of the $m(m-1)/2$ coefficients). To generate data, we used the R package GenOrd developed by Barbiero and Ferrari [1] that allows to generate samples from ordinal/discrete random variables with pre-specified correlation (Pearson/Spearman) matrix and marginal distributions. Before showing the results of the simulation study, let us briefly outline the data generation procedure.

12.5 Simulating Ordinal Data

The procedure focuses on ordinal variables, and can simulate data from discrete random variables with any finite support and with a dependence structure specified in terms of Pearson or Spearman's correlation matrix [2].

When ordinal variables are observed, the strength of the association between two variables is usually measured by Spearman's correlation coefficient, defined as the usual Pearson correlation coefficient between the two variables converted to ranks, assigning equal rank to tied categories (see, e.g., [12]). Spearman's rho is sometimes preferred to Pearson's correlation, calculated over a point scale $(1, 2, \ldots, k)$, because whereas Pearson's correlation catches and measures the linear relationship between two variables, Spearman's rho can catch any monotonic relationship.

The method is based on the transformation of a multivariate normal variable into a multivariate ordinal variable with assigned marginal distributions. It is developed in two steps: the first step finds the correlation matrix $\mathbf{R}^C = [\rho_{ij}^C]$ for the multivariate normal variable ensuring the desired $\mathbf{R}^O = [\rho_{ij}^O]$ for the correlated ordinal random variables; the second step is devoted to the very generation of samples and ensures the desires marginal distribution through the customary inverse transform method. Here, variables' association is measured through Pearson's correlations. Let $\mathbf{R}^{O*}$ be the target correlation matrix, and let us consider a normal random variable $Z \sim N\left(0, \mathbf{R}^C\right)$, and in the first stage $\mathbf{R}^C = \mathbf{R}^{O*}$. The original variable $\mathbf{Z}$ is transformed into variable $\boldsymbol{X}$ with categorical components as follows. On the basis of the k_{i-1} probabilities $0 < F_{i1} < F_{i2} < \cdots < F_{il} < \cdots < F_{i(k_{i-1})} < 1$ of the marginal distribution of the i-th component X_i of X, the corresponding normal quantiles $q_{i1} < q_{i2} < \cdots < q_{il} < \cdots < q_{i(k_i-1)}$ of Z_i are defined. The values of Z_i are then converted into integer numbers X_i as follows:

$$\begin{aligned}
&\text{if } Z_i < q_{i1} \rightarrow X_i = 1 \\
&\text{if } q_{i1} \leqslant Z_i < q_{i2} \rightarrow X_i = 2 \\
&\vdots \\
&\text{if } q_{i(k_i-1)} \leqslant Z_i \rightarrow X_i = k_i.
\end{aligned} \tag{12.1}$$

An m-dimensional point scale variable $X = (X_1, X_2, \ldots, X_k)$ is thus settled. The single components X_i of X have a different number of categories and different marginal probabilities, according to the number k_i and the values of F_{il} chosen.

This procedure meets the desired marginal distributions F_i for each component X_i, but the correlation matrix $\mathbf{R}^O$ related to vector X may sensibly differ from the chosen matrix $\mathbf{R}^C = \mathbf{R}^{O*}$ because of the discretization process (12.1) that alters the correlation coefficients. In order to overcome this problem it is necessary to determine a continuous correlation matrix $\mathbf{R}^{C*}$ able to assure the target correlation matrix $\mathbf{R}^{O*}$ for the transformed m-dimensional variable X. This is resolved by an iterative algorithm that alternates the updating of $\mathbf{R}^C$ to the discretization of Z into X according to Eq. (12.1), until the correlation matrix of X converges to $\mathbf{R}^{O*}$ (see [2] for details).

The final continuous correlation matrix $\mathbf{R}^{C*}$ is used to generate m-variate samples of size n from the target m-variate random variable $\boldsymbol{X}$, resorting again to the discretization (12.1) from an m-variate standard normal

12.6 Results and Comments

In this section the results of the simulation study for each of the different settings are described. First of all a simulation under H_0 has been considered. In particular, 4,000 random samples of sizes $n = 100$ have been generated from a trivariate ordinal random variable with 5 categories (from 1 to 5) with uniform distribution, i.e. all categories have the same probability, and a correlation matrix:

$$\mathbf{R} = \begin{pmatrix} 1 & 0 & 0 \\ 0 & 1 & 0 \\ 0 & 0 & 1 \end{pmatrix}$$

The comparative simulation study has been performed considering the permutation procedure (introduced in Sect. 12.3) the Jennrich's test and the Larntz and Perlman test, introduced in the first section. Permutation tests are performed with $B = 4{,}000$ permutations.

In particular, in the following figures with "Permutation" we refer to permutation test performance, and with terms "Jennrich" and "L&P" we refer to Jennrich and to Larntz and Perlman tests, respectively.

As it can be seen in Fig. 12.1, all procedures have substantially a similar behaviour and respect the nominal α-level.

In what follows we present all the results of the settings under H_1.

We consider $m = 3$ random variables with discrete uniform, symmetrical (non-uniform) and asymmetrical marginal distributions with 5 and 7 categories. We consider the following correlation matrices:

$$\mathbf{R}_1 = \begin{pmatrix} 1 & 0.7 & 0 \\ 0.7 & 1 & 0 \\ 0 & 0 & 1 \end{pmatrix}, \mathbf{R}_2 = \begin{pmatrix} 1 & 0.3 & 0.3 \\ 0.3 & 1 & 0.3 \\ 0.3 & 0.3 & 1 \end{pmatrix}, \mathbf{R}_3 = \begin{pmatrix} 1 & 0.3 & 0.5 \\ 0.3 & 1 & 0.7 \\ 0.5 & 0.7 & 1 \end{pmatrix}$$

From the simulation results, we can see how generally all procedures have a similar behaviour in particular with large sample size: the results with $n = 100$ were approximately the same for each procedure, with a rejection rate very close to 1.

It is worth noting that, considering a correlation matrix $\mathbf{R}_1$ the permutation test always has a greater power than the other two tests even when the sample size is small and for any considered distribution. The differences are in particular relevant at level $\alpha = 0.01$. In fact, whereas the permutation test and L&P test have very close rejection rates, Jennrich's test shows a lower power.

With correlation matrix $\mathbf{R}_2$ the L&P test loses power whereas the Jennrich test shows systematically more power in particular for $\alpha = 0.01$. Permutation test shows again the best power in all situations except for level $\alpha = 0.01$ where Jennrich test presents the best power.

When we consider a mixed correlation matrix as $\mathbf{R}_3$ the power of the three tests increases in all situations with respect to the same situation with $\mathbf{R}_2$. The permutation test has always the best power followed by Jennrich test and L&P test. The behaviour is respected either with variables with 5 or 7 categories.

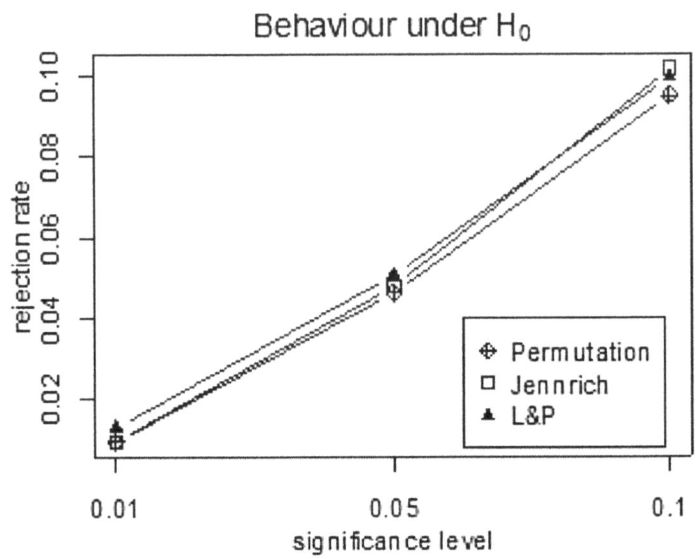

Fig. 12.1 Rejection rate of each procedure at different values of α under H_0

The simulation results are synthesized in Figs. 12.2 and 12.3, corresponding to the cases with $k = 5, n = 20$ and $k = 5, n = 50$, respectively. In Fig. 12.3, results for matrices $\mathbf{R}_1$ and $\mathbf{R}_3$ are not displayed, because the powers for all the three tests are always practically equal to 1. For the sake of brevity, even the results for $k = 7$ are not reported here, since we noted that passing from 5 to 7 categories, *coeteris paribus*, hardly affects the power of the three tests.

Conclusion

We proposed a nonparametric methodology based on partial permutation tests and NPC methodology, to test if data come from an m-dimensional random variable with correlation matrix, $\mathbf{R}_0 = \mathbf{I}_m$ where $\mathbf{I}_m$ is the identity matrix. In particular, we carried out a simulation study in order to compare the performance of the proposed procedures. In this regard we considered simulations from multivariate ordinal variables, generated through a new method recently introduced by [1].

We note that, neither passing from 5 to 7 categories nor passing from a uniform to a symmetric or an asymmetric non-uniform distribution, impacts significantly on the behaviour of the procedures. On the contrary, the structure of the correlation matrix may impact on the performance of the tests, but in most cases the permutation test seems to have the best performance, in particular when the significance level alpha is equal to 0.05 and the sample size is small.

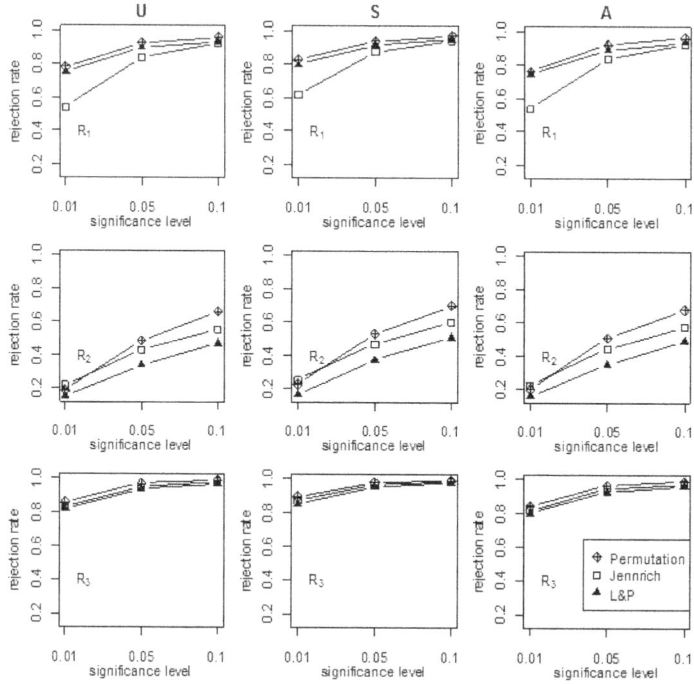

Fig. 12.2 Rejection rate of each procedure at different values of α, for uniform (U), symmetrical (S), asymmetrical (A) marginal distributions with $k = 5$ categories and $n = 20$, under $\mathbf{R}=\mathbf{R}_1$, $\mathbf{R}_2$, $\mathbf{R}_3$

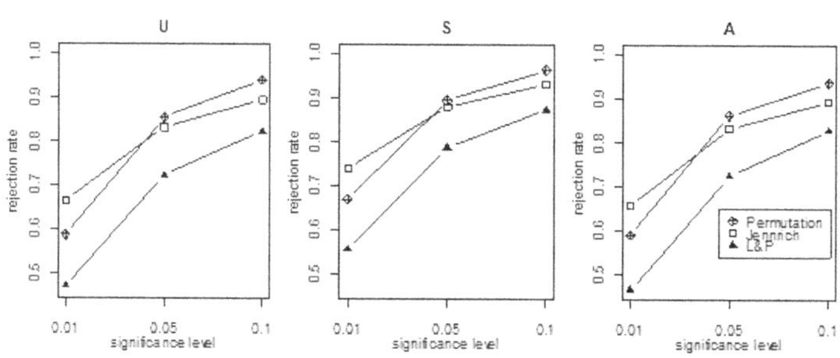

Fig. 12.3 Rejection rate of each procedure at different values of α, for uniform (U), symmetrical (S), asymmetrical (A) marginal distributions with $k = 5$ categories and $n = 50$, under $\mathbf{R}=\mathbf{R}_2$

References

1. Barbiero, A., Ferrari, P.A.: GenOrd: simulation of ordinal and discrete variables with given correlation matrix and marginal distributions. R package version 1.2.0 (2014). http://CRAN.R-project.org/package=GenOrd
2. Ferrari, P.A., Barbiero, A.: Simulating ordinal data. Multivar. Behav. Res. **47**(4), 566–589 (2012)
3. Folks, J.L.: Combination of independent tests. In: Krishnaiah, P.R., Sen, P.K. (eds.) Handbook of Statistics, Chapter 6, vol. 4, pp.113–121. Elsevier Science Publishers, New York (1984)
4. Hansen, B.B., Bowers, J.: Covariate balance in simple, stratified and clustered comparative studies. Stat. Sci. **23**(2), 219–236 (2008)
5. Hotelling, H.: The generalization of student's ratio. Ann. Math. Stat. **2**, 360–378 (1931)
6. Jackman, S.: Bayesian Analysis for the Statistical Science. Wiley, Chichester (PDF ebook) (2009)
7. Jennrich, R.I.: An asymptotic χ^2 test for the equality of two correlation matrices. J. Am. Stat. Assoc. **65**, 904–912 (1965)
8. Larntz, K., Perlman, M.D.: A simple test for the equality of correlation matrices. Unpublished report, Department of Statistics. University of Minnesota, St. Paul (1985)
9. Pesarin, F., Salmaso, L.: Permutation Tests for Complex Data: Theory, Applications and Software. Wiley, Chichester (2010)
10. Rosembaum, P.R.: Coherence in observational studies. Biometrics **50**(2), 568–574 (1994)
11. Rosembaum, P.R.: Signed rank statistics for coherent predictions. Biometrics **53**(2), 556–566 (1997)
12. Ruscio, J.: Constructing confidence intervals for Spearman's rank order correlation with ordinal data. J. Mod. Appl. Stat. Methods **7**, 416–434 (2008)
13. Timm, N.H.: Applied Multivariate Analysis. Springer, New York (2002)

Chapter 13
A Method for Selection of the Optimal Bandwidth Parameter for Beran's Nonparametric Estimator

Victor Demin and Ekaterina Chimitova

13.1 Introduction

The most popular parametric regression models in reliability are the AFT (Accelerated Failure Time) model and the proportional hazards model. The construction of any parametric model requires knowledge of the lifetime distribution and the kind of dependence of reliability function on the observed covariates. In practice, however, this information is usually absent. In such a situation it is advisable to use nonparametric methods, which enable not only to estimate the reliability function for different values of the covariate, but also can be used to construct a goodness-of-fit test for some parametric reliability model.

One of the most popular approaches to nonparametric estimation of the regression reliability model is the estimator, proposed by Beran [1]. The investigation of statistical properties of this estimator in the case of random plans, when the value of covariates are not fixed, is presented in [3, 5, 8, 9]. In [10], the properties of Beran's estimator are studied, when the values of covariate are defined in advance.

Nowadays, a great number of publications are devoted to the problem of kernel smoothing; the main attention is usually paid on the problem of selecting the optimal smoothing parameter. In the context of this problem, it is important to understand, that such methods as reference heuristic methods, substitution methods, and cross-validation are not applicable for the nonparametric Beran estimator, as in this case the kernel function determines only the weight of each observation according to the value of the covariate.

However, it is known that the quality of the Beran estimator essentially depends on the chosen value of the bandwidth parameter. In [10], a theoretical method of

V. Demin • E. Chimitova (✉)
Novosibirsk State Technical University, Novosibirsk, Russia
e-mail: ekaterina.chimitova@gmail.com

V.B. Melas et al. (eds.), *Topics in Statistical Simulation*, Springer Proceedings in Mathematics & Statistics 114, DOI 10.1007/978-1-4939-2104-1_13

selection of the optimal bandwidth parameter is suggested, however, it is extremely difficult to implement this method in practice, as it uses several functions, which are usually unknown. In [7], the method of selection of the optimal bandwidth parameter, based on the bootstrap procedure is offered, however, this approach is applicable only to the case of the random plan. Thus, it is necessary to develop the method of calculation of the optimal value of the bandwidth parameter for the Beran estimator. In [4], we proposed the idea of selecting the optimal bandwidth parameter, which is based on the minimization of the distance of failure times from kernel estimate of the inverse reliability function. So, the purpose of this paper is to investigate the statistical properties of the Beran estimator and to give some recommendations on the way of application of the proposed method.

13.2 Nonparametric Beran Estimator

Denote by T_x the lifetime of the considered technical product, which depends on a scalar covariate. The reliability function is denoted by

$$S(t|x) = P(T_x \geqslant t) = 1 - F(t|x), \tag{13.1}$$

where $F(t|x)$ is the conditional distribution function of the random variable T_x.

The main feature of the lifetime data is the presence of right censored observations, which can be represented as

$$(Y_1, x_1, \delta_1), (Y_2, x_2, \delta_2), \ldots, (Y_n, x_n, \delta_n),$$

where n is the sample size, x_i is the value of covariate for i-th object, Y_i is the failure time or censoring time, and δ_i is the censoring indicator, which is equal to 1, if the i-th observation is complete, and 0 if it is censored.

The Beran estimator is defined as follows [1]:

$$\tilde{S}_{h_n}(t|x) = \prod_{Y_{(i)} \leqslant t} \left\{ 1 - \frac{W_n^i(x; h_n)}{1 - \sum_{j=1}^{i-1} W_n^j(x; h_n)} \right\}^{\delta_i}, \tag{13.2}$$

where x is the value of the covariate, for which reliability function is estimated, $W_n^i(x; h_n)$, $i = 1, \ldots, n$ are the Nadaraya–Watson weights, which are defined as follows [9]:

$$W_n^i(x; h_n) = K\left(\frac{x - x_i}{h_n}\right) \Big/ \sum_{j=1}^{n} K\left(\frac{x - x_j}{h_n}\right),$$

where $K\left(\frac{x-x_i}{h_n}\right)$ is the kernel function, satisfying to the regularity conditions: $K(y) = K(-y)$, $0 \leqslant K(y) < \infty$, $\int_{-\infty}^{\infty} K(y)dy = 1$; $h_n > 0$ is the bandwidth parameter, which satisfies to the conditions: $\lim_{n\to\infty} h_n = 0$, $\lim_{n\to\infty} nh_n = \infty$.

13.3 The Choice of Bandwidth Parameter

The choice of the bandwidth parameter determines the values of the weights $W_n^i(x; h_n)$, which in turn determine which observations will participate in the construction of the estimate of the conditional reliability function (13.1). Thus, varying the bandwidth parameter, in a certain way, it is possible to drop "bad" observations.

In this paper, we consider the method for selecting an optimal parameter, which is based on the minimization of the mean deviation failure times $Y_1, Y_2, \ldots, Y_n$ from nonparametric estimation of the inverse reliability function $S_x^{-1}(p)$ [4]. We denote the inverse reliability function through $g(p|x)$. Then, the model (13.1) can be rewritten in the form:

$$T_x = g\,(p|x) + \varepsilon, \tag{13.3}$$

where $p \in (0, 1)$, ε is the error of observation, which, in general, may depend on p and x.

Kernel estimator for the model (13.3) can be written as

$$\hat{g}\,(\hat{p}_i|x_i) = \frac{1}{n}\sum_{j=1}^{n} \omega_n^j\,(\hat{p}_i) \cdot Y_j, \tag{13.4}$$

where ω_n^j is a certain weight, which can be calculated using various weighting functions. In particular, we consider the Nadaraya–Watson weights of the first order

$$\omega_n^j\,(\hat{p}_i) = K\left(\frac{\hat{p}_i - \hat{p}_j}{b_n}\right) \Big/ \sum_{k=1}^{n} K\left(\frac{\hat{p}_i - \hat{p}_k}{b_n}\right)$$

and the Priestley–Chao weights of the second order [1]:

$$\omega_n^j\,(\hat{p}_i) = \left\{\hat{p}_{(i)} - \hat{p}_{(i-1)}\right\} K\left(\frac{\hat{p}_i - \hat{p}_j}{b_n}\right),$$

where the smoothing parameter b_n can be selected using one of the methods proposed for kernel smoothing [1,6]. Probabilities $\hat{p}_i$ are calculated using the Beran estimates: $\hat{p}_i = \tilde{S}_{h_n}\,(Y_i|x_i)$.

Thus, the optimal value of the bandwidth parameter can be obtained by solving the following optimization problem:

$$h_n^{\text{opt}} = \arg\min_{h_n} \frac{1}{n} \sum_{i=1}^{n} \delta_i \cdot |\hat{g}\,(\hat{p}_i|x_i) - Y_i|. \tag{13.5}$$

13.4 Choice of Weights and Smoothing Parameter

As we consider the problem, involving the use of kernel smoothing, we can use pre-developed approaches for the optimal bandwidth parameter for the kernel estimator of regression. Let us consider the following method of minimal mean of integrated error according to the smoothing parameter, which is calculated as:

$$b_{\text{NS}} = \left[\frac{8\pi^{1/2}R(K)}{3\mu_2(K)^2 n}\right]^{1/5} \hat{\sigma},$$

where $\mu_2(K) = \int x^2 K(x)dx$, $R(K) = \int K^2(x)dx$, $\hat{\sigma}$ is the estimate of the variance, which can be calculated in various ways, most often used for this purpose, for example, the sample variance:

$$\hat{\sigma}^2 = S_n^2 = \frac{1}{n-1} \sum_{i=1}^{n} \left(\hat{p}_i - \bar{\hat{p}}\right)^2.$$

However, firstly, this estimate is not robust, and secondly, has "good" properties only if the distribution is close to normal. Therefore, in this paper we shall also consider the robust estimate of the variance:

$$\hat{\sigma} = S_{\text{rob}} = \operatorname*{med}_{i=1..n} \left| \hat{p}_i - \operatorname*{med}_{j=1..n,\,k=j..n} \left(\frac{\hat{p}_j + \hat{p}_k}{2}\right)\right|.$$

Let us investigate the statistical properties of the Beran estimator using the optimal bandwidth parameter (13.5). The investigation of the properties of the Beran estimates is carried out by the Monte Carlo simulations. The following statistic is used as the distance between the Beran estimates and the true conditional reliability function:

$$D_{h_n} = \sup_{j=1..k,\; t<\infty} \left|\tilde{S}_h(t|x_j) - S_{x_j}(t)\right|. \tag{13.6}$$

It is obvious that the quality of estimates (13.4) directly influences on that, how well the bandwidth parameter will be chosen. So, let us compare different weights ω_n^j for the kernel estimator $\hat{g}(\hat{p}_i|x_i)$, as well as different methods of choosing smoothing parameter from the point of view of the accuracy of the Beran estimation.

As the true reliability model we consider the parametric Cox proportional hazards model [2]:

$$S_x(t) = (S_0(t))^{r(x;\beta)}, \tag{13.7}$$

with the covariate function $r(x;\beta) = \ln(1 + e^{\beta x})$ and the lognormal baseline distribution with the density function:

$$f_0(t) = \frac{1}{\sqrt{2\pi}\theta_1 t} \exp\left(-\frac{1}{2\theta_1^2} \ln^2\left(\frac{t}{\theta_2}\right)\right).$$

Let us take the following notations for the weight functions, methods of variance estimation, and true values of parameter β used in simulation study:

- 1—the Priestley–Chao weights, variance estimate S_{rob}, $\beta = 2$;
- 2—the Priestley–Chao weights, variance estimate S_n^2, $\beta = 2$;
- 3—the Nadaraya–Watson weights, variance estimate S_{rob}, $\beta = 2$;
- 4—the Nadaraya–Watson weights, variance estimate S_n^2, $\beta = 2$;
- 5—the Priestley–Chao weights, variance estimate S_{rob}, $\beta = 5$;
- 6—the Priestley–Chao weights, variance estimate S_n^2, $\beta = 5$;
- 7—the Nadaraya–Watson weights, variance estimate S_{rob}, $\beta = 5$;
- 8—the Nadaraya–Watson weights, variance estimate S_n^2, $\beta = 5$.

We consider the case, when the covariate takes the values from the set {0, 0.11, 0.22, 0.33, 0.44, 0.56, 0.67, 0.78, 0.89, 1}, the sample size $n = 100, 200, 300$, and the number of observations corresponding to different values of the covariate is equal to each other. The samples were generated according to the model (13.7) with parameters: $\theta_1 = 21.5, \theta_2 = 1.6, \beta = 2$ or $\beta = 5$. The values of the distance (13.6) are given in Fig. 13.1; the average values of chosen bandwidth parameter h_n^{opt} and smoothing parameter b_{NS} are presented in Figs. 13.2 and 13.3, correspondingly.

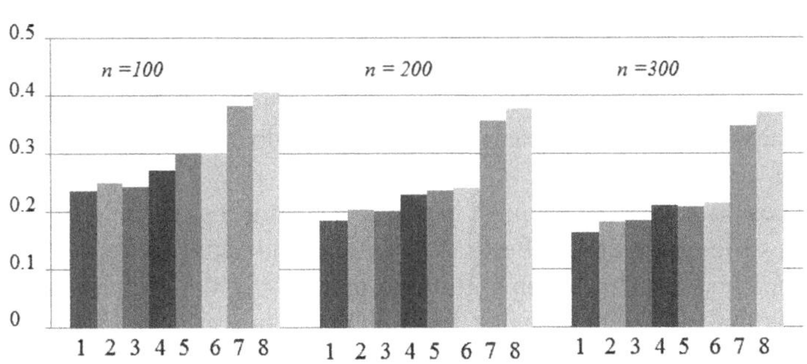

Fig. 13.1 The distance D_n for different sample sizes

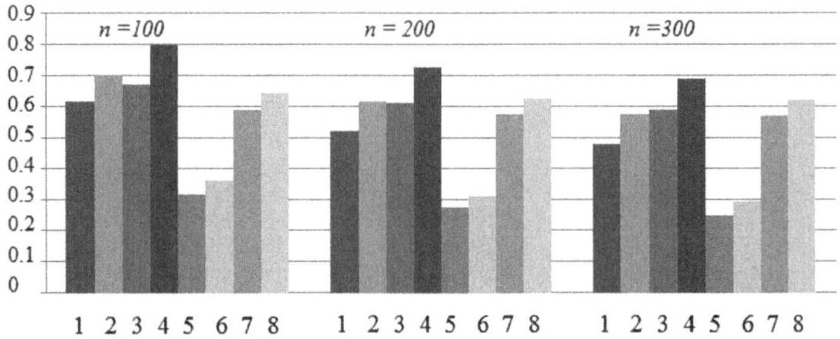

Fig. 13.2 Average values of the bandwidth parameter h_n^{opt} for different sample sizes

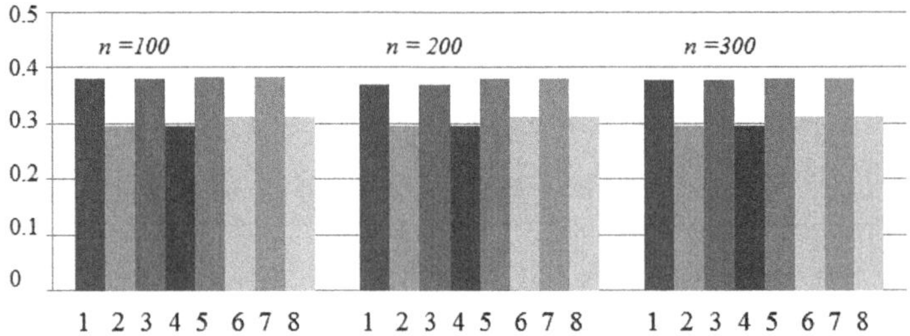

Fig. 13.3 Average values of the smoothing parameter b_n for different sample sizes

As can be seen from Fig. 13.1, the Priestley–Chao weight function allows to get more accurate Beran estimates. Thus, when the sample size is equal to 100, the value of distance (13.6) in the case of using Prestly–Chao weights is less by 3 % in comparison with the case of using Nadaraya–Watson weights; if $n = 200$ the winning is 8 % and when $n = 300$ the winning is 11 %. Moreover, the usage of robust estimator S_{rob} in calculation of the smoothing parameter b_n gives better accuracy, and accuracy of the Beran estimates increases with the sample size growth.

Figure 13.2 shows the average values of the chosen bandwidth parameter h_n^{opt}. It is seen that when the sample size increases, the value of optimal bandwidth parameter reduces; it is quite natural, since the number of observations in groups increases, and hence the number of "bad" observations increases.

Figure 13.3 illustrates the average values of smoothing parameter b_n. It is curious that the value of the smoothing parameter practically does not depend on the sample size and the weight function.

Similar results have been obtained in experiments for the parameter value $\beta = 5$ (i.e., with a stronger covariate effect). As in the considered case, the application of the robust method in conjunction with the usage of Priestley–Chao weights result in better accuracy of Beran estimates. It is interesting to consider apart the behavior of optimal bandwidth parameter h_n^{opt}: when the influence of the covariate on the

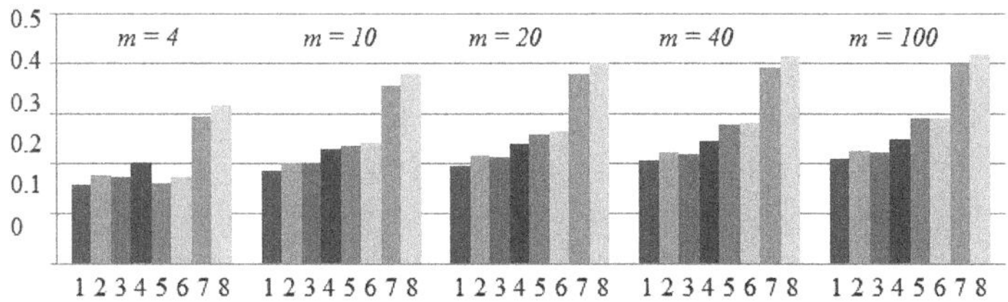

Fig. 13.4 The distance D_n for different numbers of groups

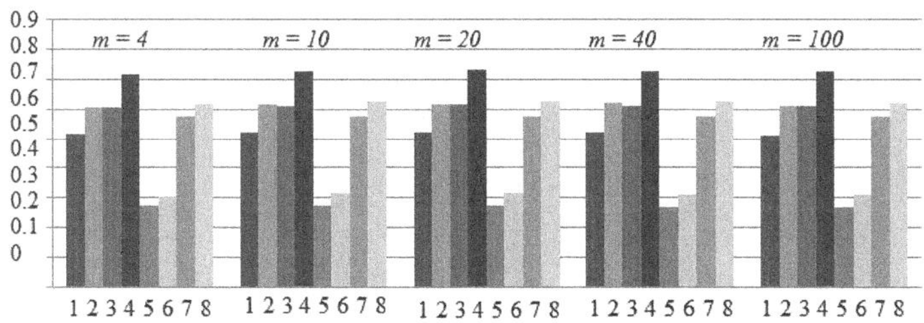

Fig. 13.5 Average values of the bandwidth parameter h_n^{opt} for different numbers of groups

reliability function increased, the average value of h_n^{opt} decreased almost twice in the case of using Priestley–Chao weights; however, in the case of Nadaraya–Watson weights such a change is not observed.

A discrete plan of experiment depends on the number of values m of the covariate. The following part of investigations is devoted to the study of the dependence of the Beran estimates on m. Simulation results for the fixed sample size $n = 200$ and for different numbers m are presented in Figs. 13.4 and 13.5.

As can be seen from Fig. 13.4, when the number of groups increases for the fixed sample size, the accuracy of the Beran estimator decreases, but this fall is not significant. This result can be explained as follows: the number of observations in a group decreases, therefore, the amount of information for each covariate value also becomes less, what leads to the loss of accuracy. However, the average values of bandwidth parameter h_n^{opt} (see Fig. 13.5) practically do not change. Similar result was observed for the smoothing parameter b_{NS}. This property of the optimal parameters h_n^{opt} and b_{NS} extends to the case of $\beta = 5$, when the degree of influence of covariate on the reliability function was increased.

Similar investigation has been carried out for the Cox proportional hazards model with exponential baseline distribution. The revealed regularities were almost the same, so specific numerical results are not given here.

Conclusions

In this paper, we have investigated the selection method of the bandwidth parameter for the Beran estimator, which is based on minimization of the distance between failure times and the kernel estimator of the inverse reliability function. It has been shown that the accuracy of the Beran estimator is influenced by the sample size, the number of different values of the covariate, as well as the weight function and the method of variance estimation used in calculation of optimal smoothing and bandwidth parameters.

The obtained results enable us to formulate a number of recommendations for calculation of Beran's estimator of conditional reliability function. It has been shown that it is preferable to use the Priestley–Chao weight function and to calculate the value of smoothing parameter by the method of minimal mean of integrated error with the robust estimator of variance, when calculating the kernel estimator of the inverse reliability function.

It should be noted that the methods considered in the paper do not cover all the variety of approaches to nonparametric estimation of reliability regression models. In particular, the development of bootstrap technique and adaptive algorithms for the choice of optimal bandwidth parameter seem to be interesting for future research.

Acknowledgements This research has been supported by the Russian Ministry of Education and Science (project 2.541.2014K).

References

1. Beran, R.: Nonparametic regression with randomly censored survival data. Technical report, Department of Statistics, University of California, Berkeley (1981)
2. Cox, D.R.: Regression models and life tables (with discussion). J. R. Stat. Soc. Ser. B **34**, 187–220 (1972)
3. Dabrowska, D.M.: Nonparametric quantile regression with censored data. Sankhya: Indian J. Stat. Ser. A **54**, 252–259 (1992)
4. Demin, V.A., Chimitova, E.V.: Choice of optimal smoothing parameter for nonparametric estimation of regression reliability model. Tomsk state university. J. Contr. Comput. Sci. **1**(22), 50–59 (2012)
5. Gonzalez, M.W., Cadarso, S.C.: Asymptotic properties of a generalized Kaplan–Meier estimator with some application. J. Nonparametr. Stat. **4**, 65–78 (1994)
6. Hardle, W.: Applied Nonparametric Regression. Cambridge University Press, Cambridge (1992)
7. Li, G., Datta, S.: A bootstrap approach to nonparametric regression for right censored data. Technical Report, Department of Statistics, University of Georgia, Athens, 30602 (1999)

8. McKeague, I.W., Utikal, K.J.: Inference for a nonlinear counting process regression model. Ann. Stat. **18**, 1172–1187 (1990)
9. Van Keilegom, I., Akritas, M.G., Veraverbeke, N.: Estimation of the conditional distribution in regression with censored data: a comparative study. Comput. Stat. Data Anal. **35**, 487–500 (2001)
10. Van Keilegom, I.: Nonparametric estimation of the conditional distribution in regression with censored data. Dissertation, Limburgs Universitair Centrum te verdedigen door (1998)

Chapter 14
Simulating from a Family of Generalized Archimedean Copulas

Fabrizio Durante

14.1 Introduction

The search for flexible multivariate statistical models has stimulated the investigations about new families of copulas that can capture stylized facts of multivariate data like tail dependence, non-exchangeability, asymmetries. See, for instance, the number of different copula families discussed in [14–16, 18, 24] and the references therein.

In such a variety of different models, it is hardly questionable that one of the most studied (and used) models is represented by the Archimedean class of copulas. Such copulas can be expressed in terms of a one-dimensional generating function $\varphi\colon [0,1] \to [0,+\infty]$ by means of the expression

$$C_{\varphi}(\mathbf{u}) = \varphi^{-1}(\varphi(u_1) + \cdots + \varphi(u_d)), \tag{14.1}$$

for all $\mathbf{u} \in [0,1]^d$. Conditions under which Eq. (14.1) describes a genuine copula are discussed in detail, for instance, in [1, 21].

Now, despite its popularity, Archimedean copulas suffers from some limitations that have been long recognized in the literature. In fact, their expression is symmetric in its arguments, so that, if a multivariate model H is constructed from an Archimedean copula C and a univariate distribution function F via the expression

$$H(\mathbf{x}) = C(F(x_1), \dots, F(x_d)), \tag{14.2}$$

F. Durante (✉)
Faculty of Economics and Management, Free University of Bozen-Bolzano, Bolzano, Italy
e-mail: fabrizio.durante@unibz.it

V.B. Melas et al. (eds.), *Topics in Statistical Simulation*, Springer Proceedings in Mathematics & Statistics 114, DOI 10.1007/978-1-4939-2104-1_14

it turns out that H is symmetric in its arguments. Thus, H may be used only when identically distributed random variables are supposed to be exchangeable. Several ways have been proposed in the literature in order to overcome the exchangeability issue by employing, for instance, ad-hoc asymmetrization procedures [5, 17] or hierarchical structures [12, 25].

Here, instead, we focus our attention on another possible extension of Archimedean copulas that is more related to the issue of shock models (or models for joint defaults). To fix ideas, suppose that (X, Y) represents a pair of identically distributed random variables (on a suitable probability space) having the meaning of lifetimes. Consider, for instance, X and Y as the lifetimes of two components of the same (engineering) system; or as time-to-default of firms or time-to-payment of some insurance contracts (i.e., house insurance against natural catastrophic events). It is likely that a suitable requirement for a parametric model for (X, Y) would be that the event $\{X = Y\}$ may happen with a non-zero probability, so that it includes the possibility of joint default for X and Y. However, models of this type are not absolutely continuous and, hence, are not often considered in the literature despite their potential interest (see, for instance, [19]).

A modification of bivariate Archimedean copulas that is able to take into account joint default (under identical marginals) has been proposed in [10]. Here, this new class of copulas is revisited and presented as distortion of the family of semilinear copula, which are copulas originated for a specific shock model. Moreover, by using a recent construction method presented in [23], a procedure is given to sample random variates from such copulas (under some additional assumptions). Such procedures are expected to be useful in multivariate models of lifetimes when, for instance, the effects of a shock are relevant for the behaviour of a system. For more details about possible use in credit risk, see also [2, 18].

14.2 The Generalized Archimedean Class of Copulas

Here we introduce the class of generalized Archimedean copulas starting with a shock model, which will be interpreted, just for the sake of presentation, as a model trying to describe random losses in a bivariate credit portfolio.

Suppose that some random losses X, Y are independent and identically distributed with distribution function F that is supported on $[0, 1]$ (basically, we are considering fractions of losses over a theoretical upper maximal loss). Following a Marshall–Olkin mechanism to construct a multivariate model [20], suppose that (X, Y) is subject to a shock represented by a random variable Z, which has distribution function $G(t) = t/F(t)$ on $(0, 1)$. Such a Z may be interpreted as another loss that can hit the system. Since an external random shock occurs, we may suppose now that the total vector of losses can be more conveniently represented in the form

$$(U, V) = (\max\{X, Z\}, \max\{Y, Z\}),$$

i.e. each individual loss tends to increase in view of the presence of random loss determined by Z. Now, it can be shown that, under the given assumptions, (U, V) is distributed according to the copula

$$C_F(u, v) = \min(u, v)F(\max(u, v)). \tag{14.3}$$

The copula C_F is a semilinear copula (see, e.g., [4, 11]). It describes positive association between two random variables and, up to the case of the independence copula $\Pi_2(u, v) = uv$, its corresponding measure admits a singular component along the main diagonal of $[0, 1]^2$. As evident, in general semilinear copulas are not Archimedean, but may serve as a basis to build a generalized Archimedean model.

In fact, consider a distortion function $h\colon [0, 1] \to [0, 1]$, i.e. an increasing and concave bijection of $[0, 1]$. Following a general construction principle, each copula C can be transformed into another copula C_h, by means of the formula

$$C_h(u, v) = h^{-1}(C(h(u), h(v))). \tag{14.4}$$

For more details, see [6, 22, 27]. By applying a distortion h to a semilinear copula C_F, we obtain the copula

$$(C_F)_h(u, v) = h^{-1}(h(\min(u, v))F(h(\max(u, v)))). \tag{14.5}$$

Setting $h(t) = \exp(-\varphi(t))$ and $F(h(t)) = \exp(-\psi(t))$ for suitable functions φ and ψ, the previous expression may be rewritten in the form

$$C_{\varphi,\psi}(u, v) = \varphi^{-1}(\varphi(\min(u, v)) + \psi(\max(u, v))). \tag{14.6}$$

Copulas of type (14.6) have been introduced in [10]. Here, they will be called GA copulas (GA stands for generalized Archimedean). Simple sufficient conditions that ensure that (14.6) defines a *bona fide* copula are provided in [10] and are reproduced below. Notice that, for a continuous function f we denote by f^{-1} it pseudo-inverse, which coincides with the standard inverse when f is strictly monotone.

Theorem 14.1. *Let* $\varphi : [0, 1] \to [0, +\infty]$ *be continuous and decreasing. Let* $\psi : [0, 1] \to [0, +\infty]$ *be continuous, decreasing and such that* $\psi(1) = 0$. *If* φ *is convex and* $(\psi - \varphi)$ *is increasing in* $[0, 1]$, *then* $C_{\varphi,\psi}$ *of Eq.* (14.6) *is a copula.*

Obviously, in the case $\varphi = \psi$, copulas of type (14.6) coincide with Archimedean copulas. However, the family of GA copulas is more general. For instance, both Cuadras-Augé copulas [3] and MT-copulas [9] are GA copulas, but not Archimedean copulas. Now, it can be proved that, in general, GA copulas are not absolutely continuous, and their singular component may be often identified, as the following result shows.

Theorem 14.2. *Let $C_{\varphi,\psi}$ be a copula of type* (14.6) *such that $\varphi \neq \psi$. Then C contains a singular component along the main diagonal $\{(x,x) : x \in [0,1]\}$.*

Proof. Since the functions φ and ψ are monotone, they are differentiable almost everywhere in $[0, 1]$. Assume $C(x, y) := C_{\varphi,\psi}(x, y) > 0$ at the point $(x, y) \in [0, 1]^2$. According to [10, Theorem 4.1],

$$\varphi'(C(x,y))\frac{\partial C}{\partial x}(x,y) = \varphi'(x), \quad x < y$$
$$\varphi'(C(x,y))\frac{\partial C}{\partial x}(x,y) = \psi'(x), \quad x > y.$$

It follows that $t \mapsto \frac{\partial C}{\partial x}(x, y)$ has some jump discontinuity along the main diagonal of the unit square. Thus, in view of [16, Theorem 1.1], there is a singular component of the probability mass associated with $C_{\varphi,\psi}$.

It turns out that copulas of type (14.6) form a class where models with singular components and Archimedean models can be joined together.

Example 14.1. Take $\varphi(t) = 1 - t$ and, for every $\alpha \in [0, 1]$,

$$\psi(t) = \begin{cases} \alpha/2, & t \in [0, \alpha/2]; \\ \alpha - t, & t \in (\alpha/2, \alpha); \\ 0, & t \in [\alpha, 1]. \end{cases}$$

Then $C_{\varphi,\psi} = C_\alpha$ is a GA copula which is an ordinal sum (see, e.g., [7]) of the counter-monotonic copula $W(u, v) = \max(u + v - 1, 0)$ and the comonotone copula $M(u, v) = \min(u, v)$ with respect to the partition $\{(0, \alpha), (\alpha, 1)\}$. It is a singular copula that spreads the probability mass along the segments with endpoints $(0, \alpha), (\alpha, 0)$ and $(\alpha, \alpha), (1, 1)$.

14.3 Sampling Generalized Archimedean Copulas

In view of the importance of considering copulas with singular components in application [19], it can be useful to have sampling procedures for such copulas. However, to the best of our knowledge, this problem has not a general solution, being sampling procedure of these copulas limited to the classical conditional distribution method [18]. Here, following a novel construction principle for copulas proposed in [23], we present a sampling method for GA copulas. As it will be noted, the stochastic mechanism is similar to the semilinear copula construction presented above. In fact, both methods are inspired by the Marshall–Olkin mechanism [20] of shock models (see also [8]).

Let X, Y be identically distributed random variables with distribution function F supported on $[0, 1]$. Suppose that (X, Y) is subject to a shock represented by a random variable Z, which has distribution function G on $[0, 1]$. Moreover, suppose that both the original random variables and the shock are not independent, but are coupled by a trivariate Archimedean copula C_φ with additive generator φ. Since an external random shock occurs, we may suppose that the new system is more conveniently represented in the form

$$(U, V) = (\max\{X, Z\}, \max\{Y, Z\}).$$

Now, it can be shown that the distribution function of (U, V) is given by

$$\begin{aligned} H(u, v) &= \mathbb{P}(U \leqslant u, V \leqslant v) \\ &= \varphi^{-1}(\varphi \circ F(u) + \varphi \circ F(v) + \varphi \circ G(\min(u, v)). \end{aligned}$$

In order to ensure that such a H is a copula, consider $G(t) = \varphi^{-1}(\varphi(t) - \varphi \circ F(t))$ for all $t \in (0, 1)$. Then it follows that G is a distribution function if, and only if, $t \mapsto (\varphi \circ F(t) - \varphi(t))$ is increasing. Under this additional assumption, H can be rewritten as

$$H(u, v) = \varphi^{-1}(\varphi(\min(u, v) + \varphi \circ F(\max(u, v))). \tag{14.7}$$

It is not difficult to show that H of Eq. (14.7) satisfies the assumption of Theorem 14.2, so that it is a GA copula. Thanks to the previous stochastic construction, the following algorithm follows.

The inputs for the algorithm are: an additive generator φ of a trivariate Archimedean copula, and the distribution functions F (supported on $[0, 1]$).

(1) Simulate (U, V, W) from the Archimedean copula C_φ.

(2) Set

$$\begin{aligned} U_1 &= F^{-1}(U), \\ V_1 &= F^{-1}(V), \\ W_1 &= G^{-1}(W), \qquad \text{where } G(t) = \varphi^{-1}(\varphi(t) - \varphi \circ F(t)) \text{ for all } t \in (0, 1). \end{aligned}$$

(3) Return

$$(S, T) = (\max\{U_1, W_1\}, \max\{V_1, W_1\}).$$

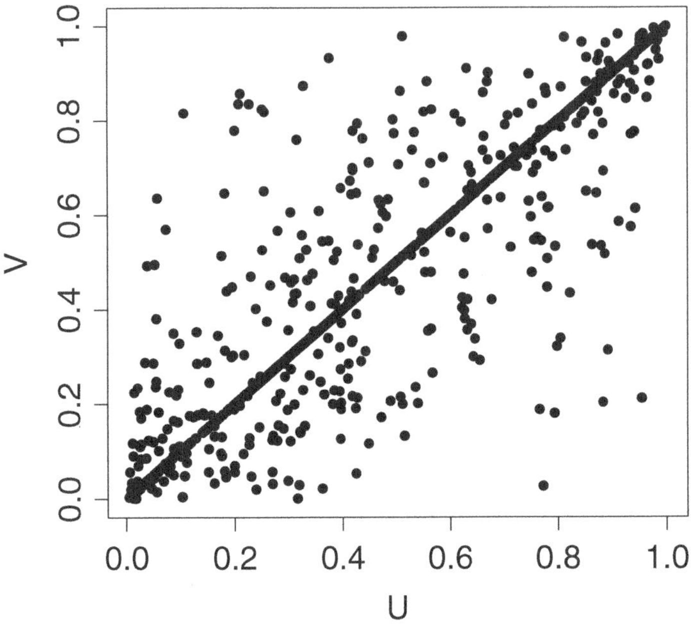

Fig. 14.1 Scatter plot from a copula of type (14.7) where φ is a generator of a Gumbel copula with Kendall's τ equal to 0.5, while $F(t) = \sqrt{t}$

Figures 14.1 and 14.2 illustrate the previous algorithms in two examples. From the pictures, it should be noted the presence of a singular component along the main diagonal of the unit square. In practice, the algorithm has been implemented in R [26] by using also some useful functions from the copula package [13] (in particular, generators and inverse generators of Archimedean copulas).

Finally, copulas H of type (14.7) represent a sub-class of GA copulas. In fact, while the former are constructed by using generators of trivariate Archimedean copulas, the latter are constructed via the larger class of generators of bivariate Archimedean copulas.

Acknowledgements The author thanks Sabrina Mulinacci (University of Bologna) for useful comments and discussions about the topic of this manuscript. The author acknowledges the support of Free University of Bozen-Bolzano, via the project MODEX.

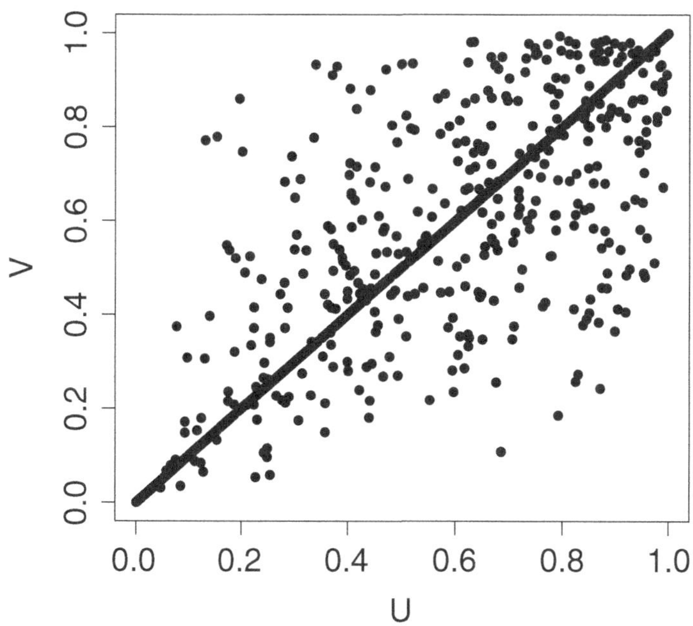

Fig. 14.2 Scatter plot from a copula of type (14.7) where φ is a generator of a Clayton copula with Kendall's τ equal to 0.5, while $F(t) = \sqrt{t}$

References

1. Alsina, C., Frank, M.J., Schweizer, B.: Associative Functions. Triangular Norms and Copulas. World Scientific Publishing, Hackensack (2006)
2. Cherubini, U., Mulinacci, S., Gobbi, F., Romagnoli, S.: Dynamic copula methods in finance. In: Wiley Finance Series. Wiley, Chichester (2012)
3. Cuadras, C.M., Augé, J.: A continuous general multivariate distribution and its properties. Commun. Stat. Theory Methods **10**(4), 339–353 (1981)
4. Durante, F.: A new class of symmetric bivariate copulas. J. Nonparametr. Stat. **18**(7–8), 499–510 (2006/2007)
5. Durante, F.: Construction of non-exchangeable bivariate distribution functions. Stat. Pap. **50**(2), 383–391 (2009)
6. Durante, F., Sempi, C.: Copula and semicopula transforms. Int. J. Math. Math. Sci. **2005**(4), 645–655 (2005)
7. Durante, F., Fernández-Sánchez, J., Sempi, C.: Multivariate patchwork copulas: a unified approach with applications to partial comonotonicity. Insur. Math. Econ. **53**, 897–905 (2013)
8. Durante, F., Hofert, M., Scherer, M.: Multivariate hierarchical copulas with shocks. Methodol. Comput. Appl. Probab. **12**(4), 681–694 (2010)
9. Durante, F., Mesiar, R., Sempi, C.: On a family of copulas constructed from the diagonal section. Soft. Comput. **10**(6), 490–494 (2006)
10. Durante, F., Quesada-Molina, J., Sempi, C.: A generalization of the Archimedean class of bivariate copulas. Ann. Inst. Stat. Math. **59**(3), 487–498 (2007)
11. Durante, F., Kolesárová, A., Mesiar, R., Sempi, C.: Semilinear copulas. Fuzzy Set. Syst. **159**(1), 63–76 (2008)

12. Hering, C., Hofert, M., Mai, J.F., Scherer, M.: Constructing hierarchical Archimedean copulas with Lévy subordinators. J. Multivar. Anal. **101**(6), 1428–1433 (2010)
13. Hofert, M., Kojadinovic, I., Maechler, M., Yan, J.: Copula: multivariate dependence with copulas (2013). http://CRAN.R-project.org/package=copula. R package version 0.999-7
14. Jaworski, P., Durante, F., Härdle, W.K. (eds.): Copulae in mathematical and quantitative finance. Lecture Notes in Statistics: Proceedings, vol. 213. Springer, Berlin/Heidelberg (2013)
15. Jaworski, P., Durante, F., Härdle, W.K., Rychlik, T. (eds.): Copula theory and its applications. Lecture Notes in Statistics: Proceedings, vol. 198. Springer, Berlin/Heidelberg (2010)
16. Joe, H.: Multivariate models and dependence concepts. In: Monographs on Statistics and Applied Probability, vol. 73. Chapman & Hall, London (1997)
17. Liebscher, E.: Construction of asymmetric multivariate copulas. J. Multivar. Anal. **99**(10), 2234–2250 (2008)
18. Mai, J.F., Scherer, M.: Simulating copulas: stochastic models, sampling algorithms, and applications. In: Series in Quantitative Finance, vol. 4. Imperial College Press, London (2012).
19. Mai, J.F., Scherer, M.: Simulating from the copula that generates the maximal probability for a joint default under given (inhomogeneous) marginals. Technical report (2013)
20. Marshall, A.W., Olkin, I.: A multivariate exponential distribution. J. Am. Stat. Assoc. **62**, 30–44 (1967)
21. McNeil, A.J., Nešlehová, J.: Multivariate Archimedean copulas, d-monotone functions and ℓ_1-norm symmetric distributions. Ann. Stat. **37**(5B), 3059–3097 (2009)
22. Morillas, P.M.: A method to obtain new copulas from a given one. Metrika **61**(2), 169–184 (2005)
23. Mulinacci, S.: Archimedean-based Marshall–Olkin distributions and related copula functions. Technical report (2013)
24. Nelsen, R.B.: An introduction to copulas, 2nd edn. In: Springer Series in Statistics. Springer, New York (2006)
25. Okhrin, O., Okhrin, Y., Schmid, W.: On the structure and estimation of hierarchical Archimedean copulas. J. Econom. **173**(2), 189–204 (2013)
26. R Core Team: R: A language and environment for statistical computing. R Foundation for Statistical Computing, Vienna. http://www.R-project.org/ (2013)
27. Valdez, E.A., Xiao, Y.: On the distortion of a copula and its margins. Scand. Actuar. J. **4**, 292–317 (2011)

Chapter 15
PS-Algorithms and Stochastic Computations

Sergej M. Ermakov

15.1 Introduction

The need to process huge volumes of data (big data analysis), on the one hand, imparts the necessity to create a computer with a very large number of processors and (or) quantum computers and, on the other hand it imparts the necessity to develop stochastic calculations that allow to draw conclusions on the basis of a sample data of a relatively small volume. We consider a special class of algorithms with parallel structure, including stochastic ones (the Monte Carlo methods) and discuss their specificity and possible applications.

It is a well-known fact that effective use of the modern supercomputers imposes certain requirements on the choice of algorithms for solving different problems. Of course, there exists a possible approach when each algorithm is analyzed by programmer or compiler, and the available parallel structures are used in calculations. However this approach in most cases leads to inefficient use of equipment. It often turns out that an algorithm that has worse properties in terms of the sequential computational structures of von Neumann is more effective if the multiprocessor technology is used. Obviously, it is interesting to study "bad" (as defined above) algorithms that can efficiently load the multiprocessor hardware, and to work out new algorithms of this type.

Further we will consider one class of algorithms (parametrically separated algorithms) that can effectively load the system with distributed memory, consisting of a central (manager) processor and k processors with autonomous memory. Clusters and clouds can serve as examples of such systems. It is assumed that the time of exchange between processors is significantly greater (by several orders of

S.M. Ermakov (✉)
Mathematics and Mechanics Faculty, St. Petersburg State University,
Universitetsky pr., 28, Stary Peterhof, Russia
e-mail: sergej.ermakov@gmail.com; nadya@statmod.ru

V.B. Melas et al. (eds.), *Topics in Statistical Simulation*, Springer Proceedings in Mathematics & Statistics 114, DOI 10.1007/978-1-4939-2104-1_15

magnitude) than the run-time of operations by each processor. In connection with this last assumption, we note the following.

Let the optimal sequential algorithm (requiring minimum average number of operations) for solving a problem requires time T_0. As an alternative we can use an algorithm that requires to be performed in a sequential version of time T, $T > T_0$, and attract k processors, each of which is occupied during the time T/k. This case, however, will require significant additional time t for exchange of information in the process of solving the problem.

The obvious requirement is fulfillment of the inequality

$$T_0 \geqslant \frac{T}{k} + t,$$

so an alternative algorithm must fulfill

$$T \leqslant k(T_0 - t).$$

It is clear that there should be $t < T_0$. Otherwise multiprocessor improves nothing. We can also see that the growth of k and reduction of t permit involving in solving the problem a wide range of algorithms with properties that are far from optimal in the traditional sense. Algorithms can be "bad" but have a parallel structure.

One class of algorithms with a parallel structure and few exchanges between processors are parametrically separable (PS)-algorithms [5].

PS-algorithm consists of three parts executed sequentially

$$B_1 \to A(\theta) \to B_2,$$

and has the following properties

(1) Algorithm $A(\theta)$ depends on the parameter θ (it has θ as an input data). θ takes values from discrete set Θ of a fairly general nature.
(2) k independent processors can be charged for execution of algorithms $A(\theta_i)$, $\theta_i \in \Theta$, $i = 1, \ldots, k$.
(3) Algorithm B_1 gives tasks to algorithm A, and B_2 computes solution of the original problem, with results received by $A(\theta_i)$, $i = 1, \ldots, k$.

If the execution times of algorithms $A(\theta_i)$ with different i vary only slightly, then the PS-algorithm is called *homogeneous*.

As one of the simplest examples of PS-algorithms we can refer to the VVR-algorithms [1]. An algorithm of this type calculates (for example) independently each digit of the number π.

Here $\Theta = N$ is the set of natural numbers. The computing system consisting of k processors independently computes k digits or k groups of digits, the results come in one of the processors (exchange), and the approximate value of π is formed with a given accuracy.

Many algorithms in the computational mathematics are PS-algorithms. For example, if the computation of values $f(X)$, where $X \in D \subset \mathbf{R}^s$, is associated with considerable difficulties, then the algorithm of interpolating polynomial construction is the following $P(X) = \sum_{i=1}^{n} l_i(X) f(X_i)$, $l_i(X)$ are polynomials, $X_i \in D$, and calculations of the integral with the use of the cubature sum

$$K_n[f] = \sum_{i=1}^{n} A_i f(X_i)$$

are PS-algorithms, where the set of points X_i constitutes the parameter set Θ.

The Newton's method of solving $f(x) = 0$, $x \in \mathbf{R}^1$ is an example of algorithm without the property of parametric separability, although it is possible that its modification obtains this property, as we will see further.

In the light of the above it becomes clear that the study of PS-algorithms may be of considerable interest, and further we will discuss a number of problems of the computational mathematics in terms of their PS-properties. For some algorithms that do not have the PS-property we will also point out ways to build their modifications with the PS-property.

15.2 Solving Linear Systems

As a very simple and very important class of algorithms we consider the iteration algorithms for solving systems of linear algebraic equations. Iteration algorithms in their general form are not PS-algorithms. Under certain conditions, asynchronous iterative methods have the PS-property. If a given system is

$$X = AX + F,\ X = (x_1, \ldots, x_n)^T,\ A = \|a_{i,j}\|_{i,j=1}^n,\ F = (f_1, \ldots, f_n)^T \tag{15.1}$$

and the majorant iterative process converges

$$\overline{X} = |A|\overline{X} + |F|, \tag{15.2}$$

then the original system also has the iterative solution $\tilde{X}$, and for each vector $H = (h_1, \ldots, h_n)$ the scalar product $(H, \tilde{X})$ can be represented as an integral over trajectories of a homogeneous Markov chain $\overrightarrow{p}^0, \mathscr{P}$ with n states $\overrightarrow{p}^0 = (p_1^0, \ldots, p_n^0)$, $\mathscr{P} = \|p_{i,j}\|_{i,j=1}^n$. Here $\overrightarrow{p}^0$ is the initial probability distribution of states, and $\mathscr{P}$ is the transfer (substochastic) matrix whose elements satisfy the following conditions

$$\sum_{i=1}^{n} p_{i,j} = 1 - g_i, \quad 0 \leqslant g_i \leqslant 1, \quad i = 1, \ldots, n. \tag{15.3}$$

The Markov chain is subject to the concordance conditions (the absolute continuity of measures), which depend on the type of selected estimate of scalar product $(H, \tilde{X})$—the function of trajectories. One of the simplest estimates is the following [4]

$$\xi_t = \frac{h_{i_0} a_{i_0,i_1}, \ldots, a_{i_{t-1},i_t} f_{i_t}}{p^0_{i_0} p_{i_0,i_1}, \ldots, p_{i_{t-1},i_t} g_{i_t}}, \tag{15.4}$$

where $i_0, i_1, \ldots, i_t$ is the trajectory of the Markov chain which requires fulfillment of conditions

$$\begin{array}{llll} p^0_i > 0, & h_i \neq 0, & g_i > 0, & f_i \neq 0, \\ p_{i,j} > 0, & a_{i,j} \neq 0, & i, j = 1, \ldots, n. & \end{array} \tag{15.5}$$

There are an infinite number of functions (estimates) which can be used to represent the scalar product $(H, \tilde{X})$ as a path integral. They require appropriate concordance conditions [4]. The integral can be calculated either with the use of the Monte Carlo method or by using the deterministic (quasi-Monte Carlo) methods. In the first case ξ_t is treated as a random variable whose expectation is the computed integral. The algorithm consists in modeling the trajectories $i_0 \to i_1 \to \cdots \to i_t$ of the Markov chain, which is selected so as to satisfy not only the concordance conditions but also the requirement of smallness of the second moment of the estimate. t is assumed finite with probability 1. ξ_t is calculated on the trajectories. Calculation of the independent groups of trajectories can be realized by different processors. The result is the mean value of the obtained ξ_t. Parameter set here is a set of pseudorandom numbers that should be divided between the processors.

A widely recognized disadvantage of the Monte Carlo method is its slow convergence. After having modeled N pathes we reduce the error $\sqrt{N}$ times. This disadvantage can be considered as a payment for unlimited parallelism of the method. It must be noted that for a certain class of problems the sequential version of the method can also be beneficial.

Indeed, the calculation of the sum $(H, X_M) = \left(H, \sum_{l=0}^{M} A^l F\right)$ requires $\sim MnK$ operations, where K is an average number of nonzero elements in the row of the matrix A. The Monte Carlo estimate of the same sum requires $\sim N(2 + \ln^2 K)M$ operations provided application of the bisection method, and $\sim 4NM$ provided application of the Walker method [12]. It is easy to see that with the growth of n, for fully filled matrices in particular, the application of the Monte Carlo method can be justified to get the results with low ($\sim 1\,\%$) accuracy. Calculations with high-precision, obviously, require different approaches.

If the system (15.1) has an iterative solution but the iterative process for (15.2) diverges, then $(H, \tilde{X})$ cannot be represented as a path integral. Therefore we cannot specify the PS-algorithm based on modeling the trajectories of the Markov chain.

In our considered case the sum $\sum_{l=0}^{\infty} |A|^l F$ is infinite, but the sum $\sum_{l=0}^{\infty} |A^l F|$ is finite. An iterative algorithm for computing solution of the system $X = AX + F$ consists in sequential calculation of vectors $\tilde{X}_1 = AF + F$, $\tilde{X}_2 = A(AF + F)$ and so on. Here it is sufficient to satisfy the condition $\|A\| < 1$.

Separately, we should mention the case when the matrix A is triangular. Difference analogue of the evolutionary net equations is usually reduced to this occasion. The largest eigenvalue of the modulus of this matrix is equal to the largest diagonal element. If it is less than unity, then it is possible to formally represent solution of the problem as a path integral. However, if other elements are large, then the variance of estimates usually grows exponentially with the order n of the matrix, and computations are unstable (stochastic instability).

The algorithm for calculation of matrix–vector product is the PS-algorithm (by line number), but this cannot be said for general algorithm for computing the sum of the Neumann series. It is required to memorize vector $\tilde{X}_m$ and synchronization is necessary in the multi-processor case. Let's recall that the necessary convergence condition for asynchronous iterations is finiteness of the sum $\sum_{l=0}^{\infty} |A|^l F$ [2]. It is possible to build PS-algorithms here if one uses methods of the numerical integration. In particular, with the use of the Monte Carlo method one can compute a sequence of unbiased estimates Ξ_m of vectors $\tilde{X}_m$. One of the simplest estimates of components of $\tilde{X}_{m+1}$ is the following. Distribution $\overrightarrow{p} = (p_1, \ldots, p_m)$ is given to select number i—the number of component of the previously computed vector Ξ_m, and stochastic matrix $\mathscr{P} = \|p_{i,j}\|_{i,j=1}^n$ is given to select element from the i-th row of the matrix A

$$\xi_l^{m+1} = \frac{\xi_i a_{i,l}}{p_i p_{i,l}}. \tag{15.6}$$

Under the concordance conditions (15.5) the following equality holds true

$$\mathrm{E}\Xi_{m+1} = \tilde{X}_{m+1}, \quad \Xi_{m+1} = (\xi_1^{m+1}, \ldots, \xi_n^{m+1}). \tag{15.7}$$

After having calculated ξ_1^{m+1} repeatedly (N_l times) with independent random numbers, we obtain with the required accuracy the estimate Ξ_{m+1} and go to the estimate Ξ_{m+2} (and so on).

Constructed algorithm is already the PS-algorithm. Estimate Ξ_M for sufficiently large M can be charged to different computers (processors). Then the expectation of Ξ_M is estimated by known methods. Simultaneously one can build the confidence interval. Analysis of the behavior of the resulting error with $N_l = N$, independent of l, can be found in [4]. N must be large enough so that the algorithm might be stochastically stable. For small N variance of estimates Ξ_m can increase indefinitely with increasing m. Here we do not discuss details of the algorithms for solving difference equations. We only note that taking into account the property of smoothness of the original problem solutions can significantly reduce the complexity of the algorithm.

For calculating the estimates ξ_l^m, as well as for calculating estimates ξ_t defined by (15.4), one can use the quasi-random numbers and grid integration along with the pseudo-random numbers. It means that we use numerical integration methods for solving systems of linear algebraic equations in the general case. As noted above, the Monte Carlo error is of order $O\left(N^{-1/2}\right)$.

The error of the quasi Monte Carlo method decreases as $O\left(\frac{\ln^s N}{N}\right)$, where s is the dimension of the computed integral.

The expectation of the estimate ξ_t is a sum of integrals of increasing multiplicity. As a rule, the quasi Monte Carlo method is used to calculate several first components. Its use for very large s is not effective. However, we can see that the sequential procedure for calculating the estimates (15.6) corresponds to $s = 1$, which allows to take full advantage of the quasi-random sequences.

If the system (15.1) appeared as a result of sampling a problem with smooth data and decision, then it is possible to use methods with significantly more rapid decrease of the error (the method of Korobov optimal coefficients for calculating the sums [7]).

These are general features of the (quasi) stochastic algorithms for solving systems of linear algebraic equations. It is quite obvious that appearance of the calculators with a very large number of processors, those that have the type of SIMD [11] in particular, makes these algorithms very attractive. Easy realization of the programs may be an additional advantage. The multigrid methods, for example, do not have this advantage.

It should also be noted that the vast majority of the results related to systems of linear algebraic equations can be transferred to equations of the form

$$\varphi(x) = \int k(x, y)\varphi(y)\mu(dy) + f(x) \qquad (\text{mod } \mu), \tag{15.8}$$

where μ—σ-finite measure. In this case the algorithm consists in simulating the Markov chain with transition density associated with kernel $k(x, y)$ by concordance conditions [4]. Application problems of particular interest are those in which $k(x, y)$ is the transition density (queueing problems, radiation transport problems, etc.).

15.3 Nonlinear Problems

Some features of the PS-algorithms for nonlinear problems can be traced by the simplest example of a quadratic equation (or rather, certain types of quadratic equations). If quadratic equation

$$x = ax^2 + bx + c \tag{15.9}$$

satisfies $|a| + |b| + |c| \leqslant 1$, then iterations $x_m = ax_m^2 + bx_m + c,\ m = 1, 2, \ldots$ converge to the smallest solution of the Eq. (15.9).

If we compare Eq. (15.9) to a branching process, determined by probability distribution p_2, p_1, p_0 and evolving in the discrete time $t = 0, 1, \ldots$ so that at $t = 0$ there is one particle, and for at $t = 1$:

it dies with probability p_0 (breakage of the trajectory);
it remains unchanged with probability p_1;
Two particles with identical properties are formed with probability p_2;

then the solution of the equation can be calculated by modeling this process. It is sufficient to calculate functional on its trajectories that is analogous to (15.4).

Everything becomes more transparent if we note that (15.9) is equivalent to infinite system of equations [9]

$$y_{s+1} = ay_{s+2} + by_{s+1} + cy_s, \quad y_0 = 1, \quad s = 0, 1, \ldots. \tag{15.10}$$

Formal application of the methods described above to the system provides an algorithm that can also be interpreted as a simulation of a branching process. The difference consists in choosing the state space of the process.

Thus we have indicated the PS-algorithm for solving (15.9). This is probably the most difficult (bad) known algorithm for solving quadratic equation. However, it can be generalized to some complex multi-dimensional equations of the form

$$\begin{aligned} \varphi(x) = f(x) + \sum_{l=1}^{\infty} \int \mu(dx_1) \ldots \int \mu(dx_l) k_l(x, x_1, \ldots, x_l) \\ \times \prod_{j=1}^{l} \varphi(x_j) \quad (\mathrm{mod}\ \mu), \end{aligned} \tag{15.11}$$

under condition of convergence of the majorant iterative process

$$\begin{aligned} \overline{\varphi}_{m+1}(x) = |f(x)| + \sum_{l=1}^{\infty} \int \mu(dx_1) \ldots \int \mu(dx_l)\, |k_l(x, x_1, \ldots, x_l)| \\ \times \prod_{j=1}^{l} \overline{\varphi}_m(x_j) \quad (\mathrm{mod}\ \mu) \end{aligned} \tag{15.12}$$

it can be very effective in this case.

The condition (15.12) is rather restrictive, but a number of techniques, including various types of changes of variables allow to achieve its fulfillment for several important applications—the Navier–Stokes equations, the Boltzmann equation.

If the nonlinearity is not polynomial, but has a more complex nature, generally speaking we can't reduce the problem to the computation of a path integral. Different types of sequential linearization are used here. At each stage to solve the linear problem, one can use the Monte Carlo and the Quasi Monte Carlo methods. The essence of the problems emerging here can be illustrated by the following example of system of equations

$$f_i(x_1, \dots, x_n) = 0, \quad i = 1, \dots, n. \tag{15.13}$$

Under certain assumptions regarding the smoothness of f_2 and others, the system can be solved by the Newton's method—linearization by expansion in the Taylor series. For each $m = 0, 1, \dots$ we solve the following system of linear equations

$$\sum_{l=1}^{n} \frac{\partial f_i(X^m)}{\partial x_l}(x_l^{m+1} - x_l^m) = -f_i(X^m), \tag{15.14}$$

$X^m = (x_1^m, \dots, x_n^m), \quad i = 1, 2, \dots, n$. It is possible that for large n the system can be successfully solved by the Monte Carlo method, especially if the occupancy of the matrix system is great. X^m, with previously computed X^{m-1}, is estimated as the average of N independent estimates $(\xi_{1,j}^m, \dots, \xi_{n,j}^m) = \Xi_j^m$ of components X^m. In contrast to the linear case

$$\mathrm{E}F\left(\frac{1}{N}\sum_{j=1}^{N} \Xi_j^m\right) \neq F(\mathrm{E}\Xi_j^m),$$

for any function F with non-zero nonlinear part. However, provided existence of the second partial derivatives with $N \to \infty$ [4] we have

$$\mathrm{E}F\left(\frac{1}{N}\sum_{j=1}^{N} \Xi_j^m\right) = F(\mathrm{E}\Xi_j^m) + O\left(\frac{1}{N}\right). \tag{15.15}$$

Therefore, for large N the arising bias can be neglected, although it is possible to conduct more accurate analysis, which requires knowledge of the second partial derivatives in the solutions neighborhood.

Analysis of the second moments behavior is more complicated [4], but if with the growth of m, they do not increase exponentially (stochastic stability), the computation can be entrusted to k independent processors and then the results are averaged. If N is independent of m and of the number of processors, then the number of processors order should not exceed $O\left(\frac{1}{\sqrt{N}}\right)$.

Thus, we have pointed out a PS-variant of the Newton's method, though with some reservations.

It is theoretically interesting, as in the linear case, to estimate the smallest N which ensure the stochastic stability [4, 6, 10].

Thus, randomization and the use of the quasi-random (deterministic) sequences allow to build PS-algorithms for solving a wide range of problems in computational mathematics. Stochastic procedures (such as the method of stochastic approximation) can obviously be a source of PS-algorithms for solving equations and determination of the extreme points of functions [3]. It is interesting to study the question—how much "worse" are they in comparison with classical algorithms, and what are the features of the quasi-random approach in these procedures.

Many randomized extremum searching algorithms have PS-properties. This is especially true for global extremum searching algorithms for functions of a large number of variables. Among such algorithms are so called genetic algorithms, algorithms for simulated annealing etc.

When solving optimization problems one frequently decides to uses many processors. Each of them is able to use its own, different from the others method. During the data exchange it is necessary to decide about the nature of the obtained results—whether an approximation obtained by the given processor is an approximation of some extremum or an approximation of the global extremum, and then the calculations are continued and completed depending on the results. Thus, the organized algorithm is is a PS-algorithm.

The most theoretically reasonable algorithms of the global extremum search are algorithms of determination of the distribution mode. If $f(X)$ is bounded and nonnegative on the set D, that does not detract from the community, and reaches its maximum value on the set $Y \subset D$ [3], then it is easy to prove that

$$F(x) = \lim_{m\to\infty} \left(f^m(X) \Big/ \int f^m(X)\, dX \right) \tag{15.16}$$

is the distribution density concentrated on the set Y.

The problem of obtaining points of the set Y is solved by modeling the density $F_m(X) = f^m(X)/\int f^m(X)\, dX$ for sufficiently large m. The only known method that allows to do this without calculating the normalization constant is the Metropolis method [8]. Obtaining independent realizations of $F_m(X)$ can be assigned to different processors—parametric separability takes place. The described method is widely used in problems of discrete optimization and it is, apparently, the only reasonable method that can numerically solve problems which require to find many (perhaps infinitely many) equal extreme points.

Acknowledgements This research has been supported by the grant of RFBR No. 11-01-00769.

References

1. Bailey, D.H., Borwein, P.B., Plouffe, S.: On the rapid computation of various polylogarithmic constants. Math. Comput. **66**(218), 903–913 (1997) (in Russian)
2. Baudet, G.M.: Asynchronous iterative methods for multiprocessors. J. ACM **25**(2), 226–244 (1978)
3. Ermakov, S.M.: The Monte Carlo Method and Related Topics. 2nd edn, pp. 472. Nauka, Moscow (1975) (in Russian)

4. Ermakov, S.M.: Monte Carlo Methods in Computational Mathematics. Introductory Course, pp. 192. Nevsky Dialect, St. Petersburg; Binom, Moscow (2009) (in Russian)
5. Ermakov, S.M.: Parametrically separated algorithms. Vestnik St.Petersburg University **1**(4), 25–31 (2010) (in Russian)
6. Ermakov, S.M., Timofeev, K.A.: On a class of Monte Carlo methods of solving equations with quadratic nonlinearity. Vestnik St.Petersburg University **10**(3), 105–109 (2008) (in Russian)
7. Korobov, N.M.: Number-Theoretic Methods in Approximate Analysis, 2nd edn. rev., Ext. Moscow: MCCME, pp. 288 (2004) (in Russian)
8. Metropolis, N., Rosenbluth, A.W., Rosenbluth, M.N., Teller, A.H., Teller, E.: Equations of state calculations by fast computing machines. J. Chem. Phys. **21**, 1087–1091 (1953)
9. Nekrutkin, V.V.: Direct and conjugate Neumann Ulam scheme for solutions non-linear integral equations. J. Comput. Math. Math. Phys. **14**(6), 409–1415 (1974) (in Russian)
10. Vidyaeva, K.O., Ermakov, S.M.: Spectrum assessment of linear operators. Vestnik St. Petersburg University **1**(1), 11–17 (2012) (in Russian).
11. Voevodin, V.V., Kapitonov, A.P.: Methods of the Description and Classification of CS, pp. 79. Moscow State University Press, Moscow (1994) (in Russian)
12. Walker, A.J.: New fast method for generating discrete random numbers with arbitrary friguency distributions. Elektronic Lett. **10**, 127–128 (1974)

Chapter 16
Laws of Large Numbers for Random Variables with Arbitrarily Different and Finite Expectations Via Regression Method

Silvano Fiorin

16.1 Assumptions and Basic Concepts

Given an arbitrary sequence of real random variables $\{Y_j : j \geqslant 1\}$ satisfying the following assumptions

A1 the Y_j's are uniformly bounded, i.e. there exists $M > 0$ such that $|Y_j| \leqslant M$, $\forall j \geqslant 1$;
A2 the Y_j's are totally independent or pairwise uncorrelated;
A3 the Y_j's have probability distributions and finite expectations which are arbitrarily different.

The basic idea in order to construct a strong law of large numbers for the sequence $\{Y_j : j \geqslant 1\}$ is that of embedding each Y_j as a conditional random variable belonging to a random vector (X, Y) in such a way that we can write

$$Y_j = (Y|X = x_j) \qquad \forall j \geqslant 1 \tag{16.1}$$

where X denotes a random element taking values into some space $\mathscr{X}$ with σ-field $\mathscr{B}_{\mathscr{X}}$. Of course the equality $Y_j = (Y|X = x_j)$ implies that the probability distribution of Y_j is completely identified by the element x_j, moreover, if we consider a different probability distribution for each Y_j, a reasonable choice for x_j is that of denoting the probability distribution function (p.d.f. hereafter) of Y_j, i.e.

$$x_j(t) = P(Y_j \leqslant t) = F_{Y_j}(t) \qquad \forall t \in \mathrm{R}^1. \tag{16.2}$$

S. Fiorin (✉)
Department of Statistical Sciences, University of Padua, Via Cesare Battisti 241, 35121 Padova, Italy
e-mail: fiorin@stat.unipd.it

V.B. Melas et al. (eds.), *Topics in Statistical Simulation*, Springer Proceedings in Mathematics & Statistics 114, DOI 10.1007/978-1-4939-2104-1_16

Furthermore, if Assumption A3 holds, a very general context has to be chosen for the elements x_j's; in fact, for instance, it may happen that x_5 and x_{10} denote, respectively, any assigned discrete and continuous type probability distribution. The problem is that of labelling by $\{x_j : j \geqslant 1\}$ the distributions of the arbitrarily and countable set of r.v.'s $\{Y_j : j \geqslant 1\}$ and a natural solution consists in giving to each x_j the meaning of probability distribution function F_{Y_j}. Recalling the sequence of random variables $\{Y_j : j \geqslant 1\}$ as a preliminary element in our analysis, then choosing the class of functions $\{x_j : j \geqslant 1\}$ as values taken by a random element X, the random vector (X, Y) will be constructed via definition over the x_j's of a metric d which is directly derived from Skorohod distance and then setting $\mathscr{X}$ equal to the closure of the set $\{x_j : j \geqslant 1\}$ under the d topology

$$\mathscr{X} = \overline{\{x_j : j \geqslant 1\}}. \tag{16.3}$$

With $\mathscr{X}$ a separable metric space of p.d.f.'s was introduced where the respective Borel σ-field $\mathscr{B}_{\mathscr{X}}$ is defined. For each fixed $x \in \mathscr{X}$ let us denote the corresponding real random variable having x as its p.d.f. by $Y(x)$, then the family of random variables $\{Y(x) : x \in \mathscr{X}\}$, through the monotone class theorem, allows us to derive a function $P(x, B)$ defined for each $x \in \mathscr{X}$ and each $B \in \mathscr{B}^1$ (the usual Borel σ-field over R^1) and satisfying the below properties

P1 for each fixed $x \in \mathscr{X}$, $P(x, \cdot)$ is a probability measure over $\mathscr{B}^1$, i.e. $P(x, \cdot)$ denotes the measure defined by the p.d.f. x;

P2 for each fixed $B \in \mathscr{B}^1$, $P(\cdot, B)$ is a Borel measurable function with respect to $\mathscr{B}_{\mathscr{X}}$.

If a marginal probability measure $P_{\mathscr{X}}$ is defined over $B_{\mathscr{X}}$, the product measure theorem can be applied to $P(x, B)$ and $P_{\mathscr{X}}$ and the existence is proved for *product measure* $P_{\mathscr{X} \times \mathrm{R}^1}$ over the product σ-field satisfying the property

P3 $P_{\mathscr{X} \times \mathrm{R}^1}(F) = \int_{\mathscr{X}} P(x, F(x)) dP_{\mathscr{X}}(x)$ where $F(x) = \{y \in \mathrm{R}^1 : (x, y) \in F\}$, for any fixed $F \in \mathscr{B}_{\mathscr{X}} \times \mathscr{B}^1$.

16.2 The Limit Value for SLLN

The product space $\mathscr{X} \times \mathrm{R}^1$ with σ-field $\mathscr{B}_{\mathscr{X}} \times \mathscr{B}^1$ and product measure $P_{\mathscr{X} \times \mathrm{R}^1}$ is the suitable context where the limit value for the sequence $\frac{1}{n} \sum_{j=1}^{n} Y_j$ can be defined. Given the product space $\mathscr{X} \times \mathrm{R}^1$, a random vector (X, Y) is easily defined through the marginals

$$X(x, y) = x \text{ and } Y(x, y) = y, \forall (x, y) \in \mathscr{X} \times \mathrm{R}^1 \tag{16.4}$$

in such a way that with $Y|X = x$ a conditional random variable exists having x as its p.d.f. and

$$(Y|X = x_j) = Y_j, \quad \forall x_j = F_{Y_j}. \tag{16.5}$$

The function $P(x, B)$ in P3 can be now rewritten adopting a new notation:

$$P(Y \in B|X = x) = P(x, B), \quad \forall \text{ fixed } B \in \mathscr{B}^1 \text{ and } x \in \mathscr{X}, \tag{16.6}$$

and the main result consists in proving the convergence

$$\lim_{n\to\infty} \frac{1}{n} \sum_{j=1}^{n} Y_j = E(Y) \text{ almost surely} \tag{16.7}$$

with respect to the infinite product probability measure P having each P_j as a marginal, noting that P_j denotes the probability measure defined by F_{Y_j}.

The below proof is based on the version of $E(Y)$ defined via Fubini theorem

$$E(Y) = \int_{\mathscr{X}} \int_{\mathrm{R}^1} y dP(Y|X = x) dP_{\mathscr{X}}(x) = \int_{\mathscr{X}} E(Y|X = x) dP_{\mathscr{X}}(x) \tag{16.8}$$

thus it is much more important the convergence to the value $\int_{\mathscr{X}} E(Y|X = x) dP_{\mathscr{X}}(x)$, where we compute the integral of the regression function $E(Y|X = x)$ with respect to the marginal measure $P_{\mathscr{X}}$.

16.3 The Measure $P_{\mathscr{X}}$

Before dealing with the technicalities of the rigorous proofs, some comments are made on measure $P_{\mathscr{X}}$ under an intuitive point of view. The connection of the regression function $x \to E(Y|X = x)$ with the set of random variables $\{Y_j : j \geqslant 1\}$ is intuitively evident: in fact $\{E(Y|X = x_j) = E(Y_j), \forall j \geqslant 1\}$ is a countable and dense subset of the set of values taken by the regression function, whereas it may appear not to so clear why, in order to study the limit for $\frac{1}{n}\sum_{j=1}^{n} Y_j$, we need a probability measure $P_{\mathscr{X}}$ over $\mathscr{B}_{\mathscr{X}}$. We will show below that for any n fixed $\frac{1}{n}\sum_{j=1}^{n} Y_j$ can be written by means of *pseudo-empirical measure* (p.e.m. hereafter) ν_n, where the term *pseudo* for ν_n is due to its arguments $\{x_j : j = 1, 2, \ldots, n\}$ which are not observations but the p.d.f. of the assigned random variables $\{Y_j : j = 1, 2, \ldots, n\}$.

Let us define the p.e.m.

$$\nu_n(B) = \frac{1}{n} \sum_{j=1}^{n} I_B(x_j) \tag{16.9}$$

for an assigned $B \in \mathscr{B}_{\mathscr{X}}$ and where

$$I_B(x_j) = \begin{cases} 1 & \text{if} \quad x_j \in B \\ 0 & \text{if} \quad x_j \notin B \end{cases}. \tag{16.10}$$

Then $\nu_n(B)$ depends directly on the position of the x_j's inside the space $\mathscr{X}$ and on the permutations of the x_j's. It will be shown below that a crucial argument for SLLN is the asymptotic behaviour for the sequences $\{\nu_n(B)\}$.

16.4 The Space $\mathscr{X}$

Given the set $\{Y_j \ : \ j \ \geqslant \ 1\}$ of random variables, having each Y_j an arbitrary probability distribution giving mass 1 to the interval $[-M, M]$ because of Assumption A1 above, the Y_j's can be parametrized taking the corresponding p.d.f. $F_j(t) = P(Y_j \leqslant t), \forall j \geqslant 1$.

Definition 16.1. A probability distribution function is a real valued function F that is increasing, right continuous, with left hand limits over $(-\infty, +\infty)$ and satisfying $\lim_{t \to \infty} F(t) = 1$, $F(+\infty) = 1$, $F(-\infty) = 0$.

Our aim consists in defining over the space $\mathscr{F}$ of all p.d.f.'s a metric which is directly derived from the Skorohod distance over the space $D[0, 1]$ of real valued functions f over $[0, 1]$ which are right continuous and with left hand limits. Let us introduce the following strictly increasing function

$$\varphi_1(t) = \begin{cases} \frac{t}{1+t} & \text{if} \quad t \geqslant 0 \\ \frac{-|t|}{1+|t|} & \text{if} \quad t < 0 \\ 1 & \text{if} \quad t = +\infty \\ -1 & \text{if} \quad t = -\infty \end{cases}$$

mapping $\bar{\mathrm{R}}^1$ onto $[-1, 1]$ and then $\varphi_2(t) = \frac{\varphi_1(t)+1}{2}$ is a continuous strictly increasing function mapping R^1 onto $[0, 1]$. With the help of $\varphi_2(t)$ a transformation is defined for any assigned p.d.f. F over $\bar{\mathrm{R}}^1$ into a function defined over $[0, 1]$

$$\varphi(F)(s) = F(\varphi_2^{-1}(s)) \quad s \in [0, 1]. \tag{16.11}$$

The transformation defined by φ on F is a very simple one: the transition from F to $\varphi(F)$ is performed only taking the transformation of t into $\varphi_2(t) = s$ and then $\varphi(F)(s) = F(t)$. Several interesting properties can be directly checked: $\varphi(F)$ is increasing, right continuous with left hand limits and $\varphi(F)(0) = 0, \varphi(F)(1) = 1$

Definition 16.2. A metric d is defined over the class $\mathscr{F}$ of p.d.f.'s by the below equality $d(F_1, F_2) = d_s(\varphi(F_1), \varphi(F_2))$ where d_s is the Skorohod distance over $D[0, 1]$, F_1 and F_2 belong to $\mathscr{F}$ and φ is the map defined by (16.11).

The distance $d(F_1, F_2)$ of two assigned p.d.f.'s is computed via the corresponding functions $\varphi(F_1), \varphi(F_2) \in D[0,1]$ and thus a simple method in order to get the topology τ_d and the Borel σ-field $\mathscr{B}_d$ generated by the metric d over $\mathscr{F}$ is that of giving τ_{d_s} and the Borel σ-field $\mathscr{B}_{d_s}$ defined through the Skorohod distance d_s over $\varphi(\mathscr{F}) \subset D[0,1]$ and then $\tau_d = \varphi^{-1}(\tau_{d_s}), \mathscr{B}_d = \varphi^{-1}(\mathscr{B}_{d_s})$ where $\varphi : \mathscr{F} \to \varphi(\mathscr{F})$, defined by (16.11) is a bijection from $\mathscr{F}$ onto $\varphi(\mathscr{F})$. Analogously the definition of the space $\mathscr{X}$ on the base of the sequence of p.d.f.'s $\{x_j : j \geqslant 1\}$ is introduced via the corresponding set $\{\varphi(x_j) : j \geqslant 1\} \subset \varphi(\mathscr{F}) \subset D[0,1]$ and the closure with respect to the d_s metric

$$\mathscr{X} = \varphi^{-1}(\overline{\{\varphi(x_j) : j \geqslant 1\}}) \tag{16.12}$$

Moreover the Borel σ-field $\mathscr{B}_d$ defined through the metric d over $\mathscr{X}$ can be assigned as preimage by φ of the Borel σ-field $\mathscr{B}_{d_s}$ over $\varphi(\mathscr{X})$

$$\mathscr{B}_d = \varphi^{-1}(\mathscr{B}_{d_s})) \tag{16.13}$$

Going back to notations adopted above, notice that we take $\mathscr{B}_{\mathscr{X}} = \mathscr{B}_d$.

16.5 The Product Space $\mathscr{X} \times \mathbf{R}^1$

For any fixed p.d.f. $x \in \mathscr{X}$ let us assign a real random variable $Y(x)$ having x as its p.d.f.; the two sets $\mathscr{X}$ and $\{Y(x) : x \in \mathscr{X}\}$ can be thought as the elements of a regression scheme: $\mathscr{X}$ denotes the set of values taken by a regressor X and each $Y(x)$ could be a conditional random variable $(Y|X = x)$ for some random variable Y. Let us denote as $P(x, B)$ the function defined for each $x \in \mathscr{X}$ and $B \in \mathscr{B}^1$ such that $P(x, \cdot)$ is the probability measure generated by the p.d.f. x. Then our purpose consists in proving the following result.

Lemma 16.1. *For each fixed $B \in \mathscr{B}^1$ $P(\cdot, B)$ is a Borel measurable function with respect to the Borel σ-field $\mathscr{B}_{\mathscr{X}}$ over $\mathscr{X}$.*

Proof. The first step concerns the measurability with respect to $\mathscr{B}_{\mathscr{X}}$ for the function $x \to P(x, (-\infty, t]) = x(t)$, with fixed $t \in \mathrm{R}^1$, which may be written as

$$\pi_t(x) = x(t) \tag{16.14}$$

i.e. as the function which assigns the value $x(t)$ to any p.d.f. $x \in \mathscr{X}$. Given the subset $A \subseteq D[0,1]$ and any fixed $s \in [0,1]$, the measurability for the map $\pi_s(f) = f(s), \forall f \in A$, with respect to $\mathscr{B}_{d_s}(A)$, the Borel σ-field on A in the Skorohod topology, is proved by Billingsley [2] (see p. 121). Then, if $A = \varphi(\mathscr{X})$ with φ defined as in (16.11), the measurability for π_s holds true for any fixed $s \in [0,1]$, i.e. $\pi_s^{-1}(\mathscr{B}^1) \subset \mathscr{B}_{d_s}$. Recalling now (see (16.12) and (16.13) above) that over $\mathscr{X}$ we have $\mathscr{B}_{\mathscr{X}} = \mathscr{B}_d$ and $\mathscr{B}_{\mathscr{X}} = \varphi^{-1}(\mathscr{B}_{d_s})$ then

$\varphi^{-1}(\pi_s^{-1}(\mathscr{B}^1)) \subset \varphi^{-1}(\mathscr{B}_{d_s}) = \mathscr{B}_{\mathscr{X}}$. Moreover, applying (16.11), we have $\varphi^{-1}(\pi_s^{-1}(\mathscr{B}^1)) = \pi_t^{-1}(\mathscr{B}^1)$ with $t = \varphi_2^{-1}(s)$, proving the inclusion $\pi_t^{-1}(\mathscr{B}^1) \subset \mathscr{B}_{\mathscr{X}}$, which shows the measurability of π_t with respect to $\mathscr{B}_{\mathscr{X}}$. Thus the $\mathscr{B}_{\mathscr{X}}$-measurability is proved for the class of functions $P(\cdot, (-\infty, s])$ and $P(\cdot, [-\infty, s])$ for all fixed $s \in \mathrm{R}^1$. Consequently the $\mathscr{B}_{\mathscr{X}}$-measurability is obtained for any function $P(\cdot, (a, b]) = P(\cdot, (-\infty, b]) - P(\cdot, (-\infty, a])$ and also for each function $P(\cdot, B)$ with B belonging to the field $\mathscr{F}_0$ of all finite disjoint unions of intervals $(a, b]$ and $[-\infty, b]$. The monotone class theorem is now applied introducing the class $\mathscr{C} = \{B \in \mathscr{B}^1 : P(\cdot, B) \text{ is a } \mathscr{B}_{\mathscr{X}}\text{-measurable function}\}$. $\mathscr{C}$ is a monotone class including the field $\mathscr{F}_0$ and then the inclusion $\mathscr{B}^1 = \sigma(\mathscr{F}_0) \subset \mathscr{C}$ holds true where $\sigma(\mathscr{F}_0)$ denotes the σ-field generated by $\mathscr{F}_0$. And this proves Lemma 16.1.

If the marginal probability measure $P_{\mathscr{X}}$ is defined over $\mathscr{B}_{\mathscr{X}}$, all the assumptions are fulfilled for the product measure theorem and then there exists a unique probability measure $P_{\mathscr{X} \times \mathrm{R}^1}$ over the product σ-field $\mathscr{B}_{\mathscr{X}} \times \mathscr{B}^1$ such that

$$P_{\mathscr{X} \times \mathrm{R}^1}(k) = \int_{\mathscr{X}} P(x, k(x)) dP_{\mathscr{X}}(x), \quad \forall k \in \mathscr{B}_{\mathscr{X}} \times \mathscr{B}^1 \tag{16.15}$$

where $k(x) = \{y \in \mathrm{R}^1 : (x, y) \in k\}$ is the section of k at point x.

16.6 The Limit for SLLN

The product space $\mathscr{X} \times \mathrm{R}^1$ with product measure $P_{\mathscr{X} \times \mathrm{R}^1}$ is the setting in which to define the main tool of our analysis.

If for each point $(x, y) \in \mathscr{X} \times \mathrm{R}^1$ the two maps are defined

$$X(x, y) = x \text{ and } Y(x, y) = y \tag{16.16}$$

the random vector (X, Y) is given where $(Y|X = x_j) = Y_j, \forall j$.

Of course Y is a marginal random variable and its expectation $E(Y)$, which is finite under Assumption A1, plays a key role. In fact, it will be shown below that $E(Y)$ is the limit for our SLLN and the proof developed in the sequel, applying Fubini theorem as in (16.8), deals with convergence to the integral $\int_{\mathscr{X}} E(Y|X = x) dP_{\mathscr{X}}(x)$.

16.7 The Convergence Technique

A slight modification is needed avoiding some complication in proving the convergence of $\frac{1}{n}\sum_{j=1}^{n} Y_j$ to $\int_{\mathscr{X}} E(Y|X = x) dP_{\mathscr{X}}(x)$ and this because the function $\psi(x) = E(Y|X = x)$ has to be integrated over the infinite dimensional space $\mathscr{X}$; a preferable context can be found by means of the composition function

$$I \circ E(Y|X = x) = I(E(Y|X = x)) \tag{16.17}$$

where $I(v) = v$ is the identity map for all $v \in [-M, M]$. Observing that, because of A1 and Fubini theorem, $\psi(x) = E(Y|X = x)$ is a $\mathscr{B}_{\mathscr{X}}$-measurable map from $\mathscr{X}$ into $[-M, M] \subset \mathrm{R}^1$, the induced probability measure

$$P_E(B) = P_{\mathscr{X}}(\psi^{-1}(B)) \tag{16.18}$$

is defined over the Borel σ-field $\mathscr{B}[-M, M]$ on the interval $[-M, M]$. Thus, by integration theorem for the composition map $I \circ E(Y|X = x)$ the equality holds true

$$\int_{-M}^{M} I(v)dP_E(v) = \int_{\mathscr{X}} I(E(Y|X = x))dP_{\mathscr{X}}(x) = \int_{\mathscr{X}} E(Y|X = x)dP_{\mathscr{X}}(x)$$

and the convergence is studied for $\frac{1}{n}\sum_{j=1}^{n} Y_j$ to $\int_{-M}^{M} I(v)dP_E(v)$.

The basic idea is now shown for proving convergence of SLLN. Given $\epsilon_m > 0$, a partition of $[-M, M]$ into subintervals $\{H_{im} : im = 1, 2, \ldots, h_m\}$ can be chosen such that $H_{im} = [-M, a]$, if $im = 1$, with $a \in (-M, M)$ and $H_{im} = (c, d] \subset [-M, M], \forall im = 2, \ldots, h_m$ where each H_{im} is an interval with length at most equal to ϵ_m. For n fixed $\frac{1}{n}\sum_{j=1}^{n} Y_j$ involves the random variables $\{Y_j : j = 1, 2, \ldots, n\}$ which are partitioned into subsets: for each fixed interval H_{im} let us denote by

$$\{Y_{j_{im}} : j_{im} = 1, 2, \ldots, n(H_{im})\} \tag{16.19}$$

the set of random variables Y_j such that $E(Y_j) \in H_{im}$ and where

$$n(H_{im}) \tag{16.20}$$

is the cardinality of random variables Y_j with $j = 1, 2, \ldots, n$ and $E(Y_j) \in H_{im}$. The decomposition is then obtained

$$\sum_{j=1}^{n} Y_j = \sum_{im=1}^{h_m} \sum_{j_{im}=1}^{n(H_{im})} Y_{j_{im}} \tag{16.21}$$

and then for each fixed $im = 1, 2, \ldots, h_m$ we write

$$\frac{\sum_{j_{im}=1}^{n(H_{im})} Y_{j_{im}}}{n} = \frac{\sum_{j_{im}=1}^{n(H_{im})} Y_{j_{im}}/n(H_{im})}{n/n(H_{im})} = \frac{n(H_{im})}{n} \cdot \frac{\sum_{j_{im}=1}^{n(H_{im})} Y_{j_{im}}}{n(H_{im})} \tag{16.22}$$

and finally

$$\frac{1}{n}\sum_{j=1}^{n} Y_j = \sum_{im=1}^{h_m} \frac{n(H_{im})}{n} \cdot \frac{\sum_{j_{im}=1}^{n(H_{im})} Y_{j_{im}}}{n(H_{im})}. \tag{16.23}$$

The formula (16.23) states the decomposition of $\frac{1}{n}\sum_{j=1}^{n} Y_j$ into h_m terms of the type $\frac{n(H_{im})}{n} \cdot \frac{\sum_{j_{im}=1}^{n(H_{im})} Y_{j_{im}}}{n(H_{im})}$ where both $\frac{n(H_{im})}{n}$ and $\frac{\sum_{j_{im}=1}^{n(H_{im})} Y_{j_{im}}}{n(H_{im})}$ are arguments for asymptotic results, when n tends to infinity. Moreover the right member of (16.23) may be thought as the integral of the simple function taking values $\frac{\sum_{j_{im}=1}^{n(H_{im})} Y_{j_{im}}}{n(H_{im})}$ over the corresponding set H_{im} with respect to the probability $\frac{n(H_{im})}{n}$ and this for each $im = 1, 2, \ldots, h_m$; this idea suggests the strategy in order to prove the convergence of $\frac{1}{n}\sum_{j=1}^{n} Y_j$ to $\int_{-M}^{M} I(v)dP_E(v)$.

In fact, via convergence for sequences $\frac{n(H_{im})}{n}$ and $\frac{\sum_{j_{im}=1}^{n(H_{im})} Y_{j_{im}}}{n(H_{im})}$, $\forall im = 1, 2, \ldots, h_m$, when n tends to infinity and the partition of $[-M, M]$ into intervals $\{H_{im} : im = 1, 2, \ldots, h_m\}$ is assigned, it will be proved, in the sequel, that the sequence of integrals defined by the right member of (16.23) is convergent to the integral of a simple function which approximates $\int_{-M}^{M} I(v)dP_E(v)$, and this under suitable assumptions for the limiting behaviour of $\frac{n(H_{im})}{n}$, $\forall im = 1, 2, \ldots, h_m$.

Let us observe that the limiting behaviour is easy to study for sequences of type $\frac{\sum_{j_{im}=1}^{n(H_{im})} Y_{j_{im}}}{n(H_{im})}$ if $n(H_{im}) \to \infty$, i.e. when there exist infinitely many values $E(Y_j)$ belonging to the interval H_{im}: under the Assumptions A1 and A2 Theorem 5.1.2 on p. 108 of Chung book [5] can be applied and then the sequence $\frac{1}{n(H_{im})}\sum_{j_{im}=1}^{n(H_{im})}(Y_{j_{im}} - E(Y_{j_{im}}))$ is almost surely convergent to zero.

Thus, because of decomposition

$$\frac{\sum_{j_{im}=1}^{n(H_{im})} Y_{j_{im}}}{n(H_{im})} = \frac{\sum_{j_{im}=1}^{n(H_{im})}(Y_{j_{im}} - E(Y_{j_{im}}))}{n(H_{im})} + \frac{\sum_{j_{im}=1}^{n(H_{im})} E(Y_{j_{im}})}{n(H_{im})} \tag{16.24}$$

we have that $\frac{\sum_{j_{im}=1}^{n(H_{im})} Y_{j_{im}}}{n(H_{im})}$ is given by the sum of $\frac{\sum_{j_{im}=1}^{n(H_{im})} E(Y_{j_{im}})}{n(H_{im})} \in H_{im}$ (because of $E(Y_{j_{im}}) \in H_{im}$) plus a sequence converging to zero.

The behaviour of each sequence $\frac{n(H_{im})}{n}$ is then a central argument for convergence of (16.22). In the sequel two different types of behaviour for $\frac{n(H_{im})}{n}$ will be considered and then different results are discussed.

In the next section we assume that, for an assigned partition of $[-M, M]$ into intervals $\{H_{im} : im = 1, 2, \ldots, h_m\}$ the convergence holds to

$$\lim_{n\to\infty} \frac{n(H_{im})}{n} = \lim_{n\to\infty} \frac{\sum_{j=1}^{n} I_{H_{im}}(Y_j)}{n} = P_E(H_{im}) \tag{16.25}$$

where $I_{H_{im}}(Y_j)$ is 1 if $E(Y_j) \in H_{im}$ and zero otherwise. And P_E is an assigned probability measure over $\mathscr{B}[-M, M]$.

It is interesting to explain the intuitive reason suggesting the introduction of convergence (16.25); the sequence of real values $\{E(Y_j) : j \geqslant 1\}$ is supposed to satisfy a property which is very similar to the case of an i.i.d. sequence of observations from the probability distribution P_E. (See the well-known Glivenko–Cantelli theorem). Someone could object that, under (16.25), the deterministic values $\{E(Y_j) : j \geqslant 1\}$ are too close to an i.i.d. sequence of observations. The answer to this objection concerns the second type of behaviour for $\frac{n(H_{im})}{n}$ (see Sect. 16.9 below) where a more general assumption will be introduced. Nevertheless the first type behaviour, even if dealing with a simplified situation, is useful for the proof technique.

16.8 A Strong Law of Large Numbers

The limit (16.25) above for a class of sets H_{im} is the content of Assumption A4 which plays a key role in the below statement.

A4 Given the set of expectations $\{E(Y_j) : j \geqslant 1\}$, the existence is assumed of positive values $\{\epsilon_m : m \geqslant 1\}$ with $\epsilon_m \downarrow 0$ and such that, for each fixed ϵ_m, there exists a finite partition of $[-M, M]$ into subintervals $\{H_{im} : im = 1, 2, \ldots, h_m\}$, where each H_{im} has length not greater than ϵ_m and satisfies $\lim_{n\to\infty} \frac{n(H_{im})}{n} = P_E(H_{im})$, where P_E is an assigned probability measure over $\mathscr{B}[-M, M]$.

Theorem 16.1. *If the family of random variables $\{Y_j : j \geqslant 1\}$ satisfies Assumptions A1–A4, then the strong law of large numbers holds true, i.e. the sequence $\frac{1}{n}\sum_{j=1}^{n} Y_j$ is almost surely convergent to $\int_{-M}^{M} I(v)dP_E(v)$ when n tends to infinity and for an assigned probability measure P_E over $\mathscr{B}[-M, M]$.*

Proof. The proof is given in case of pairwise uncorrelated and uniformly bounded Y_j's (see A1 and A2); then there exists $M > 0$ such that $|Y_j| \leqslant M, \forall j \geqslant 1$ and each Y_j defines a probability measure over $\mathscr{B}[-M, M]$. Let us consider the infinite product space $\Omega = [-M, M]^\infty$ embedded with the product σ-field $\mathscr{F}^\infty = \mathscr{B}^\infty[-M, M]$ and product probability measure P and, because of A4, for each positive ϵ_m belonging to a sequence decreasing to zero there exists a finite partition of $[-M, M]$ into subintervals $\{H_{im} : im = 1, 2, \ldots, h_m\}$. For each fixed H_{im} let

$$\{Y_{j_{im}} : j_{im} \geqslant 1\} \tag{16.26}$$

denote the set of random variables Y_j such that $E(Y_j) \in H_{im}$. If the set (16.26) contains infinitely many elements, then Theorem 5.1.2 of p. 108 of Chung is applied to the sequence of random variables $\{(Y_{j_{im}} - E(Y_{j_{im}})) : j_{im} \geqslant 1\}$ and then we obtain that the usual SLLN holds true, i.e.

$$\frac{1}{n(H_{im})}\sum_{j_{im}=1}^{n(H_{im})}(Y_{j_{im}} - E(Y_{j_{im}})) \qquad \text{is convergent to } 0 \tag{16.27}$$

when $n(H_{im}) \to \infty$ and where $n(H_{im})$ (see (16.20)) is the number of random variables belonging to the set $\{Y_j : j = 1, \ldots, n\}$ and such that $E(Y_j) \in H_{im}$; this implies the existence of a set $C_{im} \in \mathscr{B}^{\infty}[-M, M]$ with $P(C_{im}) = 1$ such that any point $\omega \in C_{im}$ makes the sequence (16.27) convergent to zero.

Iterating the above procedure for all $m \geqslant 1$ and $im = 1, 2, \ldots, h_m$, a class of sets C_{im} with $P(C_{im}) = 1$ is given such that the intersection

$$C = \cap_{m \geqslant 1; im=1,2,\ldots,h_m} C_{im} \tag{16.28}$$

satisfies $P(C) = 1$ and for any assigned $\omega \in C$ each sequence (16.27) is convergent to zero. Given $\epsilon > 0$, it will be proved that for any fixed $\omega \in C$ there exists $n_0(\omega, \epsilon)$ such that $\forall n > n_0(\omega, \epsilon)$

$$\left| \frac{1}{n}\sum_{j=1}^{n} Y_j - \int_{-M}^{M} I(v)dP_E(v) \right| < \epsilon. \tag{16.29}$$

Under Assumption A4, a value ϵ_m can be selected with $\epsilon_m < \epsilon/2$ and the associated partition of $[-M, M]$ into intervals $\{H_{im} : im = 1, 2, \ldots, h_m\}$ contains only sets whose length is at most ϵ_m. Applying (16.23) the $\frac{1}{n}\sum_{j=1}^{n} Y_j$ can be written as

$$\frac{1}{n}\sum_{j=1}^{n} Y_j = \sum_{im=1}^{h_m} \frac{n(H_{im})}{n} \frac{\sum_{j_{im}=1}^{n(H_{im})} Y_{j_{im}}}{n(H_{im})}$$

and the convergence can be proved

$$\lim_{n\to\infty} \left| \frac{n(H_{im})}{n} \frac{\sum_{j_{im}=1}^{n(H_{im})} Y_{j_{im}}}{n(H_{im})} - P_E(H_{im}) \frac{\sum_{j_{im}=1}^{n(H_{im})} E(Y_{j_{im}})}{n(H_{im})} \right| = 0 \tag{16.30}$$

$\forall im = 1, 2, \ldots, h_m$, applying Assumption A4 to each sequence $\frac{n(H_{im})}{n}$ and the decomposition (16.24) jointly with Theorem 5.1.2 of Chung or Theorems 3.1.1 and 3.1.2 in Chandra [3] to the sequence $\frac{\sum_{j_{im}=1}^{n(H_{im})} Y_{j_{im})}}{n(H_{im})}$.

Thus we have:

$$\left| \frac{1}{n}\sum_{j=1}^{n} Y_j - \int_{-M}^{M} I(v)dP_E(v) \right| = \left| \sum_{im=1}^{h_m} \frac{n(H_{im})}{n} \frac{\sum_{j_{im}=1}^{n(H_{im})} Y_{j_{im}}}{n(H_{im})} - \sum_{im=1}^{h_m} \int_{H_{im}} I(v)dPE(v) \right|$$

$$\leqslant \sum_{im=1}^{h_m} \left| \frac{n(H_{im})}{n} \frac{\sum_{j_{im}=1}^{n(H_{im})} Y_{j_{im}}}{n(H_{im})} - P_E(H_{im}) \frac{\sum_{j_{im}=1}^{n(H_{im})} E(Y_{j_{im}})}{n(H_{im})} \right|$$

$$+ \sum_{im=1}^{h_m} \left| P_E(H_{im}) \frac{\sum_{j_{im}=1}^{n(H_{im})} E(Y_{j_{im}})}{n(H_{im})} - \int_{H_{im}} I(v) dP_E(v) \right|$$

Applying the limit in (16.30) the convergence holds true

$$\lim_{n\to\infty} \sum_{im=1}^{h_m} \left| \frac{n(H_{im})}{n} \frac{\sum_{j_{im}=1}^{n(H_{im})} Y_{j_{im}}}{n(H_{im})} - P_E(H_{im}) \frac{\sum_{j_{im}=1}^{n(H_{im})} E(Y_{j_{im}})}{n(H_{im})} \right| = 0.$$

For a fixed value im the below equality

$$P_E(H_{im}) \frac{\sum_{j_{im}=1}^{n(H_{im})} E(Y_{j_{im}})}{n(H_{im})} = \int_{H_{im}} \frac{\sum_{j_{im}=1}^{n(H_{im})} E(Y_{j_{im}})}{n(H_{im})} dP_E(v)$$

is easily proved if the left hand product is thought as the intergral of a constant function over the interval H_{im} with respect to measure $P_E(H_{im})$. Thus the inequality

$$\left| P_E(H_{im}) \frac{\sum_{j_{im}=1}^{n(H_{im})} E(Y_{j_{im}})}{n(H_{im})} - \int_{H_{im}} I(v) dP_E(v) \right|$$

$$\leqslant \int_{H_{im}} \left| \frac{\sum_{j_{im}=1}^{n(H_{im})} E(Y_{j_{im}})}{n(H_{im})} - I(v) \right| dP_E(v) \leqslant \epsilon_m P(H_{im})$$

can be shown, recalling that by (16.26) $\frac{\sum_{j_{im}=1}^{n(H_{im})} E(Y_{j_{im}})}{n(H_{im})} \in H_{im}$, and this implies that

$$\sum_{im=1}^{h_m} \left| P_E(H_{im}) \frac{\sum_{j_{im}=1}^{n(H_{im})} E(Y_{j_{im}})}{n(H_{im})} - \int_{H_{im}} I(v) dP_E(v) \right| \leqslant \sum_{im=1}^{h_m} \epsilon_m P_E(H_{im}) = \epsilon_m < \frac{\epsilon}{2}$$

and this completes the proof.

16.9 A Further Result

Theorem 16.1 states an SLLN on the base of the limits $\lim_{n\to\infty} \frac{n(H_{im})}{n} = P_E(H_{im})$. Recalling the equality (16.25) $\frac{n(H_{im})}{n} = \frac{1}{n} \sum_{j=1}^{n} I_{H_{im}}(Y_j)$ where $I_{H_{im}}(Y_j)$ is 1 if $E(Y_j) \in H_{im}$ and zero otherwise, the limits above imply for the deterministic values $\{E(Y_j) : j \geqslant 1\}$ a behaviour which is very close to the case of an i.i.d. sequence of observations from a probability distribution.

The purpose consists now in giving an SLLN when any sequence $\{\frac{n(H_{im})}{n} : n \geqslant 1\}$ is not necessarily convergent to $P_E(H_{im})$ where P_E is an assigned probability measure on $\mathscr{B}[-M, M]$. Given the partition $\{H_{im} : im = 1, 2, \ldots, h_m\}$ of $[-M, M]$ and the value $L = \int_{-M}^{M} I(v)dP_E(v)$, let us introduce the class

$$M_L = \Big\{ P : P \text{ is a probability measure over } \mathscr{B}[-M, M] \tag{16.31}$$

$$\text{satisfying} \int_{-M}^{M} I(v)dP(v) = \int_{-M}^{M} I(v)dP_E(v) = L \Big\}$$

In the general case M_L contains infinitely many probability measures and an interesting condition leading to an SLLN is here provided.

Using the simplified notation

$$\mu_n(H_{im}) = \frac{n(H_{im})}{n} \tag{16.32}$$

and given the quantity

$$S(\mu_n, M_L, m) = \inf_{P \in M_L} s(\mu_n, P, m) \tag{16.33}$$

where

$$s(\mu_n, P, m) = \max_{im=1,2,\ldots,h_m} |\mu_n(H_{im}) - P(H_{im})|, \tag{16.34}$$

let us assume that the following condition holds true

A5

$$\lim_{n \to \infty} S(\mu_n, P, m) = 0 \tag{16.35}$$

for each assigned partition $\{H_{im} : im = 1, 2, \ldots, h_m\}$ of $[-M, M]$ having at most length ϵ_m for each interval H_{im} and such that $\epsilon_m \downarrow 0$.

In order to explain the meaning of A5, let us suppose that the limit (16.35) holds true for the assigned partition; then, there exists a sequence of probability measure $P_n \in M_L$ such that

$$\lim_{n \to \infty} \max_{im=1,2,\ldots,h_m} |\mu_n(H_{im}) - P_n(H_{im})| = 0. \tag{16.36}$$

A direct comparison of (16.36) with (16.35) shows that, under A5, the sequence $\{\mu_n(H_{im})\}$ is not still a convergent one: in fact we have $|\mu_n(H_{im}) - P_n(H_{im})| \to 0$ where $\{P_n(H_{im})\}$ is not necessarily convergent, and this because $\{P_n : n \geqslant 1\}$

is only a sequence of probability measure belonging to M_L and then satisfying $\int_{-M}^{M} I(v)dP_n(v) = L, \quad \forall n.$

A second argument which may be helpful in understanding the above framework consists in giving an example of a class M_L including infinitely many probability measures. As a preliminary tool the notation of symmetric measure is introduced.

Definition 16.3. A positive measure μ defined over $\mathscr{B}[-M, M]$ is said to be symmetric if it satisfies the equality $\mu(A) = \mu(-A)$ for any interval $A \subseteq [0, M]$.

For instance, if f is a positive and Borel measurable function over $[-M, M]$ such that $f(x) = f(-x)$, then the integral $\mu(B) = \int_B f(v)dv$ defines a symmetric measure. If μ is a symmetric measure it descends that $\int_{-M}^{M} I(v)d\mu(v) = 0$; thus, if P_0 and μ are, respectively, a probability measure and a symmetric measure over $\mathscr{B}[-M, M]$, $\alpha > 1$ an assigned constant and $t_0 = \left(1 - \frac{P_0([-M,M])}{\alpha}\right) \cdot \frac{1}{\mu([-M,M])}$ we have that

$$\left(\frac{P_0}{\alpha} + t_0\mu\right) \tag{16.37}$$

is a probability measure over $\mathscr{B}[-M, M]$ and it satisfies

$$\int_{-M}^{M} I(v)d\left(\frac{P_0}{\alpha} + t_0\mu\right) = \int_{-M}^{M} I(v)d\frac{P_0}{\alpha} + t_0\int_{-M}^{M} I(v)d\mu = \int_{-M}^{M} I(v)d\frac{P_0}{\alpha}.$$

The above procedure shows that, given a probability measure P_0 over $\mathscr{B}[-M, M]$ and a constant $\alpha > 1$, for any symmetric measure μ there exists a corresponding probability measure $\left(\frac{P_0}{\alpha} + t_0\mu\right)$ satisfying $\int_{-M}^{M} I(v)d\left(\frac{P_0}{\alpha} + t_0\mu\right)$ $= \int_{-M}^{M} I(v)d\frac{P_0}{\alpha} = L.$

If M_L is the class of probability measure defined in (16.31) with $L = \int_{-M}^{M} I(v)d\frac{P_0}{\alpha}$, then M_L includes the family

$$\left\{\left(\frac{P_0}{\alpha} + t_0\mu\right) : \mu \text{ is a symmetric measure over } \mathscr{B}[-M, M]\right\}.$$

This implies that M_L contains infinitely many probability measures.

Theorem 16.2. *If the family of random variables $\{Y_j : j \geqslant 1\}$ satisfies Assumptions A1–A3, A5, then the strong law of large numbers holds true, i.e. the sequence $\frac{1}{n}\sum_{j=1}^{n} Y_j$ is almost surely convergent to the value $L = \int_{-M}^{M} I(v)d\frac{P_0}{\alpha}$ where P_0 and $\alpha > 1$ are, respectively, an assigned probability measure over $\mathscr{B}[-M, M]$ and a constant.*

The proof is omitted because of its close analogy to that of Theorem 16.1.

Conclusions

The procedure shown above is a general one: starting with the p.d.f.'s F_{Y_j}, the product space $\mathscr{X} \times R^1$ with the product measure $P_{\mathscr{X} \times R^1}$ allows us to derive an SLLN for the family of random variables $\{Y_j : j \geqslant 1\}$ having the value $L = \int_{-M}^{M} I(v) dP_E(v) = \int_{\mathscr{X}} E(Y|X = x) dP_{\mathscr{X}}(x)$ as its almost sure limit, where a leading role is that of the marginal measure $P_{\mathscr{X}}$ or P_E which is the transformed measure of $P_{\mathscr{X}}$ over $\mathscr{B}[-M, M]$ through the map $x \to \psi(x) = E(Y|X = x), \quad \forall x \in \mathscr{X}$.

The measure P_E is strictly connected to the *pseudo empirical measures* $\mu_u(H_{im}) = \frac{n(H_{im})}{n}$, where $n(H_{im})$ is the number of values $E(Y_j)$, with $j = 1, \ldots, n$, and such that $E(Y_j) \in H_{im}$ for each interval H_{im} of the assigned partition $\{H_{im} : im = 1, 2, \ldots, h_m\}$ of $[-M, M]$.

The central hypothesis in proving the SLLN is concerning the asymptotic behaviour for sequences $\{\mu_n(H_{im}) : n \geqslant 1\}$ for each fixed H_{im}. The first type assumption, given in A4, is intuitively simple but it may appear as a restrictive request: in fact it consists in the convergence

$$\lim_{n \to \infty} \mu_n(H_{im}) = P_E(H_{im}) \quad \forall H_{im}.$$

The second type assumption, given by A5, looks at the class M_L of all the probability measures P on $\mathscr{B}[-M, M]$ such that $\int_{-M}^{M} I(v) dP(v) = \int_{-M}^{M} I(v) dP_E(v) = L$ and the convergence above is replaced by the existence of a sequence $P_n \in M_L$ such that

$$\lim_{n \to \infty} \max_{im=1,2,\ldots,h_m} |\mu_n(H_{im}) - P_n(H_{im})| = 0.$$

Thus the sequences $\{\mu_n(H_{im}) : n \geqslant 1\}$ are not necessarily convergent and A5 seems to be not too severe. Furthermore notice that $\mu_n(H_{im})$ depends on the following important elements:

- the position of each value $E(Y_j)$ inside $[-M, M]$;
- the permutation of Y_j's inside the series $\sum_{j=1}^{\infty} Y_j$ and then of the corresponding values $E(Y_j)$'s. In fact let us observe that even if the sequence $\{E(Y_j) : j \geqslant 1\}$ is completely known, different permutations of Y_j's and of the respective expectations $E(Y_j)$'s may produce different values for $\mu_n(H_{im})$'s and then for the limit in the SLLN. Permutations of Y_j's in the SLLN is a well-known topic; see, for instance, Chobanyan et al. [4]. Nevertheless the context adopted in this paper is consistently different from the literature.

(continued)

Finally a comparison may be interesting for the result shown above with the analogous statements available in the literature. The following proposition, referred to as Theorem 16.3, can be found with the complete proof at p. 281 of Ash and Doleans [1].

Theorem 16.3. *Let $Y_1, Y_2, \ldots$ be independent, uniformly bounded random variables. Then $\sum_{j=1}^{\infty} Y_j$ converges a.e. iff $\sum_{j=1}^{\infty} Var(Y_j) < \infty$ and $\sum_{j=1}^{\infty} E(Y_j)$ converges.*

An SLLN can be derived by Theorem 16.3: the convergence of the series $\sum_{j=1}^{\infty} Y_j$ implies that $\frac{1}{n}\sum_{j=1}^{n} Y_j$ is convergent to zero. The method discussed in this paper is not dealing with the convergence of the series; moreover, the convergence is given to a general not necessarily null value L, and this even if the series $\sum_{j=1}^{\infty} Y_j$ is not convergent.

References

1. Ash, R.B., Doleans-Dade, C.A.: Probability and Measure Theory, 2nd edn. Hartcourt Academic, New York (2000)
2. Billingsely, P.: Probability and Measure. Wiley, New York (1968)
3. Chandra, T.K.: Laws of Large Numbers. Narosa publishing house, New Dehli (2012)
4. Chobanyan, S., Levental, S., Mandrekar, V.: Prokhorov blocks and strong law of large numbers under rearrangements. J. Theor. Probab. **17**, 647–672 (2004)
5. Chung, T.K.: A Course in Probability Theory, 3rd edn. Academic, San Diego (2001)

Chapter 17
D-optimal Saturated Designs: A Simulation Study

Roberto Fontana, Fabio Rapallo, and Maria Piera Rogantin

17.1 Introduction

The optimality of an experimental design depends on the statistical model that is assumed and is assessed with respect to a statistical criterion. Among the different criteria, in this chapter we focus on D-optimality.

Widely used statistical systems like `SAS` and `R` have procedures for finding an optimal design according to the user's specifications. `Proc Optex` of `SAS/QC` [5] searches for optimal experimental designs in the following way. The user specifies an efficiency criterion, a set of candidate design points, a model and the size of the design to be found, and the procedure generates a subset of the candidate set so that the terms in the model can be estimated as efficiently as possible.

There are several algorithms for searching for D-optimal designs. They have a common structure. Indeed, they start from an initial design, randomly generated or user specified, and move, in a finite number of steps, to a better design. All of the search algorithms are based on adding points to the growing design and deleting points from a design that is too big. Main references to optimal designs include [1,4,7–9,11].

R. Fontana (✉)
Department DISMA, Politecnico di Torino, Corso Duca degli Abruzzi 24, 10127 Torino, Italy
e-mail: roberto.fontana@polito.it

F. Rapallo
Department DISIT, Università del Piemonte Orientale, Viale Teresa Michel 11, 15121 Alessandria, Italy
e-mail: fabio.rapallo@unipmn.it

M.P. Rogantin
Department DIMA, Università di Genova, Via Dodecaneso 35, 16146 Genova, Italy
e-mail: rogantin@dima.unige.it

V.B. Melas et al. (eds.), *Topics in Statistical Simulation*, Springer Proceedings in Mathematics & Statistics 114, DOI 10.1007/978-1-4939-2104-1_17

In this work, we perform a simulation study to analyze a different approach for describing D-optimal designs in the case of saturated fractions. For saturated fractions, or saturated designs, the number of points is equal to the number of estimable parameters of the model. It follows that saturated designs are often used in place of standard designs, such as orthogonal fractional factorial designs, when the cost of each experimental run is high. We show how the geometric structure of a fraction is in relation with its D-optimality, using a recent result in [3] that allows us to identify saturated designs with the points with coordinates in $\{0, 1\}$ of a polytope, being the polytope described by a system of linear inequalities. The linear programming problem is based on a combinatorial object, namely the circuit basis of the model matrix. Since the circuits yield a geometric characterization of saturated fractions, we investigate here the connections between the classical D-optimality criterion and the position of the design points with respect to the circuits.

In this way the search for D-optimal designs can be stated as an optimization problem where the constraints are a system of linear inequalities. Within the classical framework the objective function to be maximized is the determinant of the information matrix. In our simulations, we define new objective functions, which take into account the geometric structure of the design points with respect to the circuits of the relevant design matrix. We study the behavior of such objective functions and we compare them with the classical D-efficiency criterion.

The chapter is organized as follows. In Sect. 17.2 we briefly describe the results of [3] and in particular how saturated designs can be identified with $\{0, 1\}$ points that satisfy a system of linear inequalities. Then in Sect. 17.3 we present the results of a simulation study in which, using some test cases, we experiment different objective functions and we analyze their relationship with the D-optimal criterion. Concluding remarks are made in Sect. 17.4.

17.2 Circuits and Saturated Designs

As described in [3], the key ingredient to characterize the saturated fractions of a factorial design is its circuit basis. We recall here only the basic notions about circuits in order to introduce our theory. For a survey on circuits and its connections with Statistics, the reader can refer to [6].

Given a model matrix X of a full factorial design $\mathscr{D}$, an integer vector f is in the kernel of X^t if and only if $X^t f = 0$. We denote by A the transpose of X. Moreover, we denote by $\mathrm{supp}(f)$ the support of the integer vector f, i.e., the set of indices j such that $f_j \neq 0$. Finally, the indicator vector of f is the binary vector $(f_j \neq 0)$, where $(\cdot)$ is the indicator function. An integer vector f is a circuit of A if and only if:

1. $f \in \ker(A)$;
2. there is no other integer vector $g \in \ker(A)$ such that $\mathrm{supp}(g) \subset \mathrm{supp}(f)$ and $\mathrm{supp}(g) \neq \mathrm{supp}(f)$.

The set of all circuits of A is denoted by $\mathscr{C}_A$, and is named as the circuit basis of A. It is known that $\mathscr{C}_A$ is always finite. The set $\mathscr{C}_A$ can be computed through specific software. In our examples, we have used `4ti2` [10].

Given a model matrix X on a full factorial design $\mathscr{D}$ with K design points and p degrees of freedom, we recall that a fraction $\mathscr{F} \subset \mathscr{D}$ with p design points is saturated if $\det(X_{\mathscr{F}}) \neq 0$, where $X_{\mathscr{F}}$ is the restriction of X to the design points in $\mathscr{F}$. With a slight abuse of notation, $\mathscr{F}$ denotes both a fraction and its support. Under these assumptions, the relations between saturated fractions and the circuit basis $\mathscr{C}_A = \{f_1, \ldots, f_L\}$ associated with A is illustrated in the theorem below, proved in [3].

Theorem 1. *$\mathscr{F}$ is a saturated fraction if and only if it does not contain any of the supports $\{\mathrm{supp}(f_1), \ldots, \mathrm{supp}(f_L)\}$ of the circuits of $A = X^t$.*

17.3 Simulation Study

The theory described in Sect. 17.1 allows us to identify saturated designs with the feasible solutions of an integer linear programming problem. Let $C_A = (c_{ij}, i = 1, \ldots, L, j = 1, \ldots, K)$ be the matrix, whose rows contain the values of the indicator functions of the circuits $f_1, \ldots, f_L$, $c_{ij} = (f_{ij} \neq 0), i = 1, \ldots, L$, $j = 1, \ldots, K$ and $Y = (y_1, \ldots, y_K)$ be the K-dimensional column vector that contains the unknown values of the indicator function of the points of $\mathscr{F}$. In our problem the vector Y must satisfy the following conditions:

1. the number of points in the fractions must be equal to p;
2. the support of the fraction must not contain any of the supports of the circuits.

In formulae, this fact translates into the following constraints:

$$1_K^t Y = p, \tag{17.1}$$

$$C_A Y < b, \tag{17.2}$$

$$y_i \in \{0, 1\} \tag{17.3}$$

where $b = (b_1, \ldots, b_L)$ is the column vector defined by $b_i = \#\mathrm{supp}(f_i), i = 1, \ldots, L$, and 1_K is the column vector of length K and whose entries are all equal to 1.

Since $D_Y = \det(V(Y)) = \det(X_{\mathscr{F}}^t X_{\mathscr{F}})$ is an objective function, it follows that a D-optimal design is the solution of the optimization problem

$$\text{maximize } \det(V(Y))$$
$$\text{subject to } (17.1), (17.2) \text{ and } (17.3).$$

In general the objective function to be maximized $\det(V(Y))$ has several local optima and the problem of finding the global optimum is part of current research, [2]. Instead of trying to solve this optimization problem in this work we prefer to study different objective functions that are simpler than the original one but that could generate the same optimal solutions. By analogy of Theorem 1, our new objective functions are defined using the circuits of the model matrix.

For any Y, we define the vector $b_Y = C_A Y$. This vector b_Y contains the number of points that are in the intersection between the fraction $\mathscr{F}$ identified by Y and the support of each circuit $f_i \in \mathscr{C}_A, i = 1, \ldots, L$. From (17.2) we know that each of these intersections must be strictly contained in the support of each circuit. For each circuit $f_i, i = 1, \ldots, L$ it seems natural to minimize the cardinality $(b_Y)_i$ of the intersection between its support $\text{supp}(f_i)$ and Y with respect to the size of its support, b_i. Therefore, we considered the following two objective functions:

- $g_1(Y) = \sum_{i=1}^{L}(b - b_Y)_i$;
- $g_2(Y) = \sum_{i=1}^{L}(b - b_Y)_i^2$.

From the examples analyzed in Sect. 17.3.1, we observe that the D-optimality is reached with fractions that contain part of the largest supports of the circuits, although this fact seems to disagree with Theorem 1. In fact, Theorem 1 states that fractions containing the support of a circuit are not saturated, and therefore one would expect that optimal fractions will have intersections as small as possible with the supports of the circuits. On the other hand, our experiments show that optimality is reached with fractions having intersections as large as possible with such supports. For this reason we consider also the following objective function:

- $g_3(Y) = \max(b_Y)$.

As a measure of D-optimality we use the D-efficiency, [5]. The D-efficiency of a fraction $\mathscr{F}$ with indicator vector Y is defined as

$$E_Y = \left(\frac{1}{\#\mathscr{F}} D_Y^{\frac{1}{\#\mathscr{F}}}\right) \times 100$$

where $\#\mathscr{F}$ is the number of points of $\mathscr{F}$ that is equal to p in our case, since we consider only saturated designs.

17.3.1 First Case: 2^4 with Main Effects and Two-Way Interactions

Let us consider the 2^4 design and the model with main factors and two-way interactions. The design matrix X of the full design has 16 rows and 11 columns, the number of estimable parameters. As the matrix X has rank 11, we search for fractions with 11 points. A direct computation shows that there are $\binom{16}{11} = 4{,}368$ fractions with 11 points: among them 3,008 are saturated, and the remaining 1,360 are not. Notice that equivalences up to permutations of factor or levels are not considered here.

Table 17.1 Frequency tables of $b - b_Y$ for the 2^4 design with main effects and two-way interactions

Table $(b - b_Y)$					E_Y			
1	2	3	4	5		68.29	77.46	83.38
5	15	50	60	10		192	0	0
5	18	48	55	14		1,040	0	0
5	21	46	50	18		960	0	0
5	24	44	45	22		480	0	0
5	27	42	40	26		0	320	0
5	30	40	35	30		0	0	16
					Total	2,672	320	16

Table 17.2 Classification of all saturated fractions for the 2^4 design with main effects and two-way interactions

$g_1(Y)$	$g_2(Y)$	$g_3(Y)$	E_Y	n
475	1,725	9	68.29	192
475	1,739	10	68.29	960
475	1,753	10	68.29	960
475	1,739	11	68.29	80
475	1,767	11	68.29	480
475	1,781	11	77.46	320
475	1,795	11	83.38	16
			Total	3,008

The circuits are 140 and the cardinalities of their supports are 8 in 20 cases, 10 in 40 cases, 12 in 80 cases. For more details refer to [3]. This example is small enough for a complete enumeration of all saturated fractions. Moreover, the structure of that fractions reduces to few cases, due to the symmetry of the problem.

For each saturated fraction $\mathscr{F}$ with indicator vector Y we compute the vector b_Y, whose components are the size of the intersection between the fraction and the support of all the circuits, $\mathscr{F} \cap \text{supp}(f_i), i = 1, \ldots, 140$, and we consider $b - b_Y$. Recall that b is the vector of the cardinalities of the circuits. The frequency table of $b - b_Y$ describes how many points need to be added to a fraction in order to complete each circuit. All the frequency tables are displayed in the left side of Table 17.1, while on the right side we report the corresponding values of D-efficiency.

For instance, consider one of the 192 fractions in the first row. Among the 140 circuits, 5 of them are completed by adding one point to the fraction, 15 of them by adding two points, and so on. We observe that there is a perfect dependence between the D-efficiency and the frequency table of $b - b_Y$.

However, analyzing the objective functions $g_1(Y)$, $g_2(Y)$, and $g_3(Y)$, we argue that the previous finding has no trivial explanation. The values of all our objective functions are displayed in Table 17.2.

From Table 17.2 we observe that both $g_2(Y)$ and $g_3(Y)$ are increasing as D-efficiency increases. Notice also that $g_1(Y)$ is constant over all the saturated fractions. This is a general fact for all no-m-way interaction models.

Proposition 17.1. *For a no-m-way interaction model, $g_1(Y)$ is constant over all saturated fractions.*

Proof. We recall that $C_A = (c_{ij}, i = 1, \ldots, L, j = 1, \ldots, K)$ is the $L \times K$ matrix, whose rows contain the values of the indicator functions of the supports of the circuits $f_1, \ldots, f_L, c_{ij} = (f_{ij} \neq 0), i = 1, \ldots, L, j = 1, \ldots, K$. We have

$$g_1(Y) = \sum_{i=1}^{L} (b - b_Y)_i = \sum_{i=1}^{L} (b)_i - \sum_{i=1}^{L} (b_Y)_i.$$

The first addendum does not depend on Y, and for the second one we get

$$\sum_{i=1}^{L} (b_Y)_i = \sum_{i=1}^{L} \sum_{j=1}^{K} c_{ij} Y_j = \sum_{j=1}^{K} Y_j \sum_{i=1}^{L} c_{ij}.$$

Now observe that a no-m-way interaction model does not change when permuting the factors or the levels of the factors. Therefore, by a symmetry argument, each design point must belong to the same number q of circuits, and thus $\sum_{i=1}^{L} c_{ij} = q$. It follows that

$$\sum_{i=1}^{L} (b_Y)_i = q \sum_{j=1}^{K} Y_j = pq.$$

□

In view of Proposition 17.1, in the remaining examples we will consider only the functions g_2 and g_3.

17.3.2 Second Case: $3 \times 3 \times 4$ with Main Effects and Two-Way Interactions

Let us consider the $3 \times 3 \times 4$ design and the model with main factors and two-way interactions. The model has $p = 24$ degrees of freedom. The number of circuits is 17,994. In this case the number of possible subsets of the full design is $\binom{36}{24} =$ 1,251,677,700. It would be computationally unfeasible to analyze all the fractions. We use the methodology described in [2] to obtain a sample of saturated D-optimal designs. It is worth noting that this methodology finds D-optimal designs and not simply saturated designs. This is particularly useful in our case because it allows us to study fractions for which the D-efficiency is very high. The sample contains 500 designs, 380 different.

The results are summarized in Table 17.3, where the fractions with minimum D-efficiency E_Y have been collapsed in a unique row in order to save space.

Table 17.3 Classification of 380 random saturated fractions for the 3 × 3 × 4 design with main effects and two-way interactions

$g_2(Y)$	$g_3(Y)$	E_Y	n
$\leqslant$ 963,008	$\leqslant$ 21	22.27	37
962,816	21	23.6	7
962,816	22	23.6	12
963,700	22	23.6	34
965,308	22	23.6	46
966,760	22	23.6	9
967,676	22	23.6	6
970,860	24	23.6	91
970,896	24	24.41	138
		Total	380

Table 17.4 Classification of 414 random saturated fractions for the 2^5 design with main effects

$g_2(Y)$	$g_3(Y)$	E_Y	n
11,360,866	6	76.31	31
11,342,586	6	83.99	9
11,371,834	6	83.99	126
11,375,490	5	83.99	54
11,375,490	6	90.48	194
		Total	414

We observe that for 138 different designs the maximum value of D-efficiency, $E_Y = 24.41$ is obtained for both $g_2(Y)$ and $g_3(Y)$ at their maximum values $g_2(Y) = 970{,}896$ and $g_3(Y) = 24$.

17.3.3 *Third Case: 2^5 with Main Effects*

Let us consider the 2^5 design and the model with main effects only. The model has $p = 6$ degrees of freedom. The number of circuits is 353,616. As in the previous case we use the methodology described in [2] to get a sample of 500 designs, 414 different.

The results are summarized in Table 17.4. We observe that for 194 different designs, the maximum value of D-efficiency, $E_Y = 90.48$ is obtained for both $g_2(Y)$ and $g_3(Y)$ at their maximum values $g_2(Y) = 11{,}375{,}490$ and $g_3(Y) = 6$.

17.4 Concluding Remarks

The examples discussed in the previous section show that the D-efficiency of the saturated fractions and the new objective functions based on combinatorial objects are strongly dependent. The three examples suggest to investigate such connection in a more general framework, in order to characterize saturated D-optimal fractions in terms of their geometric structure. Notice that our presentation is limited to saturated fractions, but it would be interesting to extend the analysis to other kinds of fractions. Moreover, we need to investigate the connections between the new objective functions and other criteria than D-efficiency.

Since the number of circuits dramatically increases with the dimensions of the factorial design, both theoretical tools and simulation will be essential for the study of large designs.

References

1. Atkinson, A.C., Donev, A.N., Tobias, R.D.: Optimum Experimental Designs, with SAS. Oxford University Press, New York (2007)
2. Fontana, R.: Random generation of optimal saturated designs. Preprint available at arXiv: 1303.6529 (2013)
3. Fontana, R,, Rapallo, F., Rogantin, M.P.: A characterization of saturated designs for factorial experiments. J. Stat. Plann. Inference **147**, 204–211 (2014). Doi: 10.1016/j.jspi.2013.10.011
4. Goos, P., Jones, B.: Optimal Design of Experiments: A Case Study Approach. Wiley, Chichester (2011)
5. SAS Institute Inc: SAS/QC® 13.2 User's Guide, Cary, NC; SAS Institute Inc. (2014)
6. Ohsugi, H.: A dictionary of Gröbner bases of toric ideals. In: Hibi, T. (ed.) Harmony of Gröbner Bases and the Modern Industrial Society, pp. 253–281. World Scientific, Hackensack (2012)
7. Pukelsheim, F.: Optimal design of experiments. Classics in Applied Mathematics, vol. 50. Society for Industrial and Applied Mathematics, Philadelphia (2006)
8. Rasch, D., Pilz, J., Verdooren, L., Gebhardt, A.: Optimal Experimental Design with R. CRC, Boca Raton (2011)
9. Shah, K.R., Sinha, B.K.: Theory of Optimal Designs. Lecture Notes in Statistics, vol. 54. Springer, Berlin (1989)
10. 4ti2 team: 4ti2—a software package for algebraic, geometric and combinatorial problems on linear spaces. www.4ti2.de (16 October 2014)
11. Wynn, H.P.: The sequential generation of D-optimum experimental designs. Ann. Math. Stat. **41**(5), 1655–1664 (1970)

to be significant that are larger than the smallest effect for which $P_s \leqslant p_0$, where p_0 is a critical value chosen to give the test procedure the desired Type I error rate. The way in which p_0 is determined is described in Sects. 2.2 and 2.3 of [9].

We consider two approaches. The first one, referred to as the LNmod-approach, is the standard Loughin Noble testing procedure in which we suppose that the experiments are run according to a saturated D-optimal design. Let us denote by $\mathscr{F}_0$ this design. The second method, referred to as the ND-approach, is a modification of the LNmod-approach, in which reference distributions are determined using the non-isomorphic D-optimal designs instead of permutations of the vector response. Let us denote by N the number of the non-isomorphic designs and by $\mathscr{F}_s, s = 1, \ldots, N$ the non-isomorphic D-optimal designs. For the ND-approach the Loughin Noble algorithm remains the same apart from the step 3 that is modified as follows.

For each non-isomorphic D-optimal designs, $\mathscr{F}_s, s = 1, \ldots, N$.

1. Compute $\hat{\boldsymbol{\beta}}_s^\star$ from $\mathbf{y}$ using $\mathscr{F}_s$.
2. Obtain $\left|\hat{\beta}_s^\star\right|_{(1)}$ from $\hat{\boldsymbol{\beta}}_s^\star$.
3. Compute

$$W_s^\star = \left(\frac{m-1}{m-s}\right)^{\frac{1}{2}} \left|\hat{\beta}_s^\star\right|_{(1)}.$$

18.4 A Comparative Simulation Study

We consider 7 factors, each with 2 levels. We make the hypothesis that the active effects belong to the set of all the main effects and interactions. It follows that the model has $1 + 7 + \binom{7}{2} = 29$ degrees of freedom.

Let us suppose that the experiments are run according to $\mathscr{F}_0$, a saturated D-optimal design. The D-optimal deign $\mathscr{F}_0$ has been generated using Proc Optex of SAS/QC [13] with the default setting.

We consider both the LNmod-approach, i.e. the Loughin–Noble approach with $\mathscr{F}_0$ and the ND-approach, i.e. the testing procedure based on nonisomorphic D-optimal designs. In this case the methodology described in [6] provides a set of 315 non-isomorphic D-optimal designs.

$$\underbrace{\qquad\qquad}_{k \text{ times}}$$

The design matrix is

$$\mathbf{X} = \left[x_1^{\alpha_1} \cdot \ldots \cdot x_k^{\alpha_k} : (x_1, \ldots, x_k) \in \mathscr{D}, \alpha_i \in \{0, 1\}, i = 1, \ldots, k\right]$$

Being $\mathbf{X}'\mathbf{X} = m\mathbf{I}$, where I is the identity matrix of dimension m, the vector of the estimates of the unknown parameters $\hat{\boldsymbol{\beta}}$ can be computed as

$$\hat{\boldsymbol{\beta}} = \frac{1}{m}\mathbf{X}'\mathbf{y}.$$

The first element of $\hat{\boldsymbol{\beta}}$ is the mean of the observed responses, the remaining elements are contrasts corresponding to the factorial effects (main effects and interactions). The general algorithm for the entire testing procedure can be written as follows.

1. Compute $\hat{\boldsymbol{\beta}}$ from $\mathbf{y}$ and order the effects $\hat{\beta}_1, \ldots, \hat{\beta}_{m-1}$ and the $\mathbf{X}$ columns $\mathbf{x}_1, \ldots, \mathbf{x}_{m-1}$ to correspond to the ordered absolute effects (OAE) $\left|\hat{\beta}\right|_{(1)} \geqslant \left|\hat{\beta}\right|_{(2)} \geqslant \cdots \geqslant \left|\hat{\beta}\right|_{(m-1)}$.
2. At step s set $\hat{W}_s = \left|\hat{\beta}\right|_s$ and obtain $\tilde{\mathbf{y}}_s = \mathbf{y} - \hat{\beta}_1\mathbf{x}_1 - \ldots \hat{\beta}_{s-1}\mathbf{x}_{s-1}$ with $\tilde{\mathbf{y}}_1 = \mathbf{y}$.
3. Select a large number B (e.g., $B = 5{,}000$) and repeat B times.
   (a) Obtain $\tilde{\mathbf{y}}_s^{\star}$ through a random permutation of $\tilde{\mathbf{y}}_s$.
   (b) Compute $\hat{\boldsymbol{\beta}}_s^{\star}$ from $\tilde{\mathbf{y}}_s^{\star}$.
   (c) Obtain $\left|\hat{\beta}_s^{\star}\right|_{(1)}$ from $\hat{\boldsymbol{\beta}}_s^{\star}$.
   (d) Compute

$$W_s^{\star} = \left(\frac{m-1}{m-s}\right)^{\frac{1}{2}} \left|\hat{\beta}_s^{\star}\right|_{(1)}.$$

4. Compute the observed significance level (OSL), P_s, for the test as

$$P_s = 1 - \left[\frac{\#W_s^{\star} < \hat{W}_s}{B}\right]^{(m-s)/(m-1)}$$

5. Repeat steps 2–4 for testing other OAEs.

designs. It is known that D-optimal designs do not require orthogonal design matrices and, as a result, parameter estimates may be correlated. At first we analyse the behaviour of the Loughin–Nobel algorithm when a non-orthogonal design is used. Then we also describe a new algorithm that generates reference distributions using a class of non-isomorphic D-optimal designs. This algorithm generalizes the results presented in [3,7] where the use of nonisomorphic orthogonal fractional factorial designs, including orthogonal arrays, for non-parametric testing has already been studied.

The paper is organized as follows. In Sect. 18.2 we briefly describe the procedure to build a class of D-optimal non-isomorphic designs. In Sect. 18.3 we synthesize the Loughin–Nobel test procedure and we describe the new algorithm. In Sect. 18.4 we present the results of a simulation study. Concluding remarks are in "Conclusion" section.

18.2 D-optimal Non-isomorphic Designs

Efficient algorithms for searching for optimal saturated designs are widely available (see, for example, Proc Optex of SAS/QC, [13]). Nevertheless, they do not guarantee a *global* optimal design. Indeed, they start from an initial random design and find a local optimal design. If the initial design is changed the optimum found will, in general, be different. In a recent work Fontana uses discovery probability methods to support the search for globally optimal designs. The basic idea is to search for optimal designs until the probability of finding a new one is less than a given threshold. The methodology has been implemented in a software tool written in SAS. We invite the interested reader to refer to [6].

When the methodology is applied to D-optimal designs, a set of non-isomorphic designs is generated. In this work we use such set to obtain reference distributions that are a key ingredient of non-parametric testing procedures.

18.3 Non-parametric Permutation Testing

We shortly outline the procedure proposed by Loughin and Noble for the analysis of unreplicated factorials. The interested reader should refer to [9] for a detailed description.

The well-known linear model corresponding to k factors, each with 2 level, is considered:

$$\mathbf{y} = \mathbf{X}\boldsymbol{\beta} + \boldsymbol{\varepsilon}$$

18.1 Introduction

Research and applications related to permutation tests have increased in the recent years. Several books have been dedicated to these methods [1, 5, 8, 10, 11]. A recent review and some new results on multivariate permutation testing are available in [12].

Unreplicated orthogonal factorial designs are often used in sciences and engineering, and they become particularly useful for highly expensive experiments, or when time limitations impose the choice of the minimum possible number of design points.

We report here some parts of the introduction of the paper [9] since it summarizes the common approaches for the analysis of unreplicated factorial designs.

> There are two common analysis approaches recommended in many experimentale design books, [4]. The first is to make a normal or half-normal probability plot of the estimated effects.... The interpretation of the resulting plot is entirely subjective, however.... The second approach is to identify, prior to the analysis, certain effects that are known or believed to have means of 0. The variability from the estimation of these effects is then pooled to form an estimate of the inherent process variability and this is used to test the significance of all the effects remaining in the model. This also involves some subjectivity in the nomination of effects for nonsignificance,.... Finally, all of the preceding methods assume normality of the error distribution, which is difficult to verify in this problem.

R. Fontana
Politecnico di Torino, Corso Duca degli Abruzzi 24, Torino, Italy
e-mail: roberto.fontana@polito.it

L. Salmaso (✉)
University of Padua, Strada San Nicola 3, Vicenza, Italy
e-mail: luigi.salmaso@unipd.it

V.B. Melas et al. (eds.), *Topics in Statistical Simulation*, Springer Proceedings in Mathematics & Statistics 114, DOI 10.1007/978-1-4939-2104-1_18

Twenty-seven different scenarios have been built according to different values of $\boldsymbol{\beta}$ and to different distributions for the error term $\boldsymbol{\epsilon}$. In more detail each scenario has been defined as follows:

- we set $\beta_0 = 0$ and then we considered the number a of active effects equal to 5, 12 and 24. For the sake of simplicity we made the hypothesis that all the active effects have the same size, denoted by c. The value of c has been set equal to 0.5, 1 and 1.5. For example, the case $a = 5$ and $c = 0.5$ corresponds to $\beta_0 = 0, \beta_1 = \cdots = \beta_5 = 0.5$ and $\beta_6 = \cdots = \beta_{28} = 0$.
- we considered four possible distributions for the error term: standard normal, standard Cauchy, exponential with mean equal to 1 and Student's t-distribution with 3 degrees of freedom.

For each scenario we run 1,000 simulations. Each simulation is based on a vector of responses $\mathbf{y}$ defined as $\mathbf{X}\boldsymbol{\beta} + \boldsymbol{\epsilon}$ where $\boldsymbol{\epsilon}$ has been generated using normally (Cauchy/exponentially/Student's t with 3 degrees of freedom) distributed random numbers.

We set the experimentwise error rate (EER), which is the probability of making a Type I error on at least one effect in the experiment, to 0.20. All testing procedures have been calibrated according to the method suggested in [9].

For each simulation $i = 1, \ldots, 1{,}000$, we run both the LNmod-approach and the ND-approach. We registered the number of active effects correctly detected, B_i, and the number of nonactive effects correctly ignored, A_i. We measured the performance of the algorithms using the ratio R between the total number of active effects correctly detected in all the simulations $\sum_{i=1}^{1000} B_i$ and the total number of active effects for that scenario in all the simulations, $1{,}000 * a$, i.e.:

$$R = \frac{\sum_{i=1}^{1000} B_i}{1000 * a}.$$

We considered an analogous ratio S for the nonactive effects, i.e. $S = \frac{\sum_{i=1}^{1000} A_i}{1000 * (28 - a)}$.

Tables 18.1, 18.2, 18.3 and 18.4 summarize the results of the simulations. We observe that both the approaches perform quite well when the size of the effects is at least one, the number of active effects is relatively small and the error terms are not Cauchy distributed. For example, from Table 18.1, we observe that when we have $a = 5$ active effects with size $c = 1$ and the errors are normally distributed both procedures obtain a value of R equal to 0.94 that means that 94 % of active effects have been correctly detected. We further investigated the case in which error terms are exponentially distributed considering also $a = 8$ and $a = 16$ active effects with size $c = 1$. The results are presented in Fig. 18.1. In the case of exponential distributions the ND-approach is very effective. Hence, in general it could be a useful procedure to take into account when a suitable catalogue of inequivalent matrices is available.

Table 18.1 Error distribution: standardized normal

Number of active effects a	Size of active effects c	R for LNmod	R for ND
5	0.5	0.27	0.26
5	1	0.94	0.94
5	1.5	1	1
12	0.5	0.09	0.09
12	1	0.71	0.69
12	1.5	1	0.99
24	0.5	0.02	0.02
24	1	0.01	0.01
24	1.5	0	0

Table 18.2 Error distribution: standard Cauchy

Number of active effects a	Size of active effects c	R for LNmod	R for ND
5	0.5	0.04	0.04
5	1	0.09	0.08
5	1.5	0.14	0.14
12	0.5	0.04	0.03
12	1	0.05	0.04
12	1.5	0.07	0.06
24	0.5	0.03	0.03
24	1	0.03	0.03
24	1.5	0.03	0.03

Table 18.3 Error distribution: exponential with mean equal to 1

Number of active effects a	Size of active effects c	R for LNmod	R for ND
5	0.5	0.39	0.44
5	1	0.91	0.94
5	1.5	0.99	1
12	0.5	0.11	0.14
12	1	0.6	0.69
12	1.5	0.94	0.97
24	0.5	0.01	0.02
24	1	0	0.01
24	1.5	0	0

Table 18.4 Error distribution: 3df Student's T

Number of active effects a	Size of active effects c	R for LNmod	R for ND
5	0.5	0.13	0.13
5	1	0.51	0.51
5	1.5	0.86	0.87
12	0.5	0.06	0.05
12	1	0.22	0.2
12	1.5	0.61	0.6
24	0.5	0.03	0.02
24	1	0.01	0.01
24	1.5	0.01	0.01

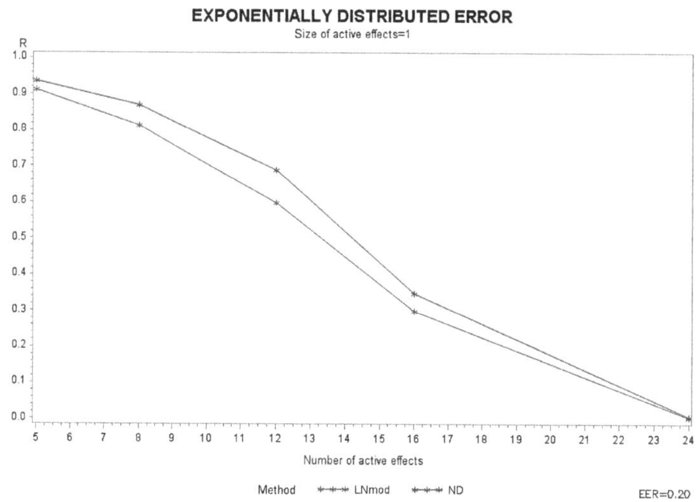

Fig. 18.1 $R = \frac{\sum_{i=1}^{1000} B_i}{1000*a}$ vs the number of active effects a

Conclusion

We introduced a new permutation test as a modification of the well-known Loughin and Noble test by also taking into account an innovative permutation mechanism based on non-isomorphic designs which can be often available for a given model and a given number of runs by means of the algorithm proposed by Fontana [6].

As it is pointed out in [2] it is quite difficult to construct powerful permutation tests for unreplicated factorial designs. In particular the original Loughin and Noble test loses power when the number of active effects and/or the error distribution is heavy tailed. Our modified procedures seem to have a general better behaviour in terms of power especially in presence of heavy tailed error distributions.

References

1. Basso, D., Pesarin, F., Salmaso, L., Solari, A.: Permutation Tests for Stochastic Ordering and ANOVA. Springer, New York (2009)
2. Basso, D., Salmaso, L.: A discussion of permutation tests conditional to observed responses in unreplicated 2 m full factorial designs. Commun. Stat. **35**(1), 83–97 (2006)
3. Basso, D., Salmaso, L., Evangelaras, H., Koukouvinos, C.: Nonparametric testing for main effects on inequivalent designs. In: mODa 7—Advances in Model-Oriented Design and Analysis, pp. 33–40. Springer, New York (2004)
4. Box, G.E., Hunter, W.G., Hunter, J.S.: Statistics for Experimenters: An Introduction to Design, Data Analysis, and Model Building. Wiley, New York (1978)
5. Edgington, E.S., Onghena, P.: Randomization tests, vol. 191. CRC, New York (2007)
6. Fontana, R.: Random generation of optimal saturated designs. ArXiv e-prints (2013)
7. Giancristofaro, R.A., Fontana, R., Ragazzi, S.: Construction and nonparametric testing of orthogonal arrays through algebraic strata and inequivalent permutation matrices. Commun. Stat. **41**(16–17), 3162–3178 (2012)
8. Good, P.: Permutation, Parametric and Bootstrap Tests of Hypotheses. Springer, New York (2005)
9. Loughin, T.M., Noble, W.: A permutation test for effects in an unreplicated factorial design. Technometrics **39**(2), 180–190 (1997)
10. Paul Jr, W., Berry, K.J.: Permutation Methods: A Distance Function Approach. Springer, New York (2007)
11. Pesarin, F., Salmaso, L.: Permutation Tests for Complex Data: Theory, Applications and Software. Wiley, Chichester (2010)
12. Pesarin, F., Salmaso, L.: A review and some new results on permutation testing for multivariate problems. Stat. Comput. **22**(2), 639–646 (2012)
13. SAS Institute, Inc.: SAS/QC 9.2 User's Guide, 2nd edn. SAS Institute, Cary (2010)

Chapter 19
Timely Indices for Residential Construction Sector

Attilio Gardini and Enrico Foscolo

19.1 Introduction

When we deal with the study of economic variables the first problem is related to the empirical evidence. We sometimes have little direct and lagged evidence and no wide standard databases to observe our phenomena are available. The problem is connected with the considerable time for collecting, processing, and releasing data. A possible solution is represented by moving to Internet data.

In the last decades the Internet has assumed a key role in representing fashions and consumer trends. One reason for this claim lies in the possibility of managing (mostly free) up-to-the-minute data. The World Wide Web therefore becomes the preferred channel for who wants to understand market dynamics. Thus, exploiting the World Wide Web we can mimic official statistics and provide new more timely indices. The most important contribution by Internet data consists of improving future predictions and allowing better understanding for current unobserved dynamics. Although the Internet provides an answer to this thirst for knowledge, its amount of data may be misleading. One requires management tools and filters can indeed help to separate the signal from noise. A simple and easy solution to represent trends comes from the most popular and used search engine: Google. This mechanism enables to forecast economic trends by examining the repetitive sequences that occur in search engine-based queries. We already recognize some contributors that

---

A. Gardini
Department of Statistical Sciences, University of Bologna, via delle Belle Arti 41, 40126 Bologna, Italy
e-mail: attilio.gardini@unibo.it

E. Foscolo (✉)
Faculty of Economics and Management, Free Univeristy of Bozen-Bolzano, Univeristätsplatz 1-Piazza Universitá 1, 39100 Bozen-Bolzano, Italy
e-mail: enrico.foscolo@unibz.it

V.B. Melas et al. (eds.), *Topics in Statistical Simulation*, Springer Proceedings in Mathematics & Statistics 114, DOI 10.1007/978-1-4939-2104-1_19

exploit Google data as exogenous variable. Choi and Varian [5] first explored the possibility of adding the search indices to a simple autoregressive model in order to improve the forecasting of new home sales. Askitas and Zimmermann [1], D'Amauri and Marcucci [6] tested the relevance of a Google job-search index as an indicator for unemployment dynamics in the USA, finding that it fruitfully increases the precision of the forecasts. Vosen and Schmidt [11, 12] introduced monthly consumer indicators based on Google search activity data, providing significant benefits to forecasts compared to common survey-based counterparts. Ginsberg et al. [8] monitored health-seeking behavior in the form of on-line Google search queries.

In this paper we give a general framework in order to exploit search engine based data. Our aim does not consist of replacing official statistics, but only to give proxies for better understanding current unobserved dynamics. In order to do so, we extract latent factors from query data and we provide a dynamic specification based on a cointegrated Vector Error Correction Model (hereafter, VECM; see [10]) for assessing the linkage with the reference series. As an illustration, we provide Internet indicators for the Italian Construction Production index. Lags from three to 6 months are common for quarterly indices and many of these indicators are subject to serious and time-consuming revisions. Nevertheless, since the construction sector has been played a central role in the Italian economy since 1999, having grown more than twice as fast as GDP until the 2008 financial crises, updated information are important for policy makers, firms, and investors.

The strength of the approach consists of providing new more timely indices for target economic time series by means of Google search engine query data. In this work, however, Google time series do not play the role of exogenous variables. With respect to the other cited approach, for the first time the exogeneity of extracted factors dealing with the new indices is not assumed a priori: Google indicators and official statistics are considered as endogenous variables.

The paper is organized as follows. Section 19.2 is devoted to the presentation of Google data and methods in order to obtain the search engine based indicators. To evaluate the performance of Internet indices in connection with the Construction Production index we estimate a cointegrated VECM in Sect. 19.3. Finally, some concluding remarks are outlined in Sect. 19.4.

19.2 Managing Query Data

Google Insights for Search[1] is the system provided by Google in order to analyze portions of worldwide Google web searches from all Google domains starting from January 2004. This mechanism computes how many searches have been sent for the entered keywords, relative to the total number of searches processed by Google

---

[1] http://www.google.com/insights/search/?hl=en-US.

Table 19.1 The selected Google categories for the Italian residential construction production

Google categories	Google subcategories
Real Estate	Property Inspections & Appraisals
	Property Management
	Property Development
	Real Estate Agencies
	Real Estate Listings
	Timeshares & Vacation Properties
	Apartments & Residential Rentals
	Commercial & Investment Real Estate
Construction & Maintenance	Building Materials & Supplies
	Civil Engineering
	Construction Consulting & Contracting
	Urban & Regional Planning

over time. Insights for Search generates normalized not seasonally adjusted weekly indexed series with values between 0 and 100. Updating is provided once a day. Moreover, it eliminates repeated queries from a single user over a short period of time. To determine the context of the terms, some Categories are provided. Categories refer to a classification of industries or markets provided by an automated classification engine.[2] When filters by Category are applied, Google system only evaluates queries that are related to that category. For our aims, the system provides several options, such as Real Estate Agencies or Property Development (cf. Table 19.1). In this sense, filters may be an additional guarantee for reducing the noise generated by searches not connected with the residential construction sector. Moreover, specific keywords are included in our queries for detecting the construction production dynamics.

In order to investigate the structure of Google time series, we provide the following procedure. Let q_t^j be the j-th query evaluated at time t obtained by fitting together the j-th group of keywords and selected categories. Extract common unobserved factors from $\left\{q_t^j : j = 1, \ldots, J, t = 1, \ldots, T\right\}$ as in [11, 12]. To identify residential construction production factors we exploit asymptotically distribution-free estimation methods in order to overwhelm some distributional assumptions on latent variables; see [2, 3]. Then, we select the number of factors by means of the parallel analysis (see [9]); i.e., we compare the decision about the number of factors to that of random data with the same properties as the real dataset.

For our purpose, parallel analysis suggests that 3 factors for construction sector might be most appropriate; see Fig. 19.1. Nevertheless, we only choose the latent

[2]See http://support.google.com/insights/?hl=enforacomprehensivedescription.

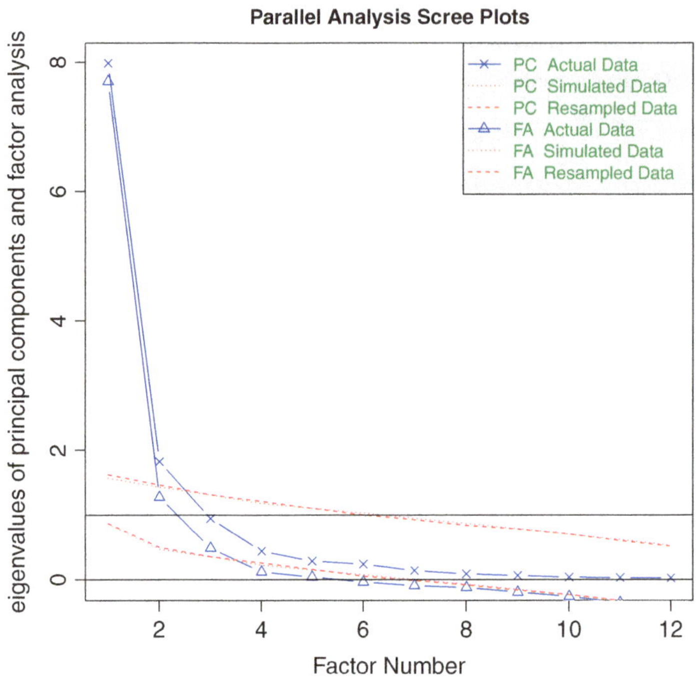

Fig. 19.1 Number of factors for the Italian residential construction production

variable with higher factor loadings in correspondence with Real Estate Agencies and Real Estate Listings subcategories, since they are, in our opinion, the best matches with the construction production.

19.3 From Query Data to Econometric Framework

The underlying nature of the data needs to be carefully analyzed before starting to handle the indices. Checking the connection between Google factors and official related data is the necessary step in order to consider these indices as housing market dynamics proxies. In order to do that, we exploit the vector autoregressive specification (hereafter, VAR) and we test the presence of stochastic common trends. Cointegrated variables are driven by the same persistent shocks. Thus, when cointegration is detected, the involved variables will show a tendency to co-move over time. Such cointegrated relations can often be interpreted as long-run economic relations and are therefore of considerable economic interest for our purpose. In this connection we use the VECM which gives a convenient reformulation of VARs in terms of differences, lagged differences, and levels of variables.

Table 19.2 The asterisks indicate the best values of the respective information criteria; i.e., Akaike criterion (shortly, AIC), Schwarz Bayesian criterion (shortly, BIC), and Hannan–Quinn criterion (shortly, HQC)

Lags	Log-likelihood	AIC	BIC	HQC
1	−327.4027	7.6431	8.4056	7.9509
2	−320.3550	7.5775	8.4490	7.9294
3	−305.0129	7.3336*	8.3140*	7.7295*
4	−304.4843	7.4083	8.4976	7.8481
5	−302.3366	7.4481	8.6463	7.9319
6	−297.8367	7.4373	8.7445	7.9651

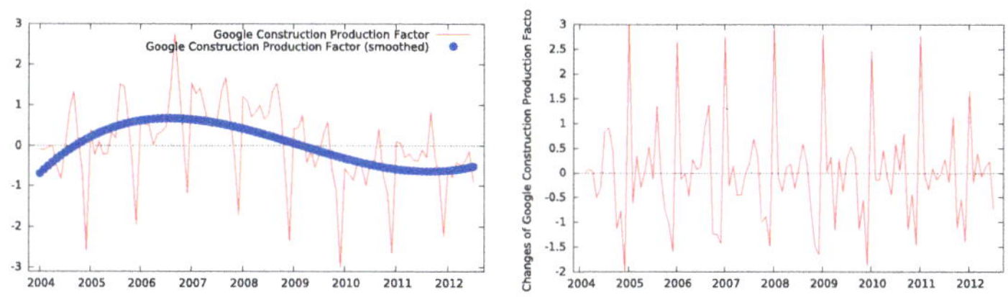

Fig. 19.2 Plot of the Google Construction factor and its first differences (the *right panel*)

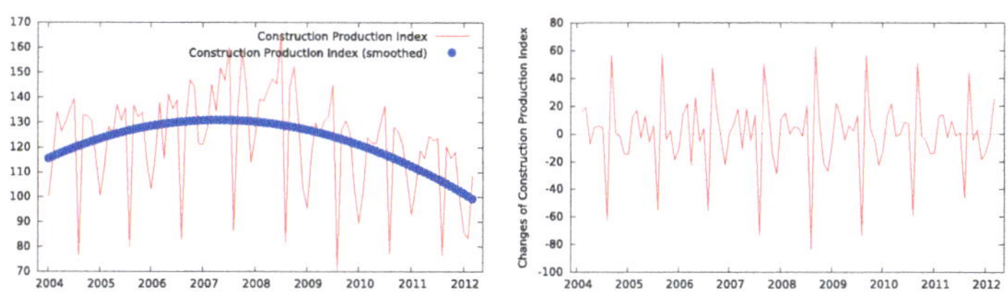

Fig. 19.3 Plot of the construction production index provided by Eurostat and its first differences (the *right panel*)

The proposed application refers to monthly Google Italian activities over the period 2004:04–2012:07 ($T = 103$), hereafter denoted by S_t, and monthly Construction Production index (not seasonally adjusted; shortly, cpi_t) provided by Eurostat, the Statistical Office of the European Union.

For notational reasons we include these $p = 2$ variables in the vector $X_t = (S_t, \mathrm{cpi}_t)^{\top}$. We fit the unrestricted VAR with $k = 3$ lags (see Table 19.2 for the validity of the chosen lag specification) by maximum likelihood, here in the corresponding VEC formulation,

$$\Delta X_t = \Pi X_{t-1} + \sum_{i=1}^{k-1} \Gamma_i \Delta X_{t-1} + \mu + \epsilon_t \tag{19.1}$$

where ϵ_t is a Gaussian white noise process with covariance matrix Ω.

Inspection of the data in Figs. 19.2 and 19.3 shows that involved variables are clearly nonstationary, while the differences, however, look like stationary processes. In order to take account different seasonality we also include in matrix μ centered seasonal dummies and an unrestricted drift term which creates a linear trend in the

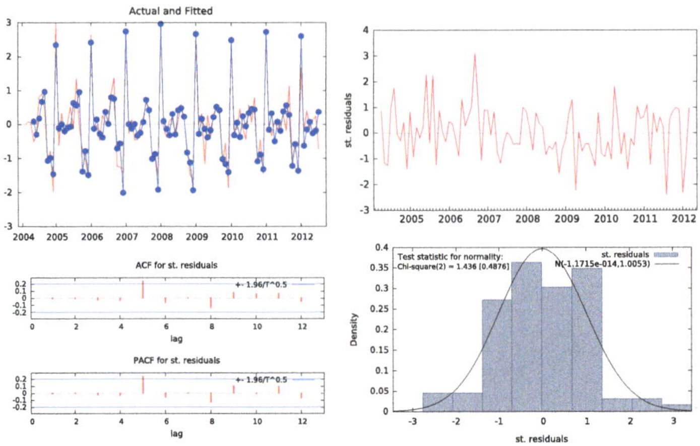

Fig. 19.4 Plot of actual and fitted values of ΔS_t, the standardized residuals of ΔS_t, the standardized residuals autocorrelation function of ΔS_t, and the histogram of standardized residuals of ΔS_t (with normality assumption check)

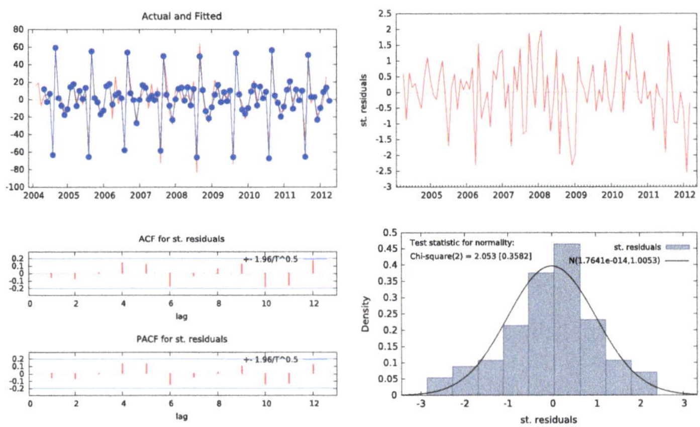

Fig. 19.5 Plot of actual and fitted values of Δcpi_t, the standardized residuals of Δcpi_t, the standardized residuals autocorrelation function of Δcpi_t, and the histogram of standardized residuals of Δcpi_t (with normality assumption check)

processes. The choice of linearity in trend is of course just a simplification of the reality. The adequacy of the model is checked by residual analysis in Figs. 19.4 and 19.5, and Table 19.3 where it can be seen that there is no autocorrelation in the residuals and the normal distribution assumption is not rejected.

The primary hypothesis of interest is to test the presence of unit roots (i.e., stochastic trends) which leads to a reduced rank condition on the impact matrix $\Gamma = \alpha\beta^\top$ in Eq. (19.1), where α and β are $p \times r$ matrices, and r $(0 < r < p)$ is the cointegration rank of the system. If accepted, it would establish cointegration.

Table 19.3 Residual diagnostic tests (p-values in the brackets)

Multivariate tests		
Normality	$\chi_4^2 = 2.9622$	(0.5642)
Residual autocorrelation (lag order = 16)	$\chi_{54}^2 = 64.9297$	(0.1466)
ARCH(12)	$\chi_{45}^2 = 53.5860$	(0.1781)
Univariate tests		
	$\Delta\, S_t$	$\Delta\, \text{cpi}_t$
Normality	$\chi_2^2 = 0.5890$	$\chi_2^2 = 1.8660$
	(0.7449)	(0.3934)
Residual autocorrelation (lag order = 16)	$\chi_{16}^2 = 13.2187$	$\chi_{16}^2 = 26.0279$
	(0.6570)	(0.0536)
ARCH(16)	$\chi_{16}^2 = 10.4250$	$\chi_{16}^2 = 18.7309$
	(0.8435)	(0.2829)

Table 19.4 Rank determination of $\Gamma = \alpha\beta^\top$ in case of unrestricted constant and periodic dummies (p-values in the brackets)

Rank	Eigenvalue	Trace test	$\lambda_{\max}$ test	Trace test corrected for sample ($df = 78$)
0	0.1107	11.5560	11.2640	11.5560
		(0.1818)	(0.1429)	(0.1910)
1	0.0030	0.2926	0.2926	0.2926
		(0.5886)	(0.5886)	(0.5952)

Estimation period: 2004:04–2012:03 ($T = 96$)

We involve testing for the cointegration rank r according to the [10] approach. The rank of Γ is investigated by computing the eigenvalues of a closely related matrix whose rank is the same as Γ. Two Johansen tests for cointegration are used to quantify r; i.e., the $\lambda_{\max}$ test (for hypotheses on individual eigenvalues) and the Trace test (for joint hypotheses). In Table 19.4 we present the results including sample size corrected Trace test statistic; see [7]. The findings are that both $r = 0$ and $r = 1$ should be not rejected for all tests. Nevertheless, the graphical inspections and the reciprocal roots of the unrestricted VAR(3) given in Fig. 19.6 suggest to accept the null hypothesis of one cointegration relation. Moreover, the right panel in Fig. 19.6 suggests that the two indices move together around the identify line for the whole period. Thus, the analysis indicates that our variables are nonstationary but cointegrate. The most obvious choice becomes $r = 1$.

The error-correction term is given by

$$ect_{1t} = -cpi_t + \underset{(3.4686)}{19.5781}\, S_t \tag{19.2}$$

and the remaining parameter estimates of model in Eq. (19.1) are

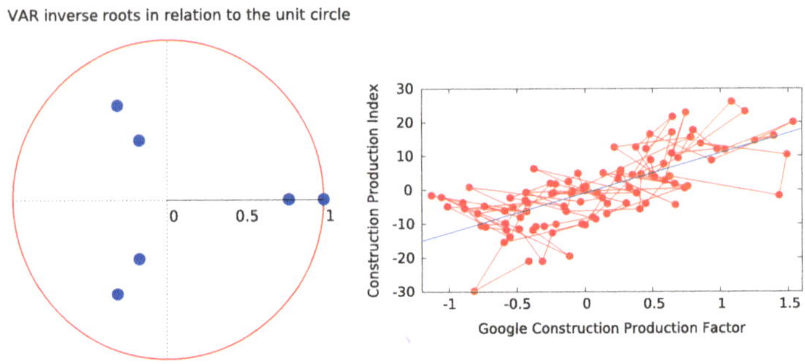

Fig. 19.6 The inverse $pk = 6$ roots of the unrestricted VAR(3) model for the Italian construction production and the scatter plot between S_t and cpi_t

$$\begin{aligned}
\Delta S_t = {} & \underset{(0.0053)}{0.0080}\, ect_{1t-1} \\
& - \underset{(0.1342)}{0.2652}\, \Delta S_{t-1} - \underset{(0.0058)}{0.0019}\, \Delta cpi_{t-1} - \underset{(0.1150)}{0.0635}\, \Delta S_{t-2} - \underset{(0.0052)}{0.0130}\, \Delta \mathrm{cpi}_{t-2} \\
& - \underset{(0.6513)}{0.9849} + \underset{(0.2119)}{3.7922}\, M_{1t} + \underset{(0.4453)}{2.3591}\, M_{2t} + \underset{(0.4609)}{2.1961}\, M_{3t} + \underset{(0.2496)}{1.9661}\, M_{4t} \\
& + \underset{(0.2210)}{2.0714}\, M_{5t} + \underset{(0.2318)}{1.7038}\, M_{6t} + \underset{(0.2271)}{2.2762}\, M_{7t} + \underset{(0.2638)}{2.5550}\, M_{8t} \\
& + \underset{(0.4337)}{3.2157}\, M_{9t} + \underset{(0.4979)}{0.5611}\, M_{10t} + \underset{(0.3556)}{1.5386}\, M_{11t}, \quad \sigma_1^2 = 0.0850 \\
\Delta \mathrm{cpi}_t = {} & - \underset{(0.0980)}{0.2552}\, ect_{1t-1} \\
& + \underset{(2.4699)}{1.3897}\, \Delta S_{t-1} - \underset{(0.1070)}{0.5491}\, \Delta \mathrm{cpi}_{t-1} + \underset{(2.1172)}{2.0409}\, \Delta S_{t-2} - \underset{(0.0963)}{0.5333}\, \Delta \mathrm{cpi}_{t-2} \\
& + \underset{(11.9883)}{30.6396} + \underset{(3.9008)}{2.5277}\, M_{1t} + \underset{(8.1963)}{1.6524}\, M_{2t} + \underset{(8.4830)}{22.8772}\, M_{3t} + \underset{(4.5948)}{23.9567}\, M_{4t} \\
& + \underset{(4.0673)}{28.5055}\, M_{5t} + \underset{(4.2661)}{18.5787}\, M_{6t} + \underset{(4.1804)}{30.5768}\, M_{7t} - \underset{(4.8547)}{42.1111}\, M_{8t} \\
& + \underset{(7.9828)}{19.4161}\, M_{9t} + \underset{(9.1634)}{7.6849}\, M_{10t} + \underset{(6.5455)}{40.5303}\, M_{11t}, \quad \sigma_2^2 = 28.8090
\end{aligned}$$

where standard errors are included in parentheses. The estimates of this model suggest that cpi_t reacts to the disequilibrium between variables in X_t as measured by the disequilibrium error in Eq. (19.2).

To conclude, we check parameter constancy throughout the sample period. The largest eigenvalue which is also used in the cointegration rank tests have been computed recursively from 2008:01 and approximate 99 % confidence intervals are plotted in Fig. 19.7. The Chow forecast (shortly, CF) test is also considered. The test checks for a structural break in 2008:09 (we identify September 2008 as the beginning of the 2008 financial crises). The CF test tests against the alternative

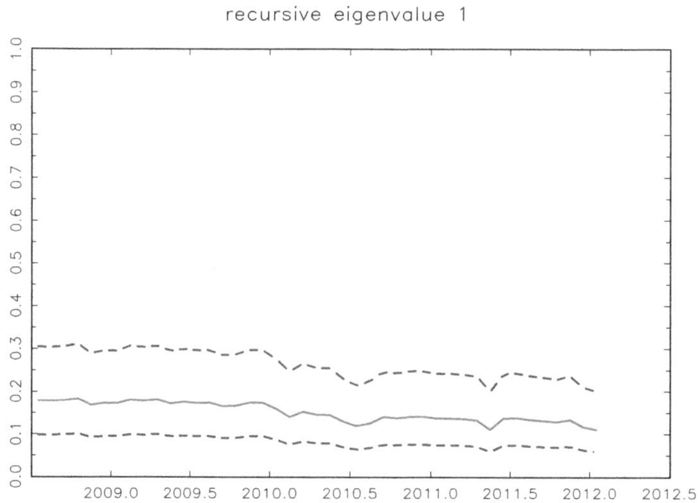

Fig. 19.7 Plot of the largest recursively estimated eigenvalue

that all coefficients including the residual covariance matrix Ω may vary. Because the sample distribution of the test statistic under the null hypothesis may be quite different from the asymptotic distribution, we compute bootstrap p-values obtained by means of 1,000 replications; see [4]. Anyway the null hypothesis of constant parameters is not rejected; i.e., $F\,(86, 67) = 1.1987$ with bootstrapped p-value $=$ 0.1580 and asymptotic p-value $= 0.2204$.

19.4 Concluding Remarks

In this paper we have proposed new timelier indices based on weekly Google search activities in order to anticipate official reports. An illustration on the Italian residential construction production is provided in order to show how Internet data may be useful. We recall that our aim is not to replace official statistics, but only to give proxies for our phenomenon to better understand current unobserved dynamics. Time series provided by statistical agencies has been showed to be related with Google data by means of a cointegrated VAR. The long-run equation significantly contributes to the short-run movements of both Google factor and construction production index. Moreover, the construction production short dynamics have been shown to be related with the Google indicator. Thus, the analysis confirms the use of the Google index as proxy of construction production trend-cycle. The findings permit us to monitor residential construction developments without expensive surveys and before official data are published.

One possible limitations for this approach may be recognized. It is related with some nonnegligible variations between samples drawn on different weeks.

As [11, 12] pointed out, this may pose a potential problem for identifying model parameters. We addressed this problem by considering the whole period at disposal and by performing recursive tests for parameter constancy. We found out that the full sample period defines a constant parameter regime and the assumed break point (i.e., September 2008 that we identify as the beginning of the 2008 financial crises) did not suggest a change in the structure.

Acknowledgements We acknowledge an anonymous reviewer for the careful reading and useful comments. We also acknowledge Luca Fanelli (University of Bologna) for some comments and useful remarks. The second author acknowledges the support of Free University of Bozen-Bolzano, School of Economics and Management, via the projects "Handling High-Dimensional Systems in Economics."

References

1. Askitas, N., Zimmermann, K.F.: Google econometrics and unemployment forecasting. Appl. Econ. Q. **55**, 107–120 (2009)
2. Browne, M.W.: Generalized least squares estimators in the analysis of covariance structures. S. Afr. Stat. J. **8**, 1–24 (1974)
3. Browne, M.W.: Asymptotically distribution-free methods for the analysis of covariance structures. Br. J. Math. Stat. Psychol. **37**, 62–83 (1984)
4. Candelon, B., Lütkephol, H.: On the reliability of chow type tests for parameter constancy in multivariate dynamic models. Econ. Lett. **73**, 155–160 (2001)
5. Choi, H., Varian, H.: Predicting the present with Google trends. Technical report, Google Inc. (2009)
6. D'Amauri, F., Marcucci, J.: "Google it!" Forecasting the US unemployment rate with a Google job search index. Technical report, MPRA Paper 18248, University Library of Munich, Germany (2009)
7. Doornik, J.A.: Approximations to the asymptotic distributions of cointegration tests. J. Econ. Surv. **12**, 573–593 (1998)
8. Ginsberg, J., Mohebbi, M.H., Patel, R.S., Brammer, L., Smolinski, M.S., Brilliant, L.: Detecting influenza epidemics using search engine query data. Nature **457**, 1012–1014 (2009)
9. Horn, J.L.: A rational test for the number of factors in factor analysis. Psychometrika **30**, 179–185 (1965)
10. Johansen, S.: Likelihood-Based Inference in Cointegrated Vector Autoregressive Models, 2nd edn. Advanced Text in Econometrics, Oxford University Press, Oxford (1996)
11. Vosen, S., Schmidt, T.: Forecasting private consumption: survey-based indicators vs. Google trends. J. Forecast. **30**, 565–578 (2011)
12. Vosen, S., Schmidt, T.: A monthly consumption indicator for Germany based on internet search query data. Appl. Econ. Lett. **19**, 683–687 (2012)

Chapter 20
Measures of Dependence for Infinite Variance Distributions

Bernard Garel

20.1 Introduction

One of the first things a statistician does when he begins a data analysis is to compute the empirical matrix of covariance or correlation. Indeed, the independence and also the level of dependence constitute a fundamental information for the modelisation of data. In the standard case, the correlation coefficient is the usual measure of dependence. When the involved random variables have infinite variance this coefficient does not exist and to find a good substitute continues to be of interest. Indeed, many physical phenomena or financial data exhibit a very high variability, showing heavy tailed distributions. Among distributions with infinite variance we find stable distributions introduced by Lévy. Mandelbrot [6] suggested the stable laws as possible models for the distributions of income and speculative prices. Excellent contributions were given by Samorodnitzky and Taqqu [14] about stable non-Gaussian random processes and by Uchaikin and Zolotarev [15] who gave examples in telecommunications, physics, biology, genetic, and geology.

A practical guide to heavy tails was published by Adler et al. [1] with a lot of interesting papers. Nolan [9–11] also did a huge contribution to stable laws.

In the case of stable distributions a few measures of dependence have been proposed. Here we start from notion around covariation. Then we generalize the contribution to arbitrary distributions with a first order moment and infinite variance.

This first section recalls a number of concepts around stability. Then in the second section we introduce the signed symmetric covariation coefficient and give its fundamental properties. The third and last section is devoted to the new measure of dependence and its estimation.

B. Garel (✉)
University of Toulouse, INP-ENSEEIHT and Mathematical Institute
of Toulouse, Toulouse, France
e-mail: garel@enseeiht.fr

V.B. Melas et al. (eds.), *Topics in Statistical Simulation*, Springer Proceedings
in Mathematics & Statistics 114, DOI 10.1007/978-1-4939-2104-1_20

We denote the law of a stable random variable by $S_\alpha(\gamma, \beta, \delta)$, with $0 < \alpha \leqslant 2$, $\gamma \geqslant 0, -1 \leqslant \beta \leqslant 1$ and δ a real parameter.

A random variable X has a stable distribution $S_\alpha(\gamma, \beta, \delta)$ if its characteristic function has the form

$$\varphi_X(t) = E \exp i\,tX = \exp\Big\{ - \gamma^\alpha |t|^\alpha \left[1 + i\,\beta \mathrm{sign}(t) w(t, \alpha)\right] + i\,\delta t \Big\},$$

where

$$w(t, \alpha) = \begin{cases} -\tan \frac{\pi\alpha}{2} & \text{if } \alpha \neq 1, \\ \frac{2}{\pi} \ln |t| & \text{if } \alpha = 1, \end{cases}$$

with t a real number, and $\mathrm{sign}(t) = 1$ if $t > 0$, $\mathrm{sign}(t) = 0$ if $t = 0$ and $\mathrm{sign}(t) = -1$ if $t < 0$.

The parameter α is the characteristic exponent or index of stability, β is a measure of skewness, γ is a scale parameter and δ is a location parameter.

The special cases $\alpha = 2$, $\alpha = 1$ and $\alpha = 0.5$ correspond, respectively, to the Gaussian, Cauchy and Lévy distributions and it is only in these cases that stable laws have a closed form expression for the density.

When $\beta = \delta = 0$, the distribution is symmetric (i.e. X and $-X$ have the same law) and is denoted $S\alpha S(\gamma)$ or for short $S\alpha S$.

Let $0 < \alpha < 2$. The characteristic function of a stable random vector $\mathbf{X} = (X_1, X_2)$ is given by

$$\varphi_{\mathbf{X}}(\mathbf{t}) = \exp\left\{ - \int_{S_2} |\langle \mathbf{t}, \mathbf{s} \rangle|^\alpha \left[1 + i\,\mathrm{sign}(\mathbf{t}, \mathbf{s}) w(\langle \mathbf{t}, \mathbf{s} \rangle, \alpha)\right] \boldsymbol{\Gamma}(d\mathbf{s}) + i\,\langle \mathbf{t}, \mathbf{d} \rangle \right\},$$

where $\boldsymbol{\Gamma}$ is a finite measure on the unit circle $S_2 = \{\mathbf{s} \in \mathbb{R}^2 : \|\mathbf{s}\| = 1\}$ and $\mathbf{d}$ is a vector in $\mathbb{R}^2$. Here $\langle \mathbf{t}, \mathbf{s} \rangle$ denotes the inner product of $\mathbb{R}^2$ and $\|.\|$ stands for the Euclidian norm in $\mathbb{R}^2$. The measure $\boldsymbol{\Gamma}$ is called the spectral measure of the α-stable random vector $\mathbf{X}$ and the pair $(\boldsymbol{\Gamma}, \mathbf{d})$ is unique. The vector $\mathbf{X}$ is symmetric if, and only if, $\mathbf{d} = 0$ and $\boldsymbol{\Gamma}$ is symmetric on S_2. In this case, its characteristic function is given by

$$\varphi_{\mathbf{X}}(\mathbf{t}) = \exp\left\{ - \int_{S_2} |\langle \mathbf{t}, \mathbf{s} \rangle|^\alpha \boldsymbol{\Gamma}(d\mathbf{s}) \right\}. \tag{20.1}$$

If necessary, we also denote the spectral measure of $\mathbf{X}$ by $\boldsymbol{\Gamma}_{\mathbf{X}}$.

The spectral measure carries essential information about the vector, in particular the dependence structure between the coordinates. So, it is not surprising that measures of dependence rely on this spectral measure. In the sequel, unless specified otherwise, we assume $\alpha > 1$ and consider symmetric stable random variables or vectors. Miller [7] introduced the covariation as follows.

Definition 20.1. Let X_1 and X_2 be jointly $S\alpha S$ and let $\boldsymbol{\Gamma}$ be the spectral measure of the random vector (X_1, X_2).

The covariation of X_1 on X_2 is the real number defined by

$$[X_1, X_2]_\alpha = \int_{S_2} s_1 s_2^{\langle \alpha-1 \rangle} \boldsymbol{\Gamma}(d\mathbf{s}). \tag{20.2}$$

where for real numbers s and a:

- if $a \neq 0$, $s^{\langle a \rangle} = |s|^a \text{sign}(s)$
- and if $a = 0$, $s^{\langle a \rangle} = \text{sign}(s)$.

Although the covariation (20.2) is linear in its first argument, it is, in general, not linear in its second argument and not symmetric in its arguments. We also have

$$[X_1, X_1]_\alpha = \int_{S_2} |s_1|^\alpha \boldsymbol{\Gamma}(d\mathbf{s}) = \gamma_{X_1}^\alpha,$$

where γ_{X_1} is the scale parameter of the $S\alpha S$ random variable X_1.

The covariation norm is defined by

$$\|X_1\|_\alpha = ([X_1, X_1]_\alpha)^{1/\alpha}. \tag{20.3}$$

When X_1 and X_2 are independent, $[X_1, X_2]_\alpha = 0$.

The covariation coefficient of X_1 on X_2 is the quantity:

$$\lambda_{X_1,X_2} = \frac{[X_1, X_2]_\alpha}{\|X_2\|_\alpha^\alpha}. \tag{20.4}$$

It is the coefficient of the linear regression $E(X_1|X_2)$. This coefficient is not symmetric and may be unbounded. We see it easily by setting $X_2 = cX_1$, where $c \neq 0$ and $c \neq ś1$. In this case we have

$$\lambda_{X_1,X_2} = \frac{1}{c} \quad \text{and} \quad \lambda_{X_2,X_1} = c.$$

Paulauskas [12, 13] introduced the alpha-correlation. This coefficient is applicable to all symmetric stable random vectors in $\mathbb{R}^2$ and has all the properties of the ordinary Pearson correlation coefficient.

Let (X_1, X_2) be $S\alpha S$, $0 < \alpha \leqslant 2$ and $\boldsymbol{\Gamma}$ its spectral measure on the unit circle S_2. Let (U_1, U_2) be a random vector on S_2 with probability distribution $\tilde{\boldsymbol{\Gamma}} = \boldsymbol{\Gamma}/\boldsymbol{\Gamma}(S_2)$. Due to the symmetry of $\boldsymbol{\Gamma}$, one has $EU_1 = EU_2 = 0$. The alpha-correlation is defined as:

$$\tilde{\rho}(X_1, X_2) = \frac{EU_1U_2}{(EU_1^2 EU_2^2)^{1/2}}.$$

It is a measure of dependence of (X_1, X_2).

20.2 Signed Symmetric Covariation Coefficient

20.2.1 Definition and First Properties

Definition 20.2. Let (X_1, X_2) be a bivariate $S\alpha S$ random vector with $\alpha > 1$. The signed symmetric covariation coefficient between X_1 and X_2 is the quantity:

$$\text{scov}(X_1, X_2) = \varkappa_{(X_1, X_2)} \left| \frac{[X_1, X_2]_\alpha [X_2, X_1]_\alpha}{\|X_1\|_\alpha^\alpha \|X_2\|_\alpha^\alpha} \right|^{\frac{1}{2}},$$

where

$$\varkappa_{(X_1, X_2)} = \begin{cases} \text{sign}([X_1, X_2]_\alpha) & \text{if} \quad \text{sign}([X_1, X_2]_\alpha) = \text{sign}([X_2, X_1]_\alpha), \\ \\ -1 & \text{if} \quad \text{sign}([X_1, X_2]_\alpha) = -\text{sign}([X_2, X_1]_\alpha). \end{cases}$$

Remark: The value of $\varkappa_{(X_1, X_2)}$ above is natural in the first case. In fact, if (X_1, X_2) was a random vector with finite variance, the equality $\text{sign}([X_1, X_2]_\alpha) = \text{sign}([X_2, X_1]_\alpha)$ would always be true, because $[X_1, X_2]_2 = \frac{1}{2}\text{Cov}(X_1, X_2)$.

But in the case of stable non-gaussian random vectors, we can have $\text{sign}([X_1, X_2]_\alpha) = -\text{sign}([X_2, X_1]_\alpha)$, see Garel et al. [4]. If it is so, we set $\varkappa_{(X_1, X_2)} = \text{sign}([X_1, X_2]_\alpha \times [X_2, X_1]_\alpha) = -1$.

The signed symmetric covariation coefficient has the following properties:

- $-1 \leqslant \text{scov}(X_1, X_2) \leqslant 1$ and if X_1, X_2 are independent, then $\text{scov}(X_1, X_2) = 0$;
- $|\text{scov}(X_1, X_2)| = 1$ if and only if $X_2 = \lambda X_1$ for some $\lambda \in \mathbb{R}$;
- let a and b be two non-zero reals, then

$$\text{scov}(aX_1, bX_2) = \begin{cases} \text{sign}(ab)\text{scov}(X_1, X_2) & \text{if } \text{sign}([X_1, X_2]_\alpha) = \text{sign}([X_2, X_1]_\alpha), \\ \\ \text{scov}(X_1, X_2) & \text{if } \text{sign}([X_1, X_2]_\alpha) = -\text{sign}([X_2, X_1]_\alpha); \end{cases} \tag{20.5}$$

- for $\alpha = 2$, $\text{scov}(X_1, X_2)$ coincides with the usual correlation coefficient.

A more detailed study of this coefficient has been done by Kodia and Garel [5].

20.2.2 Sub-Gaussian Case

Definition 20.3. Let $0 < \alpha < 2$, let $\mathbf{G} = (G_1, G_2, \ldots, G_d)$ be zero mean jointly normal random vector and let A be a positive random variable such that $A \sim S_{\alpha/2}\big((\cos\frac{\pi\alpha}{4})^{2/\alpha}, 1, 0\big)$, independent of $\mathbf{G}$, then $\mathbf{X} = A^{1/2}\mathbf{G} = (A^{1/2}G_1, A^{1/2}G_2, \ldots, A^{1/2}G_d)$ is a sub-Gaussian random vector with underlying Gaussian vector $\mathbf{G}$.

The characteristic function of $\mathbf{X}$ has the particular form:

$$\varphi_{\mathbf{X}}(\mathbf{t}) = E\exp\Big\{i\sum_{m=1}^{d} t_m X_m\Big\} = \exp\Big\{-\Big|\frac{1}{2}\sum_{j=1}^{d}\sum_{k=1}^{d} t_j t_k R_{jk}\Big|^{\alpha/2}\Big\}, \tag{20.6}$$

where $R_{jk} = EG_jG_k$, $j, k = 1, \ldots, d$ are the covariances of $\mathbf{G}$.

Theorem 20.1. *Let $1 < \alpha < 2$ and $\mathbf{X}$ a sub-Gaussian random vector with characteristic function (20.6).*

- *Then the signed symmetric covariation coefficient matrix of $\mathbf{X}$ coincides with the correlation matrix of the underlying Gaussian random vector $\mathbf{G}$.*
- *Then the signed symmetric covariation coefficient matrix coincides with the matrix of α-correlation.*

20.3 The New Coefficient of Dependence

Now we assume only that the distributions admit a finite 1st-order moment. Then the variance may be infinite.

Definition 20.4. We define the coefficient R_G by:

$$R_G(X_1, X_2) = \varkappa(X_1, X_2)\frac{|E(X_1\text{sign}(X_2))E(X_2\text{sign}(X_1))|^{1/2}}{(E|X_1|E|X_2|)^{1/2}}$$

$$\varkappa_{(X_1,X_2)} = \begin{cases} \text{sign}(E(X_1\text{sign}(X_2))) & \text{if } \text{sign}(E(X_1\text{sign}(X_2))) = \text{sign}(E(X_2\text{sign}(X_1))), \\ -1 & \text{if } \text{sign}(E(X_1\text{sign}(X_2))) = -\text{sign}(E(X_2\text{sign}(X_1))). \end{cases} \tag{20.7}$$

It is easy to show that the coefficient R_G has the following properties:

- $-1 \leqslant R_G(X_1, X_2) \leqslant 1$ and if X_1, X_2 are independent, then $R_G(X_1, X_2) = 0$;
- If $X_2 = \lambda X_1$ for some $\lambda \in R$, $|R_G(X_1, X_2)| = 1$;
- If (X_1, X_2) is a $S\alpha S$ random vector with $1 < \alpha < 2$, $R_G(X_1, X_2)$ coincides with the signed symmetric covariation coefficient.

This last property results from the following one. If (X_1, X_2) follows a stable bivariate distribution, then the covariation coefficient of X_1 on X_2:

$$\lambda_{X_1,X_2} = \frac{[X_1, X_2]_\alpha}{\|X_2\|_\alpha^\alpha}.$$

satisfies for all $1 \leqslant p < \alpha$

$$\lambda_{X_1,X_2} = \frac{E(X_1 X_2^{\langle p-1 \rangle})}{E(|X_2|^p)} = \frac{E X_1 \operatorname{sign}(X_2)}{E|X_2|}.$$

The last equality is obtained by taking $p = 1$ in the first equality. That has been suggested by Nikias and Shao [8] and also by Gallagher [3].

20.3.1 Estimation of this Coefficient

Let X_{1j} and X_{2j} $1 \leqslant j \leqslant n$ be independent copies of X_1 and X_2, respectively. We estimate R_G by:

$$\widehat{R_G}(X_1, X_2) = \hat{\varkappa}_{(X_1,X_2)} \frac{\left| \left(\sum_{j=1}^n X_{1j} \operatorname{sign} X_{2j} \right) \left(\sum_{j=1}^n X_{2j} \operatorname{sign} X_{1j} \right) \right|^{1/2}}{\left[\left(\sum_{j=1}^n |X_{1j}| \right) \left(\sum_{j=1}^n |X_{2j}| \right) \right]^{1/2}}$$

where $\hat{\varkappa}_{(X_1,X_2)} =$

$$\begin{cases} \operatorname{sign}\left(\sum_{j=1}^n X_{1j} \operatorname{sign} X_{2j} \right) & \text{if } \operatorname{sign}\left(\sum_{j=1}^n X_{1j} \operatorname{sign} X_{2j} \right) = \operatorname{sign}\left(\sum_{j=1}^n X_{2j} \operatorname{sign} X_{1j} \right), \\ -1 & \text{if not.} \end{cases}$$

This estimator is convergent. Here we give some results of simulation in the sub-Gaussian case.

Estimates of R_G and $\tilde{\rho}$ for $n = 1600$ sub-Gaussian data vectors with $\alpha = 1.5$, $\gamma_1 = 5$, $\gamma_2 = 10$ and $A \sim S_{\alpha/2}\left(\left(\cos \frac{\pi\alpha}{4}\right)^{2/\alpha}, 1, 0\right)$,

$G = (G_1, G_2)$, $(X_1, X_2) = A^{1/2}\mathbf{G} = (A^{1/2}G_1, A^{1/2}G_2)$. Number of replications: 100.

True value	−1.00	−0.60	−0.40	−0.20	0.00	0.10	0.30	0.50	0.90
$\hat{R}_G$	−1.00	−0.60	−0.38	−0.20	−0.01	0.10	0.31	0.50	0.90
	0.00	0.04	0.05	0.07	0.06	0.07	0.06	0.05	0.02
$\tilde{\rho}$.est	−0.80	−0.55	−0.36	−0.18	−0.00	0.09	0.27	0.46	0.81
	0.00	0.02	0.02	0.03	0.02	0.02	0.02	0.02	0.00

Programmed by Bernédy KODIA

The first number is the mean calculated over 100 replications. The second number (in tiny) is the mean absolute deviation from the mean above. In order to estimate $\tilde{\rho}$, we used an estimation of the spectral measure $\boldsymbol{\Gamma}$ and then a formula given by Paulauskas [12, p. 364], having replaced the weights by their estimates. See Kodia and Garel [5].

20.4 Stable Linear Processes

For a definition, see Brockwell and Davis [2]. In the same spirit that [3] we introduce the signed symmetric autocovariation function (and the coefficient R_G) in this context. We assume $\alpha > 1$ and $h \in \mathbb{N}$.

Definition 20.5. Let $\{X_t,\ t \in T\}$ be a stationary $S\alpha S$ process. The signed symmetric autocovariation function at level h is defined by:

$$\text{scov}(h) = \text{scov}(X_{t+h}, X_t) = \varkappa_{(X_{t+h}, X_t)} \frac{|E(X_{t+h}\text{sign}X_t).E(X_t\text{sign}X_{t+h})|^{1/2}}{E|X_t|}.$$

with $\varkappa_{(X_{t+h}, X_t)}$ defined as in Definition 4.

Then we obtain the following characterization:

Proposition 20.1. *Let $\{X_t\}$ be a causal linear stationary $S\alpha S$ process, with signed symmetric autocovariation function* scov(·). *Then $\{X_t\}$ is a MA(q) process if and only if* $\text{scov}(h) = 0$ *for $h > q$ and* $\text{scov}(q) \neq 0$. *This means that there exists a $S\alpha S$ white noise $\{Z_t\}$ such that:*

$$X_t = Z_t + \theta_1 Z_{t-1} + \cdots + \theta_q Z_{t-q}.$$

The neccessary condition is rather easy to prove. For the converse, we start from the causal representation of the process $X_t = \sum_{j=0}^{\infty} \psi_j Z_{t-j}$ and we show that for all $j > q$ we have $\psi_j = 0$.

Acknowledgements I would like to thank Professor Subir Ghosh for his invitation to the Rimini Conference and Bernedy Kodia for his help for simulations. I also thank the Referee for his suggestions.

References

1. Adler, R., Feldman, R., Taqqu, M.S.: A Practical Guide to Heavy Tails. Birkhäuser, Boston (1999)
2. Brockwell, P.J., Davis, R.A.: Time Series: Theory and Methods. Springer, New York/Berlin (1991)
3. Gallagher, C.M.: Estimating the autocovariation function with application to heavy-tayled ARMA modeling. Technical Report (2001)
4. Garel, B., d'Estampes, L., Tjøstheim, D.: Revealing some unexpected dependence properties of linear combinations of stable random variables using symmetric covariation. Commun. Stat. **33**(4), 769–786 (2004)
5. Kodia, B., Garel, B.: Estimation and comparison of signed symmetric covariation coefficient and generalized association parameter for alpha-stable dependence modeling. Commun. Stat. **54**, 252–276(2014)
6. Mandelbrot, B.: The variation of certain speculative prices. J. Bus. **36**, 394–419 (1963)
7. Miller, G.: Properties of certain symmetric stable distributions. J. Multivar. Anal. **8**(3), 346–360 (1978)
8. Nikias, C.L., Shao, M.: Signal Processing with Alpha-Stable Distributions and Applications. Wiley, New York (1995)
9. Nolan, J.P.: Multivariate stable distributions: approximation, estimation, simulation and identification. In: Adler, R.J., Feldman, R.E., Taqqu, M.S. (eds.) A Practical Guide to Heavy Tails. Statistical Techniques and Applications, pp. 509–525 (1998)
10. Nolan, J.P.: Multivariate stable densities and distribution functions: general and elliptical case. Deutsche Bundesbank's 2005 Annual Fall Conference. http://www.researchgate.net/publication/246910601_Multivariate_stable_densities_and_distribution_functions_general_and_elliptical_case (2005)
11. Nolan, J.P., Panorska, A.K., McCulloch, J.H.: Estimation of stable spectral measures. Math. Comput. Model. **34**, 1113–1122 (2001)
12. Paulauskas, V.J.: Some remarks on multivariate stable distributions. J. Multivar. Anal. **6**, 356–368 (1976)
13. Paulauskas, V.J.: On α-covariance, long, short and negative memories for sequence of random variables with infinite variance. Preprint (2013)
14. Samorodnitsky, G., Taqqu, M.S.: Stable non-Gaussian random processes: Stochastic Models with Infinite Variance. Stochastic Modeling. Chapman & Hall, New York/London (1994)
15. Uchaikin, V.V., Zolotarev, V.M.: Chance and Stability. Stable Distributions and their Applications. de Gruyter, Berlin/New York (1999)

Chapter 21
Design of Experiments Using R

Albrecht Gebhardt

21.1 Overview of Existing DOE Implementations

21.1.1 CRAN Task View

The R language, see [2], has become a widely used software toolkit for statisticians and statistical applications in many fields of science. One of its advantages is the extensibility through add-on packages, the current number (at the time of writing, 2013/2014) of available packages has passed 5000.

For a better overview of available packages, CRAN[1] delivers so-called task views, these are moderated collections of packages belonging to different tasks. The task view on design of experiments[2] at CRAN (maintained by Ulrike Grömping) mentions a large number of general and specialized design of experiments packages which cover:

- general purpose design of experiments,
- agricultural design of experiments,
- industrial design of experiments,
- experimental designs for computer experiments,
- and more.

[1]Comprehensive R Archive Network, http://cran.r-project.org, the central web archive of the R language.

[2]http://cran.r-project.org/web/views/ExperimentalDesign.html.

A. Gebhardt (✉)
University Klagenfurt, Universitätsstr. 65-67, 9020 Klagenfurt, Austria
e-mail: albrecht.gebhardt@aau.at

V.B. Melas et al. (eds.), *Topics in Statistical Simulation*, Springer Proceedings in Mathematics & Statistics 114, DOI 10.1007/978-1-4939-2104-1_21

Beside this collection other R packages for special design of experiments applications exist, e.g. for spatial statistics: `edesign` (Entropy based design of monitoring networks, also developed at university Klagenfurt, see [1]). The library optimal design of experiments (OPDOE) (see Sect. 21.4) will be partly introduced here, covering ANOVA and some special kind of sequential tests.

21.2 Designs for ANOVA

The task of designing ANOVA experiments consists in determining sample sizes in order to fulfill several demands regarding the risks of 1st and 2nd kind. The tests can be generally represented as testing a factor for no effects. For instance, in a single factor model with a factor A at levels a_i the test with the hypothesis

$$H_0 : \forall i \; a_i = 0 \quad H_A : \exists i \; a_i \neq 0$$

leads to an F-test with degrees of freedom f_1, f_2. The characteristics of this test can be described by

- α, the risk of 1st kind,
- $1-\beta$, the power of the F-test, risk of 2nd kind,
- σ_y^2, the population variance,
- δ, a minimum difference between levels of the factor to be detected.

The optimal size for a given set of accuracy parameters α, β and δ can be determined by solving

$$F(f_1, f_2, 0, 1-\alpha) = F(f_1, f_2, \lambda, \beta) \tag{21.1}$$

with a non-centrality parameter $\lambda = C \cdot \frac{1}{\sigma_y^2} \sum_{i=1}^{q} (E_i - \overline{E})^2$ where E_i denote the effects of the main factor, C is a constant depending on the model. The solution is found by iteration, see, e.g., [4].

21.2.1 R Implementation

ANOVA models can be classified according to the number of factors involved and the type of interaction of these factors as cross, nested or mixed classification. Additionally some of the factors can be treated as random. The function `size.anova()` is called for all possible models, the parameter `model` describes the ANOVA model using the characters "x" and ">" for cross and nested classification, "`()`" for mixed effects, small letters `a,b,c` for fixed and capital letters `A,B,C` for random effects, e.g.

```
size.anova(model="a" ,...)      # One-Way
size.anova(model="axb" ,...)    # Two-Way cross
size.anova(model="a>B" ,...)    # Two-Way nested, B random
size.anova(model="(axb)>c" ,...)   # Three-Way mixed
size.anova(model="(a>B)xc" ,...)   # B random
```

The parameter `hypothesis` selects the desired null hypothesis. It is only needed in some cases where it is not as obvious as $H_0 : \forall j : a_j = 0$ which is the default.

```
# Two-Way cross, test for interactions
size.anova(model="axb" , hypothesis="axb", ...)
# Three-Way mixed, test for effects of A
size.anova(model="(axb)>c" , hypothesis="a", ...)
# Three-Way mixed, test for interactions AxB
size.anova(model="(axb)>c" , hypothesis="axb", ...)
```

Some tests need additional assumptions, e.g. given as

```
# Three-Way cross, test for effects of A
size.anova(model="axBXC" , hypothesis="a",
           assumption="sigma_AC=0,b=c", ...)
```

The sizes a, b, c and n have to be given, omitting just the size to be determined. Additionally the accuracy parameters α, β, δ and the optimization strategy `cases` (choosing `maximin` or `minimin`) have to be specified.

21.2.2 *One-Way Classification*

Model types for one-way ANOVA are

- Type I: factor A fixed,
- Type II: factor A random, this is not covered here.

The model equation for a type I one-way ANOVA can be given as follows, bold symbols are associated with random terms in contrast to fixed terms in normal font:

$$\mathbf{y}_{ij} = E(\mathbf{y}_{ij}) + \mathbf{e}_{ij} = \mu + a_i + \mathbf{e}_{ij} \quad (i = 1, \ldots, a; j = 1, \ldots, n) \quad (21.2)$$
$$H_0 : \forall i \; a_i = 0 \quad H_A : \exists i \; a_i \neq 0$$

An example call for one-way ANOVA looks like

```
> size.anova(model="a",a=4,
      alpha=0.05,beta=0.1, delta=2, case="maximin")
n
9
> size.anova(model="a",a=4,
      alpha=0.05,beta=0.1, delta=2, case="minimin")
n
5
```

In this case n is omitted in the arguments which means that it should be calculated, in this simple case this is the only possible question.

21.2.3 Two-Way Classification

Possible model types for two-way classification $A \times B$ (cross classification) and $A \succ B$ (nested classification) are (`model` parameter notation for `size.anova()` given in parenthesis):

- Cross classification type I: All factors fixed $A \times B$ (`model ="axb"`)
- Type II: All factors random, not covered here.
- Cross classification type III: $A \times \mathbf{B}$, B random (`model ="axB"`)
- Nested classification type I: All factors fixed $A \succ B$ (`model ="a>b"`)
- Nested classification type III: $A \succ \mathbf{B}$, $\mathbf{A} \succ B$, A or B random (`model ="a>B"`)

Taking a two-way cross-classification, model I, $A \times B$ yields the model equation and hypothesis

$$\mathbf{y}_{ijk} = \mu + a_i + b_j + (ab)_{ij} + \mathbf{e}_{ijk} \tag{21.3}$$

$$H_0 : \forall i \ a_i = 0 \quad H_A : \exists i \ a_i \neq 0$$

A sample call with $a = 6$, $b = 4$, accuracy requirements $\alpha = 0.05$, $\beta = 0.1$, $\delta = 1$ asking for the size n gives for the `"minimin"` case:

```
> size.anova(model="axb", hypothesis="a", a=6, b=4,
             alpha=0.05,beta=0.1, delta=1, cases="minimin")
n
4
```

21.2.4 Three-Way Cross Classification

Model types for three-way cross classification $A \times B \times C$ are

- Type I: $A \times B \times C$ All factors fixed (`model ="axbxc"`).
- Type II: All factors random, not covered here.
- Type III: $A \times B \times \mathbf{C}$ mixed, A and B fixed, C random (`model ="axbxC"`).
- Type IV: $A \times \mathbf{B} \times \mathbf{C}$ mixed, A fixed, B and C random (`model ="axBxC"`).

Picking a three-way cross classification, model IV, $A \times \mathbf{B} \times \mathbf{C}$ as an example leads to the equation

$$\mathbf{y}_{ijkl} = \mu + a_i + \mathbf{b}_j + \mathbf{c}_k + (\mathbf{ab})_{ij} + (\mathbf{ac})_{ij} + (\mathbf{bc})_{jk} + (\mathbf{abc})_{ijk} + \mathbf{e}_{ijkl} \tag{21.4}$$

$$\sum_{i=1}^{a} a_i = 0, \forall j,k \ \sum_{i=1}^{a} (\mathbf{ab})_{ij} = \sum_{i=1}^{a} (\mathbf{ac})_{ij} = \sum_{i=1}^{a} (\mathbf{abc})_{ijk} = 0$$

$$H_0 : \forall i \ a_i = 0 \quad H_A : \exists i \ a_i \neq 0$$

This model needs additional assumptions about σ_{AB} and the sizes b and c, see [5]. In the `R` call fix $a = 6$, $n = 2$, assume $\sigma_{AB} = 0$ and $b = c$, take the precision requirements $\alpha = 0.05$, $\beta = 0.1$, $\delta = 0.5$ and get:

```
> size.anova(model="axBxC",hypothesis="a",
            assumption="sigma_AC=0,b=c",a=6,n=2,
            alpha=0.05, beta=0.1, delta=0.5, cases="maximin")
b c
9 9
```

21.2.5 Three-Way Nested Classification

For three-way nested classification $A \succ B \succ C$ we get even more model types (omitting type II again):

- Type I: All factors fixed $A \succ B \succ C$, in OPDOE: `model="a>b>c"`
- Type III: $\mathbf{A} \succ B \succ C$, A random, (`model ="A>b>c"`)
- Type IV: $A \succ \mathbf{B} \succ C$, B random, (`model ="a>B>c"`)
- Type V: $A \succ B \succ \mathbf{C}$, C random, (`model ="a>b>C"`)
- Type VI: $A \succ \mathbf{B} \succ \mathbf{C}$, A fixed, (`model ="a>B>C"`)
- Type VII: $\mathbf{A} \succ B \succ \mathbf{C}$, B fixed, (`model ="A>b>C"`)
- Type VIII: $\mathbf{A} \succ \mathbf{B} \succ C$, C fixed (`model ="A>B>c"`)

Taking a three-way nested classification, model IV, B random, $A \succ \mathbf{B} \succ C$ as example we can test for no effects of C:

$$\mathbf{y}_{ijkl} = \mu + a_i + \mathbf{b}_{j(i)} + c_{k(ij)} + \mathbf{e}_{ijkl} \tag{21.5}$$

$$H_0 : \forall i \; c_i = 0 \quad H_A : \exists i \; c_i \neq 0$$

In R again fix $a = 6$, $c = 4$, try $b = 2$ and $b = 20$, set precision requirements $\alpha = 0.05$, $\beta = 0.1$, $\delta = 0.5$ and calculate n:

```
> size.anova(model="a>B>c", hypothesis="c",a=6, b=2, c=4,
            alpha=0.05, beta=0.1, delta=0.5, case="maximin")
  n
262
```

21.2.6 Three-Way Mixed Classification

Model types for three-way mixed classification of type $(A \times B) \succ C$ are:

- Type I: All factors fixed $(A \times B) \succ C$ (`model ="(axb)>c"`)
- Type VI: $(A \times \mathbf{B}) \succ C$, B random (`model ="(axB)>c"`)
- Type V: $(A \times \mathbf{B}) \succ C$, C random (`model ="(axb)>C"`)
- Type VI: $(A \times \mathbf{B}) \succ \mathbf{C}$, B, C random (`model ="(axB)>C"`)
- Type VII: $(\mathbf{A} \times B) \succ \mathbf{C}$, A, C random (`model ="(Axb)>C"`)
- Type VIII: $(\mathbf{A} \times \mathbf{B}) \succ C$, A, B random (`model ="(AxB)>c"`)

Model II (all factors random) is not handled by the function, model III (only A random) can be achieved using model IV by exchanging A and B. Similar types can be given for $(A \succ B) \times C$:

- Type I: All factors fixed $(A \succ B) \times C$ (`model="(a>b)xc"`)
- Type III: $(\mathbf{A} \succ B) \times C$, A random (`model ="(A>b)xc"`)
- Type IV: $(A \succ \mathbf{B}) \times C$, B random (`model ="(a>B)xc"`)
- Type V: $(A \succ B) \times \mathbf{C}$, C random (`model ="(a>b)xC"`)
- Type VI: $(A \succ \mathbf{B}) \times \mathbf{C}$, B, C random (`model ="(a>B)xC"`)
- Type VII: $(\mathbf{A} \succ B) \times \mathbf{C}$, A, C random (`model ="(a>B)xC"`)
- Type VIII: $(\mathbf{A} \succ \mathbf{B}) \times \mathbf{C}$, A, B random (`model ="(A>B)xc"`)

As example take a three-way mixed classification, model I, $(A \times B) \succ C$ with model equation and hypothesis

$$\begin{aligned} \mathbf{y}_{ijkl} &= \mu + a_i + b_j + (ab)_{ij} + c_{k(ij)} + \mathbf{e}_{ijkl} \\ \forall i, j &\sum_{k=1}^{c} c_{k(ij)} = 0 \\ H_0 : \forall i \; a_i &= 0 \quad H_A : \exists i \; a_i \neq 0 \end{aligned} \tag{21.6}$$

For a sample call take $a = 6$, $b = 5$, $c = 4$, precision requirements $\alpha = 0.05$, $\beta = 0.1$, $\delta = 0.5$ test for no effect of A and get:

```
> size.anova(model="(axb)>c", hypothesis="a",a=6, b=5, c=4,
             alpha=0.05, beta=0.1, delta=0.5, case="minimin")

n
3
```

21.3 Sequential Designs

The idea of sequential designs is to start with a small sample, add step by step new data and to decide if the desired accuracy demands are fulfilled at this step so that a decision regarding the null hypothesis can be found. The tests can be one-sided or two-sided, one-sample or two-sample, for means or proportions.

The principle can be sketched as follows:

- An initial sample is taken (with size $\geqslant 1$)
- A stopping rule determines whether or not to continue sampling, sampling stops if the actual sample size fulfills the accuracy demands, expressed in terms of risks of 1st and 2nd kind.
- After stopping, a decision rule is applied (the test is performed).

Continuation can be done as a single step or by adding larger batches of data (bearing the risk of overshooting the minimum required sample size).

21.3.1 Triangular Tests

A triangular test is a sequential test which allows for early stopping of trials, it was introduced by John Whitehead, see [7]. The sample size is increased step by step until a decision can be made. The decision rule can be interpreted with a graphical representation: A derived quantity falls into a triangular shaped region (it means the test can not yet be finished) or leaves this region on the upper or lower side (H_0 is rejected or not). Triangular tests finish after a finite number of steps because of the finite size of the triangular shaped continuation region.

For a test for the mean of a normal distributed population with σ^2 unknown, a one-sided hypothesis $H_0 : \theta = \theta_0$ versus $H_A : \theta = \theta_1$ the continuation region is given by

$$\begin{array}{ll} -a + 3bv_n < z_n < a + bv_n & \text{if } \theta > \theta_0 \\ -a + bv_n < z_n < a + 3bv_n & \text{if } \theta < \theta_0 \end{array} \qquad (21.7)$$

with

$$v_n = n - \frac{z_n^2}{n},\ z_n = \frac{\sum_{i=1}^n y_i}{\sqrt{\frac{1}{n}\sum_{i=1}^n y_i^2}},\ a = 2\ln\left(\frac{1}{2\alpha}\right)/\theta_1,\ b = \frac{\theta_1}{4}$$

For an example initialize some heights data, taken from a male sample

```
> male <- c( 183, 187, 179, 190, 184, 192, 198, 182, 188,186)
```

Now perform a test $H_0 : \mu = \mu_0 = 180$ versus $H_1 > 180 + \delta$ with $\delta = 5$, $\alpha = 0.01$, $\beta = 0.1$. Assume a prior $\sigma^2 = 16$. Take a subset of the first 8 elements and generate a plot, see Fig. 21.1:

```
> tt <- triangular.test.norm(x=male[1:8], mu0=180, mu1=185,
                             alpha=0.01, beta=0.1,sigma=4)
```

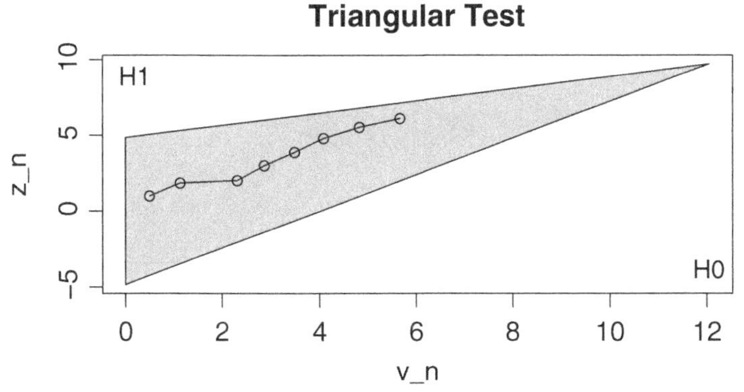

Fig. 21.1 Triangular test, yet unfinished

```
Triangular Test for normal distribution
Sigma known: 4

H0: mu= 180  versus H1: mu> 185
alpha: 0.01  beta: 0.1

Test not finished, continue by adding single data via update()
current sample size for x:  8
```

By applying further updates to the object `tt` we continue the test:

```
> tt <- update(tt,x=male[9])
Triangular Test for normal distribution
Sigma known: 4

H0: mu= 180  versus H1: mu> 185
alpha: 0.01  beta: 0.1

Test not finished, continue by adding single data via update()
current sample size for x:  9
```

It is still not finished, have a look at the plot again (Fig. 21.2). Then again add another value

```
> tt <- update(tt,x=male[10])
Triangular Test for normal distribution
Sigma known: 4

H0: mu= 180  versus H1: mu> 185
alpha: 0.01  beta: 0.1

Test finished: accept H1
Sample size for x:  10
```

Now the test finishes, H_0 is rejected, the needed sample size was 10, see Fig. 21.3.

The triangular testing procedure may be reminiscent of quality control charts, but despite of keeping the score between the borders as long as possible in case of

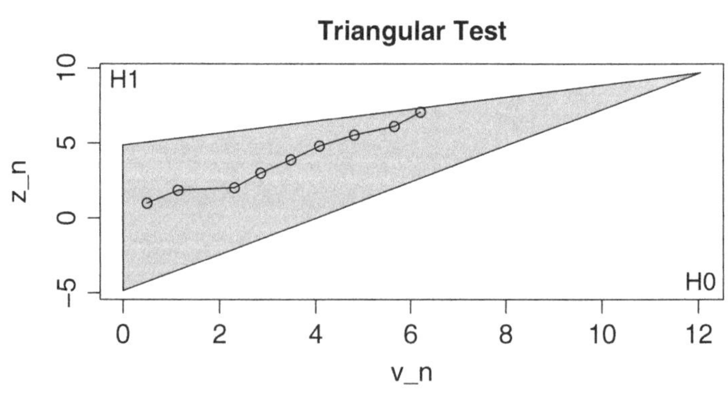

Fig. 21.2 Triangular test, still not finished

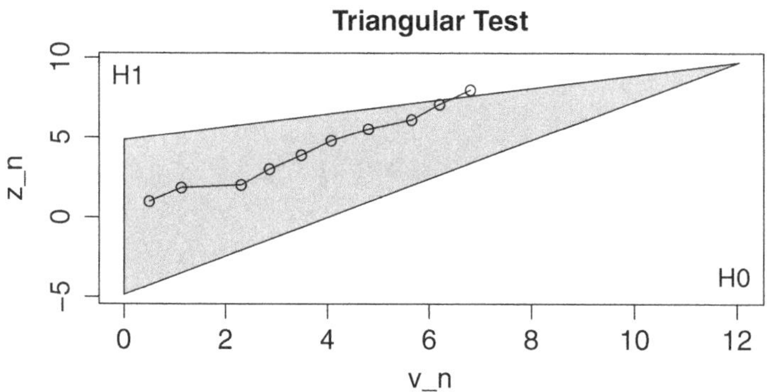

Fig. 21.3 Triangular test, finished, H_0 is rejected

quality control charts the desire of the experimenter is to finish the trials as soon as possible by leaving the continuation region with a minimal sample size.

The implementation makes use of object oriented `R` programming. Several methods like `update`, `print` and `plot` had to be written for the new object class `triangular.test`.

- First a `triangular.test` object is created with the `triangular.test.norm()` function.
- Then the `update` method of that object is used to add new data.
- Plots are generated on the fly or afterwards with the `plot` method of that object.

In the case of a two-sided test, two triangular shaped continuation regions overlap, see the plots in the forthcoming examples. If the acceptance region for some reason gets modeled without symmetry, also the triangles lose their symmetry.

The triangular test principle is now applied to two groups, leading to the double triangular test, see [6]. In this case the size of the groups is increased alternately until a decision is found, again allowing for early stopping of the trials. For this example first initialize two other male/female heights data sets for this two-sided, two-sample test, start with three measurements in each group, assume $\sigma^2 = 49$ known, the test is not finished at this state, see Fig. 21.4:

```
> heights.male <- c(179, 180, 188, 174, 185, 183, 179)
> heights.female <- c(165, 168, 168, 173, 167, 169, 162)
> tt <- triangular.test.norm(x=heights.female[1:3],
    y=heights.male[1:3], mu1=170,mu2=176,mu0=164,
    alpha=0.05, beta=0.2,sigma=7)
Triangular Test for normal distribution
Sigma known: 7

H0: mu1=mu2= 170  versus H1: mu1= 170  and mu2>= 176  or
mu2<= 164 alpha: 0.05  beta: 0.2

Test not finished, continue by adding single data via update()
current sample size for x:  3
current sample size for y:  3
```

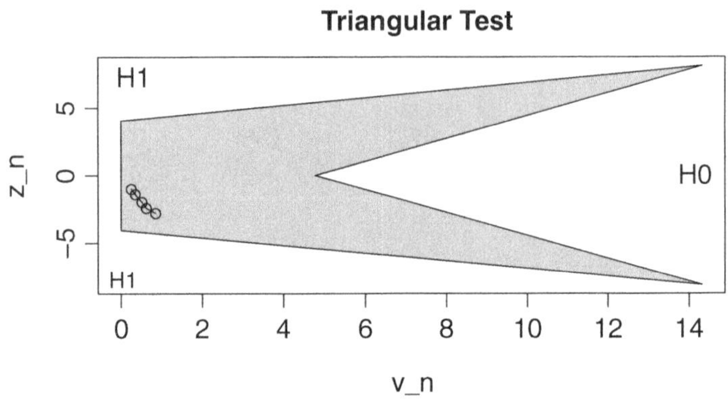

Fig. 21.4 Triangular test, two-sided, not finished

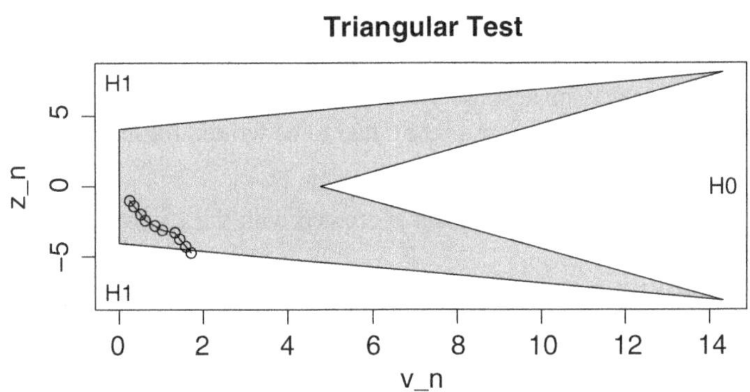

Fig. 21.5 Triangular test, two-sided, finished, H_o is rejected

Then continue with more data, the best practice would be to add samples one by one into each group. To save some lines of output produced by intermediate steps let's just add the next 4 samples (elements 4–7) into both groups:

```
> tt <- update(tt,x=heights.female[4:7], y=heights.male[4:7])
Triangular Test for normal distribution
Sigma known: 7

H0: mu1=mu2= 170  versus H1: mu1= 170  and mu2>= 176  or
mu2<= 164 alpha: 0.05  beta: 0.2

Test finished: accept H1
Sample size for x:  6
Sample size for y:  5
```

It turns out that 3 more samples for group 1 and 2 more samples in group 2 would have been sufficient, the resulting sizes are $n_x = 6$ and $n_y = 5$ (Fig. 21.5)

21.4 Summary

The previous sections give a short overview of some functions of the `R` library OPDOE which is the companion package to the book [3]. It is built as a collection of recipes for several tasks of design of experiments. It implements its own functions and reuses existing design of experiments functions and packages and tries to introduce a common naming scheme for these functions.

21.4.1 OPDOE Installation

Early releases of OPDOE had to be downloaded separately[3] and installed manually. Recent versions are part of the package collection at CRAN and can be installed the standard way using the `install.packages()` function. This also involves the automatic installation of some other needed libraries:

- `conf.design` for symmetric confounded factorial designs.
- `orthopolynom` for Legendre polynomials used in design of experiments for polynomial regression.
- `crossdes`, `gmp` for BIBD (not yet finished).
- `mvtnorm`, `nlme` for implementing Bechhofers selection rules.

21.4.2 OPDOE Contents

The package covers functions for completely randomized designs, ANOVA, sequential testing, regression analysis and more. It also contains some helper functions needed, e.g., for balanced block designs like a Hadamard matrix generator.

Acknowledgements Thanks go to the co-authors of [3], Minghui Whang, who wrote most of the ANOVA functions, and Petr Simeček who wrote the initial version of the library.

References

1. Baume, O.P., Gebhardt, A., Gebhardt, C., Heuvelink, G.B.M., Pilz, J.: Network optimization algorithms and scenarios in the context of automatic mapping. Comput. Geosci. **37**(3), 289–294 (2011)
2. R Development Core Team: R: A Language and Environment for Statistical Computing. R Foundation for Statistical Computing, Vienna (2013)
3. Rasch, D., Pilz, J., Verdooren, R., Gebhardt, A.: Optimal Experimental Design with R. Chapman and Hall/CRC, Boca Raton (2011)

[3]http://wwwu.aau.at/agebhard/OPDOE/.

4. Rasch, D., Wang, M., Herrendörfer, G.: Determination of the size of an experiment for the F-test in the analysis of variance. model i. In: Advances in Statistical Software. The 9th Conference on the Scientific Use of Statistical Software, vol. 6. Heidelberg (1997)
5. Rasch, D., Spangl, B., Wang, M.: Minimal experimental size in the three way ANOVA cross classification model with approximate F-tests. J. Stat. Plann. Inference **52**, 219-231 (2011)
6. Whitehead, J., Bruniera, H.: The double triangular test: a sequential test for the two-sided alternative with early stopping under the null hypothesis. Seq. Anal. **9**(2), 117–136 (1990)
7. Whitehead, J.: The Design and Analysis of Sequential Clinical Trials, revised 2nd edn. Wiley, Chichester (1997)

Chapter 22
The Influence of the Dependency Structure in Combination-Based Multivariate Permutation Tests in Case of Ordered Categorical Responses

Rosa Arboretti Giancristofaro, Eleonora Carrozzo, Iulia Cichi, Vasco Boatto, and Luigino Barisan

22.1 Introduction

The comparison of two multivariate populations via hypothesis testing is a considerable task in many applied research fields. For example, in some biostatistical problem such as shape analysis the goal is at comparing two populations considering a possible large set of two- or three-dimensional coordinates called landmarks [2]; in a similar way, quite often in genomics we want to compare two populations using a large set of microarray data. Finally, the multivariate two-sample location problem is the main methodological background of the multivariate control charts, which represents one of the most important tools of Statistical Process Control techniques [1]. In such multidimensional applications a quite important problem occurs when the analyzed variables are correlated and their associated regression forms are different (linear, quadratic, exponential, general monotonic, etc.).

The NonParametric Combination of a finite number of dependent permutation tests (NPC; [5]) is a suitable approach to cover almost all real situations of practical interest since the dependence relations among partial tests are implicitly captured by the combining procedure itself.

One open problem related to NPC-based tests is the possibility for the experimenter to manage with the impact of the dependency structure on the possible significance of combined tests.

R.A. Giancristofaro (✉) • I. Cichi • V. Boatto • L. Barisan
Department of Land, Environment, Agriculture and Forestry, University of Padova, Padova, Italy
e-mail: rosa.arboretti@unipd.it

E. Carrozzo
Department of Management and Engineering, University of Padova, Padova, Italy

V.B. Melas et al. (eds.), *Topics in Statistical Simulation*, Springer Proceedings in Mathematics & Statistics 114, DOI 10.1007/978-1-4939-2104-1_22

The aim of this work is to investigate the influence of the dependency structure in combination-based permutation tests for the multivariate two-sample location problem.

The present paper is organized as follows: Sect. 22.2 provides a short overview on multivariate permutation tests and the NPC methodology. In Sect. 22.3, the main results of a simulation study are shown and discussed, and finally section "Conclusions" deals with conclusions, final remarks, and future perspectives.

22.2 Multivariate Permutation Tests and Nonparametric Combination Methodology

For any general testing problem, in the null hypothesis (H_0), which usually assumes that data come from only one (with respect to groups) unknown population distribution P, the whole set of observed data $\mathbf{Y}$ is considered to be a set of exchangeable observations, taking values on sample space $\mathcal{Y}^n$, where $\mathbf{Y}$ is one observation of the n-dimensional sampling variable and where this random sample does not necessarily have independent and identically distributed (i.i.d.) components. We note that the observed data set $\mathbf{Y}$ is always a set of sufficient statistics under H_0 for any underlying distribution [5]. Since, in the null hypothesis and assuming exchangeability, the conditional probability distribution of a generic point $\mathbf{Y}' \in \mathcal{Y}^n$, for any underlying population distribution $P \in \mathcal{P}$, is distribution-independent, permutation inferences are invariant with respect to the underlying distribution in H_0. Some authors, emphasizing this invariance property, prefer to give them the name of invariant tests. However, due to this invariance property, permutation tests are distribution-free and nonparametric. Permutation tests have general good properties such as exactness, unbiasedness, and consistency (see [4,5]).

In order to provide details on the construction of multivariate permutation tests by the NPC approach, let us consider two multivariate populations and the related two-sample multivariate hypothesis testing problem where p (possibly dependent) variables are considered. We focus on ordered categorical variables, but any of the presented procedures could be applied to continuous or binary data or multivariate data that consists of some continuous/binary and some other ordered categorical responses.

The main difficulties when developing a multivariate hypothesis testing procedure arise because of the underlying dependence structure among variables, which is generally unknown. Moreover, a global answer involving several dependent variables is often required, hence the main question is how to combine the information related to the p variables into one global test. In order to better explain the proposed approach let us denote the $n \times p$, $n = n_1 + n_2$, data set with $\mathbf{Y} = [\mathbf{Y}_1, \mathbf{Y}_2]$, where $\mathbf{Y}_1$ and $\mathbf{Y}_2$ are the $n_1 \times p$ and the $n_2 \times p$ samples drawn from the first and second population, respectively. In the framework of the NPC of Dependent Permutation Tests we suppose that, if the global null hypothesis $H_0 : Y_1 \stackrel{d}{=} Y_2$ of equality in distribution of the two populations is true, the hypothesis of exchangeability of random errors holds. Hence, the following set of mild conditions should be jointly satisfied:

(a) we suppose that for $\mathbf{Y} = [\mathbf{Y}_1, \mathbf{Y}_2]$ an appropriate p-dimensional distribution exists, $P_j \in \mathscr{P}$, $j = 1, 2$, belonging to a (possibly non-specified) family $\mathscr{P}$ of non-degenerate probability distributions;

(b) the null hypothesis H_0 states the equality of the mean vectors of the p variables in the two groups:

$$H_0 : Y_1 \stackrel{d}{=} Y_2$$

The null hypothesis H_0 implies the exchangeability of the individual data vector with respect to the two groups. Moreover H_0 is supposed to be properly decomposed into p sub-hypotheses H_{0k}, $k = 1, \ldots, p$, each appropriate for partial (univariate) tests, thus H_0 (multivariate) is true if all the H_{0k} (univariate) are jointly true:

$$H_0 : \left[\bigcap_{k=1}^{p} Y_{1k} \stackrel{d}{=} Y_{2k} \right] = \left[\bigcap_{k=1}^{p} H_{0k} \right].$$

H_0 is called the *global* or *overall null hypothesis*, and H_{0k}, $k = 1, \ldots, p$, are called the *partial null hypotheses*. It is worth noting that the decomposition of the global null hypothesis into a set of partial null hypotheses does not mean that the equality of all marginal means implies the equality of the mean vectors, but it should be interpreted as a way to express an overall hypothesis in an equivalent form [5]. Substantially, this approach corresponds to a method of analysis carried out in two phases: the first focusing on p partial location aspects, and the second on their combination that should be referred to the global location aspect;

(c) the alternative hypothesis H_1 can be represented by the union of partial H_{1k} sub-alternatives:

$$H_1 : \left[\bigcup_{k=1}^{p} H_{1k} \right].$$

hence, H_1 is true if at least one of sub-alternatives is true. In this context, H_1 is called the *global* or *overall alternative*, and H_{1k}, $k = 1, \ldots, p$, are called the *partial alternatives*. Note that each univariate sub-alternatives can be expressed in the form $H_{1k} : Y_{1k} \stackrel{d}{\neq} Y_{2k}$ or $H_{1k} : Y_{1k} \stackrel{d}{>} Y_{2k}$ $H_{1k} : Y_{1k} \stackrel{d}{<} Y_{2k}$;

(d) let $\mathbf{T} = T(\mathbf{Y})$ represent a p-dimensional vector of test statistics, $p \geqslant 1$, whose components T_k, $k = 1, \ldots, p$, represent the partial univariate and non-degenerate *partial tests* appropriate for testing the sub-hypothesis H_{0k}

against H_{1k}. In case of ordered categorical responses (with S ordered categories) and one-sided alternatives, a suitable test statistic is the Anderson–Darling, i.e.:

$$T_k = \sum_{s=1}^{S-1} N_{1sk} \left[N_{\cdot sk}(n_1 + n_2 - N_{\cdot sk})\right]^{-\frac{1}{2}},$$

where $N_{\cdot sk} = N_{1sk} + N_{2sk}$ are the cumulative frequencies. Without loss of generality, all partial tests are assumed to be marginally unbiased, consistent, and significant for large values [5].

At this stage, in order to test the global null hypothesis H_0 and the p univariate hypotheses H_{0k}, the key idea comes from the partial (univariate) tests which are focused on the kth component variable, and then combining them through an appropriate combining function, to test the global (multivariate) test which is referred to as the global null hypothesis H_0.

However, we should observe that in most real problems when the sample sizes are large enough, there is a clash over the problem of computational difficulties in calculating the conditional permutation space. Hence, it is not possible to calculate the exact p-value λ_k of observed statistic T_{k0}. This is usually overcome by using the CMCP (Conditional Monte Carlo Procedure). The CMCP on the pooled data set $\mathbf{Y}$ is a random sampling from the set of all possible permutations of the same data under H_0. Hence, in order to obtain an estimate of the permutation distribution under H_0 of all test statistics, a CMCP can be used. Every resampling without replacement $\mathbf{Y}^*$ from the data set $\mathbf{Y}$ actually consists of a random attribution of the individual block data vectors to the two treatments. In every $\mathbf{Y}_b^*$ resampling, $b = 1, \ldots, B$, the k partial tests are calculated to obtain the set of values $\left[\mathbf{T}_b^* = \mathbf{T}(\mathbf{Y}_{bk}^*), k = 1, \ldots, p; b = 1, \ldots, B\right]$, from the B independent random re-samplings. It should be emphasized that CMCP only considers permutations of individual data vectors, so that all underlying dependence relations which are present in the component variables are preserved.

Without loss of generality, let us suppose that partial tests are significant for large values. More formally, the steps of the CMC procedure are described as follows:

1. calculate the p-dimensional vectors of statistics, each one related to the corresponding partial tests from the observed data:

$$\mathbf{T}^{\text{obs}}_{p\times 1} = \mathbf{T}(\mathbf{Y}) = \left[T_k^{\text{obs}} = T_k(\mathbf{Y}), k = 1, \ldots, p\right],$$

2. calculate the same vectors of statistics for the permuted data:

$$\mathbf{T}_b^* = \mathbf{T}(\mathbf{Y}_b^*) = \left[\mathbf{T}_{bk}^* = T_k(\mathbf{Y}_b^*), k = 1, \ldots, p\right],$$

3. repeat the previous step B times independently. We denote with $\{\mathbf{T}_b^*, b = 1, \ldots, B\}$ the resulting sets from the B conditional resamplings. Each element represents a random sample from the p-variate permutation c.d.f. $F_T(z|\mathbf{Y})$ of the test vector $\mathbf{T}(\mathbf{Y})$.

The resulting estimates are:

$$\hat{\lambda}_k = \left[\frac{1}{2} + \sum_{b=1}^{B} I\left(\mathbf{T}^*_{bk} \geqslant T_k^{\text{obs}}\right)\right] \Big/ (B+1), k = 1, \ldots, p,$$

where $I(\cdot)$ is the indicating function and where with respect to the traditional EDF estimators, 1/2 and 1 have been added, respectively, to the numerators and denominators in order to obtain estimated values in the open interval (0,1), so that transformations by inverse CDF of continuous distributions are continuous, i.e. are always well defined.

Hence, if the null hypothesis corresponding to the kth variable (H_{0k}) is rejected at significance level equal to α.

Moreover, choice of partial tests has to provide that:

1. all partial tests T_k are marginally unbiased, formally:

$$P\{T_k \geqslant z|\mathbf{Y}, H_{0k}\} \leqslant P\{T_k \geqslant z|\mathbf{Y}, H_{1k}\}, \forall z \in R^1$$

2. all partial tests are consistent, i.e.

$$P\{T_k \geqslant T_{k\alpha}|H_{1k}\} \rightarrow 1, \forall \alpha > 0 \quad \text{as} \quad n \rightarrow \infty$$

   where $T_{k\alpha}$ is a finite α-level for T_k.

Let us now consider a suitable continuous non-decreasing real function, $\varphi : (0,1)^p \rightarrow P^1$, that applied to the p-values of partial tests T_k defines the second order global (multivariate) test T'',

$$T'' = \varphi(\lambda_1, \ldots, \lambda_p)$$

provided that the following conditions hold:

- φ is non-increasing in each argument: $\varphi(\ldots, \lambda_k, \ldots) \geqslant \varphi(\ldots, \lambda'_k, \ldots)$, if $\lambda_k \leqslant \lambda'_k, k = 1, \ldots, p$;
- φ attains its supremum value $\bar{\varphi}$, possibly not finite, even when only one argument attains zero:

$$\varphi(\ldots, \lambda_k, \ldots) \rightarrow \bar{\varphi} \quad \text{if} \quad \lambda_k \rightarrow 0, k = 1, \ldots, p;$$

- φ attains its infimum value $\underline{\varphi}$, possibly not finite, even when only one argument attains one:

$$\varphi(\ldots, \lambda_k, \ldots) \rightarrow \underline{\varphi} \quad \text{if} \quad \lambda_k \rightarrow 1, k = 1, \ldots, p;$$

- $\forall \alpha > 0$, the acceptance region is bounded: $\underline{\varphi} < T''_{\alpha/2} < T'' < T''_{1-\alpha/2} < \bar{\varphi}$.

Frequently used combining function are:

- Fisher combination: $\varphi_F = -2\sum_k \log(\lambda_k)$;

- Tippet combination: $\varphi_T = \max_{1 \leqslant k \leqslant p} (1 - \lambda_k)$;
- Liptak combination: $\varphi_L = \sum_k \Phi^{-1}(1 - \lambda_k)$;

where $k = 1, \ldots, p$ and Φ is the standard normal c.d.f. It can be seen that under the global null hypothesis the CMC procedure allows for a consistent estimation of the permutation distributions, marginal, multivariate and combined, of the k partial tests. Usually, Fisher's combination function is considered mainly for its finite and asymptotic good properties. Of course, it would be also possible to take into consideration any other combining function (Lancaster, Mahalanobis, etc.; see [3, 5]). The combined test is also unbiased and consistent.

It is worth noting that NPC Tests overcome some limitations of traditional multivariate hypothesis testing procedures, such as the ability to include a large number of variables, and offer several advantages: (1) it is always an exact inferential procedure, for whatever finite sample size; (2) it is a robust solution with respect to the true underlying random error distribution; (3) it implicitly takes into account the underlying dependence structure of response variables and (4) it is not affected by the problem of the loss of the degrees of freedom when keeping fixed the number of observations, and the number of informative variables or aspects increases.

22.3 A Simulation Study

In order to investigate the influence of the dependency structure in combination-based permutation tests for the multivariate two-sample location problem, we performed a Monte Carlo simulation study. The rationale of the simulation study was focused on investigating how power of the NPC tests is affected by the different strength of dependence for random errors in case of balanced design with small sample sizes commonly used in real applications. More specifically, the simulation study considered 4,000 independent data generation of samples and was designed to take into account for different settings.

Let us consider two multivariate populations and the related two-sample multivariate hypothesis testing problem where three variables with different dependence structure are considered. The number of ordered categories for each variable equals to 6. In order to generate ordered categorical variables we rounded continuous values to the nearest integer [6, 8].

We referred as test statistic to the Anderson–Darling permutation test and the Fisher's combining function with the hypotheses:

$$H_0 : \{\mathbf{X}_A \overset{d}{=} \mathbf{X}_B\}, \quad H_1 : \{\mathbf{X}_A \overset{d}{>} \mathbf{X}_B\},$$

The null permutation distribution was estimated by $B = 4{,}000$ CMC iterations.

Tables 22.1, 22.2, and 22.3 display the simulation results in terms of rejections rates under the alternatives. Bold values are referred to results involving correlated variables. Rejection rates have been calculated setting $\alpha = 0.05$.

Table 22.1 Simulation results: homoscedastic linear dependence

	Setting 1: $\sigma_{12} = 0.75$			Setting 2: $\sigma_{12} = 0.50$			Setting 3: $\sigma_{12} = 0.25$			Setting 0: I_3		
	Partial tests			Partial tests			Partial tests			Partial tests		
α	T_1	T_2	T_3	T_1	T_2	T_3	T_1	T_2	T_3	T_1	T_2	T_3
0.01	0.008	0.096	0.009	0.01	0.094	0.011	0.01	0.099	0.007	0.009	0.087	0.011
0.05	0.045	0.266	0.045	0.047	0.264	0.053	0.049	0.272	0.044	0.047	0.253	0.045
0.10	0.097	0.394	0.095	0.096	0.399	0.111	0.099	0.404	0.091	0.09	0.387	0.095
	Combined tests			Combined tests			Combined tests			Combined tests		
	(pairs)			(pairs)			(pairs)			(pairs)		
α	T_{12}	T_{13}	T_{23}	T_{12}	T_{13}	T_{23}	T_{12}	T_{13}	T_{23}	T_{12}	T_{13}	T_{23}
0.01	**0.038**	0.01	0.061	**0.044**	0.011	0.063	**0.056**	0.007	0.065	0.059	0.01	0.061
0.05	**0.156**	0.048	0.193	**0.164**	0.049	0.201	**0.183**	0.043	0.192	0.18	0.05	0.18
0.10	**0.264**	0.094	0.306	**0.268**	0.105	0.311	**0.295**	0.092	0.307	0.297	0.095	0.289
	Global			Global			Global			Global		
	Combined test			Combined test			Combined test			Combined test		
α		T_{123}			T_{123}			T_{123}			T_{123}	
0.01		**0.035**			**0.045**			**0.047**			0.045	
0.05		**0.138**			**0.155**			**0.151**			0.154	
0.10		**0.238**			**0.248**			**0.249**			0.254	
	$\rho_{12} = 0.66$			$\rho_{12} = 0.43$			$\rho_{12} = 0.22$			$\rho_{12} = 0.00$		

Table 22.2 Simulation results: heteroscedastic non linear dependence

	Setting 1 $b = 0.85, \sigma_{12} = 0.75$			Setting 2 $b = 0.45, \sigma_{12} = 0.50$			Setting 3 $b = 0.18, \sigma_{12} = 0.25$		
	Partial tests			Partial tests			Partial tests		
α	T_1	T_2	T_3	T_1	T_2	T_3	T_1	T_2	T_3
0.01	0.009	0.081	0.013	0.010	0.135	0.013	0.010	0.160	0.010
0.05	0.048	0.242	0.053	0.049	0.341	0.058	0.051	0.369	0.050
0.10	0.103	0.368	0.099	0.099	0.481	0.107	0.098	0.515	0.090
	Combined tests (pairs)			Combined tests (pairs)			Combined tests (pairs)		
α	T_{12}	T_{13}	T_{23}	T_{12}	T_{13}	T_{23}	T_{12}	T_{13}	T_{23}
0.01	**0.041**	0.010	0.060	**0.059**	0.012	0.089	**0.090**	0.010	0.109
0.05	**0.139**	0.056	0.180	**0.208**	0.056	0.253	**0.253**	0.052	0.279
0.10	**0.234**	0.103	0.290	**0.332**	0.107	0.387	**0.378**	0.098	0.402
	Global Combined test			Global Combined test			Global Combined test		
α		T_{123}			T_{123}			T_{123}	
0.01		**0.036**			**0.058**			**0.080**	
0.05		**0.135**			**0.188**			**0.216**	
0.10		**0.224**			**0.297**			**0.334**	
	$\rho_{12} = 0.69$			$\rho_{12} = 0.48$			$\rho_{12} = 0.22$		

Table 22.3 Simulation results: quadratic dependence

	Setting 1 $c = 0.82, \sigma_{12} = 0.75$			Setting 2 $c = 0.42, \sigma_{12} = 0.50$			Setting 3 $c = 0.15, \sigma_{12} = 0.20$		
	Partial tests			Partial tests			Partial tests		
α	T_1	T_2	T_3	T_1	T_2	T_3	T_1	T_2	T_3
0.01	0.009	0.049	0.009	0.011	0.075	0.009	0.01	0.094	0.01
0.05	0.047	0.177	0.042	0.049	0.228	0.047	0.044	0.264	0.044
0.10	0.098	0.282	0.098	0.097	0.352	0.098	0.093	0.396	0.1
	Combined tests (pairs)			Combined tests (pairs)			Combined tests (pairs)		
α	T_{12}	T_{13}	T_{23}	T_{12}	T_{13}	T_{23}	T_{12}	T_{13}	T_{23}
0.01	**0.034**	0.009	0.036	**0.053**	0.01	0.05	**0.061**	0.01	0.062
0.05	**0.135**	0.044	0.129	**0.168**	0.05	0.172	**0.191**	0.047	0.193
0.10	**0.232**	0.089	0.221	**0.270**	0.098	0.276	**0.307**	0.092	0.309
	Global Combined test			Global Combined test			Global Combined test		
α		T_{123}			T_{123}			T_{123}	
0.01		**0.030**			**0.041**			**0.048**	
0.05		**0.110**			**0.146**			**0.158**	
0.10		**0.196**			**0.245**			**0.264**	
	$\rho_{12} = 0.69$			$\rho_{12} = 0.48$			$\rho_{12} = 0.22$		

The first four settings concerned to an homoscedastic linear dependence among two of the three variables, with $n_A = n_B = 20$ and $\mathbf{Y}_A \sim \mathcal{N}_3(\boldsymbol{\mu}_A, \Sigma)$, $\boldsymbol{\mu}_A = (0, 0.35, 0)$, $\mathbf{Y}_B \sim \mathcal{N}_3(\boldsymbol{\mu}_B, \Sigma)$, $\boldsymbol{\mu}_B = (0, 0, 0)$. The correlation index σ_{12} increased from 0.25 to 0.75 (Table 22.1). In this situation we have one variable under the alternative and the other two variables under the null hypothesis.

The results show how there is a decreasing power when dependence increases. If we consider at first the combination among pairs of tests, we can see that T_{12} has a lower power with respect to T_{23} for example, where T_{23} is a combined test between two independent variables, while T_{12} is a combined test between two dependent variables with different degrees of dependence showed in the table. The global combined test which combines all the variables is less affected by the dependence with respect to the combination in pairs. When the correlation between two of the three variables is high then the power decreases, but when the correlation is not so high, for example when the estimated correlation is 0.22, the power is more or less comparable with the case of three independent partial tests. So there is an influence of the correlation if both variables are correlated, but combining more than two variables adding uncorrelated variables helps in decreasing the intensity of the problem of the correlation. For the global combined test T_{123}, the power is less affected with respect to the power of T_{12}.

The second set of simulation considered an homoscedastic non-linear dependence among two of the three variables, with $n_A = n_B = 20$ and the following configurations:

$$\begin{aligned}
&Y_{1A}, Y_{1B}, Z_{2A}, Z_{2B} \sim \mathcal{N}(0, 1), \text{ i.i.d.};\\
&Y_{3A}, Y_{3B} \sim \mathcal{N}(0, 0.5^2), \text{ i.i.d.}, \boldsymbol{\mu}_A = (0, 0, 0), \boldsymbol{\mu}_B = (0, 0.35, 0);\\
&Y_{2A} = \mu_{2A} + bY_{1A} + f(Y_{1A})Z_{2A}, \; f(Y_{1A}) = \exp(-\beta|Y_{1A}|); \; \beta = 1/2;\\
&Y_{2B} = \mu_{2B} + bY_{1B} + f(Y_{1B})Z_{2B}, \; f(Y_{1B}) = \exp(-\beta|Y_{1B}|); \; \beta = 1/2;
\end{aligned}$$

The correlation index σ_{12} increased from 0.25 to 0.75 and the parameter b varied assuming the values: 0.18, 0.45 and 0.85 (Table 22.2).

When we consider one variable under the alternative hypothesis and the other variables under the null hypothesis and there is a correlation between two variables, the combined test of this two variables T_{12} presents a decreasing power. The global combined test T_{123} also shows a decreasing power, more strong when the correlation is high, less strong when the correlation is low. The correlation influences a little bit more the global combined test with respect the previous case showed in Table 22.1.

The next settings are related to a quadratic dependence between two of the three variables, with $n_A = n_B = 20$ and the following configurations:

$$\begin{aligned}
&Y_{1A}, Y_{1B}, Y_{3A}, Y_{3B}, Z_{2A}, Z_{2B} \sim \mathcal{N}(0, 1), \text{ i.i.d.};\\
&Y_{2A} = \mu_{2A} + c(Y_{1A})^2 + Z_{2A}, \; Y_{2B} = \mu_{2B} + c(Y_{1B})^2 + Z_{2B};\\
&\boldsymbol{\mu}_A = (0, 0, 0), \; \boldsymbol{\mu}_B = (0, 0.35, 0);
\end{aligned}$$

The correlation index σ_{12} increased from 0.20 to 0.75 and the parameter c varied assuming the values: 0.15, 0.42, and 0.82 (Table 22.3).

The results of the simulations show that the combined test T_{12} loses power when the correlation is intermediate or moderate. The problem in this situation is mostly related to the global combined test which is affected by the dependence even if the correlation is low. With quadratic dependence the problem of correlation affects not only the combined test of the two correlated variables but also the global combined test.

Focusing our attention on the role played by the correlation between variables, the results of the simulations show how as correlation among variables increases, the performance of NPC tests is negatively affected and this is explained by the fact that in the observed dataset we expect to have a relatively less amount of information useful to detect the difference between the location parameters.

Conclusion
The goal of this paper was to verify the influence of the dependency structure among variables on the power of multivariate combination-based permutation tests in multidimensional applications.

On the basis of the results of the simulation study, correlation seems to affect combination-based permutation tests by reducing power of multivariate tests.

Future prospects concern some specific procedures aimed at possibly improving power of multivariate combination-based permutation tests. The application of special forms of combination function known as truncated product method [7] is under investigation to verify if it is possible to mitigate the negative effect on the power of combination-based multivariate permutation tests produced by an increasing level of correlation/association among responses.

References

1. Bersimis, S., Psarakis, S., Panaretos, J.: Multivariate statistical process control charts: an overview. Qual. Reliab. Eng. Int. **23**, 517–543 (2007)
2. Brombin, C., Salmaso, L.: Multi-aspect permutation tests in shape analysis with small sample size. Comput. Stat. Data Anal. **53**, 3921–3931 (2009)
3. Folks, J.L.: Combinations of independent tests. In: Krishnaiah, P.R., Sen, P.K. (eds.) Handbook of Statistics,vol. 4, pp. 113–121. North-Holland, Amsterdam (1984)
4. Hoeffding, W.: The large-sample power of tests based on permutations of observations. Ann. Math. Stat. **23**, 169–192 (1952)
5. Pesarin, F., Salmaso, L.: Permutation Tests for Complex Data: Theory, Applications and Software. Wiley, Chichester (2010)
6. Vale, C., Maurelli, V. Simulating multivariate nonnormal distributions. Psychometrika **48**, 465–471 (1983)
7. Zaykin, D.V., Zhivotovsky, L.A., Westfall, P.H., Weir, B.S.: Truncated product method for combining p-values. Genet. Epidemiol. **22**, 170–185 (2002)
8. Zopluoglu, C.: Applications in R: generating multivariate non-normal variables. University of Minnesota, http://www.tc.umn.edu/~zoplu001/resim/gennonnormal.pdf (2011)

Chapter 23
Potential Advantages and Disadvantages of Stratification in Methods of Randomization

Aenne Glass and Guenther Kundt

23.1 Motivation

Clinical trials are an established method to evaluate the effectiveness and safety of a new medication to diagnose or treat a disease. To reduce the risk of randomization-associated imbalance between treatment groups for known factors which might influence therapy response, patients are randomized in strata. Besides, stratification may help to prevent type I and type II errors via reduction of variance in several trial constellations, e.g. protection against trial site drop out.

On the other hand, stratification requires administrative effort, and an increasing number of strata decreases the sample size in each stratum.

Against this background the following cost-benefit-questions rise: How useful is stratified randomization really, compared to the unstratified case? According to which criteria should one decide whether to stratify randomization or not? Are these criteria the prevalence of a prognostic factor, the trial size, or others? How is each criterion to be weighted?

23.2 Methods

To investigate a shortlist of potential advantages and disadvantages of stratification in methods of randomization, firstly, CR was considered to quantify the basical risk of imbalance due to chance [1]. Secondly, restricting this chance by using PBR(B), the risk of imbalance of success rates π under H_0: $\pi_1 = \pi_2$ was

---

A. Glass (✉) • G. Kundt
Institute for Biostatistics and Informatics in Medicine and Ageing Research, University Medicine Rostock, Rostock, Germany
e-mail: aenne.glass@uni-rostock.de; guenther.kundt@uni-rostock.de

V.B. Melas et al. (eds.), *Topics in Statistical Simulation*, Springer Proceedings in Mathematics & Statistics 114, DOI 10.1007/978-1-4939-2104-1_23

Table 23.1 Simulated probability P of exceeding a clinically relevant prognostic imbalance $I = 10$ pps of two treatment groups after CR, depending on trial size N and prevalence of a prognostic factor ($P \leqslant 0.05$ marked bold, underlined values are referred to in the text)

Probability of reaching $I > 10\,\%$	Prevalence of a prognostic factor					
Trial size N	10 %	15 %	25 %	30 %	40 %	50 %
N = 30 patients	0.347	0.436	0.531	0.550	0.589	0.587
N = 50 patients	0.235	0.320	0.419	0.445	0.476	0.487
N = 100 patients	0.094	0.160	0.243	0.276	0.307	0.320
N = 200 patients	**0.019**	**0.048**	0.101	0.121	0.152	0.156
N = 500 patients	**<0.001**	**0.002**	**0.010**	**0.014**	**0.023**	**0.025**
N = 1,000 patients	**<0.001**	**<0.001**	**<0.001**	**<0.001**	**0.0014**	**0.0018**

simulated [2], and compared for the stratified vs. the unstratified case. Thus, the effects of stratification could be discussed from different angles. Differently designed hypothetical trials were computer simulated (at least 1,000 times) for two therapy groups and two strata. We used two different simulation approaches and calculated the probability of observing

1. a clinically relevant imbalance of a prognostic factor of more than 10 pps between two treatment groups, caused by complete (unstratified) randomization, cf. Table 23.1,

as well as

2. clinically relevant, cf. Tables 23.2 and 23.3, or statistically significant, cf. Table 23.4, differences between the endpoint success rates of two treatments after unstratified/stratified PBR(B), when both treatments were equally effective.

Now we can quantify the impact of stratification on the risk of imbalance for particular trial situations. Thus the investigator is supported in his decision whether stratification could be a valuable feature for the current clinical trial.

23.3 Results

1. The risk of randomization-associated imbalance that two therapy groups will differ for the prognostic factor by more than 10 pps after CR is topping out at almost 59 % (587 of 1,000 trials), depending on trial size N and prevalence of a prognostic factor. Table 23.1 shows the risk of prognostic imbalance between therapy groups for a broad range of trial constellations, including small ($N = 30$ and 50), middle-sized ($N = 100$ and 200) and large ($N = 500$ and 1,000) trials and different factor prevalences (10 %, 15 %, 25 %, 30 %, 40 %, 50 %).

   The risk of imbalance is minimum for large trials ($N = 1{,}000$ patients) and/or small factor prevalence (10 %), but multiplies according to a more prevalent

Table 23.2 Simulated frequencies of observing a clinically relevant difference of more than 10 pps in success rates between two treatment groups after PBR(10), for the stratified vs. the unstratified case, under the null hypothesis ($\pi_1 = \pi_2$)

Success rates in strata (%)	Difference of success rates between strata	Small trials $N = 100$ patients		Middle-sized trials $N = 400$ patients	
		Stratif. rand.	Unstratif. rand.	Stratif. rand.	Unstratif. rand.
50 vs. 50	Difference 0	321	296	41	33
40 vs. 60	Difference 20	279	267	36	35
30 vs. 50		287	262	29	43
20 vs. 40		258	222	28	26
10 vs. 30		179	150	11	6
30 vs. 60	Difference 30	274	270	33	47
20 vs. 50		263	258	27	32
10 vs. 40		199	202	17	23
30 vs. 70	Difference 40	257	287	26	41
20 vs. 60		254	292	28	30
10 vs. 50		218	235	16	30
20 vs. 80	Difference 60	212	255	14	25
10 vs. 70		181	267	10	34
10 vs. 90	Difference 80	101	264	1	32

Reductions of frequencies by stratification are highlighted in grey, underlined values are referred to in the text

Table 23.3 The benefit of stratification, based on simulated frequencies of observing a clinically relevant difference in endpoint success rates between two treatment groups after PBR(10), for the stratified vs. the unstratified case, under the null hypothesis ($\pi_1 = \pi_2$), $N = 100$

Average success rates (%)	Success rates between strata (%)	Differences of success rates between strata (%)	Stratif. rand.	Unstratif. rand.	Benefit of stratification (pps)
50	50 vs. 50	Difference 0	321	296	−2.5
	40 vs. 60	Difference 20	279	267	−1.2
	30 vs. 70	Difference 40	257	287	3
	20 vs. 80	Difference 60	212	255	4.3
	10 vs. 90	Difference 80	101	264	16.3
45	30 vs. 60	Difference 30	274	270	−0.4
	20 vs. 70	Difference 50	243	262	1.9
	10 vs. 80	Difference 70	158	256	9.8
40	30 vs. 50	Difference 20	287	262	−2.5
	20 vs. 60	Difference 40	254	292	3.8
	10 vs. 70	Difference 60	181	267	8.6
35	20 vs. 50	Difference 30	263	258	−0.5
	10 vs. 60	Difference 50	207	264	5.7
30	20 vs. 40	Difference 20	258	222	−3.6
	10 vs. 50	Difference 40	218	235	1.7
25	10 vs. 40	Difference 30	199	202	0.3
20	10 vs. 30	Difference 20	179	150	−2.9

Reductions of frequencies by stratification are highlighted in grey, underlined values are referred to in the text

Table 23.4 Simulated frequencies of observing a statistically significant difference ($P \leqslant 0.05$) in success rates between two treatment groups after PBR(10), for the stratified vs. the unstratified case, under the null hypothesis ($\pi_1 = \pi_2$)

Success rates in strata (%)	Difference of success rates between strata	Trial size $N = 100$ patients		Trial size $N = 400$ patients	
		Stratif. rand.	Unstratif. rand.	Stratif. rand.	Unstratif. rand.
50 vs. 50	Difference 0	67	55	54	53
30 vs. 30		51	46	41	49
5 vs. 5		50	58	48	59
40 vs. 60	Difference 20	67	52	45	44
30 vs. 50		44	57	51	51
20 vs. 40		49	54	40	45
10 vs. 30		53	44	42	36
30 vs. 60	Difference 30	36	54	52	37
20 vs. 50		38	44	39	51
10 vs. 40		30	55	33	49
30 vs. 70	Difference 40	36	59	25	58
20 vs. 60		38	58	33	53
10 vs. 50		29	50	27	60
20 vs. 80	Difference 60	17	64	22	50
10 vs. 70		14	53	22	46
10 vs. 90	Difference 80	1	57	4	55

The two-sided (unadjusted) chi-square test was used. Reductions of frequencies by stratification are highlighted in grey, underlined values are referred to in the text prognostic factor (up to 50 %), and/or for smaller trials with down to $N = 30$ patients. It is at least 23.5 % in small trials ($N \leqslant 50$), independently of factor prevalence, and up to 2.5 % in large trials ($N \geqslant 500$).

In particular, for $N = 30$ patients the risk almost doubles (35–59 %) for increasing factor prevalence from 10–50 %, and in larger trials with $N = 100$ or 200 patients it ranges between 1.9–32 %. In case of low prevalence $\leqslant$ 15 % and $N \geqslant 200$ patients the risk is acceptably small ($\leqslant$ 5 %), similarly to large trials with $N \geqslant 500$ patients for any factor prevalence.

Since the risk of imbalance is highest for a high factor prevalence of 50 %, for greater insight we give some probabilities for only this prevalence for trials of $N = 400$ patients ($P = 4.6$ %) and $N = 300$ ($P = 8.3$ %). Obviously, the risk is acceptable small for 400 patients, and accepting a risk of 10 %, a trial of even $N = 300$ patients (150 per trial arm) is not necessarily to be stratified.

2. Since the prevalence of a prognostic factor in clinical trials is given, we focus on other potential influencing values that could be modified to reduce the risk of imbalance, when planning a trial. As known and depictable from Table 23.1 this could obviously be the trial size. However more interesting with respect to economic aspects may be the reduction of risk by pre-stratification of randomization, processing randomization separately for each stratum.

   We show in Tables 23.2, 23.3, and 23.4 the frequencies per 1,000 trials that observed differences between treatment groups exceeded clinically relevant and statistically significant differences, respectively, even though in populations no difference exists (π_1=π_2). Results are given for several success rates and differences between strata. We compare the stratified vs. the unstratified case for PBR(B) with a block size $B = 10$, and thus, illustrate the impact of stratification to reduce the error risk for specific trial constellations. Unstratified permuted-block randomization can lead to a clinically relevant imbalance ($I > 10$ %) in up to 296 of 1,000 hypothetical trials ($P = 30$ %), if both the trial size is small ($N = 100$ patients), and population success rate is large (50 %), and success rates in strata do not differ, cf. Table 23.2. The number of hypothetical trials exceeding a clinically relevant difference was diminished by stratification between minimum 0.3 pps (stratified: 199 vs. unstratified: 202 per 1,000), and up to 16.3 pps (101 vs. 264). The reduction is proportional to differences of success rates between strata, and occurs for differences between (30–80 pps) for trials of $N = 100$. We demonstrated and specify the conclusions by [1,2] that stratification reduces type I error rates for clinical differences, if differences between stratum success rates are large (at least 30 pps), in small (101 vs. 264) and middle-sized (1 vs. 32) trials as well, even if the reduction seems to be of relevance rather in small trials. Since the error rate increases for smaller differences, we recommend to not stratify in that case in small and middle-sized trials.

   We present the frequencies for error I again to gain a more detailed insight into the impact of stratification. This time, we change the order of the rows of Table 23.2 and show the frequencies based on the average success rates in Table 23.3, instead of differences between strata. This way, we grasp the relationship between potential influences on the impact of stratification. The effect of stratified randomization depends both on the average success rate and differences between stratum success rates. It can obviously be maximized to 16.3 pps for a maximum success rate (50 %) and a maximum difference between stratum success rates (80 pps). In fact, high differences in success rates between strata can only be expected for high average success rates. Thus we conclude that the impact of stratification in randomization is related to the average success rates, and hence to the differences of success rates between strata as stated in [2].

The frequencies of type I error for statistically significant differences are about 50 per 1,000, as expected, cf. Table 23.4. The benefit of stratification in terms of the risk of exceeding a statistically significant difference between treatment groups was shown by a maximum reduction of type I error for a difference of 80 pps: from 55 per 1,000 (unstratified) to 4 per 1,000 (stratified) in trials of $N = 400$ patients, and from 57 to 1 per 1,000 in trials of $N = 100$. Although the risk of imbalance under H_0 via stratification could be reduced by 5.6 pps (57 per 1,000 to 1), the impact of stratification (very large strata differences of 80 pps in small trials with $N = 100$ patients) is rather of minor practical relevance.

Conclusion

The risk of randomization-associated prognostic imbalance > 10 pps between therapy groups of a clinical trial could be quantified in simulation studies with maximum 59 % for complete randomization, and thus, is highly important, cf. Table 23.1. In larger trials, and/or with a factor of less prevalence this risk decreases. Compared to the straightforward range of trial constellations investigated by [1] we show the risk to exceed $I = 10\,\%$ for even border-lined trial situations of very large trials and both very small and high prevalence of a prognostic factor. For large trials $\geqslant$ 500 patients, the risk will never exceed 3 %, independently of any factor prevalence, and thus, this trial situation can comfortably be conducted.

Restricted randomization as (unstratified) PBR(B) reveals results comparable to CR, concerning the risk of imbalance $P = 32\,\%$ (50 % factor prevalence, cf Table 23.1) vs. $P = 29.6\,\%$ (50 % average success rate, cf. Table 23.2) in small studies ($N = 100$).

For large superiority trials with $N > 400$ patients, a relevant risk for a prognostic imbalance was not observed, independently of any factor prevalence, and hence, it is not necessary to stratify. Our results confirm recommendations of [1, 2] to stratify in trials with $N < 400$ patients. The risk for an imbalance is <5 % in trials of $N = 400$ patients, and <9 % in trials of $N = 300$ patients. If trialists accept even a 9 % risk of imbalance, less effort has to made when planning a trial of $N = 300$, neither by enlarging the trial to $N > 150$ patients per arm, nor by pre-stratifying it.

(continued)

Otherwise, in small trials, stratification of randomization can be helpful to provide comparable groups with higher probability, for certain trial constellations and for clinically relevant differences. Reduced probabilities of type I error rates by maximum 16 pps due to stratification were detected for large differences of success rates between strata (80 pps) in small trials ($N = 100$), cf. Table 23.2. A reduction effect of stratification on type I error in small trials is not detectable unless differences between strata are $\geqslant$30 pps, so that for smaller differences should not be stratified.

From Table 23.3 we get more insight into the effect of stratification. The average success rate primarily influences the impact of stratification, rather than the differences in success rates between strata as presented in [2]. This is caused by the fact that high differences in success rates between strata are expected for high average success rates, if at all.

The detected effect of stratification on the frequencies of statistically significant differences in endpoint success rates is rated less relevant, cf. Table 23.4.

Taken together, if both trials are small (<150 patients per arm) and success rates between strata differ by 30 pps or more, stratification is recommended to reduce the expected risk of error I rate for clinically relevant differences under the assumption of H_0.

References

1. Kernan, W.N., et al.: Stratified randomization for clinical trials. J. Clin. Epidemiol. **52**(1), 19–26 (1999)
2. Feinstein, A.R., Landis, J.R.: The role of prognostic stratification in preventing the bias permitted by random allocation of treatment. J. Chronic Dis. **29**(4), 277–284 (1976)

Chapter 24
Additive Level Outliers in Multivariate GARCH Models

Aurea Grané, Helena Veiga, and Belén Martín-Barragán

24.1 Introduction

The correlation structure of security returns is the keystone of both portfolio allocation and risk management decisions. In the literature, there are several models to estimate correlations. They often belong to the class of multivariate GARCH models. In the univariate setting it is well known that extreme observations caused by jumps or the presence of outliers affect the estimation of GARCH parameters [10, 18, 19], the tests of conditional homoscedasticity [4, 13], and the out-of-sample volatility forecasts [3, 5, 11, 12, 15]. Moreover, when there are extreme returns standard GARCH models tend to overestimate the volatility the days following the presence of these extreme observations. Similar biases are expected to occur when the correlations are estimated using multivariate GARCH-type models.

The first objective of this paper is to study the effect of additive level outliers on the estimated correlations of three well-known multivariate GARCH models. The second aim is to propose an outlier detection procedure for multivariate GARCH models based on wavelets that can be interpreted as a misspecification test for the model. The procedure is based on the multivariate series of residuals and if outliers are detected in these series this implies a rejection of the model.

A. Grané (✉)
Universidad Carlos III de Madrid, c/ Madrid 126, 28903 Getafe, Spain
e-mail: aurea.grane@uc3m.es; agrane@est-econ.uc3m.es

H. Veiga
Instituto Flores de Lemus and Financial Research Center/UNIDE,
Universidad Carlos III de Madrid, Getafe, Spain
e-mail: mhveiga@est-econ.uc3m.es

B. Martín-Barragán
University of Edinburgh Business School, 29 Buccleuch Place, Edinburgh EH8 9JS, UK
e-mail: Belen.Martin@ed.ac.uk

V.B. Melas et al. (eds.), *Topics in Statistical Simulation*, Springer Proceedings in Mathematics & Statistics 114, DOI 10.1007/978-1-4939-2104-1_24

The organization of this paper is as follows. In Sect. 24.2 we present the volatility models under study and review the concept of additive level outlier in Sect. 24.3. The effects of outliers on the estimated correlations are analyzed in Sect. 24.4 via an intensive simulation study. In Sect. 24.5 we propose an outlier detection algorithm and evaluate its performance.

24.2 Models Under Study

The models under consideration are the diagonal Baba–Engle–Kraft–Kroner (D-BEKK) model defined in Engle and Kroner [9], the constant conditional correlation (CCC) model by Bollerslev [2], and the dynamic conditional correlation (DCC) model by Engle [8] because they are often applied empirically to many fields such as portfolio management, asset allocation, volatility spillover transmission, contagion, etc. (see [1] and [17] for excellent surveys on these models). However, the methodology developed in this paper is not restricted to these models.

Let $\{\mathbf{y}_t\}$ be a vector stochastic process with dimension $N \times 1$ such that $E(\mathbf{y}_t) = \mathbf{0}$ and $\mathscr{F}_{t-1}$ is the information set till time $t-1$. We consider that

$$\mathbf{y}_t = \mathbf{H}_t^{1/2} \boldsymbol{\eta}_t,$$

where $\mathbf{H}_t$ is the conditional covariance matrix of $\mathbf{y}_t$ and $\boldsymbol{\eta}_t$ is an iid vector error process such that $E(\boldsymbol{\eta}_t \boldsymbol{\eta}_t') = \mathbf{I}$, the identity matrix of order N. We assume that there is no linear dependence in $\mathbf{y}_t$. Different approaches in the literature propose different models for the dependence of $\mathbf{H}_t$ on past information $\mathscr{F}_{t-1}$.

In the D-BEKK, this dependence of $\mathbf{H}_t$ on past information is modeled directly. In contrast, in the CCC and DCC models, which belong to a subclass of the multivariate GARCH models called conditional correlation models, first the conditional variances and correlations are modeled using univariate specifications and then $\mathbf{H}_t$ is obtained by using these conditional standard deviations and correlations.

24.3 Additive Level Outliers

Additive level outliers (ALOs)[1] can be caused by institutional changes or market corrections that do not affect volatility. Then, the conditional mean equation is:

$$\mathbf{y}_t = \boldsymbol{\omega} \cdot I_T(t) + \mathbf{H}_t^{1/2} \boldsymbol{\eta}_t,$$

[1] We refer to the concept of ALO that appears in [14].

where $\boldsymbol{\eta}_t$ is defined as before, $\boldsymbol{\omega} = (\omega_1, \ldots, \omega_N)'$ is a vector containing the ALOs' sizes and $I_T(t) = 1$ for $t \in T$ and 0 otherwise, representing the presence of ALOs at a given set of times T. ALOs can occur simultaneously at the same time t or not and their sizes can coincide or not. The equation of the conditional variance–covariance remains the same, since ALOs only affect the level of the series. Regarding the conditional correlation models, the situation is similar. ALOs affect separately each conditional mean equation, supposing that each series of financial returns is modeled by a univariate GARCH–type model.

In the simulation study below ALOs are set in the same positions in the $N = 2$ simulated series to reproduce the scenario of contagion, usual in financial markets.

24.4 Effects of ALOs on the Correlations: A Simulation Study

In this section we implement an intensive simulation study to assess the impact of outliers on the estimated correlations. The frequency of the simulations is daily, outliers are placed randomly across the series and each scenario involves 1,000 replications.[2] We consider the following situations: Single or multiple isolated ALOs of two different sizes ($5\boldsymbol{\sigma}_{\mathbf{y}}$ and $10\boldsymbol{\sigma}_{\mathbf{y}}$) in simulated series from a CCC, DCC, and a D-BEKK models with errors following, respectively, univariate or multivariate Normal distributions. For each outlier size, the sample sizes considered are $n = 1{,}000, 3{,}000, 5{,}000$.

From Table 24.1 and Fig. 24.1 we observe that the estimated correlations are affected by the presence of ALOs and the relative errors are higher the higher is the ALO size, the higher the number of ALOs included in the simulated series and the smaller the sample sizes of the simulated time series. Moreover, the biases in the correlations are higher for the DCC model in comparison with the CCC and D-BEKK models. In particular, this latter model seems to be more robust to the presence of ALOs since the correlations present small relative errors over the sample size.

[2]Parameters used are: $\{\mathbf{C} = (0.053, 0.042, 0.020), \mathbf{A} = (0.161, 0.164), \mathbf{B} = (0.983, 0.981)\}$ for the D-BEKK; $\{\boldsymbol{\alpha}_0 = (0.010, 0.013), \boldsymbol{\alpha}_1 = (0.049, 0.067), \boldsymbol{\beta}_1 = (0.940, 0.926), \boldsymbol{\rho} = (1, -0.606)\}$ for the CCC and $\{\boldsymbol{\alpha}_0 = (0.010, 0.013), \boldsymbol{\alpha}_1 = (0.049, 0.067), \boldsymbol{\beta}_1 = (0.940, 0.926), \alpha = 0.015, \beta = 0.981\}$ for the DCC, which were chosen by fitting the models to real time series of financial returns.

Table 24.1 Average relative biases of the estimated CCC correlations for different sample sizes

	n	Estimated correlation	Relative bias		n	Estimated correlation	Relative bias		n	Estimated correlation
1 ALO	1,000	−0.5987	−0.012	3 ALOs	1,000	−0.5892	−0.028	No	1,000	−0.6060
$5\sigma_y$	3,000	−0.6042	−0.003	$5\sigma_y$	3,000	−0.6007	−0.009	ALOs	3,000	−0.6060
	5,000	−0.6051	−0.002		5,000	−0.6017	−0.008		5000	−0.6064
1 ALO	1,000	−0.5872	−0.031	3 ALOs	1,000	−0.5545	−0.085			
$10\sigma_y$	3,000	−0.5970	−0.015	$10\sigma_y$	3,000	−0.5810	−0.041			
	5,000	−0.6012	−0.009		5,000	−0.5902	−0.027			

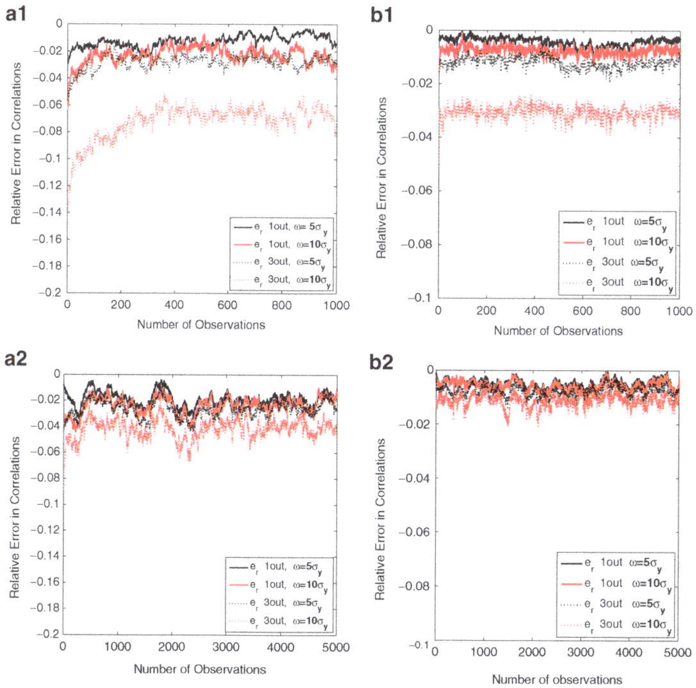

Fig. 24.1 Average relative biases of the estimated (**a**) DCC correlations and (**b**) BEKK correlations, for different sample sizes. (**a1**) DCC, $n = 1{,}000$. (**a2**) DCC, $n = 5{,}000$. (**b1**) BEKK, $n = 1{,}000$. (**b2**) BEKK, $n = 5{,}000$

24.5 Wavelet-Based Detection Procedure

In Grané and Veiga [12] a general outlier detection method based on wavelets was introduced. The method was designed for univariate time series and was proven to be very reliable, since it detects a significantly smaller number of false outliers compared to other competitive methods. Although in this work we face to multivariate time series, it is of our interest to develop a procedure with as good properties as the univariate one, effectiveness and reliability, and also of feasible implementation in large data sets. A possible way to proceed is to translate the multivariate problem to a univariate setting. This is achieved by applying the random projection method. In Cuesta-Albertos et al. [6, 7] some theoretical results were developed in the context of functional data (also of application whenever the data can be considered as independent and identically distributed draws of a stochastic process taking values in a Hilbert space). In practice, the number of random projections used is low (1 or 2), which is exactly contrary to the Projection Pursuit paradigm, avoiding implementation problems due to high dimensionality.

24.5.1 The Procedure

The procedure we propose is based on detail coefficients resulting from the discrete wavelet transform (DWT) of a univariate series of (standardized) residuals. The procedure starts with fitting a multivariate GARCH model and obtaining the series of multivariate residuals. The next step consists in transforming the multivariate series of residuals into univariate series to which DWT will be applied. Here we consider two different cases. Conditional correlation models, such as CCC and DCC, are based on the decomposition of the conditional covariance matrix. Hence, for these models, the decomposition property suggests that it is enough to consider only the univariate marginals. However, for models that do not have this property, as it is the case of the D-BEKK model, in addition to the marginals, we consider one randomly chosen projection [6]. DWT is applied to each of the univariate series under consideration and outliers are identified as those observations in the original series whose detail coefficients are greater (in absolute value) than a certain threshold.

In the context of financial return time series it is quite common to assume an underlying model for the data. Then, if the fitted model has captured the structure of the data, the residuals are supposed to be independent and identically distributed random variables following a specified (usually standard normal) distribution. Hence, our aim is to check whether a univariate series of (standardized) residuals follows a standard normal distribution. Our proposal is to use the following test statistic: the maximum of the detail wavelet coefficients (in absolute value) resulting from the DTW of a univariate series of (standardized) residuals. If the univariate series under consideration is obtained as the marginal of the multivariate one, the distribution of the test statistic reported in Grané and Veiga [12] for the univariate case is still valid. For the case in which the univariate series is obtained as a random projection the distribution is obtained via Monte Carlo, analogously. In all cases, threshold values are obtained as percentiles of the distribution of the test statistic computed on 20,000 Monte Carlo samples of size n. In practice, we find that in order to detect isolated ALOs it suffices to work with the first level detail wavelet coefficients and from the simulation study (see Sect. 24.5.2) we recommend the 95th percentile as a reasonable threshold to use in the detection of isolated ALOs.[3] Since in the multivariate case we are considering more than one series, the thresholds proposed in Grané and Veiga (2010) [12] for the univariate case are not directly applicable and the union-intersection principle [16] with Bonferroni correction is applied.

[3]Other percentiles can be used leading to more conservative results.

Table 24.2 Percentage of correct detection of ALOs (and percentage of false ALOs) in 1,000 replications of size n for a multivariate GARCH model with errors following a normal distribution

	n	D-BEKK	CCC	DCC		n	D-BEKK	CCC	DCC		n	D-BEKK	CCC	DCC
1 ALO	1,000	43.8	77.1	77.2	3 ALOs	1,000	36.7	69.6	68.9	No	1,000	–	–	–
$5\sigma_y$		(0.004)	(0.005)	(0.005)	$5\sigma_y$		(0.003)	(0.004)	(0.004)	ALOs		(0.004)	(0.006)	(0.005)
	3,000	38.7	76.2	75.3		3,000	36.5	71.2	71.1		3,000	–	–	–
		(0.001)	(0.001)	(0.001)			(0.001)	(0.001)	(0.001)			(0.001)	(0.001)	(0.001)
	5,000	36.1	69.8	70.8		5,000	36.1	71.5	71.4		5,000	–	–	–
		(0.001)	(0.001)	(0.001)			(0.001)	(0.001)	(0.001)			(0.001)	(0.001)	(0.001)
1 ALO	1,000	99.1	100.0	100.0	3 ALOs	1,000	96.5	97.8	97.8					
$10\sigma_y$		(0.004)	(0.004)	(0.004)	$10\sigma_y$		(0.002)	(0.005)	(0.005)					
	3,000	99.3	99.9	99.9		3,000	97.8	98.9	98.8					
		(0.001)	(0.001)	(0.001)			(0.001)	(0.001)	(0.001)					
	5,000	99.3	99.9	99.8		5,000	97.8	99.1	98.9					
		(0.001)	(0.001)	(0.001)			(0.001)	(0.001)	(0.001)					

Threshold values ($\alpha = 0.05$) for the D-BEKK model: $k(1000) = 4.1386$; $k(3000) = 4.3885$; $k(5000) = 4.5027$. For the CCC and DCC models: $k(1000) = 4.0595$; $k(3000) = 4.2995$; $k(5000) = 4.4062$

24.5.2 Performance of the Procedure: A Simulation Study

Here we present the results of an intensive simulation study to assess the performance of our detection proposal. The measures used in the performance study are the percentage of times that the localization of the outliers is correctly detected and the percentage of false outliers.

These results are shown in Table 24.2, where we observe that when the magnitude of the outlier is $10\sigma_y$, the procedure detects more than 96% of the outliers, reaching the 100% in two cases. When the magnitude of the outlier is relatively small, $5\sigma_y$, the detection rate goes from 36% to 43% for the BEKK model and from 68% and 77% for the CCC and DCC models. The method is very reliable, since the percentage of false outliers is at most 0.006%.

Concluding Remarks: The main conclusions are: First, outliers affect the estimated correlations and the effects are stronger for the conditional correlation models. Second, our detection procedure is effective and reliable, since the percentage of correct detections is high and the number of false outliers is very low.

References

1. Bauwens, L., Laurent, S., Rombout, J.: Multivariate GARCH models: a survey. J. Appl. Econom. **21**, 79–109 (2006)
2. Bollerslev, T.: Modeling the coherence in short-run nominal exchange rates: a multivariate generalized ARCH model. Rev. Econ. Stat. **42**, 498–505 (1990)
3. Boudt, K., Daníelsson, J., Laurent, S.: Robust forecasting of dynamic conditional correlation GARCH models. Int. J. Forecasting **29**, 244–257 (2013)
4. Carnero, M., Peña, D., Ruiz, E.: Effects of outliers on the identification and estimation of GARCH models. J. Time Ser. Anal. **28**, 471–497 (2007)
5. Chen, C., Liu, L.: Joint estimation of model parameters and outlier effects. J. Am. Stat. Assoc. **88**, 284–297 (1993)
6. Cuesta-Albertos, J., Fraiman, R., Ransford, T.: Random projections and goodness-of-fit tests in infinite-dimensional spaces. Bull. Braz. Math. Soc. **37**, 1–25 (2006)
7. Cuesta-Albertos, J., del Barrio, E., Fraiman, R., Matrán, C.: The random projection method in goodness of fit for functional data. Comput. Stat. Data Anal. **51**, 4814–4831 (2007)
8. Engle, R.: Dynamic conditional correlation—a simple class of multivariate GARCH models. J. Bus. Econ. Stat. **20**, 339–350 (2002)
9. Engle, R., Kroner, K.: Multivariate simultaneous generalized ARCH. Econom. Theory **11**, 122–150 (1995)
10. Fox, A.: Outliers in time series. J. R. Stat. Soc. B **34**, 350–363 (1972)
11. Franses, P., Ghijsels, H.: Additive outliers, GARCH and forecasting volatility. Int. J. Forecasting **15**, 1–9 (1999)
12. Grané, A., Veiga, H.: Wavelet-based detection of outliers in financial time series. Comput. Stat. Data Anal. **54**, 2580–2593 (2010)
13. Grossi, L., Laurini, F.: A robust forward weighted lagrange multiplier test for conditional heteroscedasticity. Comput. Stat. Data Anal. **53**, 2251–2263 (2009)

14. Hotta, L., Tsay, R.: Outliers in GARCH processes. Manuscript. Graduate School of Business, University of Chicago (1998)
15. Ledolter, J.: The effect of additive outliers on the forecasts from ARIMA models. Int. J. Forecasting **5**, 231–240 (1989)
16. Roy, S.: On a Heuristic Method of Test Construction and its use in multivariate analysis. Ann. Math. Stat. **24**, 220–238 (1953)
17. Silvennoinen, A., Teräsvirta, T.: Multivariate GARCH models. In: Handbook of Financial Time Series, pp. 201–226. Springer, Berlin (2009)
18. Van Dijk, D., Franses, P., Lucas, A.: Testing for ARCH in the presence of additive outliers. J. Appl. Econom. **14**, 539–562 (1999)
19. Verhoeven, P., McAleer, M.: Modelling outliers and extreme observations for ARMA-GARCH processes. Working Paper, University of Western Australia (2000)

Chapter 25
A Comparison of Efficient Permutation Tests for Unbalanced ANOVA in Two by Two Designs and Their Behavior Under Heteroscedasticity

Sonja Hahn, Frank Konietschke, and Luigi Salmaso

25.1 Introduction

In many biological, medical, and social trials, data are collected in terms of a two by two design, e.g. when male and female patients are randomized to two different treatment groups (placebo and active treatment). The data is often analyzed by assuming linear treatment effects and ANOVA procedures. These approaches rely on rather strict model assumptions like normally distributed error terms and variance homogeneity. However, these model assumptions can rarely be justified. In particular, heteroscedastic variances occur frequently in a variety of disciplines, e.g. in genetic data. It is well known that the classical ANOVA F-test tends to result in liberal or conservative decisions, depending on the underlying distribution, the amount of variance heterogeneity, and unbalance. Thus, asymptotic (or approximate) procedures, which allow the data to be heteroscedastic, are a robust alternative to the classical ANOVA F-test. An asymptotic testing procedure is the Wald-type statistic (see, e.g., [2, 10]), which is based on the asymptotic distribution of an appropriate quadratic form. It is even valid without the assumptions of normality and variance homogeneity. However, very large sample sizes are necessary to achieve accurate test results (see, e.g., [2] and the references therein). As an approximate solution, [2] propose the so-called ANOVA-type statistic (ATS), which is based

S. Hahn
Department of Psychology, University of Jena, Jena, Germany
e-mail: hahn.sonja@uni-jena.de

F. Konietschke
Department of Mathematical Sciences, The University of Texas at Dallas, USA
e-mail: frank.konietschke@utdallas.edu

L. Salmaso (✉)
Department of Management and Engineering, University of Padova, Padova, Italy
e-mail: luigi.salmaso@unipd.it

V.B. Melas et al. (eds.), *Topics in Statistical Simulation*, Springer Proceedings in Mathematics & Statistics 114, DOI 10.1007/978-1-4939-2104-1_25

on an Box-type approximation approach. The ATS, however, is an approximate test and its asymptotically exactness is unknown (see, e.g., [10]). On the other hand, permutation approaches are known to be very robust under non-normality. In particular, under certain model assumptions, permutation tests are exact level α tests. Usual permutation tests assume that the data is exchangeable, which particularly implies homogeneous variances. Recently, Pauly et al. [10] propose asymptotic permutation tests, which are asymptotically exact even under non-normality and possibly heteroscedastic variances.

Various permutational approaches for factorial designs have been developed within the last years, but a comparison of the different permutational approaches for unbalanced factorial designs with variance heterogeneity remains.

The aim of the present paper is to investigate different parametric and permutation tests for factorial linear models. For simplicity, we focus on two by two designs within this paper.

The paper is organized as follows: After some notational issues we summarize different existing approaches that were developed for unbalanced ANOVA designs. Afterwards we investigate the behavior of these procedures in a simulation study. Here we focus on small sample sizes, heterogeneity of variances, and different error term distributions. Finally, we discuss the results of the simulation study and add further considerations about the procedures.

25.1.1 Notation and Hypotheses

We consider the two way factorial crossed design

$$X_{ijk}=\mu+\alpha_i+\beta_j+(\alpha\beta)_{ij}+\epsilon_{ijk},\ i=1,2;\ j=1,2;\ k=1,\ldots,n_{ij}, \qquad (25.1)$$

where α_i denotes the effect of level i from factor A, β_j denotes the effect of level j from factor B, and $(\alpha\beta)_{ij}$ denotes the (ij)th interaction effect from $A \times B$. Here, ϵ_{ijk} denotes the error term with $E(\epsilon_{ijk}) = 0$ and $\mathrm{Var}(\epsilon_{ijk}) = \sigma_{ij}^2 > 0$. Under the assumption of equal variances, we simply write $\mathrm{Var}(\epsilon_{ijk}) = \sigma^2$. It is our purpose to test the null hypotheses

$$\begin{aligned} H_0^{(A)} &: \alpha_1 = \alpha_2 \\ H_0^{(B)} &: \beta_1 = \beta_2 \\ H_0^{(A\times B)} &: (\alpha\beta)_{11} = \ldots = (\alpha\beta)_{22} \end{aligned} \qquad (25.2)$$

For simplicity, let $\mu_{ij} = \mu + \alpha_i + \beta_j + (\alpha\beta)_{ij}$, then, the hypotheses defined above can be equivalently written as

$$H_0^{(A)} : \mathbf{C}_A \boldsymbol{\mu} = \mathbf{0}$$
$$H_0^{(B)} : \mathbf{C}_B \boldsymbol{\mu} = \mathbf{0}$$
$$H_0^{(A\times B)} : \mathbf{C}_{A\times B} \boldsymbol{\mu} = \mathbf{0},$$

where $\mathbf{C}_L, L \in \{A, B, A \times B\}$, denote suitable contrast matrices and $\boldsymbol{\mu} = (\mu_{11}, \ldots, \mu_{22})'$. To test the null hypotheses formulated in (25.2), various asymptotic and approximate test procedures have been proposed. We will explain the current state of the art in the subsequent sections.

25.1.2 Wald-Type Statistic (WTS)

Let $\overline{\mathbf{X}}_{\cdot} = (\overline{X}_{11\cdot}, \ldots, \overline{X}_{22\cdot})'$ denote the vector of sample means $\overline{X}_{ij\cdot} = \frac{1}{n_{ij}} \sum_{k=1}^{n_{ij}} X_{ijk}$, and let $\widehat{\mathbf{S}}_N = \mathrm{diag}(\hat{\sigma}_{11}^2, \ldots, \hat{\sigma}_{22}^2)$ denote the 4×4 diagonal matrix of sample variances $\hat{\sigma}_{ij}^2 = \frac{1}{n_{ij}-1} \sum_{k=1}^{n_{ij}} (X_{ijk} - \overline{X}_{ij\cdot})^2$. Under the null hypothesis $H_0 : (L) : \mathbf{C}_L \boldsymbol{\mu} = \mathbf{0}$, the Wald-type statistic

$$W_N(L) = N \overline{\mathbf{X}}_{\cdot}' \mathbf{C}_L' (\mathbf{C}_L \mathbf{S}_N \mathbf{C}_L')^{+} \mathbf{C}_L \overline{\mathbf{X}}_{\cdot} \to \chi^2_{\mathrm{rank}(\mathbf{C}_L)} \tag{25.3}$$

has, asymptotically, as $N \to \infty$, a $\chi^2_{\mathrm{rank}(\mathbf{C}_L)}$ distribution. The rate of convergence, however, is rather slow, particularly for larger numbers of factor levels and smaller sample sizes. For small and medium sample sizes, the WTS tends to result in rather liberal results (see [2, 10] for some simulation results). However, the Wald-type statistic is asymptotically exact even under non-normality.

25.1.3 ANOVA-Type Statistic (ATS)

In order to overcome the strong liberality of the Wald-type statistic in (25.3) with small sample sizes, [2] propose the so-called ANOVA-type statistic (ATS)

$$F_N(L) = \frac{N\, \overline{\mathbf{X}}_{\cdot}' \mathbf{T}_L \overline{\mathbf{X}}_{\cdot}}{\mathrm{trace}(\mathbf{T}_L \mathbf{S}_N)}, \tag{25.4}$$

where $\mathbf{T}_L = \mathbf{C}_L' (\mathbf{C}_L \mathbf{C}_L')^{+} \mathbf{C}_L$. The null distribution of $F_N(L)$ is approximated by a F-distribution with

$$f_1 = \frac{[\mathrm{trace}(\mathbf{T}_L \mathbf{S}_N)]^2}{\mathrm{trace}[(\mathbf{T}_L \mathbf{S}_N)^2]} \text{ and } f_2 = \frac{[\mathrm{trace}(\mathbf{T}_L \mathbf{S}_N)]^2}{\mathrm{trace}(\mathbf{D}_{T_L}^2 \mathbf{S}_N^2 \boldsymbol{\Lambda})}, \tag{25.5}$$

where $\mathbf{D}_T$ is the diagonal matrix of the diagonal elements of $\mathbf{T}_L$ and $\boldsymbol{\Lambda} = \text{diag}\{(n_{ij} - 1)^{-1}\}_{i,j=1,2}$. The ATS relies on the assumption of normally distributed error terms [10]. Especially for skewed error terms the procedure tends to be very conservative [10, 16]. When the sample sizes are extremely small ($n_{ij} \approx 5$), it tends to result in conservative decisions [13]. We note that in two by two designs, the Wald-type statistics $W_N(L)$ in (25.3) and the ANOVA-type statistic $F_N(L)$ are identical. Furthermore, the ATS is even asymptotically an approximate test and its asymptotical exactness is unknown.

25.1.4 Wald-Type Permutation Test (WTPS)

Recently, [10] proposed an asymptotic permutation based Wald-test, which is even asymptotically exact when the data is not exchangeable. In particular, it is asymptotically valid under variance heterogeneity. This procedure denotes an generalization of two-sample studentized permutation tests for the Behrens–Fisher problem [5, 6, 8, 9]. The procedure is based on (randomly) permuting the data $\mathbf{X}^* = (X^*_{111}, \dots, X^*_{22n_{22}})'$ within the whole data set. Let $\overline{\mathbf{X}}^*_{\cdot} = (\overline{X}^*_{11\cdot}, \dots, \overline{X}^*_{22\cdot})'$ denote the vector of permuted means $\overline{X}^*_{ij\cdot} = n_{ij}^{-1} \sum_{k=1}^{n_{ij}} X^*_{ijk}$, and let $\widehat{\mathbf{S}}^*_N = \text{diag}(\hat{\sigma}^{2*}_{11}, \dots, \hat{\sigma}^{2*}_{22})$ denote the 4×4 diagonal matrix of permuted sample variances $\hat{\sigma}^{2*}_{ij} = \frac{1}{n_{ij}-1} \sum_{k=1}^{n_{ij}} (X^*_{ijk} - \overline{X}^*_{ij\cdot})^2$. Further let

$$W^*_N(L) = N(\overline{\mathbf{X}}^*_{\cdot})' \mathbf{C}'_L (\mathbf{C}_L \mathbf{S}^*_N \mathbf{C}'_L)^+ \mathbf{C}_L \overline{\mathbf{X}}^*_{\cdot} \tag{25.6}$$

denote the permuted Wald-type statistics $W_N(L)$. Pauly et al. [10] show that, given the data $\mathbf{X}$, the distribution of $W^*_N(L)$ is, asymptotically, the $\chi^2_{\text{rank}(\mathbf{C}_L)}$ distribution. The p-value is derived as the proportion of test statistics of the permuted data sets that are equal or more extreme than the test statistic of the original data set.

If data is exchangeable, this Wald-type permutation tests guarantees an exact level α test. Otherwise, this procedure is asymptotically exact due to the multivariate studentization. Simulation results showed that this test adheres better to the nominal α-level than its unconditional counterpart for small and medium sample sizes (see [10] and the supplementary materials therein). Furthermore, the Wald-type permutation test achieves a higher power than the ATS in general. We note that the WTPS is not restricted to two by two designs. The procedure is applicable in higher-way layouts and even in nested and hierarchical designs.

25.1.5 Synchronized Permutation Tests (CSP and USP)

Synchronized permutation tests were designed to test the different hypotheses in (25.2) of a factorial separately (e.g., testing a main effect when there is an interaction effect). There are two important differences to the WTPS approach: (1) Data is not permuted within the whole data set, but there exists a special synchronized permutation mechanism. (2) The test statistic is not studentized. This procedure assumes that the error terms are exchangeable.

Basso et al. [1], Pesarin and Salmaso [11] and Salmaso [14] propose synchronized permutation tests for balanced factorial designs. *Synchronization* means that data is permuted within blocks built by one of the factors. In addition, the number of exchanged observations in each of these blocks is equal for a single permutation. For example, when testing for the main effect A or the interaction effect, the observations can be permuted within the blocks built by the levels of factor B. Different variants of synchronized permutations have been developed (see [3] for details):

Constrained Synchronized Permutations (CSP). Here only observations on the same position within each subsample are permuted. When applied to real data set it is strongly recommended to pre-randomize the observations in the data set to eliminate possible systematic order effects.

Unconstrained Synchronized Permutations (USP). Here also observations on different position can be permuted. In this case it has to be ensured that the test statistic follows a uniform distribution.

The test statistics for the main effect A and the interaction effect are

$$T_A = (T_{11} + T_{12} - T_{21} - T_{22})^2, \text{ and}$$
$$T_{A\times B} = (T_{11} - T_{12} - T_{21} + T_{22})^2$$

with

$$T_{ij} = \textstyle\sum_k X_{ijk}.$$

Due to the synchronization and the test statistic, the effects not of interest are eliminated (e.g, when testing for main effect A, main effect B and the interaction effect are eliminated, see [1] for more background information). When testing for main effect B, the data has to be permuted within blocks built by the levels of A and the test statistics have to be adapted.

For certain unbalanced factorial designs this method can be extended [4]. In the case of CSP this leads to the situation that some observations will never be exchanged. In the case of USP the maximum number of exchanged observations equals the minimum subsample size.

A test statistic that finally eliminates the effects of interest is only available in special cases [4]. For example, when $n_{11} = n_{12}$ and $n_{21} = n_{22}$, possible test statistics are:

$$T_A = (n_{21}T_{11} + n_{22}T_{12} - n_{11}T_{21} - n_{12}T_{22})^2,$$
$$T_{A\times B} = (n_{21}T_{11} - n_{22}T_{12} - n_{11}T_{21} + n_{12}T_{22})^2.$$

For both, the balanced and the unbalanced case, the p-value is again calculated as the proportion of test statistics of permuted data sets greater or equal than the test statistic of the original data set.

These procedures showed a good adherence to the nominal α-level as well as power in simulation studies [1, 4]. However, this procedure is limited in various ways:

- It is restricted to very specific cases of unbalanced designs due to assumptions on equal sized subsamples.
- Extension to more complex factorial designs seems quite difficult (see, e.g., [1] for balanced cases with more levels).
- It assumes exchangeability. This might not be given in cases with heteroscedastic error variances.

As the behavior of this procedure under variance heterogeneity has not been investigated yet, we included it in the following simulation study.

25.1.6 Summary

We outlined various procedures that aim to compensate shortcomings of classical ANOVA. Some procedures are only valid under normality and possibly heteroscedastic variances (ATS). CSP and USP are valid under non-normally distributed error terms and homoscedastic variances. Both the WTS and WTPS are asymptotically valid even under non-normality and heteroscedasticity, respectively. Most of these procedures are intended to be used for small samples (ATS, WTPS, CSP, and USP), only the WTS requires a sufficiently large sample size.

In the following simulation we vary additionally the aspect of balanced vs. unbalanced designs, as heteroscedasticity is especially problematic in the latter one.

25.2 Simulation Study

25.2.1 General Aspects

The present simulation study investigates the behavior of the procedures described above (see Sect. 25.1) for balanced vs. unbalanced designs and homo- vs. heteroscedastic variances. A major assessment criterion for the accuracy of the procedures is their behavior when increasing sample sizes are combined with increasing variances (positive pairing) or with decreasing variances (negative pairing).

We investigate data sets that did not contain any effect, and data sets that contained an effect. In the first case we were interested if the procedures keep the nominal level; in the second case additionally the power behavior was investigated. Similar to the notation introduced above we used the following approach for data simulation:

$$X_{ijk} = \mu + \alpha_i + \beta_j + (\alpha\beta)_{ij} + \epsilon_{ijk}. \tag{25.7}$$

Specifications for the different data settings can be found below. Throughout all studies we focused on the main effect A and the interaction effect.

All simulations were conducted using the freely available software *R* (www.r-project.org), version 2.15.2 [12]. The numbers of simulation and permutation runs were $n_{\text{sim}} = 5{,}000$ and $n_{\text{perm}} = 5{,}000$, respectively. All simulations were conducted at 5 % level of significance.

25.2.2 Data Sets Containing No Effect

25.2.2.1 Description

Table 25.1 outlines the combinations of balanced vs. unbalanced designs and homo- vs. heteroscedastic variances. Larger sample sizes were obtained by adding a constant number to each of the sample sizes. Those numbers were 5, 10, 20, and 25.

There was no effect in the data (i.e., for Eq. (25.7) $\mu = \alpha_i = \beta_j = 0$). For the error terms, different symmetric and skewed distributions were used:

- Symmetrical distributions: normal, Laplace, logistic, and a "mixed" distribution, where each factor level combination has a different symmetric distribution (normal, Laplace, logistic, and uniform).
- Skewed distributions: log-normal, χ^2_3, χ^2_{10}, and a "mixed" distribution, where each factor level combination has a different skewed distribution (exponential, log-normal, χ^2_3, χ^2_{10}).

To generate variance heterogeneity, random variables were first generated from the distributions mentioned above and standardized to achieve an expected value of 0 and a standard deviation of 1. These values were further multiplied by the standard deviations given in Table 25.1 to achieve different degrees of variance heteroscedasticity.

Table 25.1 Different subsample sizes and variances considered in the simulation study

	Data setting	n_{11}	(SD)	n_{12}	(SD)	n_{21}	(SD)	n_{22}	(SD)
1	Balanced and homoscedastic	5	(1.0)	5	(1.0)	5	(1.0)	5	(1.0)
2	Differing sample sizes	5	(1.0)	7	(1.0)	10	(1.0)	15	(1.0)
3	Differing variances	5	(1.0)	5	(1.3)	5	(1.5)	5	(2.0)
4	Positive pairings	5	(1.0)	7	(1.3)	10	(1.5)	15	(2.0)
5	Negative pairings	5	(2.0)	7	(1.5)	10	(1.3)	15	(1.0)

Besides the data settings in the table, bigger samples were achieved by adding 5, 10, 20, or 25 observations to each subsample

25.2.2.2 Results

Figure 25.1 shows the behavior of the different procedures in the case of symmetric and homoscedastic error terms. Most procedures keep close to the nominal α-level of 0.05 that is indicated by the red thin line. WTS tends to be quite liberal, while ATS tends to be slightly conservative for small sample sizes.

Figure 25.2 shows the behavior for skewed but still homoscedastic error term distributions. The picture is very similar to the previous one, but the conservative behavior of the ATS procedure is more pronounced.

Figure 25.3 shows the behavior in the symmetric and heteroscedastic case. For Setting 3 with equal sample sizes there is not much difference in comparison with the previous cases. In Setting 4, the positive pairings, WTPS and ATS show a good adherence to the level and a slightly conservative behavior in the case of the

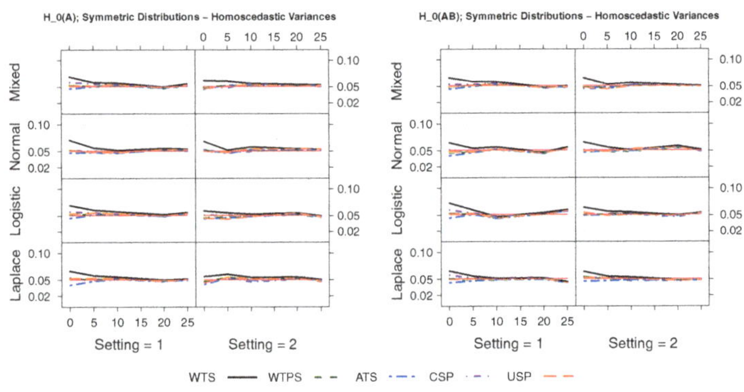

Fig. 25.1 Results for the different procedures testing main effect A (*left-hand side*) or the interaction effect (*right-hand side*) for symmetric distributions and homoscedastic variances

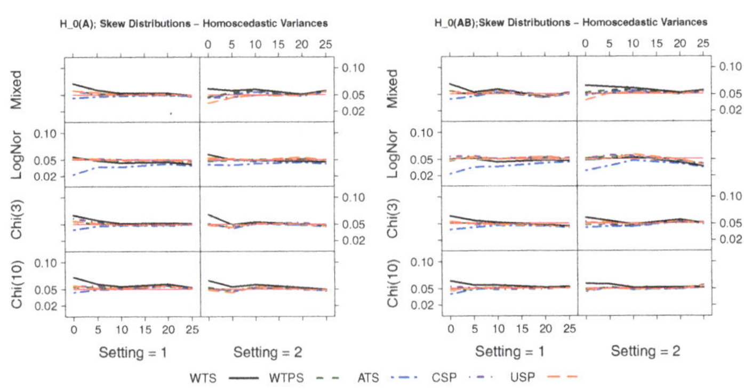

Fig. 25.2 Results for the different procedures testing main effect A (*left-hand side*) or the interaction effect (*right-hand side*) for skewed distributions and homoscedastic variances

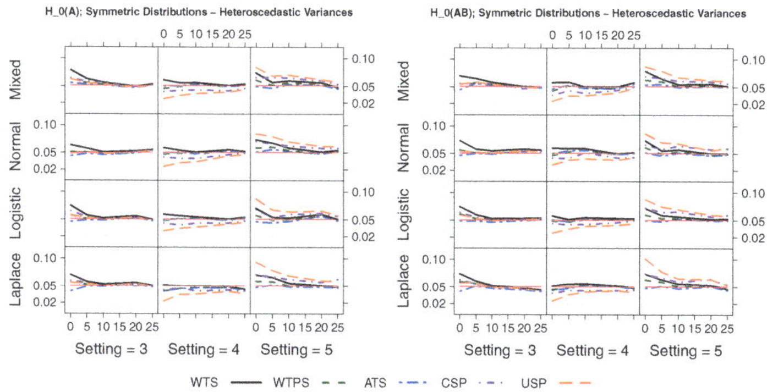

Fig. 25.3 Results for the different procedures testing main effect A (*left-hand side*) or the interaction effect (*right-hand side*) for symmetric distributions and heteroscedastic variances

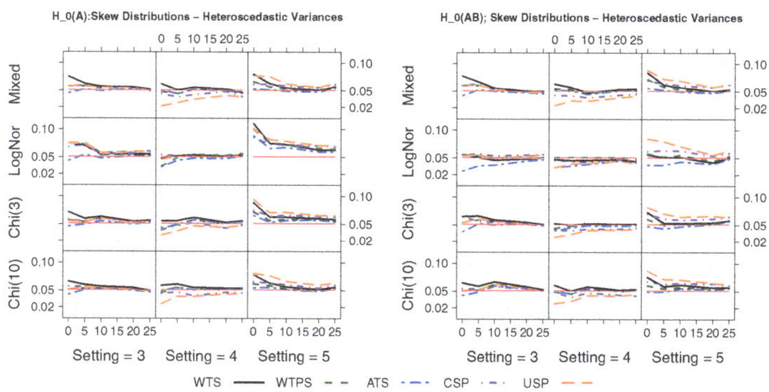

Fig. 25.4 Results for the different procedures testing main effect A (*left-hand side*) or the interaction effect (*right-hand side*) for skewed distributions and heteroscedastic variances

Laplace-distribution. WTS tends to over-reject the null in small sample size settings. Both the CSP and USP tests tend to result in conservative decisions. This is more pronounced for small sample sizes and for the USP-procedure. In Setting 5, that indicates negative pairings, all procedures unless ATS tends to result in a liberal behavior—especially for small sample sizes. USP has the strongest tendency with Type-I-error rates up to 0.08.

Figure 25.4 shows the behavior in the skewed and heteroscedastic case. In general, the same conclusions can be drawn. For the log-normal distribution there is a general tendency to get a more liberal decision than in the other cases. This means that in Setting 4 with positive pairings the procedures keep the level almost well, but in the other cases the Type-I-error rate is up to 0.10.

25.2.3 *Data Sets Containing an Effect*

25.2.3.1 Description

Table 25.2 shows the different combinations of subsample sizes and standard deviations for data sets that contained an effect. Two aspects were considered: The power behavior as well as the level of the procedures when testing an inactive effect. To ensure a valid comparison of the power behavior, the sample sizes and variance heterogeneity was chosen less extreme than in the previous simulation study (see Sect. 25.2.2).

Again different error term distributions were used:

- normal and Laplace distribution as symmetric distributions, and
- log-normal distribution and exponential distribution as skewed distributions.

Different error term variances were obtained in the same manner as described above in Sect. 25.2.2 using the standard deviations from Table 25.2. Additionally, there were active effects as described in Table 25.3 in the data with $\mu = 0$ and $\delta \in \{0, 0.2, \dots, 1\}$. The tested effects were again main effect A and the interaction effect. In some cases where only main effect B was active the aim was to test if the procedures kept the level in these cases.

25.2.3.2 Results

Figures 25.5, 25.6, 25.7, 25.8, and 25.9 show the behavior of the different procedures for data sets containing an effect. The procedures show a very similar power behavior.

Table 25.2 Different subsample sizes and standard deviations considered in the simulation study containing effects

	Data setting	n_{11}	(SD)	n_{12}	(SD)	n_{21}	(SD)	n_{22}	(SD)
1	Balanced and homoscedastic	10	(1)	10	(1)	10	(1)	10	(1)
2	Differing sample sizes	9	(1)	9	(1)	15	(1)	15	(1)
3	Differing variances	10	(1)	10	(1)	10	($\sqrt[4]{2}$)	10	($\sqrt[4]{2}$)
4	Positive pairings	9	(1)	9	(1)	15	($\sqrt[4]{2}$)	15	($\sqrt[4]{2}$)
5	Negative pairings	9	($\sqrt[4]{2}$)	9	($\sqrt[4]{2}$)	15	(1)	15	(1)

Table 25.3 Different effect in simulated data sets with $\mu = 0$ and $\delta \in \{0, 0.2, \dots, 1\}$ in the simulation study containing effects

Condition	α_1	α_2	β_1	β_2	$\alpha\beta_{11}$	$\alpha\beta_{12}$	$\alpha\beta_{21}$	$\alpha\beta_{22}$	Active effects
1	$+\delta$	$-\delta$	0	0	0	0	0	0	Main effect A
2	0	0	$+\delta$	$-\delta$	0	0	0	0	Main effect B
3	$+\frac{\delta}{2}$	$-\frac{\delta}{2}$	0	0	$+\frac{\delta}{2}$	$-\frac{\delta}{2}$	$-\frac{\delta}{2}$	$+\frac{\delta}{2}$	Main effect A and interaction effect

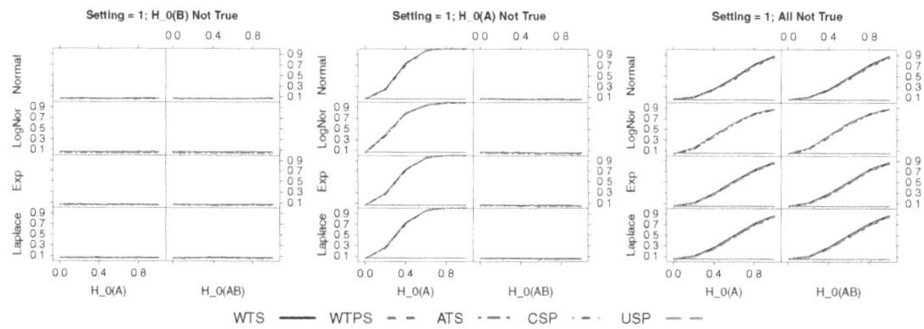

Fig. 25.5 Results for data sets containing effects (equal subsample sizes and homoscedastic variances)

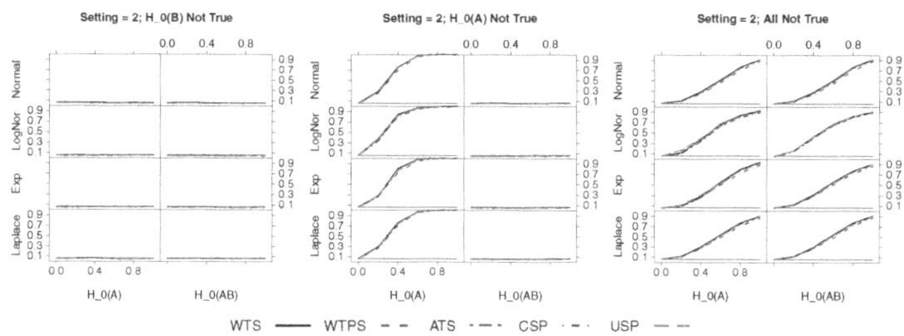

Fig. 25.6 Results for data sets containing effects (equal subsample sizes and heteroscedastic variances)

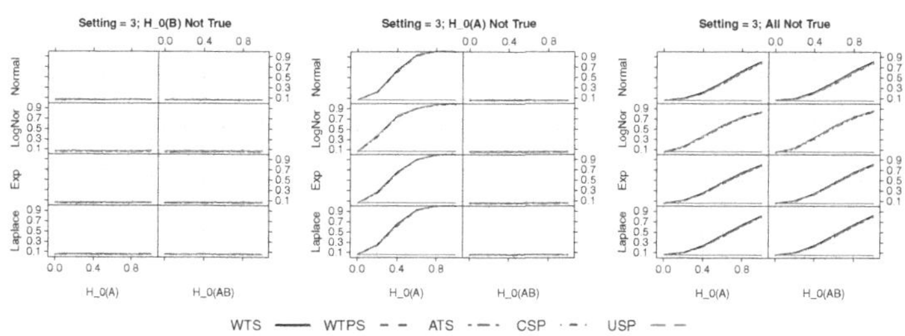

Fig. 25.7 Results for data sets containing effects (unequal subsample sizes and homoscedastic variances)

Conclusion

As the simulation study showed, the different procedures may be useful depending on the data setting and further aspects.

The ATS procedure was the only one that never exceeded the nominal level. On the other hand it may show a conservative behavior, but in the simulations containing effects this was only slightly observable. Similar to the results of previous simulation studies, the conservative behavior was higher

(continued)

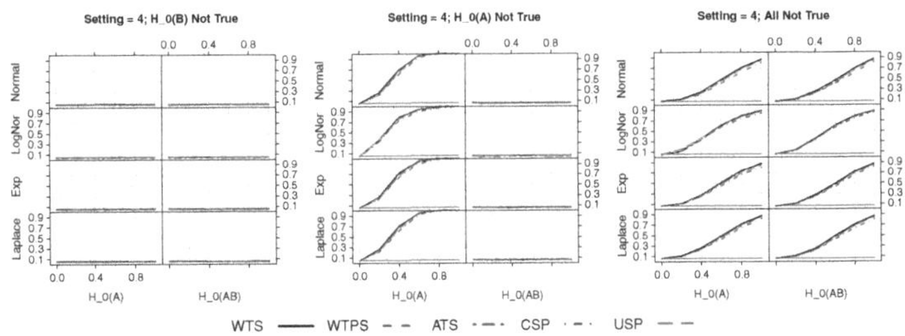

Fig. 25.8 Results for data sets containing effects (positive pairings)

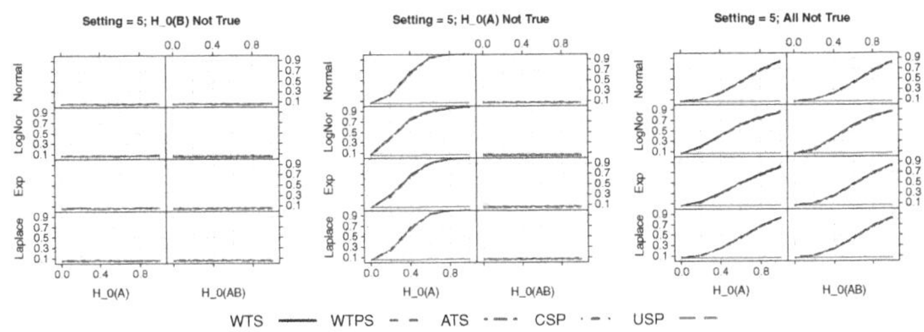

Fig. 25.9 Results for data sets containing effects (negative pairings)

for skewed distributions, especially with homoscedastic error term variances. An advantage of this procedure is that it can be adapted for very different designs and hypotheses (see [16] for more background information).

The WTS procedure showed in almost every data setting a liberal behavior for small samples. It should be only applied when sample sizes are large.

The WTPS procedure overcomes this problem. In all considered simulation settings this procedure controls the type-I error rate quite accurately. In case of positive or negative pairings, this permutation test shows better results than its competitors. Both the WTS and WTPS can be adapted to higher-way layouts and hierarchical designs.

The CSP and the USP procedures work well for all cases with equal subsample sizes or homogeneous variances. This implies cases where exchangeability of the observations might not be given due to different error term distributions (mixed distributions) or heterogeneous variances. In case of positive and negative pairings, the behavior is similar to parametric ANOVA with a conservative behavior for positive pairings and a liberal behavior for negative pairings. This is more pronounced for the USP-procedure. The power behavior of both procedures was very comparable to the other procedures. CSP showed in some cases a slightly lower power than the other procedures. The CSP and the USP procedures are restricted to certain hypothesis due to

(continued)

their construction: They assumed that all cells should get the same weight in the analysis. This corresponds to Type III sums of squares [15]. Extension of these procedures to other kind of hypotheses, unbalancedness, or more complex designs might be challenging.

References

1. Basso, D., Pesarin, F., Salmaso, L., Solari, A.: Permutation Tests for Stochastic Ordering and ANOVA. Springer, New York (2009)
2. Brunner, E., Dette, H., Munk, A.: Box-type approximations in nonparametric factorial designs. J. Am. Stat. Assoc. **92**(440), 1494–1502 (1997). http://www.jstor.org/stable/2965420
3. Corain, L., Salmaso. L.: A critical review and a comparative study on conditional permutation tests for two-way ANOVA. Commun. Stat. Simul. Comput. **36**(4), 791–805 (2007). doi:10.1080/03610910701418119
4. Hahn, S., Salmaso, L.: A comparison of different synchronized permutation approaches to testing effects in two-level two-factor unbalanced ANOVA designs (submitted)
5. Janssen, A.: Studentized permutation tests for non-i.i.d. hypotheses and the generalized Behrens–Fisher problem. Stat. Probab. Lett. **36**, 9–21 (1997)
6. Janssen, A., Pauls, T.: How do bootstrap and permutation tests work? Ann. Stat. **31**, 768–806 (2003)
7. Kherad-Pajouh, S., Renaud, D.: An exact permutation method for testing any effect in balanced and unbalanced fixed effect ANOVA. Comput. Stat. Data Anal. **54**, 1881–1893 (2010)
8. Konietschke, F., Pauly, M.: A studentized permutation test for the nonparametric Behrens–Fisher problem in paired data. Electron. J. Stat. **6**, 1358–1372 (2012)
9. Konietschke, F., Pauly, M.: Bootstrapping and permuting paired t-test type statistics. Stat. Comput. **36** (2013). doi:10.1007/s11222-012-9370-4
10. Pauly, M., Brunner, E., Konietschke, F.: Asymptotic permutation tests in general factorial designs. J. R. Stat. Soc. Ser. B (Stat Methodol.) (2014). doi:10.1111/rssb.12073
11. Pesarin, F., Salmaso, L.: Permutation Tests for Complex Data: Theory, Application and Software. Wiley, New York (2010)
12. R Core Team: R: A Language and Environment for Statistical Computing. R Foundation for Statistical Computing, Vienna (2013). http://www.R-project.org
13. Richter, S.J., Payton, M.E.: Performing two way analysis of variance under variance heterogeneity. J. Mod. Appl. Stat. Methods **2**(1), 152–160 (2003)
14. Salmaso, L.: Synchronized permutation tests in 2^k factorial designs. Commun. Stat. **32**, 1419–1437 (2003). doi:10.1081/STA-120021566
15. Searle, S.R.: Linear Models for Unbalanced Data. Wiley Series in Probability and Mathematical Statistics. Wiley, New York (1987)
16. Vallejo, G., Fernández, M.P., Livacic-Rojas, P.E.: Analysis of unbalanced factorial designs with heteroscedastic data. J. Stat. Comput. Simul. **80**(1), 75–88 (2010). doi:10.1080/00949650802482386. http://www.tandfonline.com/doi/abs/10.1080/00949650802482386, http://www.tandfonline.com/doi/pdf/10.1080/00949650802482386

Chapter 26
Likelihood-Free Simulation-Based Optimal Design: An Introduction

Markus Hainy, Werner G. Müller, and Helga Wagner

26.1 Introduction

In the past decades simulation techniques, particularly the use of Markov chain Monte Carlo (MCMC) methods, have revolutionized statistical inference (cf. [9]). There has, however, been little impact of this revolution on the experimental design literature other than the pioneering work initiated by Peter Müller (cf. [7] and [8]) and his followers.

In this contribution, we consider an adaptive design situation, where some observations have already been collected. The information obtained through these observations can be used to update the prior information on the unknown parameters. In this case, it is usually necessary to evaluate the likelihood function. If the likelihood function is intractable, we cannot perform the standard simulation-based MCMC scheme.

With the advent of the so-called *likelihood-free* (or *approximate Bayesian computation—ABC*) methods, the latter issue can be overcome, and we therefore propose to employ these techniques also for finding optimal experimental designs. There are essentially two ways of accomplishing this: the first one is to marry ABC with Müller's essentially MCMC-based methods, which is also the method we pursue in this contribution; the second one is a more basic approach that does not make use of the MCMC methodology but allows to deal with more general design criteria. A thorough review of the former is given in [4], on parts of which this article is based, while the latter has been put forward in [5]. Similar ideas have been

M. Hainy (✉) • W.G. Müller • H. Wagner
Johannes Kepler University Linz, Altenberger Straße 69, 4040 Linz, Austria
e-mail: markus.hainy@jku.at; werner.mueller@jku.at; helga.wagner@jku.at

V.B. Melas et al. (eds.), *Topics in Statistical Simulation*, Springer Proceedings in Mathematics & Statistics 114, DOI 10.1007/978-1-4939-2104-1_26

developed in [3] in order to conduct simulation-based design when the likelihood is intractable. However, they do not regard the adaptive design situation, they rather use ABC to estimate complex design criteria.

Finally, we illustrate the method outlined in this contribution on an example which can be easily related to and understood from classic optimal design theory.

26.2 Simulation-Based Optimal Design

We extend the basic simulation-based optimal design setup introduced in [7] by assuming that some observations have already been collected in the past. The additional information provided by these observations may be used to update the prior knowledge about the parameters of the model.

26.2.1 *Expected Utility Maximization*

For a chosen design $\boldsymbol{\xi} \in \Xi$ and parameters $\theta \in \Theta$, the likelihood of the observed data vector $y \in \mathscr{Y}$ is given by $p(y|\theta, \boldsymbol{\xi})$. We assume that past observations $y_{1:s} = \{y_i, i = 1, \ldots, s\}$ measured at the design points $\zeta_{1:s} = \{\zeta_i, i = 1, \ldots, s\}$ are available and that these past observations are conditionally i.i.d., i.e., the likelihood function for the past observations is $p(y_{1:s}|\theta, \zeta_{1:s}) = \prod_{i=1}^{s} p(y_i|\theta, \zeta_i)$. Furthermore, we assume that the parameters follow a prior distribution $p(\theta)$ which does not depend on the design. Thus, by using the past observations we can update the prior information and obtain the posterior distribution of the parameters: $p(\theta|y_{1:s}, \zeta_{1:s}) \propto p(\theta) \prod_{i=1}^{s} p(y_i|\theta, \zeta_i)$.

The general aim of simulation-based optimal design is to find the optimal configuration $\xi_{\max} = \arg\sup_{\xi} U(\xi)$ for the expected utility integral

$$U(\xi) = \int_{z \in \mathscr{Y}} \int_{\theta \in \Theta} u(\{z, y_{1:s}\}, \{\xi, \zeta_{1:s}\}, \theta) p(z|\theta, \xi)\, p(\theta|y_{1:s}, \zeta_{1:s}) \mathrm{d}\theta \mathrm{d}z\,, \tag{26.1}$$

where z denotes a vector of (future) observations measured at the points of the candidate design $\xi \in \Xi$. The utility function $u(.)$ may depend on the (current and past) data $\{z, y_{1:s}\}$, the (current and past) designs $\{\xi, \zeta_{1:s}\}$, and the parameters θ. This is the extended setting considered, e.g., by [8]. To simplify notation, the possible dependence of $u(.)$ on $y_{1:s}$ and $\zeta_{1:s}$ will be neglected for the remainder of this article.

26.2.2 MCMC Algorithm

A particular way to tackle this optimization problem is proposed in [7]. It combines the simulation as well as the optimization steps. To implement a stochastic search, the integrand in (26.1) is regarded as being proportional to a joint probability distribution of the variables $\vartheta = (z, \xi, \theta)$:

$$h(\vartheta) \propto u(\vartheta) p(z|\theta, \xi) p(\theta|y_{1:s}, \zeta_{1:s}) \mu(\xi) , \qquad (26.2)$$

where $\mu(\xi)$ is some (usually uniform) measure on the design region. If $u(.)$ is positive and bounded, then $h(.)$ is a proper pdf and marginalizing over θ and z yields

$$U(\xi) \propto \int_{z \in \mathscr{Y}} \int_{\theta \in \Theta} h(z, \xi, \theta) \mathrm{d}\theta \mathrm{d}z ,$$

so the marginal distribution of ξ is proportional to the expected utility function. Therefore, a strategy to find the optimum design is to sample from $h(z, \xi, \theta)$, retain the draws of ξ, and then search for the mode of the marginal distribution of ξ by inspecting the draws.

The density function h is only known up to a normalizing constant. Therefore, one option to obtain a sample from h is to perform Markov chain Monte Carlo methods such as Metropolis Hastings (MH). For a review of MCMC sampling schemes see [11]. The following proposal distribution, which generates a proposed draw $\vartheta' = (z', \xi', \theta')$ given the previous draw $\vartheta^* = (z^*, \xi^*, \theta^*)$, was suggested, e.g., by [8]:

$$q_k(\vartheta'|\xi^*) = p(z'|\theta', \xi') k(\theta'|y_{1:s}, \zeta_{1:s}) g(\xi'|\xi^*) .$$

The density function g is a random walk proposal density for ξ. The data z are sampled according to the probability model. The parameters θ are sampled from a proposal distribution k which should resemble the posterior distribution as closely as possible. Common choices for these proposals are normal or random walk or independence proposals, where the scale is proportional to the inverse of the Hessian of the log-likelihood or the unnormalized log-posterior. Specifying the proposal distribution in this way leads to the MH acceptance ratio

$$\alpha = \min\left(1, \frac{u(\vartheta')}{u(\vartheta^*)} \frac{p(y_{1:s}|\theta', \zeta_{1:s}) p(\theta')}{p(y_{1:s}|\theta^*, \zeta_{1:s}) p(\theta^*)} \frac{k(\theta^*|y_{1:s}, \zeta_{1:s})}{k(\theta'|y_{1:s}, \zeta_{1:s})} \frac{g(\xi^*|\xi')}{g(\xi'|\xi^*)}\right) .$$

Due to the particular choice of the proposal distribution, the likelihood terms $p(z'|\theta', \xi')$ and $p(z^*|\theta^*, \xi^*)$ cancel out in the acceptance ratio. However, the corresponding terms for the past observations, $p(y_{1:s}|\theta', \zeta_{1:s})$ and $p(y_{1:s}|\theta^*, \zeta_{1:s})$, do not vanish. This poses a problem if the likelihood function is intractable. Note that in the specific case where it is possible to sample from $p(\theta|y_{1:s}, \zeta_{1:s})$ directly, it is convenient to set $k(\theta|y_{1:s}, \zeta_{1:s}) = p(\theta|y_{1:s}, \zeta_{1:s})$. Then the terms $p(y_{1:s}|\theta', \zeta_{1:s})p(\theta') \propto p(\theta'|y_{1:s}, \zeta_{1:s})$ and $p(y_{1:s}|\theta^*, \zeta_{1:s})p(\theta^*) \propto p(\theta^*|y_{1:s}, \zeta_{1:s})$ would cancel out in the acceptance ratio.

26.3 ABC for Simulation-Based Optimal Design

If there is no explicit formula for the likelihood function or it is very cumbersome to evaluate, one may have to resort to *likelihood-free (LF)* methods, also called *approximate Bayesian computation (ABC)*. These methods can be applied if simulating the data from the probability model is feasible for every parameter θ. Some of the earliest applications of ABC were in the context of biogenetics (e.g., in [6]). For further examples see [10].

One possibility to incorporate likelihood-free methods into the MCMC simulation-based design algorithm is to modify and augment the target distribution (26.2) in the following way:

$$h_{\mathrm{LF}}(\vartheta, x_{1:s}) \propto u(\vartheta)p(z|\theta, \xi)p_\epsilon(y_{1:s}|x_{1:s}, \theta)p(x_{1:s}|\theta, \zeta_{1:s})p(\theta)\mu(\xi) .$$

The artificial data $x_{1:s}$, which are sampled together with ϑ, are added to the arguments of the target distribution. Integrating over $x_{1:s}$ leads to the original target distribution if $p_\epsilon(y_{1:s}|x_{1:s}, \theta)$ is a point mass at the point $x_{1:s} = y_{1:s}$. Since this event has a very small probability for higher-dimensional discrete distributions and probability zero in the case of continuous distributions, a compromise has to be found between exactness and practicality by adjusting the "narrowness" of $p_\epsilon(y_{1:s}|x_{1:s}, \theta)$. Therefore, the marginal distribution of h_{LF} with respect to ϑ, $\int h_{\mathrm{LF}}(\vartheta, x_{1:s})\mathrm{d}x_{1:s}$, is only an approximation to the true target distribution h. The function $p_\epsilon(y_{1:s}|x_{1:s}, \theta)$ is usually assumed to be a smoothing kernel density function: $p_\epsilon(y_{1:s}|x_{1:s}, \theta) = (1/\epsilon)K((\|\mathrm{T}(x_{1:s}) - \mathrm{T}(y_{1:s})\|)/\epsilon)$, where T(.) is some low-dimensional statistic of $y_{1:s}$ and $x_{1:s}$, respectively. The parameter ϵ controls the tightness of $p_\epsilon(y|x, \theta)$. The approximation error induced by ϵ being positive is often called nonparametric error.

If T is a sufficient statistic for the parameters of the probability model, integrating over $\mathrm{T}(x_{1:s})$ yields the same distribution as integrating out $x_{1:s}$. Otherwise, the application of ABC introduces a bias in addition to the nonparametric error.

The reason for augmenting the model is that using the proposal distribution

$$q_{\mathrm{LF}}(\vartheta', x'_{1:s}|\xi^*) = p(z'|\theta', \xi')p(x'_{1:s}|\theta', \zeta_{1:s})p(\theta')g(\xi'|\xi^*) ,$$

leads to the MH acceptance probability

$$\alpha = \min\left(1, \frac{u(\vartheta')}{u(\vartheta^*)} \frac{p_\epsilon(y_{1:s}|x'_{1:s}, \theta')}{p_\epsilon(y_{1:s}|x^*_{1:s}, \theta^*)} \frac{g(\xi^*|\xi')}{g(\xi'|\xi^*)}\right),$$

which does not depend on the likelihood function.

A comprehensive account of likelihood-free MCMC is given in [10].

26.4 Example

We apply the simulation-based design methodology developed in the previous sections to a standard Bayesian linear regression example. In that case the likelihood function is of a well-known and simple form, so there is no need to invoke likelihood-free methods. The purpose of our example is merely to demonstrate various important aspects one has to consider when applying simulation-based design algorithms with likelihood-free extensions. For this example the expected utility integral can also be computed analytically. This allows us to compare the results from the simulation-based optimal design algorithm to the exact results.

26.4.1 Bayesian Linear Regression

We assume that

$$z|\theta, \xi \sim \mathcal{N}(\mathrm{D}\theta, \sigma^2 \mathrm{I}_n).$$

That is, the expected value of the dependent variable is a linear combination of the parameter values $\theta \in \Theta \subseteq \mathbb{R}^k$ and depends on the design through the design matrix $\mathrm{D} = (\mathrm{f}(\xi_1), \ldots, \mathrm{f}(\xi_n))^T$, where f(.) is a k-dimensional function of the design variables $\xi_i \in [-1, 1]$, and $\xi = (\xi_1, \ldots, \xi_n)$. The n observations are assumed to be normally distributed, independent, and homoscedastic with known variance σ^2.

We assume that s previous observations $y = (y_1, \ldots, y_s)$ have been collected which follow the same distribution:

$$y|\theta, \zeta \sim \mathcal{N}(\mathrm{K}\theta, \sigma^2 \mathrm{I}_s),$$

where $\mathrm{K} = (\mathrm{f}(\zeta_1), \ldots, \mathrm{f}(\zeta_s))^T$ and $\zeta = (\zeta_1, \ldots, \zeta_s)$.

Furthermore, the parameters θ follow the prior normal distribution

$$\theta \sim \mathcal{N}(\theta_0, \sigma^2 \mathrm{R}^{-1}).$$

The posterior distribution of θ given the previous and current observations can be easily obtained for this example.

We take $u(z, \xi, \theta) = \log p(\theta|\{z, y\}, \{\xi, \zeta\}) - \log p(\theta)$ as our utility function, so that the expected utility for a specific design ξ is the expected gain in Shannon information (see [2]):

$$U(\xi) = \int_{z\in\mathbb{R}^n} \int_{\theta\in\Theta} \log\left(\frac{p(\theta|\{z, y\}, \{\xi, \zeta\})}{p(\theta)}\right) p(z|\theta, \xi)p(\theta|y, \zeta)\mathrm{d}\theta\mathrm{d}z\,.$$

For our particular model, the integral can be computed analytically and is given by $U(\xi) = -\frac{k}{2}\log(2\pi) - \frac{k}{2} + \frac{1}{2}\log\det\left(\sigma^{-2}(\mathrm{D}^T\mathrm{D} + \mathrm{K}^T\mathrm{K} + \mathrm{R})\right) + C$ for some constant C. It has the same maximum as the criterion for D_B optimality, which is $\det(\mathrm{D}^T\mathrm{D}+\mathrm{K}^T\mathrm{K}+\mathrm{R})$ (cf. [1]). Note that the D_B-optimal design does neither depend on σ^2 nor on the prior mean θ_0 nor on the previous observations y.

We choose a setting for which the exact solution can be obtained easily, and thus a comparison of the results of our design algorithms is feasible.

The following setting is used: the predictor is a polynomial of order two in one factor, i.e. $\mathrm{f}(\xi_i) = (1, \xi_i, \xi_i^2)^T$ and $\mathrm{f}(\zeta_i) = (1, \zeta_i, \zeta_i^2)^T$.

The continuous optimal design for this problem puts equal weights of 1/3 on the three design points -1, 0, and 1, see [1]. Likewise, if the number of trials of an exact design is divisible by three, then at the optimal design 1/3 of the trials are set to -1, 0, and 1, respectively. For our example, we choose the prior information matrix R in a way so that it represents prior information equivalent to one trial taken at the design point 0, i.e. $\mathrm{R} = \mathrm{f}(0)\mathrm{f}^T(0) = (1, 0, 0)^T(1, 0, 0)$. A value of 10^{-5} is added to the diagonal elements, thereby making it possible to invert R and thus to sample from the prior distribution. Furthermore, we assume that one previous observation has been collected at the design point -1, so that $\mathrm{K} = \mathrm{f}^T(-1) = (1, -1, 1)$. Therefore, if we have $n = 1$ (future) trial, it is optimal to set this trial to 1.

26.4.2 MCMC Sampler for Augmented Target Distribution

As neighborhood kernel for the likelihood-free MCMC sampler we take the uniform kernel: $p_\epsilon(y, x) \propto \mathbb{I}_{|y-x|<\epsilon}(x)$.

We use the uniform distribution on the interval $[-1, 1]$ as independence proposal distribution for ξ. For our example this is a reasonable choice because the utility surface is rather flat. Furthermore, we set $\sigma^2 = 2$, $\theta_0 = (0, 0, 0)^T$, and we assume that the previously collected observation at $\zeta = -1$ is $y = 40$. Note that these parameters should have no effect on the outcome in our example.

The algorithm was run for various values of ϵ ($\sigma, 2\sigma, 4\sigma, 8\sigma, 16\sigma$) and for various lengths of the Markov chain (10^4, 10^8, 10^9). Due to memory allocation constraints, the output of the Markov chains of length 10^8 and 10^9 was thinned, keeping every 10th and 100th element of the chain, respectively.

The utility function $u(z, \xi, \theta)$ is not non-negative everywhere. If negative utilities occur, the simulation step is repeated until the sampled utility is positive. This modification distorts the output of the estimated utility surface, but we are only

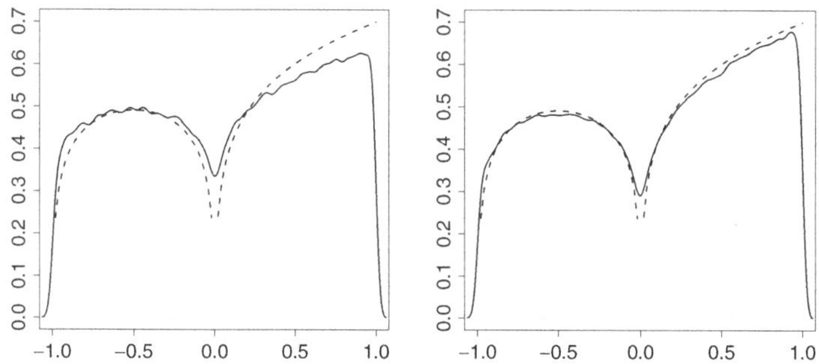

Fig. 26.1 Kernel density estimates of marginal distribution of ξ for MCMC sampler on the augmented utility (*solid lines*); 10^8 iterations; $\epsilon = \sigma$ (*left*) and $\epsilon = 16\sigma$ (*right*). For illustrative purposes, the true criterion $U(\xi)$ is also plotted (*dashed lines*; rescaled)

interested in regions of high expected utility anyway. If we use the utility function $u(z, \xi, \theta) = \log(p(\theta|\{z, y\}, \{\xi, \zeta\})) - \log(p(\theta))$, we do not observe many cases with negative utilities. If too many negative utilities were sampled, one could also add a positive constant to the utility function. We did not find it necessary to add a constant in this example.

Due to the very low acceptance rates that are usually associated with ABC sampling, a Markov chain of length 10^4 was deemed to be too short to properly represent the expected utility surface. We found that the MCMC samplers running for 10^8 iterations while keeping every 10th draw produce sufficiently long Markov chains that explore the whole design space for all values of ϵ (see Fig. 26.1) while being computationally not very demanding. Hence we will focus on the results for these samplers. On a PC with an Intel Core i3 CPU (2.10 GHz) and 4 GB RAM, they needed from 2.25 to 3 min to produce their sample.

The acceptance rate decreases from 0.017 if $\epsilon = 16\sigma$ to 0.0048 if $\epsilon = \sigma$, as would be expected. On the other hand, the integrated autocorrelation time (IAT)[1] increases from 12.28 ($\epsilon = 16\sigma$) to 44.01 ($\epsilon = \sigma$). One has to find a reasonable compromise between the accuracy of the ABC approximation and the autocorrelation of the sample, which has a negative effect on the effective sample size.

Figure 26.1 indicates that the maximum of the criterion is close to 1, which is the true optimum. It also suggests that the choice of ϵ has little impact on the marginal distribution of ξ. This might be a special feature of our example and our choice of the utility function and not the case in general. In our example the value of y does not matter for the optimal design, and hence it is irrelevant whether the simulated observations are close to the actual observations or not.

[1]The IAT of a process is defined as $\text{IAT} = 1 + 2\sum_{i=1}^{\infty} \rho_i$, where ρ_i denotes the autocorrelation of the process at lag i.

Conclusion
The integrated approach presented in this contribution is only suitable for very low-dimensional designs, where it is relatively easy to read off the mode of the target distribution from the sample output. In more complex design situations, a two-stage approach may be preferable: first obtain the posterior of the parameters using ABC, then perform simulation-based design. More details about this extension can be found in [4]. Further complications arise if the utility function $u(.)$ cannot be evaluated directly but also has to be estimated by ABC. Suggestions for dealing with that case are provided in [3] or [4].

Acknowledgements This work was partially supported by the project ANR-2011-IS01-001-01 "DESIRE" and FWF I 833-N18.

References

1. Atkinson, A.C., Donev, A.N., Tobias, R.D.: Optimum Experimental Designs, with SAS. Oxford University Press, New York (2007)
2. Chaloner, K., Verdinelli, I.: Bayesian experimental design: a review. Stat. Sci. **10**, 273–304 (1995)
3. Drovandi, C.C., Pettitt, A.N.: Bayesian experimental design for models with intractable likelihoods. Biometrics **69**, 937–948 (2013)
4. Hainy, M., Müller, W.G., Wagner, H.: Likelihood-free simulation-based optimal design. IFAS Research Paper Series 2013–64 (2013). Available via arXiv.org. http://arxiv.org/abs/1305.4273
5. Hainy, M., Müller, W.G., Wynn, H.P.: Approximate Bayesian computation design (ABCD), an introduction. In: Uciński, D., Atkinson, A.C., Patan, M. (eds.) mODa 10 - Advances in Model-Oriented Design and Analysis, pp. 135–143. Springer, Cham (2013)
6. Marjoram, P., Molitor, J., Plagnol, V., Tavaré, S.: Markov chain Monte Carlo without likelihoods. Proc. Natl. Acad. Sci. USA **100**, 15324–15328 (2003)
7. Müller, P.: Simulation based optimal design. In: Bernardo, J.M., Berger, J.O., Dawid, A.P., Smith, A.F.M. (eds.) Bayesian Statistics, vol. 6, pp. 459–474. Oxford University Press, New York (1999)
8. Müller, P., Sansó, B., De Iorio, M.: Optimal Bayesian design by inhomogeneous Markov chain simulation. J. Am. Stat. Assoc. **99**, 788–798 (2004)
9. Robert, C.P., Casella, G.: Monte Carlo Statistical Methods. Springer, New York (2004)
10. Sisson, S.A., Fan, Y.: Likelihood-free Markov chain Monte Carlo. In: Brooks, S.P., Gelman, A., Jones, G., Meng, X.-L. (eds.) Handbook of Markov Chain Monte Carlo, pp. 319–341. Chapman and Hall/CRC Press, Boca Raton (2011)
11. Tierney, L.: Markov chains for exploring posterior distributions. Ann. Stat. **22**, 1701–1728 (1994)

Chapter 27
Time Change Related to a Delayed Reflection

B.P. Harlamov

27.1 Introduction

Apparently Gihman and Skorokhod were the first who investigated reflection with delaying of one-dimensional Markov diffusion processes [1, p. 197]. They applied a method of stochastic integral equations which takes into account preserving the Markov property while reflecting. However there exist examples of interaction between a process and a boundary of its range of values, which can be interpreted like reflection, when the Markov property is being lost, although the property of continuous semi-Markov processes is preserved. Here is a simple example.

Let $w(t)$ $(t \geqslant 0)$ be Wiener process. Let us consider on the segment $[a, b]$ $(a \leqslant w(0) \leqslant b)$ the truncated process

$$\overline{w}(t) = \begin{cases} b, & w(t) \geqslant b \\ w(t), & a < w(t) < b \\ a, & w(t) \leqslant a \end{cases}$$

for all $t \geqslant 0$. It is clear that this process is not Markov. However it remains to be continuous semi-Markov [3]: the Markov property is fulfilled with respect to the first exit time from any open interval inside the segment, and also that from any one-sided neighborhood of any end of the segment.

The semi-Markov approach to the problem of reflection consists in solution of the following task: to determine a semi-Markov transition function for the process at a boundary point for the process preserving its diffusion form inside its open range of values, i.e. that up to the first exit time from the region and any time when it leaves

B.P. Harlamov (✉)
Institute of Problems of Mechanical Engineering of RAS, Saint-Petersburg, Russia
e-mail: b.p.harlamov@gmail.com

V.B. Melas et al. (eds.), *Topics in Statistical Simulation*, Springer Proceedings in Mathematics & Statistics 114, DOI 10.1007/978-1-4939-2104-1_27

the boundary. A more specific task to find reflection, preserving a global Markov property, is reduced to a problem to find a subclass of Markov reflected processes in the class of all the semi-Markov ones. Tasks of such a kind are important for applications where one takes into account interaction of diffusion particles with a boundary of a container, leading to a dynamic equilibrium of the system (see, e.g., [6]).

In paper [2] all the class of semi-Markov characteristics of reflection for a given locally Markov diffusion process is described. In paper [4] conditions for a semi-Markov characteristic to give a globally Markov process are found. In the present paper we continue to investigate processes with semi-Markov reflection. The aim of investigation is to find formulae, characterizing a time change, transforming a process with instantaneous reflection into the process with delaying reflection.

27.2 Semi-Markov Transition Function at a Boundary Point

We consider random processes on Skorokhod space $\mathscr{D} \equiv \mathscr{D}([0,\infty),\mathbb{R})$ with a natural filtration $(\mathscr{F})_0^\infty$. More special we will consider a diffusion process $X(t)$ on the half-line $t \geqslant 0$ with one boundary at zero. We assume that this process does not go to infinity and from any positive initial point it hits zero with probability one. For example, it could be a diffusion Markov process with a negative drift and bounded local variance.

Let us denote θ_t the shift operator on the set of trajectories $\mathscr{D}$; σ_Δ the operator of the first exit time from set Δ. By definition $\sigma_\Delta(\xi) = 0$, if $\xi(0) \notin \Delta$, where $\xi \in \mathscr{D}$. Semi-Markov process is a process which obeys Markov property at any time σ_Δ for an open $\Delta \subset \mathbb{R}$ (it is sufficiently to consider Δ as an open interval). We had substantiated above why it is expedient to consider semi-Markov reflection. Semi-Markov approach permits to consider from unit point of view an operation of instantaneous reflection as well as an operation of truncation, besides it opens new properties of processes interesting for applications.

In frames of semi-Markov models of reflection it is natural to assume that $X(t)$ is a semi-Markov process of diffusion type [3]. Let (P_x) $(x \geqslant 0)$ be a consistent family of measures of the process, depending on initial points of trajectories. On interval $(0,\infty)$ semi-Markov transition generating functions of the process

$$g_{(a,b)}(\lambda,x) := \mathbb{E}_x\left(e^{-\lambda\sigma_{(a,b)}};\ X(\sigma_{(a,b)}) = a\right);$$

$$h_{(a,b)}(\lambda,x) := \mathbb{E}_x\left(e^{-\lambda\sigma_{(a,b)}};\ X(\sigma_{(a,b)}) = b\right)$$

$(a < x < b)$ satisfy the differential equation

$$\frac{1}{2}f'' + A(x)f' - B(\lambda,x)f = 0,$$

with boundary conditions

$$g_{(a,b)}(\lambda,a+) = h_{(a,b)}(\lambda,b-) = 1; \quad g_{(a,b)}(\lambda,b-) = h_{(a,b)}(\lambda,a+) = 0.$$

The coefficients of the equation are assumed to be piece-wise continuous functions of $x > 0$, and for any x function $B(\lambda,x)$ is non-negative and has completely monotone partial derivative with respect to λ. First of all reflection of the process from point $x = 0$ means addition of this point to the range of values of the process. Further all the semi-closed intervals $[0,r)$ are considered what the process can only exit from open boundary. Corresponding semi-Markov transition generating functions are denoted as $h_{[0,r)}(\lambda,x)$. In this case $h_{[0,r)}(\lambda,0) > 0$. Function $K(\lambda,r) := h_{[0,r)}(\lambda,0)$ plays an important role for description of properties of reflected processes. Using semi-Markov properties of the process, we must assume

$$h_{[0,r)}(\lambda,x) = h_{(0,r)}(\lambda,x) + g_{(0,r)}(\lambda,x)\,K(\lambda,r),$$

and also

$$K(\lambda,r) = K(\lambda,r-\epsilon)(h_{(0,r)}(\lambda,r-\epsilon) + g_{(0,r)}(\lambda,r-\epsilon)K(\lambda,r)).$$

Assuming that there exist derivatives with respect to the second argument we have

$$g_{(a,b)}(\lambda,x) = 1 + g'_{(a,b)}(\lambda,a+)\,(x-a) + o(x-a),$$

$$g_{(a,b)}(\lambda,x) = -g'_{(a,b)}(\lambda,b-)\,(b-x) + o(b-x),$$

$$h_{(a,b)}(\lambda,x) = h'_{(a,b)}(\lambda,a+)\,(x-a) + o(x-a),$$

$$h_{(a,b)}(\lambda,x) = 1 - h'_{(a,b)}(\lambda,b-)\,(b-x) + o(b-x),$$

and obtain the differential equation

$$K'(\lambda,r) + K(\lambda,r)\,h'_{(0,r)}(\lambda,r-) + K^2(\lambda,r)\,g'_{(0,r)}(\lambda,r-) = 0.$$

Its family of solutions are [5]

$$K(\lambda,r) = \frac{h'_{(0,r)}(\lambda,0+)}{C(\lambda) - g'_{(0,r)}(\lambda,0+)},$$

where arbitrary constant $C(\lambda)$ can depend on λ. In order for $K(\lambda,r)$ to be a Laplace transform it is sufficient that function $C(\lambda)$ to be non-decreasing, $C(0) = 0$, and its derivative to be a completely monotone function [4]. Under our assumptions it is fair

$$K(\lambda,r) = 1 - C(\lambda)\,r + o(r) \quad (r \to 0).$$

Our next task is to learn a time change in the process with instantaneous reflection which derives the process with delayed reflection.

27.3 Time Change with Respect to the Instantaneous Reflection

For any Markov times τ_1, τ_2 (with respect to the natural filtration) on set $\{\tau_1 < \infty\}$ let us define the following operation

$$\tau_1 \dot{+} \tau_2 := \tau_1 + \tau_2 \circ \theta_{\tau_1}.$$

It is known [3] that for any open (in relative topology) sets Δ_1, Δ_2, if $\Delta_1 \subset \Delta_2$, then

$$\sigma_{\Delta_2} = \sigma_{\Delta_1} \dot{+} \sigma_{\Delta_2}.$$

Let us introduce special denotations for some first exit times and their combinations, and that for random intervals as $\epsilon > 0$

$$\alpha := \sigma_{[0,\epsilon)}, \quad \beta := \sigma_{(0,\infty)}, \quad \gamma(0) := \beta,$$

$$\gamma := \alpha \dot{+} \beta, \quad \gamma(n) := \gamma(n-1) \dot{+} \gamma \quad (n \geqslant 1),$$

$$b(0) := [0, \beta), \quad a(n) := [\gamma(n-1), \gamma(n-1) \dot{+} \alpha), \quad b(n) = [\gamma(n-1) \dot{+} \alpha, \gamma_n).$$

The random times α, $\gamma(n)$, and intervals $a(n)$, $b(n)$ $(n = 1, 2, \dots)$ depend on ϵ. In some cases we will denote this dependence by the lower index.

Let us remark that sequence $(\gamma(n))$ forms moments of jumps of a renewal process. Besides if $X(t) > 0$ then for any $t > 0$ there exist $\epsilon > 0$, and $n \geqslant 1$ such that $t \in b_\epsilon(n)$. It implies that for $\epsilon \to 0$ random set $\cup_{k=1}^{\infty} b_\epsilon(k)$ covers all the set of positive values of process X with probability one. On share of supplementary set (a limit of set $\cup_{k=1}^{\infty} a_\epsilon(k)$) there remain possible intervals of constancy and also a discontinuum of points (closed set, equivalent to continuum, without any intervals, [7, p. 158]), consisted of zeros of process X. The linear measure of it can be more than or equal to 0. This measure is included as a component in a measure of delaying while reflecting.

It is known [3, p. 111] that continuous homogeneous semi-Markov process is a Markov process if and only if it does not contain intrinsic intervals of constancy (it can have an interval of terminal stopping). This does not imply that a process with delayed deflection cannot be globally Markov. Its delaying is exceptionally at the expense of the discontinuum. A process without intervals of constancy at zero, and with the linear measure of the discontinuum of zeros which equals to zero is said to be a process with instantaneous reflection.

We will construct a non-decreasing sequence of continuous non-decreasing functions $V_\epsilon(t)$ $(t \geqslant 0)$, converging to some limit $V(t)$ as $\epsilon \to 0$ uniformly on every bounded interval.

Let $X(0) > 0$, and $V_\epsilon(t) = t$ on interval $b(0)$, and $V_\epsilon(t) = \beta$ on interval $a(1)$. On interval $b(1)$ the process V_ϵ increases linearly with a coefficient 1. On interval $a(2)$ function V_ϵ is constant. Then it increases with coefficient 1 on interval $b(2)$, and so on, being constancy on intervals $a(k)$, increasing with coefficient 1 on intervals $b(k)$. Noting that if $\epsilon_1 > \epsilon_2$, for any interval $a_{\epsilon_2}(k)$ there exists n such that $a_{\epsilon_2}(k) \subset a_{\epsilon_1}(n)$, we convince ourselves that the sequence of constructed functions does not decrease, bounded and consequently tends to a limit.

Let us define a process with instantaneous reflecting obtained from the original process X as a process, obtained after elimination of all its intervals of constancy at zero, and contraction of a linear measure of its discontinuum of zeros to zero. This process can be represented as a limit (in Skorokhod metric) of a sequence of processes $X_\epsilon(t)$, determined for all t by formula

$$X_\epsilon(t) = X(V_\epsilon^{-1}(t)),$$

where $V_\epsilon^{-1}(y)$ is defined as the first hitting time of the process $V_\epsilon(t)$ to a level y. Hence $X_\epsilon(t)$ has jumps of value ϵ at the first hitting time to zero and its iterations. Let us denote the process with instantaneous reflecting as $X_0(t)$, and the map $X \mapsto X_0$ as φ_V. Such a process is measurable (with respect to the original sigma-algebra of subsets) and continuous. Let $P_x^0 = P_x \circ \varphi_V^{-1}$ be the induced measure of this process.

Then it is clear that V is an inverse time change transforming the process X_0 into the process X, i.e. $X = X_0 \circ V$. In this case for any open interval $\Delta = (a, b)$ $(0 < a < b)$, or $\Delta = [0, r)$ $(r > 0)$ it is fair

$$\sigma_\Delta(X_0 \circ V) = V^{-1}(\sigma_\Delta(X_0)).$$

The function V^{-1} we call a direct time change, which corresponds to every "intrinsic" Markov time of the original process (in given case $X_0(t)$) the analogous time of the transformed process.

Remark that for $\epsilon_1 > \epsilon_2$ the set $\{\gamma_{\epsilon_1}(n),\ n = 0, 1, 2, \ldots\}$ is a subset of the set $\{\gamma_{\epsilon_2}(n),\ n = 0, 1, 2, \ldots\}$. That is why every Markov time $\gamma_\epsilon(n)$ is a Markov regeneration time of the process V, what permits in principle to calculate finite-dimensional distributions of this process. On the other hand, this process is synonymously characterized by its inverse, i.e. the process $V^{-1}(y) := \inf\{t \geqslant 0 : V(t) \geqslant y\}$ $(y > 0)$. This process is more convenient to deal with because Laplace transform of its value at a point y can be found as a limit of a sequence of easy calculable Laplace images of values $V_\epsilon^{-1}(y)$.

Theorem 1. *A direct time change $V^{-1}(y)$, mapping a process with instantaneous reflection into a process with delayed reflection satisfy the relation*

$$\mathbb{E}_0 \exp(-\lambda V^{-1}(y)) = \mathbb{E}_0 \exp(-\lambda\, y - C(\lambda) W(y)), \tag{27.1}$$

where $W^{-1}(t)$ *is a non-decreasing process with independent increments for which*

$$\mathbb{E}_0 \exp(-\lambda W^{-1}(t)) = \exp(g_{(0,\infty)}(\lambda, 0+)\, t). \tag{27.2}$$

Proof. Without loss of generality we suppose that $X(0) = 0$. Let $N_\epsilon(t) = n$ if and only if $\sum_{k=1}^{n-1} |b(k)| < t \leqslant \sum_{k=1}^{n} |b(k)|$ ($|a(k)|$, $|b(k)|$ are lengths of intervals $a(k)$, $b(k)$). Then

$$\mathbb{E}_0 \exp(-\lambda V^{-1}(y)) = \lim_{\epsilon \to 0} \mathbb{E}_0 \exp(-\lambda V_\epsilon^{-1}(y)) = \lim_{\epsilon \to 0} \mathbb{E}_0 \left(-\lambda y - \lambda \sum_{k=1}^{N_\epsilon(y)} |a(k)| \right).$$

We have

$$\mathbb{E}_0 \exp(-\lambda(V_\epsilon^{-1}(y) - y)) = \mathbb{E}_0 \exp\left(-\lambda \sum_{k=1}^{N_\epsilon(y)} |a(k)| \right)$$

$$= \sum_{n=0}^{\infty} \mathbb{E}_0 \exp\left(-\lambda \sum_{k=1}^{n} \alpha \circ \theta_{\gamma(k-1)};\ N_\epsilon(t) = n \right)$$

$$= P_\epsilon(\beta \geqslant y) + \sum_{n=1}^{\infty} \mathbb{E}_0 \left(\exp\left(-\lambda \sum_{k=1}^{n} \alpha \circ \theta_{\gamma(k-1)} \right);\ \sum_{k=1}^{n-1} |b(k)| < y \leqslant \sum_{k=1}^{n} |b(k)| \right)$$

$$= P_\epsilon(\beta \geqslant y) + \sum_{n=1}^{\infty} \mathbb{E}_0 \left(\exp\left(-\lambda\alpha - \lambda \sum_{k=2}^{n} \alpha \circ \theta_{\gamma(k-1)} \right);\right.$$

$$\left. \beta \circ \theta_\alpha + \sum_{k=2}^{n-1} \beta \circ \theta_\alpha \circ \theta_{\gamma(k-1)} < y \leqslant \beta \circ \theta_\alpha + \sum_{k=2}^{n} \beta \circ \theta_\alpha \circ \theta_{\gamma(k-1)} \right)$$

$$= P_\epsilon(\beta \geqslant y) + \sum_{n=1}^{\infty} \int_0^y \mathbb{E}_0 \left(\exp\left(-\lambda\alpha - \lambda \sum_{k=2}^{n} \alpha \circ \theta_{\gamma(k-1)} \right);\right.$$

$$\left. \beta \circ \theta_\alpha \in dx,\ \sum_{k=2}^{n-1} \beta \circ \theta_\alpha \circ \theta_{\gamma(k-1)} < y - x \leqslant \sum_{k=2}^{n} \beta \circ \theta_\alpha \circ \theta_{\gamma(k-1)} \right)$$

$$= P_\epsilon(\beta \geqslant y) + \sum_{n=1}^{\infty} \int_0^y \mathbb{E}_0(e^{-\lambda\alpha};\ \beta \circ \theta_\alpha \in dx) \mathbb{E}_0 \left(\exp\left(-\lambda \sum_{k=2}^{n} \alpha \circ \theta_{\gamma(k-2)} \right);\right.$$

$$\left. \sum_{k=2}^{n-1} \beta \circ \theta_\alpha \circ \theta_{\gamma(k-2)} < y - x \leqslant \sum_{k=2}^{n} \beta \circ \theta_\alpha \circ \theta_{\gamma(k-2)} \right)$$

$$= P_\epsilon(\beta \geqslant y) + \sum_{n=1}^{\infty} \int_0^y P_\epsilon(\beta \in dx) \mathbb{E}_0(e^{-\lambda\alpha}) \mathbb{E}_0 \left(\exp\left(-\lambda \sum_{k=1}^{n-1} \alpha \circ \theta_{\gamma(k-1)} \right);\right.$$

$$\left. \sum_{k=1}^{n-2} \beta \circ \theta_\alpha \circ \theta_{\gamma(k-1)} < y - x \leqslant \sum_{k=1}^{n-1} \beta \circ \theta_\alpha \circ \theta_{\gamma(k-1)} \right)$$

$$= P_\epsilon(\beta \geq y)$$
$$+ \int_0^y P_\epsilon(\beta \in dx)\mathbb{E}_0(e^{-\lambda\alpha}) \sum_{n=0}^{\infty} \mathbb{E}_0 \left(\exp\left(-\lambda \sum_{k=1}^{n} \alpha \circ \theta_{\gamma(k-1)}\right); \, N_\epsilon(y-x) = n \right)$$
$$= P_\epsilon(\beta \geq y) + \int_0^y P_\epsilon(\beta \in dx)\mathbb{E}_0(e^{-\lambda\alpha})\mathbb{E}_0 \exp(-\lambda(V_\epsilon^{-1}(y-x) - (y-x))).$$

Let us denote $Z(y) := \mathbb{E}_0 \exp(-\lambda(V_\epsilon^{-1}(y) - y))$, $F(x) := P_x(\beta < x)$, $\overline{F}(x) := 1 - F(x)$, $A := \mathbb{E}_0(e^{-\lambda\alpha})$. We obtain an integral equation

$$Z(y) = \overline{F}(x) + A \int_0^y Z(y-x)\, dF(x),$$

with a solution which can be written as follows:

$$Z(y) = \sum_{n=0}^{\infty} A^n (F^{(n)}(y) - F^{(n+1)}(y)),$$

where $F^{(n)}$ is n-times convolution of distribution F. Let us consider a sequence of independent and identically distributed random values $|b(n)|$ $(n = 1, 2, \ldots)$. Let P_ϵ^* is the distribution of a renewal process $N_\epsilon(y)$ with this sequence of lengths of intervals, and $\mathbb{E}_\epsilon^*$ is the corresponding expectation. Then

$$\mathbb{E}_\epsilon^* A^{N_\epsilon(y)} = \sum_{n=0}^{\infty} A^n P_\epsilon^*(N_\epsilon(y) = n) = \sum_{n=0}^{\infty} A^n (F^{(n)}(y) - F^{(n+1)}(y)),$$

Thus

$$\mathbb{E}_0 \exp(-\lambda V_\epsilon^{-1}(y)) = e^{-\lambda y}\, \mathbb{E}_\epsilon^* (\mathbb{E}_0 e^{-\lambda\alpha})^{N_\epsilon(y)}.$$

On the other hand, it is clear that there exists a version of the process $N_\epsilon(y)$, measurable with respect to the basic sigma-algebra, and adapted to the natural filtration of the original process, and having identical distribution with respect to measure P_0. Preserving denotations we can write

$$\mathbb{E}_\epsilon^* (\mathbb{E}_0 e^{-\lambda\alpha})^{N_\epsilon(y)} = \mathbb{E}_0 (\mathbb{E}_0 e^{-\lambda\alpha})^{N_\epsilon(y)}.$$

Moreover, measures P_0 and P_0^0 coincide on sigma-algebra F^*, generated by all the random values $\beta^\epsilon \circ \theta_{\alpha^\epsilon} \circ \theta_{\gamma(k)^\epsilon}$ $(\epsilon > 0,\ k = 1, 2, \ldots)$. From here

$$\mathbb{E}_0 (\mathbb{E}_0 e^{-\lambda\alpha})^{N_\epsilon(y)} = \mathbb{E}_0^0 (\mathbb{E}_0 e^{-\lambda\alpha})^{N_\epsilon(y)}.$$

Taking into account that α depends on ϵ and using our former denotations we can write

$$\mathbb{E}_0 e^{-\lambda\alpha} = K(\lambda,\epsilon) = 1 - C(\lambda)\,\epsilon + o(\epsilon).$$

We will show that the process $W_\epsilon(y) := \epsilon\, N_\epsilon(y)$ tends weakly to a limit $W(y)$ as $\epsilon \to 0$, which is an inverse process with independent increments with known parameters, and measurable with respect to sigma-algebra F^*. Actually, the process $W_\epsilon(y)$ does not decrease and is characterized completely by the process $W_\epsilon^{-1}(t)$. The latter has independent positive jumps on the lattice with a pitch ϵ. Hence it is a process with independent increments. Evidently a limit of a sequence of such processes, if it exists, is a process with independent increments too. Its existence follows from evaluation of Laplace transform of its increment. We have

$$\begin{aligned}\mathbb{E}_0^0 e^{-\lambda W_\epsilon^{-1}(t)} &= \mathbb{E}_0^0 \exp\left(-\lambda \sum_{k=1}^{[t/\epsilon]} |b(k)|\right)\\ &= (\mathbb{E}_\epsilon e^{-\lambda\beta})^{[t/\epsilon]}\\ &= (1 + g'_{(0,\infty)}(\lambda,0)\,\epsilon + o(\epsilon))^{[t/\epsilon]} \to e^{g'_{(0,\infty)}(\lambda,0)\,t} \quad (\epsilon \to 0).\end{aligned}$$

Using the sufficient condition of weak convergence of processes in terms of convergence of their points of the first exit from open sets [3, p. 287], we obtain

$$\mathbb{E}_0 \exp(-\lambda V^{-1}(y)) = \mathbb{E}_0^0 \exp(-\lambda\, y - C(\lambda) W(y)),$$

what can be considered as description of the direct time change in terms of the process with instantaneous reflection and the main characteristic of delaying, function $C(\lambda)$. □

We use this formula for deriving the Laplace transform of a difference between the first exit times from an one-sided neighborhood of the boundary point for processes with delayed and instantaneous reflection.

Denote

$$\beta^r := \sigma_{(0,r)}, \quad \gamma^r(0) = 0,$$

$$\gamma^r := \alpha \dot{+} \beta^r, \quad \gamma^r(n) := \gamma^r(n-1) \dot{+} \gamma^r \quad (n \geqslant 1),$$

$$b^r(n) = [\gamma^r(n-1) \dot{+} \alpha,\, \gamma^r(n)) \quad (n \geqslant 1),$$

$$M_\epsilon^r := \inf\{n \geqslant 0 :\ X(\gamma^r(n)) \geqslant r\}.$$

Hence

$$P_0(M_\epsilon^r = n) = P_0(X(\gamma^r(1)) = 0, \dots, X(\gamma^r(n-1)) = 0, X(\gamma^r(n-1)) = r)$$
$$= (p(\epsilon, r))^{n-1}(1 - p(\epsilon, r)),$$

where $p(\epsilon, r) := P_0(X(\gamma^r(1)) = 0)$.

Theorem 2. *A difference between the first exit times from a semi-closed interval $[0, r)$ for processes with delayed and instantaneous reflections obeys to the relation*

$$\mathbb{E}_0 \exp(-\lambda(\sigma_{[0,r)} - \sigma^0_{[0,r)}) = \frac{-G'_{(0,r)}(0+)}{C(\lambda) - G'_{(0,r)}(0+)}, \tag{27.3}$$

where $G_{(0,r)}(x) = g_{(0,r)}(0, x)$.

Proof. We have

$$\sigma_{[0,r)} = \gamma^r(M_\epsilon^r) = \sum_{n=1}^{\infty} \gamma^r(n) I(M_\epsilon^r = n)$$
$$= \sum_{n=1}^{\infty} \left(\sum_{k=1}^{n-1} (|a_\epsilon(k)| + |b_\epsilon(k)|) + |a_\epsilon(n)| + |b_\epsilon^r(1)| \right) I(M_\epsilon^r = n);$$
$$\sigma^o_{[0,r)} = \sum_{n=1}^{\infty} \left(\sum_{k=1}^{n-1} (|a_\epsilon^o(k)| + |b_\epsilon(k)|) + |a_\epsilon^o(n)| + |b_\epsilon^r(1)| \right) I(M_\epsilon^r = n);$$

where $a_\epsilon^o(k)$ is the first hitting time at the level ϵ by the process with instantaneous reflection after a recurrent first hitting time at 0. By definition the sum of such times up to the first hitting time at level r tends to zero as $\epsilon \to 0$ P_0-almost sure. From here it follows that

$$\sigma_{[0,r)} - \sigma^o_{[0,r)} = \lim_{n\to\infty} \sum_{n=1}^{\infty} \sum_{k=1}^{n} |a_\epsilon(n)|\, I(M_\epsilon^r = n).$$

Hence

$$E_0 e^{-\lambda(\sigma_{[0,r)} - \sigma^o_{[0,r)})} = \lim_{n\to\infty} \sum_{n=1}^{\infty} E_0 \exp\left(-\lambda \sum_{k=1}^{n} |a_\epsilon(n)|\, I(M_\epsilon^r = n) \right).$$

On the other hand,

$$F_\epsilon(\lambda) := E_0 \exp\left(-\lambda \sum_{k=1}^{M_\epsilon^r} |a_\epsilon(k)|\right) = \sum_{n=1}^{\infty} E_0 \exp\left(-\lambda \sum_{k=1}^{n} |a_\epsilon(k)|;\ M_\epsilon^r = n\right)$$

$$= \sum_{n=1}^{\infty} E_0 \exp\left(-\lambda \sum_{k=1}^{n}\right.$$

$$\left. \times\, \alpha \circ \theta_{\gamma^r(k-1)};\, X(\gamma^r(1)) = 0, \ldots, X(\gamma^r(n-1)) = 0, X(\gamma^r(n)) = r\right)$$

$$= \sum_{n=1}^{\infty} E_0 \exp\left(-\lambda\alpha - \lambda \left(\sum_{k=2}^{n} \alpha \circ \theta_{\beta^r \dot{+} \gamma^r(k-2)}\right) \circ \theta_\alpha;\right.$$

$$\left.\theta_\alpha^{-1}\left(X(\beta^r) = 0, \ldots, X(\beta^r \dot{+} \gamma^r(n-2)) = 0, X(\beta^r \dot{+} \gamma^r(n-1)) = r\right)\right)$$

$$= E_0 e^{-\lambda\alpha} \sum_{n=1}^{\infty} P_\epsilon(X(\beta^r) = 0) E_0 \exp\left(-\lambda \sum_{k=1}^{n-1} \alpha \circ \theta_{\gamma^r(k-1)};\right.$$

$$\left. X(\gamma^r(1)) = 0, \ldots, X(\gamma^r(n-2)) = 0, X(\gamma^r(n-1)) = r\right)$$

$$= E_0 e^{-\lambda\alpha}\, P_\epsilon(X(\beta^r) = r) + E_0 e^{-\lambda\alpha}\, P_\epsilon(X(\beta^r) = 0) E_0 \exp\left(-\lambda \sum_{k=1}^{M_\epsilon^r} |a_\epsilon(k)\right).$$

Hence

$$F_\epsilon(\lambda) = \frac{E_0 e^{-\lambda\alpha}\, P_\epsilon(X(\beta^r) = r)}{1 - E_0 e^{-\lambda\alpha}\, P_\epsilon(X(\beta^r) = 0)}.$$

Taking into account that

$$E_0 e^{-\lambda\alpha} = K(\lambda, \epsilon) = 1 - C(\lambda)\epsilon + o(\epsilon),$$

$$P_\epsilon(X(\beta^r) = r) := H_{(0,r)}(\epsilon) = h_{(0,r)}(0, \epsilon) = H'_{(0,r)}(0+)\epsilon + o(\epsilon),$$

$$P_\epsilon(X(\beta^r) = 0) := G_{(0,r)}(\epsilon) = 1 - H_{(0,r)}(\epsilon),$$

we obtain

$$F_\epsilon(\lambda) \to \frac{H'_{(0,r)}(0+)}{C(\lambda) + H'_{(0,r)}(0+)}$$

as $\epsilon \to 0$ □

It is interesting to note that for a linear function $C(\lambda) = k\lambda$, when a reflecting locally Markov process is globally Markov [4], the difference between the first exit times from a semi-closed interval $[0, r)$ for processes with delayed and instantaneous reflections has the exponential distribution with parameter $H'_{(0,r)}(0+)/k$. Evidently this difference is the time when the process has zero value. Taking into account that a continuous Markov process has no intervals of constancy (excepting an infinite final interval of constancy if any) we obtain that in the latter case the set of points when the process has zero value is a Cantor discontinuum with a positive linear measure.

References

1. Gihman, I.I., Skorokhod, A.V.: Stochastic Differential Equations. Naukova dumka, Kiev (1968, in Russian)
2. Harlamov, B.P.: Diffusion process with delay on edges of a segment. Zapiski nauchnyh seminarov POMI **351**, 284–297 (2007, in Russian)
3. Harlamov, B.P.: Continuous Semi-Markov Processes. ISTE & Wiley, London (2008)
4. Harlamov, B.P.: On a Markov diffusion process with delayed reflection on boundaries of a segment. Zapiski nauchnyh seminarov POMI **368**, 231–255 (2009, in Russian)
5. Harlamov, B.P.: On delay and asymmetry points of one-dimensional diffusion processes. Zapiski nauchnyh seminarov POMI **384**, 292–310 (2010, in Russian)
6. Harlamov, B.P.: Stochastic model of gas capillary chromatography. Commun. Stat. Simul. Comput. **41**(7), 1023–1031 (2012)
7. Hausdorff, F.: Theory of Sets. KomKniga, Moscow (2006, in Russian)

Chapter 28
Et tu "Brute Force"? No! A Statistically Based Approach to Catastrophe Modeling

Mark E. Johnson and Charles C. Watson Jr.

28.1 Introduction

Catastrophe modeling is complex and inherently multi-disciplinary drawing upon atmospheric science (hurricanes, nor'easters, and tornadoes), geophysics (earthquakes, volcanoes, and sinkholes) and hydrology (tsunamis, storm surge, flooding). Each natural peril exerts various pressures on building exposures, bringing wind and structural engineering into the analyses. The conversion of physical damage to economic losses requires actuarial science to provide an insured loss perspective. Throughout the catastrophe modeling process, uncertainty abounds which requires the attention of the statistician. Finally, for implementation, computer science and software engineering comes in to play.

Prior to Hurricane Andrew (1992), much of the insurance industry based its hurricane peril premiums on econometric models of historical losses from hurricanes. Some of these insurance companies who combined such econometric models with market forces that had driven premiums to historical low levels became insolvent. In response to the imminent exodus of insurance underwriters from Florida, the Florida Commission on Hurricane Loss Projection Methodology was established to develop standards for computer models that generate annual losses due to the hurricane wind peril. Since 1996, this Commission has been reviewing submitted models submitted for use in insurance rate filings. The models submitted to the Commission have the same basic structure. Simply put, some of the key hurricane related random

M.E. Johnson (✉)
Department of Statistics, University of Central Florida, Orlando, FL 32816-2370, USA
e-mail: Mark.Johnson@ucf.edu

C.C. Watson Jr.
Enki Holdings LLC, Savannah, GA 31404, USA
e-mail: cwatson@methaz.org

V.B. Melas et al. (eds.), *Topics in Statistical Simulation*, Springer Proceedings in Mathematics & Statistics 114, DOI 10.1007/978-1-4939-2104-1_28

variables (annual frequency, intensity, track, radius of maximum winds, etc.) are fit with probability distributions and then tens or hundreds of thousands of simulated years of hurricane activity are generated and annual losses accrued. The previously outlined structure is generally attributed to [7] and followers [3], and this brute-force approach has become the typical template for hurricane wind catastrophe modeling.

By modeling the hurricane as an entity having life cycle (with the caveat that some physical features are ignored such as reconstitution of the eye wall, spin off tornadoes, and so forth), the modeling community is then forced to generate massive numbers of hypothetical events in order to account for the uncertainty in estimation due to the simulation itself. Of course, this approach does not reduce the inherent uncertainty in estimating loss costs, as the individual versions of the hurricane models cannot accommodate every historical storm perfectly (Hurricane Wilma in 2005 is a case in point with stronger winds observed on the left side of its track) nor all future events, as well. A basic argument given is that the historical record is too short (50 to 100 to 160 years depending on the acceptance of historical data sources) to achieve much better results. Consequently, the models that have been approved by the Commission for use in rate filing can vary substantially from each other with respect to average annual losses and probable maximum losses, and in a manner and magnitude that is vexing to state legislators. Although disavowing the adequacy of the historical record for direct modeling purposes, the modelers then draw upon the same record to *validate* their own results in the sense that the historical results are sufficiently close to the simulated results. We view this perspective as inverted, since the data are real while the models are approximations. This perspective also elevates the simulated results in the eyes of some insurers and re-insurers, effectively blinding them from other approaches to the problem. Finally, measurement error in observed wind speeds is a contributor to uncertainty but it should not be the lone excuse for disparities in model simulated versus actual losses.

With this background in mind, it should not be surprising that the authors have pursued an alternative line of research with respect to catastrophe modeling. The authors have developed and published over the past 15 years a number of papers that offer an alternative approach to estimating insured losses and site-specific wind distributions [12, 15, 16, 22–26]. This approach makes more direct use of the historical record that is both true to the record and provides corresponding uncertainty assessments. Instead of constructing hypothetical storms based on and resembling the historical record, the approach is to run the full set of nearly 1,700 historical events and record the maximum wind speeds at each location of interest. This set of wind speeds provides an ample data set for fitting extreme value distributions at each site. Cross validation is used to confirm that accurate forecasts can be made with this approach.

The focus of this paper is on attaining realistic estimates of the phenomena of interest (average annual losses, probable maximum loss, and maximum wind speeds). The original brute force approach of the pioneer Friedman will be compared to the statistically based approach which uses the catastrophe models for augmenting historical data sets which are then subjected to sophisticated statistical treatment. Our approach will also be shown to accommodate modular subcomponents to parts of the perils affording us the possibility of assessing model misspecification. Published papers, presentations, and reports on operational projects can be found at http://hurricane.methaz.org/ while real time tracking of global hazards can be followed at http://tracking.enkiops.org.

28.2 Building a Hurricane Model

In order to build a hurricane computer simulation model, a number of decisions must be made involving a mathematical model of a hurricane, its features and their computer representation including probability distributions to capture the stochastic nature of events [1, 2, 4–6, 8–11, 13, 14, 18, 19]. At the very least, the following characteristics must be considered:

Frequency of Occurrence The number of events in a given season (June 1 through November 30 for the Atlantic basin) is of interest. Historically, there have been approximately ten tropical cyclones per season—a portion of which strengthen to hurricane force and in turn, a portion of these that make landfall in the USA. Discrete distributions (Poisson, negative binomial, Polya) are used for this purpose with the specific distribution chosen to reflect the scope of the study. The frequency of US landfalling events is related to global climate conditions such the Atlantic Multi-decadal Oscillation (AMO) and the increasingly known Pacific phenomena el Niño Southern Oscillation (ENSO). These conditions are most relevant for short term and seasonal forecasts.

Tropical Cyclone Tracks The path of a hurricane is challenging to model since the starting point can be as far east as the Atlantic Ocean off the northwest African coast through the Caribbean Sea and Gulf of Mexico. Once formed the tracks can be rather erratic with the Atlantic forming storms generally headed west until encountering steering currents which can divert them north and northeast away from harm's way, depending on the timing. Simulated tracks are so challenging that some models resort to sampling from the historical tracks to avoid the generation of completely unrealistic movements.

Intensity Strength and duration of winds dictates the level of damage for hurricanes. Modeling intensity is challenging since storms naturally evolve from weak low pressure areas with some circulation to possibly a well-defined very strong vortex. Storms can strengthen, weaken, and strengthen again with interruptions due to passage over mountains in Hispaniola or Cuba or traversing the Florida peninsula. Intensity has been modeled by determining the distribution of maximum wind velocity or its surrogate the pressure differential (far field pressure minus central pressure). Another component of intensity is the radius to maximum winds and the overall forward movement of the storm (yielding an asymmetry in the storm strength pattern about the center). The scope of the hurricane can be modeled through a transect profile of the storm from the center of the storm (calm) to the eyewall (maximum winds) to the extent of the storm (diminishing winds farther out). Once a simulated storm is over land, weakening and filling occurs which introduces additional characteristics to be modeled.

Stochastic Storm Set Assuming an individual tropical cyclone can be modeled, the next step is to generate a season's worth of events followed by the generation of multiple seasons. Within a season, care must be taken to avoid simultaneous events striking the same structures at the same time to preserve reality and to accommodate insurance policies having provisions related to damage for a season and call for repairs after each event.

Statistical Perspective Even a very simple hurricane simulation model requires the fitting of several probability distributions and consideration of their joint distributions. With the exception of hurricane models that look at the dependence between radius of maximum winds and maximum winds, little consideration has been given for joint distributions—the fits taking place individually. To summarize, the distributions to be fit and typical number of parameters include:

- Number of storms per season—one or two parameter discrete distribution
- Track distribution—multi-parameters possibly associated with a Markov chain model or discrete distribution to sample from the historical tracks with possible probabilistic perturbations
- Maximum wind (or minimum central pressure)—two parameter continuous distribution with thresholds
- Radius of maximum winds—two parameter continuous distribution with thresholds
- Forward speed (translation velocity)—two parameter continuous distribution
- Profile factor—two parameter distribution, possibly related to strength of storm

Some of the above characteristics are temporal and spatially varying, as well but tend to be sampled at landfall and then vary according to the filling of the storm now separated from its heat source. There is uncertainty in parameter estimation, owing to the data sources supporting the fits. Frequency may be based on a sample size as large as 160 whereas there are fairly few category five storms for which radius of maximum winds are available. With all of these sources of variation in a simulation model, a duration of 100,000 years or more should be viewed as a necessary evil to control the additional source of random variation attributable to sampling error.

28.3 Direct Fitting of Wind Speeds

In contrast to the brute force modeling effort described in Sect. 28.2, a more direct fitting process can be used. Since damage to structures occurs due to wind impacts, the primary distribution of interest is the maximum wind speed at each site in the study area owing to tropical cyclones. Ideally, this requirement would translate into having the historical record of wind speeds at each exposure site. Of course, anemometers are not so conveniently placed so the next best thing is to simulate all tropical cyclones in the record and then determine the wind speeds at each site using a wind field model to provide the equivalent measurement readings. Thus, only historical storms would be used so there can be no criticism of simulated storms that do not make realistic sense. Since only 1,644 storms are simulated (the number of Atlantic basin events since 1851), multiple wind field models could be used in this exercise and the median wind speed from the generated maximum winds could be used. Such a conservative approach diminishes the possibility that a particular wind field model provides a very poor representation of a particular event. Carrying this idea out further, in addition to the choice of wind field model, one could also envision different friction models (rough terrain mitigating wind speeds while increasing turbulence) and various damage functions. Such an approach was originally developed by the authors in conjunction with a rate filing in North Carolina [23] and was subsequently documented and published in the Bulletin of the American Meteorological Society [24] and the Journal of Insurance Regulation [26]. Distribution fitting is restricted to the annual maximum winds at each site and for hurricane related winds, the authors have discovered that the Weibull distribution provides an excellent model. Other distributions considered include the lognormal, extreme value, and inverse Gaussian. Using various cross validation approaches [16], the Weibull performs best in predicting the maximum wind speed across hurricane prone sites. As an example validation calculation, the most recent 20 years of experience is predicted using all previous data and then the actual and forecasted wind speeds are compared. These calculations have been used at 30 m resolution which corresponds to three billion sites in Florida.

28.4 Discussion

Two approaches have been outlined for generating loss costs associated with hurricane events. Both the brute force (Sect. 28.2) and the statistical (Sect. 28.3) approaches rely on fitting probability distributions. A key difference in the two approaches is the choice of data sets that are the basis of the fits. Once wind speeds on structures/exposures are determined, then damage and insured losses can be subsequently estimated. Although both approaches could converge in methodology at this point, most implemented brute force models opt for a single damage function/vulnerability component which is generally considered proprietary by their developers. (Incidentally, close empirical approximations to these proprietary damage functions can be obtained using publicly available data.) The statistical approach that uses the median values from a collection of model results readily accommodates multiple choices of damage functions, to broaden the range of possible values. Which approach is more sensible from a scientific viewpoint and more importantly, which approach yields the more accurate estimates of losses—the ultimate reason this modeling exercise is taking place? We consider several criteria and the relative merits of each modeling approach.

Stability The insurance industry, the re-insurance industry, and especially consumers have difficulty with massive fluctuations in loss costs or premiums from one season to the next. Such fluctuations could occur if the modeling approach is highly sensitive to one new season of considerable activity and catastrophic losses. The statistical approach is based on over 1,600 historical events and the fitting of distributions to wind speeds across the panoply of possibilities. One category five event offers possibly a new maxima to the set of data values already being fit, but if the value is in the neighborhood of the 200-year return period value (using the 160 years of HURDAT data [5, 18]), the impact is natural and modest. In a separate analysis mentioned elsewhere [24], a complete re-run of results was made with the exclusion of the 1992 season (the year of Hurricane Andrew striking Florida) to marginal effect. In contrast, such a mega-storm can have a huge impact on the brute force models. If this storm has any unusual characteristics compared to other large storms, then various probability distributions that are fit can change considerably. For example, Hurricane Charley in 2004 had an exceptionally small radius of maximum winds which forced the Public Model to adjust this distribution which in turn reduced estimated losses considerably (smaller storms tend to have smaller damage swaths). As another instance, following the very active 2004 and 2005 seasons, some modelers developed "near-term" models evidently at the request of the re-insurance industry. Selected experts argued that these two seasons were a harbinger of things to come and the long-term frequency of events needed to be increased considerably. This approach has lost some impetus following five straight years of no landfalling hurricanes in Florida.

Impact of Model Components With the statistical approach, viable wind-field, friction, and damage components are included in combination, so no one particular component drives the results. The brute force methods choose what their developers consider the best model sub-components giving them a vested interest in their use. Swapping out a sub-component can have a very large impact on results. As a case in point, some of the brute force models under review by the Florida Commission on Hurricane Loss Projection Methodology have gradually evolved from using an inland weakening model developed by [17] to an alternative due to [21]. Since the latter has much slower filling rates than the Kaplan–de Maria method, the damage swath is larger and the losses in turn are greater.

Validation There is a fundamental difficulty in determining if the collection of stochastic storms consists of realistic events (physically possible) and if the collection as a whole provides adequate coverage to generate realistic annual loss costs and probable maximum losses (near term and long term). A common remark among reviewers of the proprietary models is "let's wait ten thousand years and see how it turns out." Although tongue in cheek, this attitude conveys the difficulty in assessing competing models against the future reality. The additional phrase, "we're doing the best we can," is an assertion that the subcomponents represent the current state of the art and that the implementation is meticulous. Evidence that a modeler can match reasonably well historical insured losses is meaningless if the model has been at all calibrated just for this purpose. At least for the statistical approach, a cross-validation approach has been used extensively [24, 25] to provide objective evidence of accuracy and reliability of the modeling approach. Although several variants have been used over the years (all leading to confirmation of the approach), a common scenario is to use only the data available for the time frame 1851–1990 to develop the statistical model and then use this model to forecast the subsequent 20-year return period winds at each site in the study area with varying levels of prediction capability (e.g., 50 %, 75 %, 90 %, 95 %) and then tally the proportion of sites falling into the forecasted categories. This exercise is objective (not using the same data to validate that was used to develop the model) and lays the groundwork for further improvements. In contrast, the brute force method to our knowledge has not been subjected to such scrutiny, since the effort would be massive. Every distribution in Sect. 28.2 would need to be re-assessed and fit, and then a new 100–300K years of simulation effort would follow based on the alternate baseline data.

28.5 Final Comments

The authors contend that the statistical approach when validated is superior to the brute force simulation approach that has become ingrained in the catastrophe modeling industry. The insurance industry does not appreciate major changes from 1 year to the next that generate large changes in premiums, reserves, or the cost of re-insurance. The brute force models can be updated each year with another tweak associated with an additional season. Effects from an active season in which there were substantial losses can take five or more years to sort out with the easy claims (readily settled anyway) coming through first and then disputed claims and public adjusters entering the fray over time. With the statistical approach, results from the previous season are incorporated as soon as the hurricane characteristics have been assessed. In an earlier study, fairly simple damage functions were found to suffice [24] and that the meteorology aspects were the most critical for the overall variability in the results. Owing to space limitations, the focus in this paper has been on hurricane perils. However, the same statistical approach has been applied to earthquakes and other perils—and most notably in developing a comprehensive approach to an insurance facility in the Caribbean [20].

References

1. Bretschneider, C.: A non-dimensional stationary hurricane wave model. In: Proceedings of 1972 Offshore Technology Conference, pp. 30–42 (1972)
2. Center CER: Shore protection manual. US Army Corps of Engineers 1 (1984)
3. Clark, K.: A formal approach to catastrophe risk assessment and management. Proc. Casual. Actuar. Soc. **73**, 69–92 (1986)
4. Dumm, R.E., Johnson, M.E., Simons, M.M.: Inside the black box: evaluating and auditing hurricane loss models. J. Insur. Regul. **27**(2), 41 (2008)

5. Friedman, D.: Hurricane best track files. HURDAT, Atlantic Tracks File (1975). http://www.nhc.noaa.gov/pastall.shtml
6. Friedman, D.: An analytic model of the wind and pressure profiles in hurricanes. Mon. Weather Rev. **108**, 1212–1218 (1980)
7. Friedman, D.: Natural hazard risk assessment for an insurance program. The Geneva Papers on Risk and Insurance (1984)
8. Georgiou, P.: Design wind speeds in tropical cyclone prone regions. Ph.D. dissertation, Dept of Civil Engineering (1985)
9. Hamid, S., Kibria, G., Gulati, S., Powell, M., Annane, B., Cocke, S., Pinelli, J.P., Gurley, K., Chen, S.C.: Predicting losses of residential structures in the state of florida by the public hurricane loss evaluation model. Stat. Methodol. **7**(5), 552–573 (2010)
10. Holland, G.: Natural hazard risk assessment for an insurance program. The Geneva Papers on Risk and Insurance (1984)
11. Holton, J.: An Introduction to Dynamic Meteorology. Academic, New York (1992)
12. Iman, R.L., Johnson, M.E., Watson, C.C., Jr.: Statistical aspects of forecasting and planning for hurricanes. Am. Stat. **60**(2), 105–121 (2006)
13. Jarvinen, B., Neumann, C., Davis, M.A.S.: A tropical cyclone data tape for the north atlantic basin, 1886–1983: contents, limitations, and uses. NOAA Tech Memo NWS NHc 22, 21 (1984)
14. Jelesnianski, C.P., Chen, J., Shaffer, W.A.: SLOSH: Sea, lake, and overland surges from hurricanes. US Department of Commerce, National Oceanic and Atmospheric Administration, National Weather Service (1992)
15. Johnson, M.E., Watson, C.C.: Hurricane Return Period Estimation. Organization of American States, Washington, DC (1999)
16. Johnson, M.E., Watson, C.C., Jr.: Fitting statistical distributions to data in hurricane modeling. Am. J. Math. Manag. Sci. **27**(3–4), 479–498 (2007)
17. Kaplan, J., DeMaria, M.: A simple empirical model for predicting the decay of tropical cyclone winds after landfall. J. Appl. Meteorol. **34**(11), 2499–2512 (1995)
18. Landsea, C.W., Anderson, C., Charles, N., Clark, G., Dunion, J., Fernandez-Partagas, J., Hungerford, P., Neumann, C., Zimmer, M.: The atlantic hurricane database re-analysis project: documentation for the 1851–1910 alterations and additions to the hurdat database. In: Hurricanes and Typhoons: Past, Present and Future, pp. 177–221 (2004)
19. Miller, B.I. Characteristics of hurricanes analyses and calculations made from measurements by aircraft result in a fairly complete description. Science **157**(3795), 1389–1399 (1967)
20. Vaughan, E.J., Vaughan, T.: Fundamentals of Risk and Insurance, vol. 3. Wiley, New York (2007)
21. Vickery, P.J.: Simple empirical models for estimating the increase in the central pressure of tropical cyclones after landfall along the coastline of the united states. J. Appl. Meteorol. **44**(12), 1807–1826 (2005)
22. Watson, C.C.J., Johnson, M.E.: Using integrated multi-hazard numerical models in coastal storm hazard planning. In: Solutions for Coastal Disasters 02 Conference Proceedings, pp. 173–177 (2002)
23. Watson, C.C.J., Johnson, M.E.: An assessment of computer based estimates of hurricane loss costs in North Carolina. North Carolina Department of Insurance Technical Report (34) (2003)
24. Watson, C.C., Johnson, M.E.: Hurricane loss estimation models: opportunities for improving the state of the art. Bull. Am. Meteorol. Soc. **85**(11), 1713–1726 (2004)
25. Watson, C.C., Jr., Johnson, M.E.: Integrating hurricane loss models with climate models. In: Climate Extremes and Society, pp. 209–224. Cambridge University Press, Cambridge (2008)
26. Watson, C.C., Jr., Johnson, M.E., Simons, M.: Insurance rate filings and hurricane loss estimation models. J. Insur. Regul. **22**(3), 39–64 (2004)

Chapter 29
Optimizing Local Estimates of the Monte Carlo Method for Problems of Laser Sensing of Scattering Media

Evgeniya Kablukova and Boris Kargin

29.1 Introduction

The current paper is devoted to one way of optimizing Monte Carlo method algorithms for solving nonstationary problems of laser sensing of natural media. Such problems are of great interest in connection with wide application of laser sensors of land, aircraft, and space basing to various practical tasks. For example, efficiently diagnosing aerosol admixtures in the atmosphere, determining the space-time transformation of microphysical properties of the atmosphere and many other problems of optical remote sensing in natural media.

A more detailed list of certain modern physical problem statements for atmosphere and ocean laser sensing and related methods of statistical modeling can be obtained from a relatively fresh overview [2]. The feasibility of applying Monte Carlo methods and developing corresponding algorithms for solving problems of optical radiation transfer theory in scattering and absorbing media have been discussed in many earlier works, summarized in [3]. The laser sensing problems under consideration differ from many other problems of atmosphere optics. One aspect is the presence of complex boundary conditions, connected to the finite size of the initial beam of radiation and small phase volume of the detector. Another

---

E. Kablukova (✉)
Institute of Computational Mathematics and Mathematical Geophysics SB RAS, pr. ak. Lavrentieva 6, Novosibirsk 630090, Russia
e-mail: jane_k@ngs.ru

B. Kargin
Institute of Computational Mathematics and Mathematical Geophysics SB RAS, pr. ak. Lavrentieva 6, Novosibirsk 630090, Russia

Novosibirsk National Research State University, Pirogova str. 2, Novosibirsk 630090, Russia
e-mail: bkargin@osmf.sscc.ru

V.B. Melas et al. (eds.), *Topics in Statistical Simulation*, Springer Proceedings in Mathematics & Statistics 114, DOI 10.1007/978-1-4939-2104-1_29

is the principally nonstationary character of the radiation transfer process being modeled. This circumstance defines characteristic requirements to the technique of statistical modeling. Local estimates (LE), which have high computational cost, present the only way to compute the properties of radiation registered by a detector with a small phase volume. To decrease computational cost of the algorithm in this paper an optimization of local estimates is proposed, based on integrating in a version of the "splitting" method.

29.2 Statement of the Problem

Consider a volume $G \in R^3$ filled with a substance that scatters and absorbs radiation with coefficients $\sigma(r)$ and $\sigma_s(r)$, correspondingly, of attenuation and scattering with scattering indicatrix $g(r, \mu)$ such that

$$\int_{-1}^{1} g(r, \mu) d\mu = 1.$$

Here $\mu = (\omega', \omega)$ is the scalar product of the vectors $\omega', \omega \in \Omega = \{\omega = (a, b, c) : a^2 + b^2 + c^2 = 1\}$—the set of directions. Denote with $q(r)$ the value $\sigma_s(r)/\sigma(r)$, which is the probability of survival of a quantum of radiation (a photon) in a collision with an element of substance and let c be the velocity of propagation of radiation in the medium.

At the point r_0 a source is located, emitting at the moment of time $t = 0$ an impulse of radiation of a unit power in a circular directions cone Ω_0 with half opening angle θ_0 with respect to the cone's axis, directed along the unit vector ω_0 (Fig. 29.1).

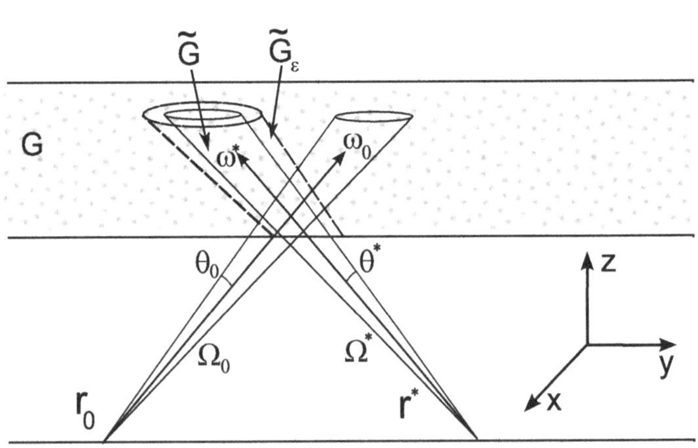

Fig. 29.1 The geometrical concept of the problem

The goal is to determine the time distribution $I_{\Omega*}(r^*, t)$ of radiation coming to the point r^* in directions ω such that $-\omega \in \Omega^*$, where Ω^* is a circular cone with half opening angle θ^* with respect to axis ω^*. Therefore the functional to be computed is $I_{\Omega*}(r^*, t) = \int_{R^3} \int_{\Omega*} I(r, -\omega, t)\delta(r - r^*)d\omega dr$, in which the function $I(r, \omega, t)$ is the radiation intensity at point r at the moment of time t in the direction ω. It's known (see, for example, [3]), that radiation intensity is connected to the collision density $\varphi(x)$ with the following relation:

$$I(r, \omega, t)\sigma(r) = \varphi(r, \omega, t).$$

Let $\sigma(r) \equiv 0$ when $r \in R^3 \backslash G$ (which corresponds to the case of absence of scattering and absorption in the complementary to G subset of the space R^3), the collision density $\varphi(x)$ in the geometrical optics approximation may be described by the integral equation [3]

$$\varphi(x) = \int_X \frac{q(r')\sigma(r)\exp(-\tau(r', r))g(r', (\omega', \omega))}{2\pi |r' - r|^2} \delta\left(\omega - \frac{r - r'}{|r - r'|}\right)$$
$$\times \delta\left(t - \left(t' + \frac{|r - r'|}{c}\right)\right) \varphi(x')dx' + f(x); \tag{29.1}$$

$x = (r, \omega, t) \in X = R^3 \times \Omega \times T,\ T = (0, \infty)$.

Here $f(x) = \sigma(r)\exp(-\tau(r_0, r))\ \Delta_{\Omega_0}(\omega)\Delta_t\left(t - \frac{|r - r_0|}{c}\right)$ is the primary collisions density of unscattered particles directly from the source, $\tau(r', r) = \int_0^{|r - r'|} \sigma (r' + s\frac{r - r'}{|r - r'|})ds$, $\Delta_{\Omega_0}(\omega) = \begin{cases} 1, & \omega \in \Omega_0, \\ 0, & \text{otherwise}, \end{cases}$ $\Delta_t(t) = \begin{cases} 1, & t > 0, \\ 0, & t \leqslant 0. \end{cases}$

29.3 Local Estimates

Let us compute the functional $I_{\Omega*}(r^*, t)$ as a histogram so that the following values are to be estimated:

$$I_{\Omega*}^{(i)}(r^*) = \int_{t_{i-1}}^{t_i} I_{\Omega*}(r^*, t)dt,\ i = 1, \ldots, n_t,$$

where t_i are the histogram nodes; $t_0 = 0$. We use the LE [3] to compute $I_{\Omega*}^{(i)}(r^*)$.

$$I_{\Omega*}^{(i)}(r^*) = \mathbf{E} \sum_{n=1}^{N} Q_n h_i^{(1)}(x_n, r^*), \tag{29.2}$$

$$h_i^{(1)}(x_n, r^*) = \frac{q(r_n)\exp(-\tau(r_n, r^*))g(r_n, (\omega_n, s))}{2\pi |r^* - r_n|^2} \Delta_{\Omega*}(s)\Delta_i(t)$$

where $\mathbf{E}$ is the mathematical expectation sign, $x_n = (r_n, \omega_n, t_n)$ is a Markov chain with the primary collision density $p_1(x)$ and transition density $p(x', x)$, N is the random number of the termination of the Markov chain, $t = t_n + |r_n - r^*|/c$, $s = (r^* - r_n)/|r^* - r_n|$, and

$$\Delta_i(t) = \begin{cases} 1, & \text{if } t \in (t_{i-1}, t_i], \\ 0, & \text{otherwise.} \end{cases} \quad \Delta_{\Omega^*}(\omega) = \begin{cases} 1, & \text{if } -\omega \in \Omega^*, \\ 0, & \text{otherwise.} \end{cases}$$

Weight multipliers Q_n in accordance with the theory of weight Monte Carlo methods [3, 4] are defined by the expressions: $Q_1 = \frac{f(x)}{p_1(x)}$, $Q_n = Q_{n-1}\frac{k(x_{n-1}, x_n)}{p(x_{n-1}, x_n)}$, $n = 2, \ldots, N$, here $k(x', x)$ is the kernel of Eq. (29.1).

It's easily seen that Eq. (29.1) is equivalent to the equation

$$\varphi(x) = \mathbf{K}^2\varphi(x) + \mathbf{K}f(x) + f(x), \tag{29.3}$$

where $\mathbf{K}$ is the integral operator with kernel $k(x', x)$. The local estimate $h_i^{(2)}$ to compute $I_{\Omega^*}^{(i)}(r^*)$ based upon the representation (29.3) is called a *double local estimate* (DLE). This estimate is defined by the formula

$$I_{\Omega^*}(r^*, t) = \int_X k_2(x', x^*)\varphi(x')dx' \tag{29.4}$$

$$\text{where } k_2(x', x^*) = \int_{X_{\tilde{G}}} k(x', \rho)k(\rho, x^*)d\rho.$$

In the latter integral integration is performed over $\tilde{G} = \{\rho(s) \in G : \rho(s) = r^* + \omega s, \omega \in \Omega^*, \ s > 0\}$. In the case when the DLE is used, instead of (29.2) we have

$$I_{\Omega^*}^{(i)}(r^*) = \mathbf{E}\xi = \mathbf{E}\left(Q_1 h_i^{(1)}(x_1, r^*) + \sum_{n=1}^{N} Q_n h_i^{(2)}(x_n, r^*, \rho_n)\right), \tag{29.5}$$

$$h_i^{(2)}(x, r^*, \rho) = \frac{q(r)q(\rho)\sigma(\rho)e^{-\tau(r,\rho)-\tau(\rho,r^*)}g(r, (\omega, \omega_\rho))g\left(\rho, \left(\omega_\rho, \frac{r^*-\rho}{|r^*-\rho|}\right)\right)}{(2\pi)^2|\rho - r|^2 p(\rho)}$$
$$\times \Delta_{\Omega^*}\left(\frac{r^* - \rho}{|r^* - \rho|}\right)\Delta_i\left(t + \frac{|\rho - r| + |r^* - \rho|}{c}\right).$$

Here $\omega_\rho = \frac{\rho - r}{|\rho - r|}$, $p(\rho)$ is an arbitrary distribution density of an intermediate random node ρ, which is chosen so that $-(r^* - \rho)/|r^* - \rho| \in \Omega^*$. The freedom of choice of $p(\rho)$ allows optimizing the estimate (29.5) to decrease the algorithm's computational cost.

The estimate (29.5) has an infinite variance, so in practice a biased intensity estimate with a finite variance is computed instead. In this case for a point of collision (r_n, ω_n, t_n) an additional random node ρ is chosen randomly in

$\tilde{G} \backslash B_\varepsilon$, $B_\varepsilon = \{\tilde{r} \in R^3 : |r_n - \tilde{r}| < \varepsilon\}$, where ε is a certain positive number chosen beforehand. Note that the relative calculation error of the value $I_{\Omega^*}(r^*)$ due to substraction of the ball B_ε may be approximately estimated (see [1]) by the value $\frac{q(r)(1-\mu_0)}{1-q(r)\mu_0}(1 - \exp(-\sigma(r)\varepsilon))$, where $\mu_0 = \int_{-1}^{1} \mu g(r, \mu) d\mu$ is the average scattering angle cosine at point r.

29.4 Modified Estimate

Consider a certain area $\tilde{G}_\epsilon \in G$, such that $\tilde{G} \in \tilde{G}_\epsilon$. For collision points $x_n = (r_n, \omega_n, t_n)$ such that $r_n \in \tilde{G}_\epsilon$ the computation of the integral (29.4) will be conducted on a certain predefined number K of integrating nodes. In this case, the function $h_i^{(2)}$ in (29.5) is substituted with

$$h_i^{(3)}(x, r^*) = \frac{1}{K} \sum_{k=1}^{K} h_i^{(2)}(x, r^*, \rho_k) \tag{29.6}$$

For x_n such that $r_n \in G \setminus \tilde{G}_\epsilon$, the contribution into the DLE is computed on a single random integrating node ρ ($K = 1$). In the current paper for test calculations the linear dimension of the area $\tilde{G}_\epsilon$ was chosen of an order of magnitude with the particle's free pass length while traveling through the area G.

To determine the optimal number of integrating nodes in the area $\tilde{G}_\epsilon \in G$, consider the problem of computing a time integral for radiation intensity incoming to the point r^* of the source in a given directions cone. Let us use the complete variance formula and the method of determining optimal parameters of the splitting method [4].

Let $h_i^{(1)}(x_1, r^*) = 0$. Let us present the modified double local estimate (MDLE) of radiation intensity in the form

$$\xi^{(K)} = \sum_{n=1}^{N} Q_n h^{(2)}(x_n, r^* | r_n \in G \setminus \tilde{G}_\epsilon) + \sum_{n=1}^{N} Q_n h^{(3)}(x_n, r^* | r_n \in \tilde{G}_\epsilon).$$

Let $\zeta = (x_1, \ldots, x_N)$ be the sequence of phase coordinates of the photon collision points with matter particles and let $\eta = (\rho_1^1, \ldots, \rho_1^s, \ldots, \rho_N^1, \ldots, \rho_N^s)$ be the random integrating nodes $\rho_i^j \in \tilde{G}$ for a sequence of collision points ζ ($s = 1$ if $r_n \in G \setminus \tilde{G}_\epsilon$; $s = K$ if $r_n \in \tilde{G}_\epsilon$). Let us use the complete variance formula for the random variable $\xi^{(K)}$ [4]:

$$\begin{aligned}\mathbf{D}\xi^{(K)} &= \mathbf{D}_\zeta \mathbf{E}_\eta(\xi^{(K)}|\zeta) + \mathbf{E}_\zeta \mathbf{D}_\eta(\xi^{(K)}|\zeta) \\ &= \mathbf{D}_\zeta \mathbf{E}_\eta\Big(\sum_{n=1}^{N} Q_n h^{(2)}(x_n, r^* | x \in \zeta, r_n \in G \setminus \tilde{G}_\epsilon)\end{aligned}$$

$$+\sum_{n=1}^{N} Q_n h^{(3)}(x_n, r^*|x \in \zeta, r_n \in \tilde{G}_\varepsilon))$$

$$+\mathbf{E}_\zeta \mathbf{D}_\eta(\sum_{n=1}^{N} Q_n h^{(2)}(x_n, r^*|x \in \zeta, r_n \in G \setminus \tilde{G}_\epsilon))$$

$$+\mathbf{E}_\zeta \mathbf{D}_\eta(\sum_{n=1}^{N} Q_n h^{(3)}(x_n, r^*|x \in \zeta, r_n \in \tilde{G}_\epsilon)).$$

The latest equality is true, because

$$\mathbf{D}_\eta(\xi^{(K)}|\zeta) = \mathbf{D}_\eta(\xi^{(K)}|\zeta, r_n \in G \setminus \tilde{G}_\epsilon) + \mathbf{D}_\eta(\xi^{(K)}|\zeta, r_n \in \tilde{G}_\epsilon)$$

for independent ρ_i^j. Using the independence and identical distributions of the components of the vector η, as well as the formula (29.6), we get $\mathbf{D}_\eta h^{(3)}(x_n, r^*|x_n \in \zeta, r_n \in \tilde{G}_\epsilon) = \frac{1}{K}\mathbf{D}_\eta h^{(2)}(x_n, r^*|x_n \in \zeta, r_n \in \tilde{G}_\epsilon)$. Denote with $D_1 = \mathbf{D}_\zeta \mathbf{E}_\eta(\xi^{(K)}|\zeta) + \mathbf{E}_\zeta \mathbf{D}_\eta(\sum_{n=1}^{N} Q_n h^{(2)}(x_n, r^*|x \in \zeta, r_n \in G \setminus \tilde{G}_\epsilon))$ and with

$$D_2 = \mathbf{E}_\zeta \mathbf{D}_\eta \sum_{n=1}^{N} Q_n h^{(2)}(x_n, r^*|x \in \zeta, r_n \in \tilde{G}_\epsilon).$$

Then $\mathbf{D}\xi^{(K)} = D_1 + D_2/K$.

Let t_1 be the average time for modeling collision points x_n, $n = 1, \ldots, N$, let t_2 be the average time for modeling one additional integrating node ρ_n for each x_n, $n = 1, \ldots, N$ and computing the value of the functional $h^{(2)}(x_n, r^*)$, l_1 and l_2 be the average ratios of the number of collision points x_n in the areas $G \setminus \tilde{G}_\epsilon$ and $\tilde{G}_\epsilon$ to their total number, $l_1 + l_2 = 1$. Then the average time required for computing one sample value $\xi_i^{(K)}$ of the random variable $\xi^{(K)}$ is equal to

$$t^{(K)} = t_1(l_1 + l_2) + t_2(l_1 + Kl_2) = t_1 + t_2 l_1 + K t_2 l_2 = \tilde{t}_1 + K\tilde{t}_2,$$
$$\tilde{t}_1 = t_1 + t_2 l_1,\ \tilde{t}_2 = t_2 l_2.$$

The optimal value of K minimizes the computational cost

$$S^{(K)} = t^{(K)}\mathbf{D}\xi^{(K)} = (\tilde{t}_1 + K\tilde{t}_2)(D_1 + D_2/K).$$

Calculating the derivative and taking into account that the values D_1, D_2, $\tilde{t}_1$, $\tilde{t}_2$ are positive, we get that the optimal number K is approximately equal to the integer number closest to the expression [4]

$$K_{\text{opt}} \approx \sqrt{\frac{D_2\tilde{t}_1}{D_1\tilde{t}_2}}. \tag{29.7}$$

The values D_1, D_2, $\tilde{t}_1$, $\tilde{t}_2$ may be approximately estimated from results of preliminary calculations, computing the intensity estimate twice: once for $K = 1$ and once for a certain given K. In this case, knowing the average time $t^{(1)}$ required for computing one sample value $\xi_i^{(1)}$ and average time $t^{(K)}$ for a value $\xi_i^{(K)}$ of random values $\xi^{(1)}$ and $\xi^{(K)}$ we have $t_2 = \frac{t^{(K)}-t^{(1)}}{(K-1)l_2}$ and $t_1 = t^{(1)} - t_2$, therefore

$$\tilde{t}_1 = \frac{t^{(1)}K - t^{(K)}}{K-1}, \quad \tilde{t}_2 = \frac{t^{(K)} - t^{(1)}}{K-1}.$$

D_1 may be estimated from the formula $D_1 = \mathbf{D}\xi^{(K)} - D_2/K$.

29.5 Results of Numeric Experiments

Further we illustrate the efficiency of the proposed algorithm by the results of calculating an estimate for radiation intensity $I_{\Omega^*}(r^*)$ and the time distribution of radiation intensity $I_{\Omega^*}^{(i)}(r^*, t)$ for a flat layer $G = \{(\tilde{x}, \tilde{y}, \tilde{z}) \in R^3 : h \leqslant \tilde{z} \leqslant H\}$ filled with an absorbing and scattering medium. The following parameters are used in the calculation: $h = 0.5\,\text{km}$, $H = 0.9\,\text{km}$, attenuation coefficient of the medium $\sigma(r) \equiv \Sigma \equiv 50\,\text{km}^{-1}$, survival probability $q = 0.95$ and the Henyey–Greenstein scattering indicatrix $g(\mu) = (1-\alpha^2)/2(1+\alpha^2-2\alpha\mu)^{3/2}, \mu \in (-1, +1)$ with various parameters α. The radiation detector was supposed to be located at the origin $r^* = (0, 0, 0)$, its axis coincides with $\omega^* = (0, 0, 1)$. The radiation source, located at point $r_0 = (-0.7\,\text{km}, 0, 0)$, emits an impulse of unit power in a circular cone with aperture $\theta_0 = 20''$ in the direction $\omega_0 = (\sqrt{2}/2, 0, \sqrt{2}/2)$.

In Table 29.1 a comparison of computational costs $S = \sigma^2 T$ is presented for the methods of DLE and MDLE for computing radiation intensity $I_{\Omega^*}(r^*)$ with the abovementioned parameters of the scattering medium with $\alpha = 0.8$ and the detector aperture $\theta^* = 1°$. Here T is the average time required for modeling one random trajectory, $\sigma = \sqrt{\mathbf{D}I_{\Omega^*}(r^*)/n}$ is the standard deviation of the estimates. When the radiation intensity was calculated by the MDLE for points $\{x_n \in X | r_n \in \tilde{G}_\varepsilon = \{r \in G : \left(\frac{r-r^*}{|r-r^*|}, \omega^*\right) \geqslant \cos 3°\}\}$ the number K of additional integration nodes ρ_k was varied from 25 to 250, $K = 1$ for all other points x_n. Radiation intensity was estimated on $n = 10^9$ trajectories, the computation time

Table 29.1 Computational cost S, standard deviation σ of the modified double local estimate of radiation intensity $I_{\Omega^*}(r^*)$ for different K

K	1	25	50	100	150	180	200	230	250
$S * 10^7$	33	3.4	4.2	3.9	4.4	2.7	1.9	2.0	1.8
$t * 10^{-4}s$	5.34	5.97	6.9	8.6	10.5	11.4	12.4	13.4	14.16
$\sigma * 10^6$	7.84	2.39	2.47	2.12	2.06	1.54	1.23	1.21	1.09

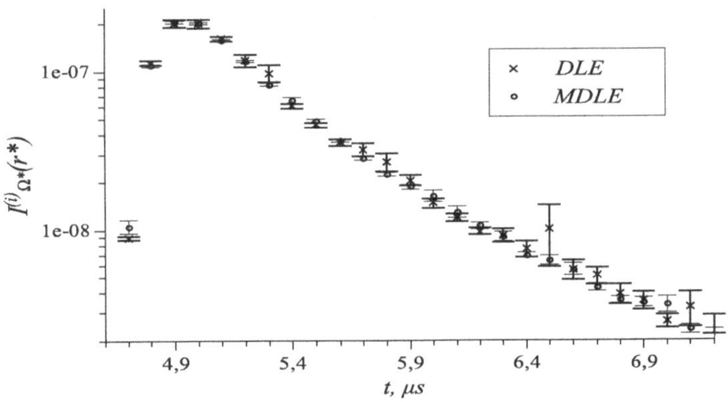

Fig. 29.2 Estimate of the time distribution of radiation intensity $I_{\Omega^*}^{(i)}(r^*, t)$, calculated with the double local and modified double local estimates. $\alpha = 0.8$, $\theta^* = 30''$

denoted by t. From the table we can see that the computational cost S is almost identical for $K = 200 - 250$ and is considerably less than the computational cost of the DLE ($K = 1$). These results are in accordance with preliminary calculations by the formula (29.7) in which the values D_1 and D_2 were estimated on $n = 10^8$ trajectories from which an optimal value of $K_{\text{opt}} \approx 205$ was obtained.

In the following calculations to construct an MDLE the area G was divided into several concentric cones: let $\nu(r) = \left(\frac{r-r^*}{|r-r^*|}, \omega^*\right)$ and $\tilde{G}_1 = (r \in G : \gamma_1 \leqslant \nu(r) \leqslant \gamma_2)$, $\tilde{G}_2 = (r \in G : \gamma_2 \leqslant \nu(r) \leqslant \gamma_3)$, $\tilde{G}_3 = (r \in G : \nu(r) \geqslant \gamma_3)$. In presented calculations, $\gamma_1 = \cos 5°, \gamma_2 = \cos 2°, \gamma_3 = \cos 0.5°$. For points $\{x_n \in X | r_n \notin \tilde{G}_1 \cup \tilde{G}_2 \cup \tilde{G}_3\}$ the functional (29.6) was estimated on one additional random node ρ ($K = 1$). For $\{x_n \in X | r_n \in \tilde{G}_1 \cup \tilde{G}_2 \cup \tilde{G}_3\}$ the choice of number of additional nodes K_1, K_2, K_3 was made in accordance with several radiation intensity $I_{\Omega^*}^{(i)}(r^*, t)$ estimates constructed on $n = 10^7 - 10^8$ trajectories.

In Fig. 29.2 the estimates of the time distribution of radiation intensity $I_{\Omega^*}^{(i)}(r^*, t)$ for indicatrix parameter $\alpha = 0.8$ and the detector with aperture $\theta^* = 30''$ in the time interval [4.6 μs, 7.2 μs] with histogram step 0.1 μs are presented. The number of additional integration nodes ρ_k was equal to $K_1 = 20$, $K_2 = 30$, $K_3 = 50$. The bold line depicts the standard deviations $\sigma_i, i = 1, \ldots, n_t$ of the DLE, the thin line—standard deviations $\sigma_i, i = 1, \ldots, n_t$ of the MDLE.

In Table 29.2 we present the computational costs S of LE for various apertures of the detector θ^* and indicatrix parameter values $\alpha = 0.7, 0.8$. For small detector apertures (of an order of magnitude with $\theta^* = 1'$ or less) the LE constructed on $n = 10^9$ trajectories does not produce a satisfactory calculation precision for the time distribution of radiation intensity $I_{\Omega^*}^{(i)}(r^*, t)$. The standard deviations and computational costs S are also large in this case. This data confirms that the computational cost S of MDLE is less than the computational cost of DLE for all presented parameters θ^*, α.

Table 29.2 Computational cost S of considered methods for $\theta^* = \{1°, 30', 1', 30''\}$, for $\alpha = 0.7$—$K_1 = 40$, $K_2 = 60$, $K_3 = 100$, for $\alpha = 0.8$—$K_1 = 20$, $K_2 = 30$, $K_3 = 50$

Method	1°	30′	1′	30″
$\alpha = 0.7$				
LE	$2.3 * 10^{-8}$	$7.0 * 10^{-9}$	–	–
DLE	$54 * 10^{-8}$	$49 * 10^{-9}$	$7.7 * 10^{-13}$	$3.4 * 10^{-14}$
MDLE	$2.2 * 10^{-8}$	$3.0 * 10^{-9}$	$0.11 * 10^{-13}$	$0.13 * 10^{-14}$
$\alpha = 0.8$				
LE	$6.9 * 10^{-8}$	$1.6 * 10^{-8}$	–	–
DLE	$1.4 * 10^{-6}$	$4.7 * 10^{-7}$	$2.8 * 10^{-13}$	$1.7 * 10^{-14}$
MDLE	$2.2 * 10^{-7}$	$2.7 * 10^{-8}$	$8.2 * 10^{-14}$	$2.5 * 10^{-15}$

Conclusion

Comparing the DLE with the modification proposed in this paper we can conclude that using the "splitting" method allows to considerably reduce the computational cost of the algorithm. It shows most prominently in problems with "smooth" scattering indicatrices ($\alpha = 0.8 - 0.6$) and big enough detector apertures θ^*. But even for elongated indicatrices and small detector apertures θ^* the proposed modifications keep the advantage in computational cost over the DLE (29.5).

Acknowledgements The work is partly supported by Integrational Project SB RAS No. 52, RAS Presidium Project No. 15.9-1 and the Program "Leading Scientific Schools" (grant SS-5111.2014.1).

References

1. Jetybayev, Ye.O.: On solving the nonstationary radiation transfer equation by the Monte Carlo method. Preprint no. 346 CC SB SAS, Novosibirsk (1982)
2. Krekov, G.A.: Monte Carlo method in problems of atmosphere optics. Optika atmosfery i okeana. **20**(9), 826–834 (2007)
3. Marchuk, G.I., Mikhailov, G.A., Nazaraliev, M.A., Darbinjan, R.A., Kargin, B.A., Elepov, B.S.: The Monte Carlo Methods in Atmospheric Optics. Springer Series in Optical Sciences, vol. 12. Springer, Berlin (1980)
4. Mikhailov, G.A., Voytishek, A.V.: Numerical statistical modeling. In: Monte Carlo Methods. Akademiya, Moscow (2006)

Chapter 30
Computer Experiment Designs via Particle Swarm Optimization

Erin Leatherman, Angela Dean, and Thomas Santner

30.1 Computer Experiments and Emulators

Computer experiments are used widely in diverse research areas such as engineering, biomechanics, and the physical and life sciences. Computer experiments use *computer simulators* as experimental tools to provide outputs $y(\boldsymbol{x})$ at specified design input points $\boldsymbol{x}$, where a computer simulator is the computer implementation of a mathematical model that describes the relationships between the input and output variables in the physical system. Computer experiments can be especially attractive when physical experiments are infeasible, unethical, or "costly to run."

For fast running codes, the output response surface can be explored by evaluating (running) the simulator at a set of inputs $\boldsymbol{x} = (x_1, \ldots, x_k)$ that are dense in the space of possible inputs, $\mathscr{X}$. For slow-running codes, an approximator (also called an "emulator" or "metamodel") is often sought for the simulator output $y(\boldsymbol{x})$; such metamodels allow, for example, the detailed (approximate) exploration of the output surface (see, for example, Santner et al. [18]).

One rapidly computable class of emulators for deterministic computer simulator output $y(\boldsymbol{x})$ assumes that $y(\boldsymbol{x})$ can be modeled as a realization of a Gaussian Stochastic Process $Y(\boldsymbol{x})$ (GaSP). In this paper, the input space $\mathscr{X}$ for the k inputs is

E.R. Leatherman
West Virginia University, PO Box 6330, Morgantown, WV 26506, USA
e-mail: erleatherman@mail.wvu.edu

A.M. Dean (✉)
University of Southampton, Southampton SO17 1BJ, UK
e-mail: dean.9@osu.edu

T.J. Santner
The Ohio State University, Columbus, OH 43210, USA
e-mail: santner.1@osu.edu

V.B. Melas et al. (eds.), *Topics in Statistical Simulation*, Springer Proceedings in Mathematics & Statistics 114, DOI 10.1007/978-1-4939-2104-1_30

rectangular and is rectangular and scaled to $[0, 1]^k$. The GaSP models are assumed to take the form

$$Y(\boldsymbol{x}) = \sum_{\ell=0}^{p} f_\ell(\boldsymbol{x})\beta_\ell + Z(\boldsymbol{x}) \quad = \quad \boldsymbol{f}'(\boldsymbol{x})\boldsymbol{\beta} \; + \; Z(\boldsymbol{x}), \tag{30.1}$$

where $\boldsymbol{f}'(\boldsymbol{x}) = \big(f_1(\boldsymbol{x}), \ldots, f_p(\boldsymbol{x})\big)$ is a vector of known regression functions, $\boldsymbol{\beta} = \big(\beta_1, \ldots, \beta_p\big)'$ is a $p \times 1$ vector of unknown regression coefficients and $Z(\boldsymbol{x})$ is a zero-mean, stationary Gaussian stochastic process on $\mathscr{X}$ with covariance

$$\text{Cov}\,(Z(\boldsymbol{x}_u), Z(\boldsymbol{x}_v)) = \sigma_Z^2 \times R(\boldsymbol{x}_u - \boldsymbol{x}_v \mid \boldsymbol{\rho}) = \prod_{j=1}^{k} \rho_j^{4\left(x_{uj} - x_{vj}\right)^2},$$

where x_{uj}, x_{vj} are the jth elements of input points $\boldsymbol{x}_u, \boldsymbol{x}_v \in \mathscr{X}$, $j = 1, \ldots, k$, $\boldsymbol{\rho} = (\rho_1, \rho_2, \ldots, \rho_k)'$, and $\rho_j \in [0, 1]$ is the correlation between two outputs whose $\boldsymbol{x}_u$ and $\boldsymbol{x}_v$ differ *only* in the jth dimension by $|x_{uj} - x_{vj}| = 1/2$, which is *half their domain*.

The design for the computer experiment is denoted by an $n \times k$ matrix $\boldsymbol{X} \in \mathscr{D}(n, k)$ whose ith row is defined by the ith design point $\boldsymbol{x}_i' = (x_{i1}, \ldots, x_{ik})$; $\mathscr{D}(n, k)$ denotes the class of all designs with n runs, k input variables, and input space $\mathscr{X}$.

Let $\boldsymbol{y}^n = (y(\boldsymbol{x}_1), \ldots, y(\boldsymbol{x}_n))'$ denote (training) data to be used for estimating the simulator output $y(\boldsymbol{x}_0)$ and let $\boldsymbol{Y}^n$ denote the corresponding random vector. When $\boldsymbol{\beta}$ is *unknown*, but the correlation parameters $\boldsymbol{\rho}$ are *known*, the best linear unbiased predictor (BLUP) of $y(\boldsymbol{x}_0)$ can be shown to be $\hat{y}(\boldsymbol{x}_0) = \boldsymbol{f}'_{x0}\widehat{\boldsymbol{\beta}} + \boldsymbol{r}'_{x0}\boldsymbol{R}^{-1}(\boldsymbol{y}^n - \boldsymbol{F}\widehat{\boldsymbol{\beta}})$, where $\widehat{\boldsymbol{\beta}} = (\boldsymbol{F}^T\boldsymbol{R}^{-1}\boldsymbol{F})^{-1}\boldsymbol{F}^T\boldsymbol{R}^{-1}\boldsymbol{y}^n$ (see, for example, [17]). Here $\widehat{\boldsymbol{\beta}}$ is the generalized least squares estimator of $\boldsymbol{\beta}$, $\boldsymbol{F}$ is an $n \times p$ matrix with uth row $\boldsymbol{f}'(\boldsymbol{x}_u)$, $\boldsymbol{R}$ is an $n \times n$ matrix whose (u, v)th element is $R(\boldsymbol{x}_u - \boldsymbol{x}_v \mid \boldsymbol{\rho})$, and $\boldsymbol{r}'_{x0} = (R(\boldsymbol{x}_0 - \boldsymbol{x}_1 \mid \boldsymbol{\rho}), \ldots, R(\boldsymbol{x}_0 - \boldsymbol{x}_n \mid \boldsymbol{\rho}))$ is a $1 \times n$ vector.

30.2 Design Criteria

Space-filling designs are popular choices for computer experiments when fitting GaSP models, (see, for example, [9] and [3]). Space-filling criteria ensure that the entire input space is sampled by preventing design points from being "close" together.

Two important space-filling criteria are the maximin (Mm) and the Average Reciprocal Distance (ARD) criteria. The *Mm* criterion specifies that a design $\boldsymbol{X}_{Mm} \in \mathscr{D}(n, k)$ that maximizes the minimum interpoint distance

$$\min_{\boldsymbol{x}_u, \boldsymbol{x}_v \in \boldsymbol{X}} q(\boldsymbol{x}_u, \boldsymbol{x}_v) \tag{30.2}$$

is optimal where $q(\boldsymbol{x}_u, \boldsymbol{x}_v)$ is the distance between $\boldsymbol{x}_u$ and $\boldsymbol{x}_v$. Here and below, we use Euclidean distance, but other metrics could equally well be used.

The *ARD* criterion is specified by a given set $\boldsymbol{J} \subset \{1, \ldots, k\}$ of sub-dimensions over which the distances are to be computed (e.g., [2, 14]). A design $\boldsymbol{X}_{\text{ARD}}$ is ARD-optimal with respect to $\boldsymbol{J}$ if it minimizes

$$av_z(\boldsymbol{X}) = \frac{1}{\binom{n}{2}\sum_{j\in J}\binom{k}{j}} \sum_{j\in J}\sum_{\ell=1}^{\binom{k}{j}} \sum_{\boldsymbol{x}_u^\star, \boldsymbol{x}_v^\star \in \boldsymbol{X}_{\ell j}} \left[\frac{j^{1/z}}{q(\boldsymbol{x}_u^\star, \boldsymbol{x}_v^\star)}\right] \tag{30.3}$$

where $\boldsymbol{X}_{\ell j}$ is the ℓth subspace of $\boldsymbol{X}$ having dimension j, $\boldsymbol{x}_u^\star$ and $\boldsymbol{x}_v^\star$ are the projections of $\boldsymbol{x}_u$, $\boldsymbol{x}_v$ onto $\boldsymbol{X}_{\ell j}$, and $q(\boldsymbol{x}_u^\star, \boldsymbol{x}_v^\star)$ is the distance between $\boldsymbol{x}_u^\star$ and $\boldsymbol{x}_v^\star$.

For prediction, [12] and [15] showed that process-based design criteria produce better designs than do space-filling criteria. Process-based criteria involve the chosen emulator rather than geometric properties. Such criteria include the *minimum integrated mean squared prediction error (IMSPE)* [16], maximum entropy [19], and maximum expected improvement [4]. For example, for a given $\boldsymbol{\rho}$, σ_Z^2 and predictor $\hat{y}(\cdot)$, the IMSPE-optimal design $\boldsymbol{X}_I \in \mathscr{D}(n, k)$ minimizes

$$\text{IMSPE}^*\left(\boldsymbol{X} \mid \boldsymbol{\rho}\right) = \frac{1}{\sigma_Z^2}\int_{\mathscr{X}=[0,1]^d} E\left[\left(\widehat{y}(\boldsymbol{w}) - Y(\boldsymbol{w})\right)^2 \mid \boldsymbol{\rho}, \sigma_Z^2\right] \mathrm{d}\boldsymbol{w} \tag{30.4}$$

where the expectation is over the joint distribution of $(Y(\boldsymbol{w}), \boldsymbol{Y}^n)$. If the values of the correlation parameters $\boldsymbol{\rho}$ cannot be specified in advance of the experiment but a distribution $\pi(\boldsymbol{\rho})$ of possible values is approximately known, an alternative criterion is to minimize the IMSPE weighted by $\pi(\boldsymbol{\rho})$, as in [12]. The examples in [12] use $\pi\left(\boldsymbol{\rho}\right) = \prod_{j=1}^{k} \pi(\rho_j)$ and independent Beta distributions for $\pi(\rho_1), \ldots, \pi(\rho_k)$. For given $\pi(\boldsymbol{\rho})$, a design $\boldsymbol{X}_A$ that minimizes *weighted* (averaged) *integrated mean squared prediction error*:

$$\text{W-IMSPE}^*\left(\boldsymbol{X}\right) = \int_{[0,1]^k} \text{IMSPE}^*\left(\boldsymbol{X} \mid \boldsymbol{\rho}\right) \pi\left(\boldsymbol{\rho}\right) \mathrm{d}\boldsymbol{\rho} \tag{30.5}$$

is said to be W-IMSPE*-optimal.

For each of the four criteria (30.2)–(30.5), Fig. 30.1 shows approximate optimal designs with $k = 2$ inputs and $n = 20$ runs constructed using particle swarm optimization (PSO) followed by a quasi-Newton optimizer. The PSO used is described in Sect. 30.3; it took $N_{\text{des}} = 4nk = 160$ "particles" and $N_{\text{its}} = 8nk = 320$ "iterations." Maximin designs tend to have design points on the boundary of the input region; as seen in the top left of Fig. 30.1, this is true in this example where 12 of the 20 points are on, or close to, the boundary. The minimum distance between the points in this design is 0.2729, which is close to the maximum achievable minimum interpoint distance of 0.2866 (http://www.packomania.com/).

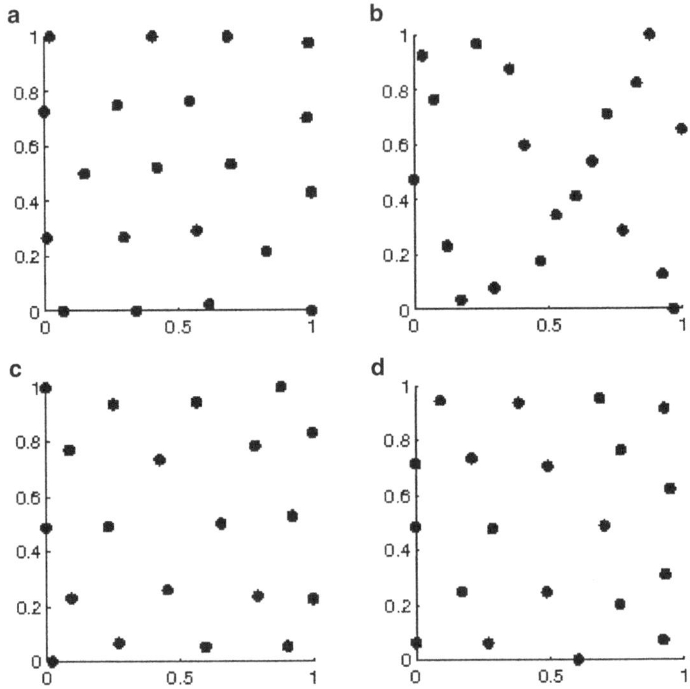

Fig. 30.1 Approximate optimal designs for $k = 2$ inputs, $n = 20$ runs, using criteria (30.2)–(30.5) (criterion value in *parentheses*): Panel (**a**) Mm design (0.2729); panel (**b**) min ARD design (2.2096); panel (**c**) min IMSPE* design (2.2827×10^{-6}); panel (**d**) min W-IMSPE* design (4.1192×10^{-4})

The minimum ARD design, shown in the top right of Fig. 30.1, used $\boldsymbol{J} = \{1, 2\}$ so that the ARD was calculated as an average over the two-dimensional input space and its two one-dimensional projections. The resulting design has more uniformly spread points in the one-dimensional subspaces than the maximin design, but at the cost of less uniformity in the two-dimensional space. A more uniform distribution of two-dimensional points would arise if $\boldsymbol{J} = \{2\}$ were to be used rather than $\boldsymbol{J} = \{1, 2\}$.

For the minimum IMSPE* design, shown in the bottom left of Fig. 30.1, the correlation parameters, ρ_1 and ρ_2 were set to 0.75 (see [12, 16]). For the minimum W-IMSPE* design, $\pi(\boldsymbol{\rho})$ took each of ρ_1 and ρ_2 to be Beta(37.96,37.96) (found by Leatherman et al. [12] to perform well for prediction). Although, visually, both of these designs seem to have more uniform two-dimensional spread than the maximin design, their minimum interpoint distances (MIPDs) are, respectively, 0.1954 and 0.2043, about 75 % of the MIPD 0.2729 for the Mm design. For more information on the prediction performances of space-filling, IMSPE*-optimal, and W-IMSPE*-optimal designs for different parameter values, see [12].

30.3 Particle Swarm Optimization

Many optimization methods have been suggested in the literature; see, for example, [7] and [20] for surveys. Some methods are most effective in local searches of the input space; for example, gradient-based methods such as the Newton and quasi-Newton algorithms (see, for example, [7]). Other optimization methods emphasize a global search over the entire input space; for example, genetic algorithms [8], simulated annealing [11], and particle swarm optimization [10]. Some methods, such as simulated annealing [11] and mesh adaptive direct search [1], have iteration-dependent parameters that enable them to search both globally and locally.

PSO algorithms introduced in [10], have had many applications including the computation of optimal designs for physical experiments using classical criteria [5, 6] and by Leatherman et al. [12] to find optimal designs for computer experiments. Leatherman et al. [12] used the output of PSO to identify starting points for a gradient-based, constrained non-linear optimizer (`fmincon.m` from the MATLAB Optimization toolbox).

In more detail, to find an $n \times k$ optimal design, PSO starts with a number (N_{des}) of $n \times k$ initial designs $\boldsymbol{X}_1, \ldots \boldsymbol{X}_{N_{\text{des}}}$. Each $\boldsymbol{X}_i$ is reshaped (column-wise) into an $nk \times 1$ vector $z_i^1 = vec\,(\boldsymbol{X}_i)$, called the ith *particle*, $i = 1, 2, \ldots, N_{\text{des}}$. To ensure wide exploration of the nk-dimensional input space, the initial set of N_{des} particles can be selected as an $N_{\text{des}} \times nk$ approximate Mm Latin Hypercube Design.

At iteration t, $t = 1, 2, \ldots, N_{\text{its}}$, every particle z_i^t is "updated," using (30.6) to z_i^{t+1}, and then evaluated under the criterion of interest. The update requires the following notation. At iteration t, let g^t denote that particle $z_i^t \in \left\{z_i^{t^\star} \mid i = 1, \ldots, N_{\text{des}}; t^\star \leqslant t\right\}$ that produces the *global best value of the criterion of interest*. Similarly, for each particle i, let p_i^t denote that $z_i^t \in \left\{z_i^{t^\star} \mid t^\star \leqslant t\right\}$ having *particle best value of the criterion*. Then

$$z_i^{t+1} = z_i^t + \boldsymbol{v}_i^{t+1}, \tag{30.6}$$

where $\boldsymbol{v}_i^{t+1} = \theta \boldsymbol{v}_i^t + \alpha \boldsymbol{\epsilon_1}_i^t \circ \left(g^t - z_i^t\right) + \beta \boldsymbol{\epsilon_2}_i^t \circ \left(p_i^t - z_i^t\right)$, $\circ$ is elementwise product of vectors, $\boldsymbol{\epsilon_1}_i^t$ and $\boldsymbol{\epsilon_2}_i^t$ are independent random vectors whose elements are independent Uniform[0,1], α and β are weights put on the step toward the global- and personal- best positions, respectively, $\theta \in [0, 1]$ is the "inertia" parameter, and $v_i^t \in [-0.25, 0.25]$.

The examples in Sect. 30.4 took $\alpha = \beta = 2$, $\theta = 0.5$, and initial velocity $\boldsymbol{v}_i^1 = \boldsymbol{0}_{nk}$, as recommended by [10] and [20]. There we describe the results of PSO in searching for a Mm design with $(n, k) = (60, 6)$ for different numbers of "particles" and different numbers of iterations, with and without final local optimization. The use of PSO for obtaining IMSPE*-optimal and W-IMSPE*-optimal designs is described in [12].

We now illustrate the working of PSO in a "toy" example with $(n, k) = (1, 2)$ so that each z_i^t is a two-dimensional vector (since $nk = 2$). Figure 30.2 shows

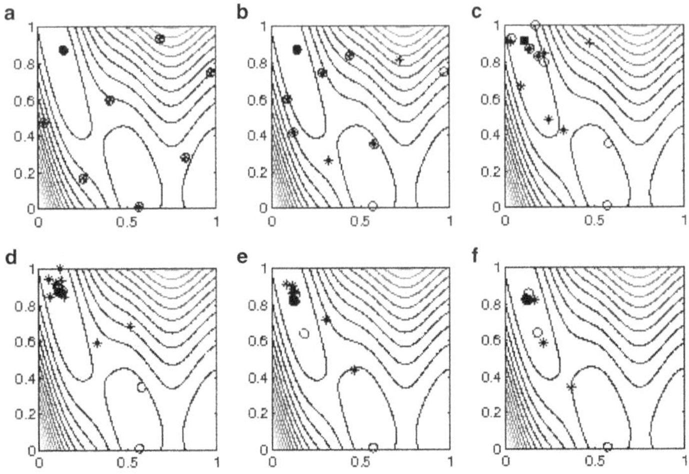

Fig. 30.2 Panel (**a**) Iteration 1, min fnc value = 3.6087; panel (**b**) Iteration 2, min fnc value = 3.6087; panel (**c**) Iteration 3, min fnc value = 3.6087; panel (**d**) Iteration 5, min fnc value = 2.0557; panel (**e**) Iteration 10, min fnc value = 1.1776; panel (**f**) Iteration 24, min fnc value = 1.0118

$N_{des} = 8$ z_i^t positions after $N_{its} = 1, 2, 3, 5, 10, 24$ iterations, together with the (unknown) contours of the design criterion, which is to be *minimized*. The optimal value is 1.0116, located at [0.1215, 0.8240].

Panel (a) of Fig. 30.2 (labeled "Iteration 1") shows the initial particle starting locations, chosen as a maximin LHD. The particle located in the top left corner of the scatterplot corresponds to the design that has the minimum criterion value (= 3.6087) when $t = 1$, so this location is g^1. At Iteration $t = 2$, the particles have taken one step towards g^1 plus a random perturbation, using (30.6). The stars denote the current particle positions z_i^2, and the open circles denote the starting positions which form the current particle-best p_i^2. An evaluation of the criterion values of the designs corresponding to the new particle positions, z_i^2, $i = 1, \ldots, 8$, finds that the global best design remains unchanged, i.e., $g^2 = g^1$. At $t = 3$, each particle i (= $1, \ldots, 8$) moves from z_i^2 towards a weighted combination of the global best particle position, g^2, and its personal best position p_i^2 resulting in z_i^3. Again the global best position is unchanged so that $g^3 = g^2 = g^1$. Some of the particle-best positions (open circles) have changed, such as that originally on the right-hand border of the picture, while others remain the same, such as that one on the bottom border. By iteration $t = 5$ (panel (d)), most of the particles are closing in on the optimum, and one particle has found a better location than g^3 with a smaller criterion value of 2.0557. This implies that the previous best particle, which had not moved in previous iterations, will now start to move towards the new best position.

By iteration $t = 10$, all but two of the z_i^{10} are in the top left corner of the figure, and one of these six has found a better location with criterion value 1.1776. The two remaining z_i^{10} are still drawn towards their previous particle-best positions further "south." One of these z_i^t particles has not found a position better than the location where it started. Because PSO requires only that one particle find the optimum, increasing the number of particles simply increases the chance that the optimum is located quickly. Here, with only eight particles in two-dimensional space, by iteration 24, the global best z_i^t is $g^{24} = [0.1211, 0.8249]$ corresponding to a criterion value of 1.0118, very close to the true optimum of 1.0116. The PSO search could be followed by a gradient-based, constrained non-linear optimizer to hone in on the exact optimum.

30.4 Quality of Designs Produced

Table 30.1 investigates the effect of varying N_{des} and N_{its} in a PSO search for a Mm design having $k = 6$ inputs and $n = 60$ runs. The running times on a Linux compute machine, having a Dual Quad Core Xeon 2.66 processor with 32GB RAM are shown, together with the achieved MIPD (to be *maximized*). The effect of following PSO by the local optimizer, `fmincon.m` starting at at $g^{N_{\text{its}}}$ is also shown.

For a given number of particles, N_{des}, the left portion of Table 30.1 shows a steady increase in the maximized MIPD of the computed design as the number of iterations, N_{its}, increases. The right portion of the table shows that an increase in MIPD could usually be achieved by following PSO with `fmincon` starting at particle $g^{N_{\text{its}}}$. The extra run time needed for additional iterations and/or use of a local optimizer is worthwhile.

Interestingly, for all four N_{des} values, running `fmincon` with starting particle g^1 produced a better design than was obtained by running $20nk = 7{,}200$ iterations of PSO alone. This suggests that a considerably larger value of N_{des} would be needed to find the optimum using only PSO. Results of a modified PSO are given by [6] for searching for maximin LHDs using approximately $N_{\text{des}} = 8000nk$ and $N_{\text{its}} = 100nk$.

Finally, Table 30.1 shows the empirical mean squared prediction error (empMSPE) obtained when using the design to fit the empirical best linear unbiased predictor obtained from (30.1) to outputs from one particular $k = 6$ output function. The values are generally, but not always, lower for designs with larger MIPD. However, maximin is not the best criterion for prediction [12, 15]. A study is currently being carried out on PSO in constructing W-IMSPE*-optimal designs for calibration [13].

Acknowledgements This research was sponsored by the National Science Foundation under Agreements DMS-0806134 and DMS-1310294 (The Ohio State University).

Table 30.1 MIPDs and computation times to find a 60 × 6 design using PSO with $N_{des} = p \times n \times k$ particles ($p \in \{.1, 1, 4, 10\}$) and $N_{its} = q \times n \times k$ PSO iterations ($q \in \{.2, 2, 8, 20\}$) optionally followed by quasi-Newton algorithm (`fmincon`)

	PSO only				PSO + fmincon				
$N_{des} = 36$; N_{its}	$0.2nk$	$2nk$	$8nk$	$20nk$	1	$0.2nk$	$2nk$	$8nk$	$20nk$
Criterion value	0.5181	0.6425	0.6881	0.7222	0.7571	0.6750	0.8412	0.8065	0.8086
Time (s)	46.4	78.5	172.5	355.7	267.9	113.0	476.4	344.7	442.8
empMSPE	5.2116	4.9567	4.7519	4.7866	5.0415	5.3442	5.0821	4.8269	4.8469
$N_{des} = 360$; N_{its}	$0.2nk$	$2nk$	$8nk$	$20nk$	1	$0.2nk$	$2nk$	$8nk$	$20nk$
Criterion value	0.5676	0.6947	0.7210	0.7230	0.7369	0.7652	0.7730	0.7235	0.7230
Time (s)	257.4	312.8	485.9	830.3	431.7	375.7	434.2	489.5	830.6
empMSPE	4.9607	5.2233	5.2379	5.3221	5.5145	5.1181	5.5151	5.2354	5.3222
$N_{des} = 1{,}440$; N_{its}	$0.2nk$	$2nk$	$8nk$	$20nk$	1	$0.2nk$	$2nk$	$8nk$	$20nk$
Criterion value	0.6004	0.6944	0.7125	0.7168	0.8216	0.7784	0.6948	0.7279	0.7168
Time (s)	2,926.5	3,043.5	3,424.3	4,182.8	3,330.2	3,138.7	3,046.0	3,458.0	4,184.4
empMSPE	5.1696	5.1995	5.0018	4.9909	4.8221	4.9231	5.0689	5.2398	4.7784
$N_{des} = 3{,}600$; N_{its}	$0.2nk$	$2nk$	$8nk$	$20nk$	1	$0.2nk$	$2nk$	$8nk$	$20nk$
Criterion value	0.6361	0.7281	0.7444	0.7637	0.7784	0.8047	0.7281	0.7444	0.7637
Time (s)	34,415.4	34,641.0	35,387.7	36,895.6	34,678.9	34,600.2	34,641.3	35,388.0	36,896.0
empMSPE	5.1640	4.6342	4.8644	5.4624	4.9741	5.1198	4.6342	4.8644	5.4624

The column labeled "1" is the MIPD for the design obtained by applying `fmincon` to the design g^1. The empMSPE for predicting one output based on this design are also listed

References

1. Audet, C., Dennis, J.E., Jr.: Mesh adaptive direct search algorithms for constrained optimization. SIAM J. Optim. **17**, 188–217 (2006)
2. Audze, P., Eglais, V.: New approach for planning out of experiments. Probl. Dyn. Strengths **35**, 104–107 (1977)
3. Bates, R.A., Buck, R.J., Riccomagno, E., Wynn, H.P.: Experimental design and observation for large systems. J. R. Stat. Soc. B **58**, 77–94 (1996)
4. Bernardo, M.C., Buck, R.J., Liu, L., Nazaret, W.A., Sacks, J., Welch, W.J.: Integrated circuit design optimization using a sequential strategy. IEEE Trans. Comput. Aided Des. **11**, 361–372 (1992)
5. Chen, R.-B., Chang, S.-P., Wang, W., Tung, H.-C., Wong, W.K.: Optimal minimax designs via particle swarm optimization methods (2013). http://www.newton.ac.uk/preprints/NI13039.pdf
6. Chen, R.-B., Hsieh, D.-N., Hung, Y., Wang, W.: Optimizing Latin hypercube designs by particle swarm. Stat. Comput. **23**, 663–676 (2013)
7. Givens, G., Hoeting, J.: Computational Statistics. Wiley, New York (2012)
8. Goldberg, D.E.: Genetic Algorithms in Search, Optimization, and Machine Learning. Addison-Wesley, Reading (1989)
9. Johnson, M.E., Moore, L.M., Ylvisaker, D.: Minimax and maximin distance designs. J. Stat. Plan. Inference **26**, 131–148 (1990)
10. Kennedy, J., Eberhart, R.: Particle swarm optimization. In: IEEE International Conference on Neural Networks, 1995. Proceedings, vol. 4, 1942–1948 (1995)
11. Kirkpatrick, S., Gelatt, C.D., Vecchi, M.P.: Optimization by simulated annealing. Science **220**, 671–680 (1983)
12. Leatherman, E.R., Santner, T.J., Dean, A.M.: Designs for computer experiments that minimize the weighted integrated mean square prediction error (2014, submitted)
13. Leatherman, E.R., Dean, A.M., Santner, T.J.: Designing combined physical and computer experiments to provide global prediction of the mean of the physical system (2014, in preparation)
14. Liefvendahl, M., Stocki, R.: A study on algorithms for optimization of Latin hypercubes. J. Stat. Plan. Inference **136**, 3231–3247 (2006)
15. Pronzato, L., Müller, W.G.: Design of computer experiments: space-filling and beyond. Stat. Comput. **22**, 681–701 (2012)
16. Sacks, J., Schiller, S.B., Welch, W.J.: Design for computer experiments. Technometrics **31**, 41–47 (1989)
17. Sacks, J., Welch, W.J., Mitchell, T.J., Wynn, H.P.: Design and analysis of computer experiments. Stat. Sci. **4**, 409–423 (1989)
18. Santner, T.J., Williams, B.J., Notz, W.I.: The Design and Analysis of Computer Experiments. Springer, New York (2003)
19. Shewry, M.C., Wynn, H.P.: Maximum entropy sampling. J. Appl. Stat. **14**, 165–170 (1987)
20. Yang, X.-S.: Engineering Optimization: An Introduction with Metaheuristic Applications, 1st edn. Wiley, New York (2010)

Chapter 31
Application of Nonparametric Goodness-of-Fit Tests for Composite Hypotheses in Case of Unknown Distributions of Statistics

Boris Yu. Lemeshko, Alisa A. Gorbunova, Stanislav B. Lemeshko, and Andrey P. Rogozhnikov

31.1 Introduction

Classical nonparametric tests were constructed for testing simple hypotheses: H_0 : $F(x) = F(x, \theta)$, where θ is known scalar or vector parameter of the distribution function $F(x, \theta)$. When testing simple hypotheses nonparametric criteria are distribution free, i.e. the distribution $G(S|H_0)$, where S is the test statistic, does not depend on the $F(x, \theta)$ when the hypothesis H_0 is true.

When testing composite hypotheses $H_0 : F(x) \in \{F(x, \theta), \theta \in \Theta\}$, where the estimate $\hat{\theta}$ of a scalar or vector parameter of the distribution $F(x, \theta)$ is calculated from the same sample, nonparametric tests lose the distribution freedom. Conditional distributions $G(S|H_0)$ of tests statistics for composite hypotheses depend on a number of factors: the type of the distribution $F(x, \theta)$, corresponding to the true hypothesis H_0; the type of the estimated parameter and the number of estimated parameters and, in some cases, the value of the parameter; the method of the parameter estimation.

31.2 Nonparametric Goodness-of-Fit Criteria for Testing Simple Hypotheses

In **Kolmogorov test** statistic the distance between the empirical and theoretical distribution is determined by

---

B.Yu. Lemeshko (✉) • A.A. Gorbunova • S.B. Lemeshko • A.P. Rogozhnikov
Department of Applied Mathematics, Novosibirsk State Technical University, K. Marx pr., 20, Novosibirsk, Russia
e-mail: Lemeshko@fpm.ami.nstu.ru

V.B. Melas et al. (eds.), *Topics in Statistical Simulation*, Springer Proceedings in Mathematics & Statistics 114, DOI 10.1007/978-1-4939-2104-1_31

$$D_n = \sup_{|x|<\infty} |F_n(x) - F(x, \theta)|,$$

where $F_n(x)$ is the empirical distribution function, n is the sample size. When $n \to \infty$, distribution of statistic $\sqrt{n} D_n$ for true hypothesis under test uniformly converges to the Kolmogorov distribution [15]

$$K(S) = \sum_{k=-\infty}^{\infty} (-1)^k e^{-2k^2 s^2}.$$

While testing hypothesis using the Kolmogorov test it is advisable to use the statistic with Bolshev correction [4] given by [5]:

$$S_K = \frac{6nD_n + 1}{6\sqrt{n}}, \tag{31.1}$$

where $D_n = \max(D_n^+, D_n^-)$,

$$D_n^+ = \max_{1 \leqslant i \leqslant n} \left\{ \frac{i}{n} - F(x_i, \theta) \right\}, D_n^- = \max_{1 \leqslant i \leqslant n} \left\{ F(x_i, \theta) - \frac{i-1}{n} \right\},$$

n is the sample size, $x_1, x_2, \ldots, x_n$ are the sample values in an increasing order. When a simple hypothesis H_0 under test is true, the statistic (31.1) converges to the Kolmogorov distribution significantly faster than statistic $\sqrt{n} D_n$.

The statistic of **Cramer–von Mises–Smirnov test** has the following form [3]:

$$S_\omega = \frac{1}{12n} + \sum_{i=1}^{n} \left\{ F(x_i, \theta) - \frac{2i-1}{2n} \right\}^2, \tag{31.2}$$

and **Anderson–Darling test** statistic [2, 3] is

$$S_\Omega = -n - 2\sum_{i=1}^{n} \left\{ \frac{2i-1}{2n} \ln F(x_i, \theta) + \left(1 - \frac{2i-1}{2n}\right) \ln(1 - F(x_i, \theta)) \right\}. \tag{31.3}$$

When testing simple hypotheses, statistic (31.2) has the following distribution $a1(s)$ and the statistic (31.3) has the distribution $a2(s)$ [5].

The **Kuiper test** [16] is based on the statistic $V_n = D_n^+ + D_n^-$. The limit distribution of statistic $\sqrt{n} V_n$ while testing simple hypothesis is the following distribution function [36]:

$$G(s|H_0) = 1 - \sum_{m=1}^{\infty} 2(4m^2 s^2 - 1)e^{-2m^2 s^2}.$$

The following modification of the statistic converges faster to the limit distribution [38]:

$$V = V_n\left(\sqrt{n} + 0.155 + \frac{0.24}{\sqrt{n}}\right),$$

or the modification that we have chosen:

$$V_n^{\mathrm{mod}} = \sqrt{n}(D_n^+ + D_n^-) + \frac{1}{3\sqrt{n}}. \tag{31.4}$$

Dependence of the distribution of statistic (31.4) on the sample size is practically negligible when $n \geqslant 30$.

As a model of limit distribution we can use the beta distribution of the third kind with the density

$$f(s) = \frac{\theta_2^{\theta_0}}{\theta_3 B(\theta_0,\theta_1)} \frac{\left(\frac{s-\theta_4}{\theta_3}\right)^{\theta_0-1}\left(1-\frac{s-\theta_4}{\theta_3}\right)^{\theta_1-1}}{\left[1+(\theta_2-1)\frac{s-\theta_4}{\theta_3}\right]^{\theta_0+\theta_1}},$$

and the vector of parameters $\boldsymbol{\theta} = (7.8624, 7.6629, 2.6927, 0.495)^T$, obtained by the simulation of the distribution of the statistic (31.4).

Watson test [41, 42] is used in the following form

$$U_n^2 = \sum_{i=1}^{n}\left(F(x_i,\theta) - \frac{i-\frac{1}{2}}{n}\right)^2 - n\left(\frac{1}{n}\sum_{i=1}^{n} F(x_i,\theta) - \frac{1}{2}\right) + \frac{1}{12n}. \tag{31.5}$$

The limit distribution of the statistic (31.5) while testing simple hypotheses is given by [41, 42]:

$$G(s|H_0) = 1 - 2\sum_{m=1}^{\infty}(-1)^{m-1}e^{-2m^2\pi^2 s}.$$

The good model for the limit distribution of the statistic (31.5) is the inverse Gaussian distribution with the density

$$f(s) = \frac{1}{\theta_2}\left(\frac{\theta_0}{2\pi\left(\frac{s-\theta_3}{\theta_2}\right)^2}\right)^{1/2} \exp\left(-\frac{\theta_0\left(\left(\frac{s-\theta_3}{\theta_2}\right)-\theta_1\right)}{2\theta_1^2\left(\frac{s-\theta_3}{\theta_2}\right)}\right)$$

and the vector of parameters $\boldsymbol{\theta} = (0.2044, 0.08344, 1.0, 0.0)^T$, obtained by the simulation of the empirical distribution of the statistic (31.5). This distribution as well as the limit one could be used in testing simple hypotheses with Watson test to calculate the achieved significance level.

Zhang tests were proposed in papers [43–45]. The statistics of these criteria are:

$$Z_K = \max_{1 \leqslant i \leqslant n} \left(\left(i - \frac{1}{2} \right) \log \left\{ \frac{i - \frac{1}{2}}{n F(x_i, \theta)} \right\} + \left(n - i + \frac{1}{2} \right) \log \left[\frac{n - 1 + \frac{1}{2}}{n \{1 - F(x_i, \theta)\}} \right] \right), \tag{31.6}$$

$$Z_A = -\sum_{i=1} n \left[\frac{\log \{F(x_i, \theta)\}}{n - i + \frac{1}{2}} + \frac{\log \{1 - F(x_i, \theta)\}}{i - \frac{1}{2}} \right], \tag{31.7}$$

$$Z_C = \sum_{i=1} n \left[\log \left\{ \frac{[F(x_i, \theta)]^{-1} - 1}{\left(n - \frac{1}{2}\right) / \left(i - \frac{3}{4}\right) - 1} \right\} \right]^2. \tag{31.8}$$

The author gives the percentage points for statistics distributions for the case of testing simple hypotheses. The strong dependence of statistics distributions on the sample size n prevents one from wide use of the criteria with the statistics (31.6)–(31.8). For example, Fig. 31.1 shows a dependence of the distribution of the statistics (31.7) on the sample size while testing simple hypotheses.

Of course, this dependence on the sample size n remains for the case of testing composite hypotheses.

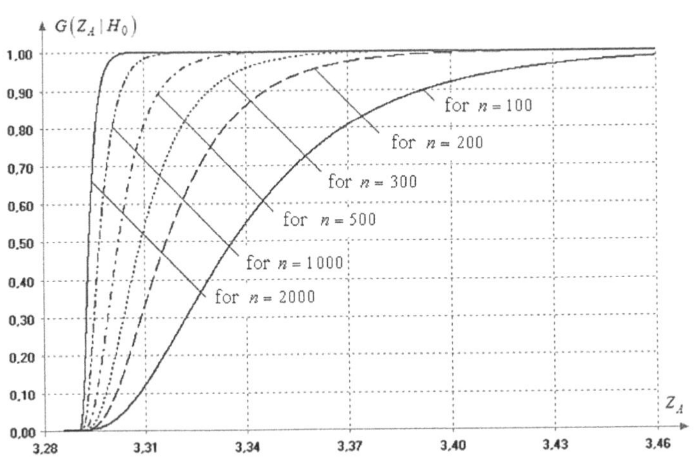

Fig. 31.1 The distribution $G_n(Z_A|H_0)$ of statistic (31.7) depending on the sample size n for testing simple hypothesis

31.3 Comparative Analysis of the Tests Power

In papers [25–27] the power of Kolmogorov (K), Cramer–von Mises–Smirnov (KMS), Anderson–Darling (AD) tests, and also χ^2 criteria was analyzed and compared for testing simple and composite hypotheses for a number of different pairs of competing distributions. In the case of testing simple hypotheses and using asymptotically optimal grouping [17] in χ^2 criterion, this test has the advantage in power compared with nonparametric tests [25, 26]. When testing composite hypotheses, the power of nonparametric tests increases significantly, and they become more powerful.

In order to be able to compare the power of Kuiper (V_n), Watson (U_n^2), and Zhang tests (Z_K, Z_A, Z_C) with the power of other goodness-of-fit tests, the power of these criteria was calculated for the same pairs of competing distributions in the paper [19] alike papers [25–27].

The first pair is the normal and logistics distribution: for the hypothesis H_0—the normal distribution with the density:

$$f(x) = \frac{1}{\theta_0\sqrt{2\pi}} \exp\left\{-\frac{(x-\theta_1)^2}{2\theta_0^2}\right\},$$

and for competing hypothesis H_1—the logistic distribution with the density:

$$f(x) = \frac{\pi}{\theta_1\sqrt{3}} \exp\left\{-\frac{\pi(x-\theta_0)}{\theta_1\sqrt{3}}\right\} \Big/ \left[1+\exp\left\{-\frac{\pi(x-\theta_0)}{\theta_1\sqrt{3}}\right\}\right]^2,$$

and parameters $\theta_0 = 1$, $\theta_1 = 1$. For the simple hypothesis H_0 parameters of the normal distribution have the same values. These two distributions are close and difficult to distinguish with goodness-of-fit tests.

The second pair was the following: H_0—Weibull distribution with the density

$$f(x) = \frac{\theta_0(x-\theta_2)^{\theta_0-1}}{\theta_1^{\theta_0}} \exp\left\{-\left(\frac{x-\theta_2}{\theta_1}\right)^{\theta_0}\right\}$$

and parameters $\theta_0 = 2$, $\theta_1 = 2$, $\theta_2 = 0$; H_1 corresponds to gamma distribution with the density

$$f(x) = \frac{1}{\theta_1\Gamma(\theta_0)} \left(\frac{x-\theta_2}{\theta_1}\right)^{\theta_0-1} e^{-(x-\theta_2)/\theta_1}$$

and parameters $\theta_0 = 2.12154$, $\theta_1 = 0.557706$, $\theta_2 = 0$, when gamma distribution is the closest to the Weibull counterpart.

Comparing the estimates of the power for the Kuiper, Watson and Zhang tests [19] with results for Kolmogorov, Cramer–von Mises–Smirnov, and

Anderson–Darling tests [25–27], the nonparametric tests can be ordered by decrease in power as follows:

- for testing simple hypotheses with a pair "normal—logistic": $Z_C \succ Z_A \succ Z_K \succ U_n^2 \succ V_n \succ AD \succ K \succ KMS$;
- for testing simple hypotheses with a pair "Weibull—gamma": $Z_C \succ Z_A \succ Z_K \succ U_n^2 \succ V_n \succ AD \succ KMS \succ K$;
- for testing composite hypotheses with a pair "normal—logistic": $Z_A \approx Z_C \succ Z_K \succ AD \succ KMS \succ U_n^2 \succ V_n \succ K$;
- for testing composite hypotheses with a pair "Weibull—gamma": $Z_A \succ Z_C \succ AD \succ Z_K \succ KMS \succ U_n^2 \succ V_n \succ K$.

31.4 The Distribution of Statistics for Testing Composite Hypotheses

When testing composite hypotheses conditional distribution $G(S\,|\,H_0)$ of the statistic depends on several factors: the type of the observed distribution for true hypothesis H_0; the type of the estimated parameter and the number of parameters to be estimated, in some cases the parameter values (e.g., for the families of gamma and beta distributions), the method of parameter estimation. The differences between distributions of the one statistic for testing simple and composite hypotheses are very significant, so we could not neglect this fact. For example, Fig. 31.2 shows the distribution of Kuiper statistic (31.4) for testing composite hypotheses for the different distributions using maximum likelihood estimates (MLE) of the two parameters.

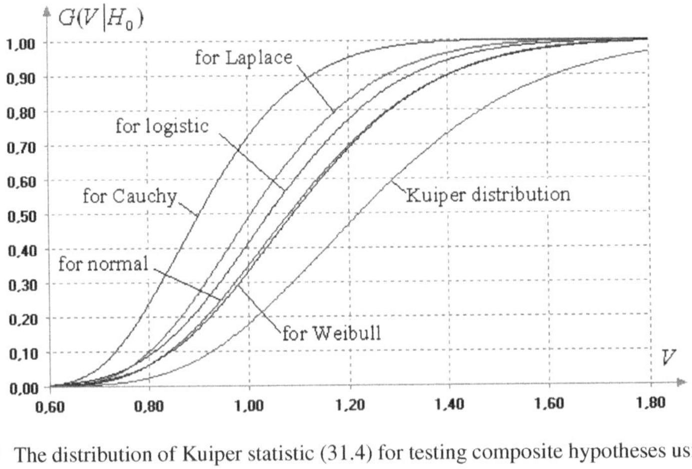

Fig. 31.2 The distribution of Kuiper statistic (31.4) for testing composite hypotheses using MLEs of the two parameters

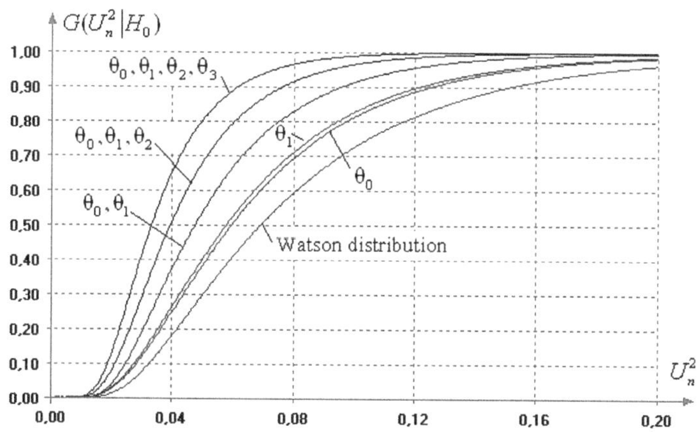

Fig. 31.3 The distribution of Watson statistic (31.5) for testing composite hypotheses using MLEs of different number of parameters of the Su-Johnson distribution

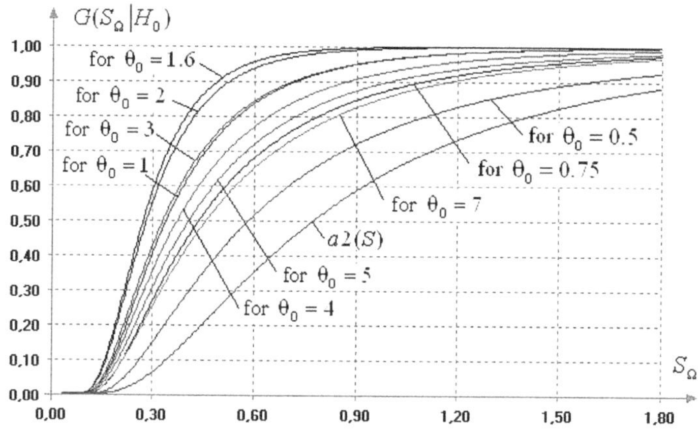

Fig. 31.4 The distribution of Anderson–Darling statistics (31.3) for testing composite hypotheses using MLEs of three parameters of the generalized normal distribution, depending on the value of the shape parameter θ_0

Figure 31.3 illustrates the dependence of the distribution of the Watson test statistic (31.5) on the type and the number of estimated parameters having as an example the *Su*-Johnson distribution with a density:

$$f(x) = \frac{\theta_1}{\sqrt{2\pi}\sqrt{(x-\theta_3)^2+\theta_2^2}} \exp\left\{-\frac{1}{2}\left[\theta_0 + \theta_1 \ln\left\{\frac{x-\theta_3}{\theta_2} + \sqrt{\left(\frac{x-\theta_3}{\theta_2}\right)^2 + 1}\right\}\right]^2\right\}.$$

Figure 31.4 shows the dependence of the distribution of Anderson–Darling test statistics (31.3) for testing composite hypotheses using MLEs of the three parameters of the generalized normal distribution depending on the value of the shape parameter θ_0.

The first work that initiates the study of limiting distributions of nonparametric goodness-of-fit statistics for composite hypotheses was [14]. Later, different approaches were used to solve this problem: the limit distribution was investigated by analytical methods [7–12, 30–34], the percentage points were calculated using statistical modeling [6,35,37,38], the formulas were obtained to give a good approximation for small values of the probabilities [39, 40].

In our studies [18–29] the distribution of nonparametric Kolmogorov, Cramer–von Mises–Smirnov, and Anderson–Darling tests statistics were studied using statistical modeling.

Further, based on obtained empirical distribution of statistics, we construct an approximate analytical model of statistics distributions.

The obtained models of limiting distributions and percentage points for Kuiper and Watson test statistics, which are required to test composite hypotheses (using MLEs), could be found in the paper [20] for the most often used in applications parametric distributions: Exponential, Seminormal, Rayleigh, Maxwell, Laplace, Normal, Log-normal, Cauchy, Logistic, Extreme-value (maximum), Extreme-value (minimum), Weibull, *Sb*-Johnson, *Sl*-Johnson, *Su*-Johnson.

Previously obtained similar models (and percentage points) for distributions of Kolmogorov, Cramer–von Mises–Smirnov, and Anderson–Darling test statistics (for distributions mentioned above) could be found in papers [21, 22, 24, 28].

The tables of percentage points and models of test statistics distributions were based on simulated samples of the statistics with the size $N = 10^6$. Such N makes the difference between the actual distribution $G(S|H_0)$ and empirical counterpart $G_N(S|H_0)$ that does not exceed 10^{-3}. The values of the test statistic were calculated using samples of pseudorandom values simulated for the observed distribution $F(x, \theta)$ with the size $n = 10^3$. In such a case the distribution $G(S_n|H_0)$ practically equal to the limit one $G(S|H_0)$. The given models could be used for statistical analysis if the sample sizes $n > 25$.

Unfortunately, the dependence of the nonparametric goodness-of-fit tests statistics distributions for testing composite hypotheses on the values of the shape parameter (or parameters) (see Fig. 31.4) appears to be for many parametric distributions implemented in the most interesting applications, particularly in problems of survival and reliability. This is true for families of gamma, beta distributions of the first, second, and third kind, generalized normal, generalized Weibull, inverse Gaussian distributions, and many others.

The limit distributions and percentage points for Kolmogorov, Cramer–von Mises–Smirnov, and Anderson–Darling tests for testing composite hypotheses with the family of gamma distributions were obtained in paper [22], with the inverse Gaussian distribution—in papers [29], with generalized normal distribution—in paper [23], with the generalized Weibull distribution—in paper [1]. It should be noted that the data in these papers were obtained only for a limited number of, generally, integer values of the shape parameter (or parameters).

31.5 An Interactive Method to Study Distributions of Statistics

The dependence of the test statistics distributions on the values of the shape parameter or parameters is the most serious difficulty that is faced while applying nonparametric goodness-of-fit criteria to test composite hypotheses in different applications.

Since estimates of the parameters are only known during the analysis, so the statistic distribution required to test the hypothesis could not be obtained in advance (before calculating estimates for the analyzed sample!). For criteria with statistics (31.6)–(31.8), the problem is harder as statistics distributions depend on the samples sizes. Therefore, statistics distributions of applied criteria should be obtained interactively during statistical analysis, and then should be used to make conclusions about composite hypothesis under test.

The implementation of such an interactive mode requires developed software that allows parallelizing the simulation process and taking available computing resources. While using parallel computing the time to obtain the required test statistic distribution $G_N(S_n|H_0)$ (with the required accuracy) and use it to calculate the achieved significance level $P\{S_n \geqslant S^*\}$, where S^* is the value of the statistic calculated using an original sample, is not very noticeable compared to a process of statistical analysis.

In the program system [13], an interactive method to research statistics distributions is implemented for the following nonparametric goodness-of-fit tests: Kolmogorov, Cramer–von Mises–Smirnov, Anderson–Darling, Kuiper, Watson, and three Zhang tests. Moreover, the different methods of parameter estimation could be used there.

The following example demonstrates the accuracy of calculating the achieved significance level depending on sample size N of simulated interactively empirical statistics distributions [13]. The inverse Gaussian distribution is widely used in reliability and in survival analysis [29]. In this case, the Γ-distribution (generalized gamma distribution) can be considered as the competing law.

Example. You should check the composite hypothesis that the following sample with the size $n = 100$ has the inverse Gaussian distribution with the density (31.9):

0.945 1.040 0.239 0.382 0.398 0.946 1.248 1.437 0.286 0.987
2.009 0.319 0.498 0.694 0.340 1.289 0.316 1.839 0.432 0.705
0.371 0.668 0.421 1.267 0.466 0.311 0.466 0.967 1.031 0.477
0.322 1.656 1.745 0.786 0.253 1.260 0.145 3.032 0.329 0.645
0.374 0.236 2.081 1.198 0.692 0.599 0.811 0.274 1.311 0.534
1.048 1.411 1.052 1.051 4.682 0.111 1.201 0.375 0.373 3.694
0.426 0.675 3.150 0.424 1.422 3.058 1.579 0.436 1.167 0.445
0.463 0.759 1.598 2.270 0.884 0.448 0.858 0.310 0.431 0.919
0.796 0.415 0.143 0.805 0.827 0.161 8.028 0.149 2.396 2.514
1.027 0.775 0.240 2.745 0.885 0.672 0.810 0.144 0.125 1.621

$$f(x) = \frac{1}{\theta_2}\left(\frac{\theta_0}{2\pi\left(\frac{x-\theta_3}{\theta_2}\right)^3}\right)^{1/2} \exp\left(-\frac{\theta_0\left(\left(\frac{x-\theta_3}{\theta_2}\right)-\theta_1\right)^2}{2\theta_1^2\left(\frac{x-\theta_3}{\theta_2}\right)}\right). \qquad (31.9)$$

The shift parameter θ_3 is assumed to be known and equal to 0.

The shape parameters θ_0, θ_1, and the scale parameter θ_2 are estimated using the sample. The MLEs calculated using the sample above are the following: $\hat{\theta}_0 = 0.7481$, $\hat{\theta}_1 = 0.7808$, $\hat{\theta}_2 = 1.3202$. Statistics distributions of nonparametric goodness-of-fit tests depend on the values of the shape parameters θ_0 and θ_1 [46, 47], do not depend on the value of the scale parameter θ_2 and can be calculated using values $\theta_0 = 0.7481$, $\theta_1 = 0.7808$.

The calculated values of the statistics S_i^* for Kuiper, Watson, Zhang, Kolmogorov, Cramer–von Mises–Smirnov, Anderson–Darling tests and achieved significance levels for these values $P\{S \geqslant S_i^*|H_0\}$ (p-values), obtained with different accuracy of simulation (with different sizes N of simulated samples of statistics) are given in Table 31.1.

The similar results for testing goodness-of-fit of a given sample with Γ-distribution with the density:

$$f(x) = \frac{\theta_1}{\theta_3 \Gamma(\theta_0)}\left(\frac{x-\theta_4}{\theta_3}\right)^{\theta_0\theta_1 - 1} e^{-\left(\frac{x-\theta_4}{\theta_3}\right)^{\theta_1}}$$

are given in Table 31.2. The MLEs of the parameters are $\theta_0 = 2.4933$, $\theta_1 = 0.6065$, $\theta_2 = 0.1697$, $\theta_4 = 0.10308$. In this case the distribution of the test statistic depends on the values of the shape parameters θ_0 and θ_1.

The implemented interactive mode to study statistics distributions enables to correctly apply goodness-of-fit Kolmogorov, Cramer–von Mises–Smirnov, Anderson–Darling, Kuiper, Watson, Zhang (with statistics Z_C, Z_A, Z_K) tests with calculating the achieved significance level (p-value) even in those cases when the statistic distribution for true hypothesis H_0 is unknown while testing composite hypothesis. For Zhang tests, this method allows us to test a simple hypothesis for every sample size.

Table 31.1 The achieved significance levels for different sizes N when testing goodness-of-fit with the inverse Gaussian distribution

The values of test statistics	$N = 10^3$	$N = 10^4$	$N = 10^5$	$N = 10^6$
$V_n^{\text{mod}} = 1.1113$	0.479	0.492	0.493	0.492
$U_n^2 = 0.05200$	0.467	0.479	0.483	0.482
$Z_A = 3.3043$	0.661	0.681	0.679	0.678
$Z_C = 4.7975$	0.751	0.776	0.777	0.776
$Z_K = 1.4164$	0.263	0.278	0.272	0.270
$K = 0.5919$	0.643	0.659	0.662	0.662
$KMS = 0.05387$	0.540	0.557	0.560	0.561
$AD = 0.3514$	0.529	0.549	0.548	0.547

Table 31.2 The achieved significance levels for different sizes N when testing goodness-of-fit with the Γ-distribution

The values of test statistics	$N = 10^3$	$N = 10^4$	$N = 10^5$	$N = 10^6$
$V_n^{\text{mod}} = 1.14855$	0.321	0.321	0.323	0.322
$U_n^2 = 0.057777$	0.271	0.265	0.267	0.269
$Z_A = 3.30999$	0.235	0.245	0.240	0.240
$Z_C = 4.26688$	0.512	0.557	0.559	0.559
$Z_K = 1.01942$	0.336	0.347	0.345	0.344
$K = 0.60265$	0.425	0.423	0.423	0.424
$KMS = 0.05831$	0.278	0.272	0.276	0.277
$AD = 0.39234$	0.234	0.238	0.238	0.237

Acknowledgements This research has been supported by the Russian Ministry of Education and Science (project 2.541.2014K).

References

1. Akushkina, K.A., Lemeshko, S.B., Lemeshko, B.Yu.: Models of statistical distributions of nonparametric goodness-of-fit tests in testing composite hypotheses of the generalized Weibull distribution. In: Proceedings Third International Conference on Accelerated Life Testing, Reliability-Based Analysis and Design, Clermont-Ferrand, 19–21 May 2010, pp. 125–132
2. Anderson, T.W., Darling, D.A.: Asymptotic theory of certain "goodness of fit"İ criteria based on stochastic processes. Ann. Math. Stat. **23**, 193–212 (1952)
3. Anderson, T.W., Darling, D.A.: A test of goodness of fit. J. Am. Stat. Assoc. **29**, 765–769 (1954)
4. Bolshev, L.N.: Asymptotic Pearson transformations. Teor. Veroyatn. Ee Primen. **8**(2), 129–155 (1963, in Russian)
5. Bolshev, L.N., Smirnov, N.V.: Tables for Mathematical Statistics. Nauka, Moscow (1983, in Russian)
6. Chandra, M., Singpurwalla, N.D., Stephens, M.A.: Statistics for test of fit for the extreme—value and Weibull distribution. J. Am. Stat. Assoc. **76**(375), 729–731 (1981)
7. Darling, D.A.: The Cramer-Smirnov test in the parametric case. Ann. Math. Stat. **26**, 1–20 (1955)
8. Darling, D.A.: The Cramer-Smirnov test in the parametric case. Ann. Math. Stat. **28**, 823–838 (1957)
9. Durbin, J.: Weak convergence of the sample distribution function when parameters are estimated. Ann. Stat. **1**(2), 279–290 (1973)
10. Durbin, J.: Kolmogorov-Smirnov tests when parameters are estimated with applications to tests of exponentiality and tests of spacings. Biometrika **62**, 5–22 (1975)
11. Durbin, J.: Kolmogorov–Smirnov test when parameters are estimated. In: Gänssler, P., Revesz, P. (eds.) Empirical Distributions and Processes. Selected Papers from a Meeting at Oberwolfach, March 28 – April 3, 1976. Series: Lecture Notes in Mathematics, **566**, pp. 33–44. Springer Berlin Heidelberg (1976)
12. Dzhaparidze, K.O., Nikulin, M.S.: Probability distribution of the Kolmogorov and omega-square statistics for continuous distributions with shift and scale parameters. J. Soviet Math. **20**, 2147–2163 (1982)

13. ISW: Program system of the statistical analysis of one-dimensional random variables. http://www.ami.nstu.ru/~headrd/ISW.htm. Accessed 25 Dec 2013
14. Kac, M., Kiefer, J., Wolfowitz, J.: On tests of normality and other tests of goodness of fit based on distance methods. Ann. Math. Stat. **26**, 189–211 (1955)
15. Kolmogoroff, A.N.: Sulla determinazione empirica di una legge di distribuzione. Giornale dell ' Istituto Italiano degli Attuari **4**(1), 83–91 (1933)
16. Kuiper, N.H.: Tests concerning random points on a circle. Proc. Konikl. Nederl. Akad. Van Wettenschappen. Ser. A. **63**, 38–47 (1960)
17. Lemeshko, B.Yu.: Asymptotically optimum grouping of observations in goodness-of-fit tests. Ind. Lab. **64**(1), 59–67(1998). Consultants Bureau, New York
18. Lemeshko, B.Yu.: Errors when using nonparametric fitting criteria. Measur. Tech. **47**(2), 134–142 (2004)
19. Lemeshko, B.Yu., Gorbunova, A.A.: Application and power of the nonparametric Kuiper, Watson, and Zhang Tests of Goodness-of-Fit. Measur. Tech. **56**(5), 465–475 (2013)
20. Lemeshko, B.Yu., Gorbunova, A.A.: Application of nonparametric Kuiper and Watson tests of goodness-of-fit for composite hypotheses. Measur. Tech. **56**(9), 965–973 (2013)
21. Lemeshko, B.Yu., Lemeshko, S.B.: Distribution models for nonparametric tests for fit in verifying complicated hypotheses and maximum-likelihood estimators. Part 1. Measur. Tech. **52**(6), 555–565 (2009)
22. Lemeshko, B.Yu., Lemeshko, S.B.: Models for statistical distributions in nonparametric fitting tests on composite hypotheses based on maximum-likelihood estimators. Part II. Measur. Tech. **52**(8), 799–812 (2009)
23. Lemeshko, B.Yu., Lemeshko, S.B.: Models of statistic distributions of nonparametric goodness-of-fit tests in composite hypotheses testing for double exponential law cases. Commun. Stat. Theory Methods **40**(16), 2879–2892 (2011)
24. Lemeshko, B.Yu., Lemeshko, S.B.: Construction of statistic distribution models for nonparametric goodness-of-fit tests in testing composite hypotheses: the computer approach. Qual. Technol. Quant. Manag. **8**(4), 359–373 (2011)
25. Lemeshko, B.Yu., Lemeshko, S.B., Postovalov, S.N.: The power of goodness of fit tests for close alternatives. Measur. Tech. **50**(2), 132–141 (2007)
26. Lemeshko, B.Yu., Lemeshko, S.B., Postovalov, S.N.: Comparative analysis of the power of goodness-of-fit tests for near competing hypotheses. I. The verification of simple hypotheses. J. Appl. Ind. Math. **3**(4), 462–475 (2009)
27. Lemeshko, B.Yu., Lemeshko, S.B., Postovalov, S.N.: Comparative analysis of the power of goodness-of-fit tests for near competing hypotheses. II. Verification of complex hypotheses. J. Appl. Ind. Math. **4**(1), 79–93 (2010)
28. Lemeshko, B.Yu., Lemeshko, S.B., Postovalov, S.N.: Statistic distribution models for some nonparametric goodness-of-fit tests in testing composite hypotheses. Commun. Stat. Theory Methods **39**(3), 460–471 (2010)
29. Lemeshko, B.Yu., Lemeshko, S.B., Akushkina, K.A., Nikulin, M.S., Saaidia, N.: Inverse Gaussian model and its applications in reliability and survival analysis. In: Rykov, V., Balakrishnan, N., Nikulin, M. (eds.) Mathematical and Statistical Models and Methods in Reliability. Applications to Medicine, Finance, and Quality Control, pp. 433–453. Birkhauser, Boston (2011)
30. Lilliefors, H.W.: On the Kolmogorov-Smirnov test for normality with mean and variance unknown. J. Am. Stat. Assoc. **62**, 399–402 (1967)
31. Lilliefors, H.W.: On the Kolmogorov-Smirnov test for the exponential distribution with mean unknown. J. Am. Stat. Assoc. **64**, 387–389 (1969)
32. Martynov, G.V.: The Omega Squared Test. Nauka, Moscow (1978, in Russian)
33. Nikulin, M.S.: Gihman and goodness-of-fit tests for grouped data. Math. Rep. Acad. Sci. R. Soc. Can. **14**(4), 151–156 (1992)
34. Nikulin, M.S.: A variant of the generalized omega-square statistic. J. Sov. Math. **61**(4), 1896–1900 (1992)

35. Pearson, E.S., Hartley, H.O.: Biometrika Tables for Statistics, vol. 2. Cambridge University Press, Cambridge (1972)
36. Stephens, M.A.: The goodness-of-fit statistic VN: distribution and significance points. Biometrika **52**(3–4), 309–321 (1965)
37. Stephens, M.A.: Use of Kolmogorov—Smirnov, Cramer—von Mises and related statistics—without extensive table. J. R. Stat. Soc. **32**, 115–122 (1970)
38. Stephens, M.A.: EDF statistics for goodness of fit and some comparisons. J.Am. Stat. Assoc. **69**(347), 730–737 (1974)
39. Tyurin, Yu.N.: On the limiting Kolmogorov—Smirnov statistic distribution for composite hypothesis. NewsAS USSR Ser. Math. **48**(6), 1314–1343 (1984, in Russian)
40. Tyurin, Yu.N., Savvushkina, N.E.: Goodness-of-fit test for Weibull—Gnedenko distribution. News AS USSR. Ser. Tech. Cybern. **3**, 109–112 (1984, in Russian)
41. Watson, G.S.: Goodness-of-fit tests on a circle. I. Biometrika **48**(1–2), 109–114 (1961)
42. Watson, G.S.: Goodness-of-fit tests on a circle. II. Biometrika **49**(1–2), 57–63 (1962)
43. Zhang, J.: Powerful goodness-of-fit tests based on the likelihood ratio. J. R. Stat. Soc. Ser. B **64**(2), 281–294 (2002)
44. Zhang, J.: Powerful two-sample tests based on the likelihood ratio. Technometrics **48**(1), 95–103 (2006)
45. Zhang, J., Wub, Yu.: Likelihood-ratio tests for normality. Comput. Stat. Data Anal. **49**(3), 709–721 (2005)

Chapter 32
Simulating from the Copula that Generates the Maximal Probability for a Joint Default Under Given (Inhomogeneous) Marginals

Jan-Frederik Mai and Matthias Scherer

32.1 Motivation

Starting from two default times with given univariate distribution functions, the copula which maximizes the probability of a joint default can be computed in closed form. This result can be retrieved from Markov-chain theory, where it is known under the terminology "maximal coupling", but typically formulated without copulas. For inhomogeneous marginals the solution is not represented by the comonotonicity copula, opposed to a common modeling (mal-)practice in the financial industry. Moreover, a stochastic model that respects the marginal laws and attains the upper-bound copula for joint defaults can be inferred from the maximal-coupling construction. We formulate and illustrate this result in the context of copula theory and motivate its importance for portfolio-credit risk modeling. Moreover, we present a sampling strategy for the "maximal-coupling copula".

In portfolio-credit risk, the modeling of dependent default times is often carried out in two subsequent steps: (1) the specification of the marginal laws, and (2) the choice of some copula connecting them, see [14]. Mathematically, this is justified by Sklar's theorem (see [15]), stating that arbitrary marginals can be connected with any copula to obtain a valid joint distribution function. The main reason for the popularity of such a modeling approach is that a dependence structure can be added to well-understood marginal models without destroying their structure. However, the danger of naïvely using this modeling paradigm is that the resulting distribution

J.-F. Mai
XAIA Investment GmbH, Sonnenstraße 19, 80331 München, Germany
e-mail: jan-frederik.mai@xaia.com

M. Scherer (✉)
Technische Universität München, Parkring 11, 85748 Garching-Hochbrück, Germany
e-mail: scherer@tum.de

V.B. Melas et al. (eds.), *Topics in Statistical Simulation*, Springer Proceedings in Mathematics & Statistics 114, DOI 10.1007/978-1-4939-2104-1_32

need not be reasonable with regard to the economic criterion in concern, as pointed out in the academic literature many times, see, e.g., [2, 6–8].

One popular misbelief is that the comonotonicity copula, which maximizes the dependence if measured in terms of concordance measures, also maximizes the probability of a joint default (or, at least, the probability of default times being quite close to each other). However, this is not the case, because for two companies' default times τ_1, τ_2 events such as $\{|\tau_1 - \tau_2| < \epsilon\}$ for small $\epsilon > 0$, or even $\{\tau_1 = \tau_2\}$, strongly depend on the marginal laws of the default times, as will be investigated in quite some detail below. Providing an example, which we adopt from [9], let τ_1 and τ_2 be exponentially distributed with rate parameters λ_1 and λ_2, respectively, and assume that they are coupled by a Gaussian copula C_ρ with parameter $\rho \in [-1, 1]$. Figure 32.1 visualizes the probability $(\lambda_1, \rho) \mapsto \mathbb{P}(|\tau_1 - \tau_2| < 1/12)$ that both default times happen within one month in dependence of the parameter ρ and rate λ_1, the rate λ_2 is fixed to 0.15. This probability can be evaluated numerically as a double integral:

$$\mathbb{P}(|\tau_1 - \tau_2| < 1/12) =$$
$$\lambda_1\, 0.15 \int_0^\infty e^{-\lambda_1 x} \int_{\max\{0, x-1/12\}}^{x+1/12} c_\rho\big(1 - e^{-\lambda_1 x}, 1 - e^{-0.15\,y}\big)\, e^{-0.15\,y}\, dy\, dx,$$

with c_ρ denoting the Gaussian copula density. This probability is not increasing in the dependence parameter, see Fig. 32.1 (left). This might be problematic if the target risk to be modeled is not the dependence per se (being measured in terms of correlation or some more general concordance measure), but rather the probability of a joint default.

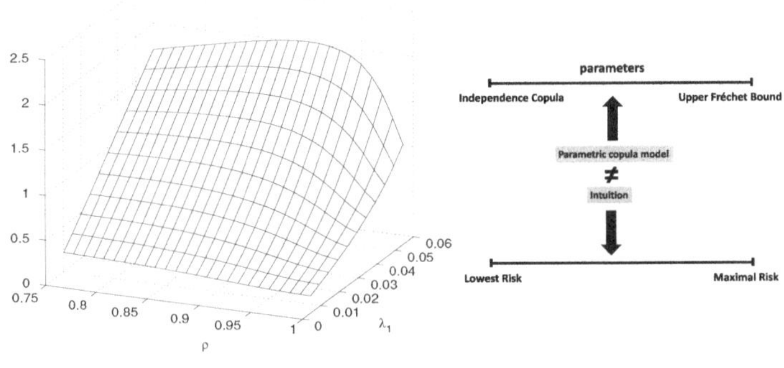

Fig. 32.1 *Left*: The probability $\mathbb{P}(|\tau_1 - \tau_2| < 1/12)$ is plotted for different ρ and λ_1, assuming $\tau_1 \sim \text{Exp}(\lambda_1)$ and $\tau_2 \sim \text{Exp}(\lambda_2 = 0.15)$, and these exponential marginals are connected with a Gaussian copula with correlation parameter ρ. Notice that (**a**) for fixed λ_1, the displayed probability is not increasing in ρ, and (**b**) in the limiting case $\rho = 1$ we have $\tau_1 = \lambda_1/\lambda_2\, \tau_2$ almost surely, i.e. $\lim_{\rho \nearrow 1} \mathbb{P}(|\tau_1 - \tau_2| < 1/12) = \mathbb{P}(\tau_1\,(1 - \lambda_1/\lambda_2) \leqslant 1/12)$. *Right*: Visualization of misleading model intuition

There are in fact various situations when the risk we truly face is the probability of a joint event, i.e. $\tau_1 = \tau_2$. One current and prominent example in a financial context is the computation of credit value adjustments (CVA) in a credit default swap (CDS). A CDS is an insurance contract between two parties. One party makes periodic premium payments to the other party. In return, the other party compensates the insurance buyer for potential losses arising from a credit event, i.e. the event that a third party (the underlying reference entity) becomes bankrupt. Regulatory requirements force the insurance buyer to compute a CVA, accounting for the counterparty credit risk arising from the fact that the insurance contract loses value upon the default of the insurance seller. The major risk for the insurance buyer with regard to this CVA is the possibility that the insurance seller defaults jointly with (or immediately before) the underlying reference entity, because in this case insurance is needed but cannot be paid by the insurance seller. It is highly plausible that the marginal survival functions of the reference entity and of the insurance seller are well understood, e.g. from credit-risk data observable in the marketplace. However, there is only limited (or even none) information available about the dependence between the two. A natural model in this situation is to estimate the margins from the observable data, and to link them with some parametric copula model. The choice of parameters for the copula is then clearly more art than science, and a great amount of intuition is required.[1] However, the intuition one might have for such a copula model can be misleading. The right graph in Fig. 32.1 visualizes the problem.

The present article is partly inspired by a series of papers dealing with the investigation of multivariate distributions under given marginals but with unknown copula. The references [10–13] study and compute lower and upper bounds for certain functionals of a multivariate law with given marginals. In comparison with these references, the present article deals with a very special functional, namely the probability of a joint default. Moreover, motivated by a financial application, [3–5] study the Value-at-Risk and related measures of a portfolio of risks with unknown copula.

32.2 An Upper Bound for the Probability of $\{\tau_1 = \tau_2\}$

To illustrate the problem we first assume identical marginal laws, i.e. $\tau_1 \sim F$, $\tau_2 \sim F$ for a univariate distribution function F. In this case (and only in this case), coupling with the comonotonicity copula $C(u, v) = \min\{u, v\}$ indeed maximizes the probability of the event $\{\tau_1 = \tau_2\}$, implying a certain joint default, i.e. $\mathbb{P}(\tau_1 = \tau_2) = 1$. This can easily be seen from a stochastic model based on the quantile transformation: simply take $U \sim \mathcal{U}(0, 1)$ and define $\tau_1 = \tau_2 := F^{-1}(U)$, where F^{-1} is the (generalized) inverse of F. Clearly, one obtains $\mathbb{P}(\tau_1 = \tau_2) = 1$

[1] In the terminology of F. Knight one might say that one is exposed to uncertainty concerning the dependence structure.

and both default times have the pre-determined marginal law F. Conversely, $\mathbb{P}(\tau_1 = \tau_2) = 1$ already implies identical default probabilities, which follows from the fact that

$$\mathbb{P}(\tau_1 \leqslant x) \overset{(*)}{=} \mathbb{P}(\tau_1 = \tau_2, \tau_1 \leqslant x) \overset{(*)}{=} \mathbb{P}(\tau_2 \leqslant x),$$

where x was arbitrary and $\mathbb{P}(\tau_1 = \tau_2) = 1$ is used in $(*)$. This implies that for inhomogeneous marginals, there does not exist a stochastic model such that the defaults take place together for sure. Moreover, it raises the following natural question: what is the dependence structure (i.e. the copula) maximizing the probability for a joint default when the marginals are fixed? An answer to this question can be retrieved by a technique called "coupling" in Markov-chain theory. Standard textbooks on the topic are [1,16]. The idea of couplings in Markov-chain theory is precisely the idea of copulas in statistical modeling, namely to define bivariate Markov chains from pre-determined univariate Markov chains. The coupling technique was initially invented to prove that Markov chains converge to a stationary law under some regularity conditions by coupling the given Markov chain with a stationary chain that shares the same transition probabilities.

The maximal-coupling construction provides a probability space supporting two default times τ_1, τ_2 with given densities f_1, f_2 on $(0, \infty)$ such that the upper bound for the joint default probability is attained. To clarify notation, recall that with $F_i(x) := \int_0^x f_i(s)\,ds$, $i = 1, 2$, the probability of a joint default can be expressed in terms of the copula C and the marginal laws F_1, F_2 as

$$\mathbb{P}(\tau_1 = \tau_2) = \iint\limits_{D(F_1,F_2)} dC(u, v), \quad D(F_1, F_2) := \big\{(u, v) \in [0, 1]^2 \,:\, F_1^{-1}(u) = F_2^{-1}(v)\big\}.$$

Formulating the result in copula language, it may be stated as follows.

Theorem 1 (A Model Maximizing $\mathbb{P}(\tau_1 = \tau_2)$). *Denote by $\mathscr{C}$ the set of all bivariate copulas and assume that τ_1, τ_2 have densities f_1, f_2 on $(0, \infty)$.*

- *One then has:*

$$\sup_{C \in \mathscr{C}} \Bigg\{ \iint\limits_{D(F_1,F_2)} dC(u, v) \Bigg\} = \int_0^\infty \min\{f_1(x), f_2(x)\} dx =: p_\infty. \qquad (32.1)$$

- *Moreover, the supremum is actually a maximum and there is a probabilistic construction for the maximizer. If $f_1 = f_2$ a.e., then $p_\infty = 1$; if the supports of f_1 and f_2 are disjoint, then $p_\infty = 0$. In all other cases we have $p_\infty \in (0, 1)$ and a maximizing copula C_{F_1,F_2}, which strongly depends on the marginals, is given by*

$$C_{F_1,F_2}(u,v) = \int_0^{\min\{F_1^{-1}(u),F_2^{-1}(v)\}} \min\{f_1(s), f_2(s)\}\, ds +$$

$$+ \frac{1}{1-p_\infty}\left(\int_0^{F_1^{-1}(u)} f_1(s) - \min\{f_1(s), f_2(s)\}\, ds\right)\left(\int_0^{F_2^{-1}(v)} f_2(s) - \min\{f_1(s), f_2(s)\}\, ds\right).$$

Proof. The proof can be retrieved from the coupling literature, e.g. [16, p. 9]. We only sketch the basic idea because it is educational and we refer to it in Example 1 below.

- If the supports of f_1 and f_2 are disjoint, then obviously $\mathbb{P}(\tau_1 = \tau_2) = 0$, irrespective of the choice of copula.
- If $f_1 = f_2$ a.e., then the distributions of τ_1 and τ_2 are identical and the comonotonicity copula $\min\{u, v\}$ provides the maximum $\mathbb{P}(\tau_1 = \tau_2) = 1$.

We define $p_\infty := \int_0^\infty \min\{f_1(x), f_2(x)\}dx$, which—excluding the two degenerate cases from above—is in $(0, 1)$. Moreover, define the densities

$$h_{\min} := \frac{\min\{f_1, f_2\}}{p_\infty}, \quad h_{f_1} := \frac{f_1 - p_\infty h_{\min}}{1-p_\infty}, \quad h_{f_2} := \frac{f_2 - p_\infty h_{\min}}{1-p_\infty}. \tag{32.2}$$

Consider a probability space $(\Omega, \mathscr{F}, \mathbb{P})$ supporting the independent random variables $H_{\min} \sim h_{\min}$, $H_{f_1} \sim h_{f_1}$, $H_{f_2} \sim h_{f_2}$, and a Bernoulli variable X with success probability p_∞. Define

$$(\tau_1, \tau_2) := (H_{\min}, H_{\min})\mathbb{1}_{\{X=1\}} + (H_{f_1}, H_{f_2})\mathbb{1}_{\{X=0\}}. \tag{32.3}$$

Whenever $X = 1$, one has $\tau_1 = \tau_2 = H_{\min}$. In the case $X = 0$, the probability for $\tau_1 = H_{f_1} = H_{f_2} = \tau_2$ is zero. Hence, we have $\mathbb{P}(\tau_1 = \tau_2) = \mathbb{P}(X = 1) = p_\infty$ and

$$\begin{aligned}\mathbb{P}(\tau_i \leqslant x) &= p_\infty\, \mathbb{P}(H_{\min} \leqslant x) + (1-p_\infty)\, \mathbb{P}(H_{f_i} \leqslant x)\\ &= \int_0^x p_\infty\, h_{\min}(s) + (1-p_\infty)\, h_{f_i}(s) ds = \int_0^x f_i(s) ds = F_i(x),\end{aligned}$$

for $i = 1, 2$. Moreover, it can be shown that p_∞ is actually an upper bound for the probability $\iint_{D(F_1,F_2)} dC(u,v)$ across all copulas $C \in \mathscr{C}$, a step which we omit in full detail for the sake of brevity. Intuitively speaking, at each point in time x we have to maintain the marginal laws, specified by $f_1(x)$ and $f_2(x)$. So an upper bound for the (local) joint default probability at time x is the minimum of the densities f_1 and f_2 at time x. Integrating over the positive half line yields the global upper bound, denoted p_∞. The claimed copula of the stochastic model (32.3) that attains the upper bound is then easily inferred from the probabilistic construction outlined above.

It is very educational to understand the idea of the proof above, because it readily implies a simulation algorithm for the maximizing copula C_{F_1,F_2}. We explicitly extract the stochastic construction idea from the proof in the following simulation algorithm for the copula C_{F_1,F_2}.

A simulation via the rejection algorithms in (1), (2), and (3) has the advantage of being very generic: only little knowledge is needed about the involved marginal laws. The runtime of each rejection step is random, the number of required runs until we accept is a Geometric random variable with success probability depending on the involved densities. Our implementation of Example 1 in `Matlab` (on a standard desktop PC) required less than two minutes to produce 100, 000 samples. In case the marginals are chosen such that X, $H_{\min}$, and H_{f_i} can be simulated without rejection algorithm, this might be accelerated further. This, however, requires p_∞ explicitly and depends on the choice of f_1 and f_2, so we cannot provide a generic recipe.

Example 1 (Illustration of the Construction). We assume that τ_1 and τ_2 have lognormal densities. More precisely, we assume that

$$f_i(x) = \frac{1}{\sqrt{2\pi}\,\sigma_i\,x}\exp\left(-\frac{(\log(x)-\mu_i)^2}{2\sigma_i^2}\right)\mathbb{1}_{\{x>0\}}, \quad i = 1, 2,$$

with $\sigma_1 = 1$, $\sigma_2 = 2$, $\mu_1 = \log(10)$, and $\mu_2 = \log(30)$. These two densities are visualized in Fig. 32.2.

Figure 32.2 (left) shows three more density functions, which are constructed from f_1 and f_2. The solid line is the density $h_{\min} : x \mapsto \min\{f_1(x), f_2(x)\}/p_\infty$, which exhibits a kink at approximately 1.4801, where f_1 and f_2 intersect. This equals the density of the random variable $H_{\min}$ in the proof of Theorem 1, which is supported on all of $(0,\infty)$. The other two densities h_{f_1} and h_{f_2} are obtained by subtraction of $h_{\min}$ from f_1 and f_2, respectively (and appropriate scaling). Hence, h_{f_1} is positive only on $(0, 1.4801)$ and h_{f_2} is positive only on the complementary interval $(1.4801,\infty)$. Finally, Fig. 32.2 (right) visualizes a scatter plot from the

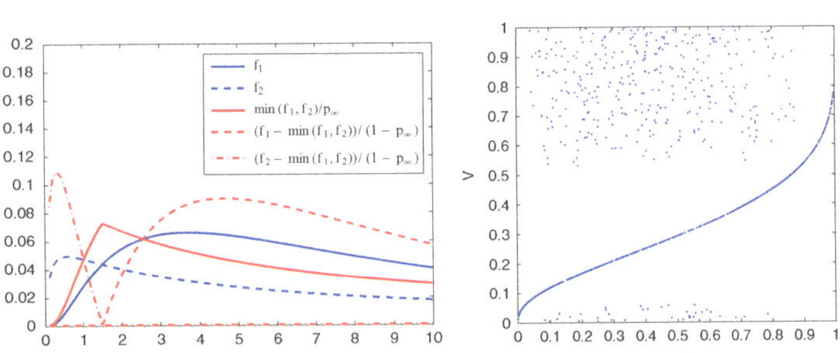

Fig. 32.2 *Left*: Two lognormal densities f_1 and f_2, together with the three further density functions constructed from them in Eq. (32.2). *Right*: Scatter plot from the copula C_{F_1,F_2} in the lognormal example. The singular component is easy to spot

Algorithm 1 Simulation of C_{F_1,F_2}

The input for the algorithm are two densities f_1, f_2. Moreover, we need to be able to simulate these two univariate distributions and to evaluate the distribution functions F_1, F_2 and the densities f_1, f_2.

(1) Simulate a Bernoulli random variable $X \sim \text{Bernoulli}(p_\infty)$. There is no need to compute p_∞ explicitly. Rather, the simulation of X can be accomplished by the following code:

```
SIMULATE Y_1 ~ F_1, U ~ 𝒰(0, 1)
IF (U ≤ min{1, f_2(Y_1)/f_1(Y_1)})
    X := 1
ELSE
    X := 0
END
```

(2) Simulate $H_{\min}$ by the following rejection acceptance algorithm:

```
SIMULATE Y_1 ~ F_1, U ~ 𝒰(0, 1)
WHILE (U > min{1, f_2(Y_1)/f_1(Y_1)})
    SIMULATE Y_1 ~ F_1, U ~ 𝒰(0, 1)
END
H_min := Y_1
```

(3) Simulate $H_{f_i}, i = 1, 2$ by the following rejection acceptance algorithm:

```
SIMULATE Y_i ~ F_i, U ~ 𝒰(0, 1)
WHILE (U > 1 − min{1, f_{3−i}(Y_i)/f_i(Y_i)})
    SIMULATE Y_i ~ F_i, U ~ 𝒰(0, 1)
END
H_{f_i} := Y_i
```

(4) Return the desired sample (U_1, U_2) from the copula C_{F_1,F_2} as follows:

```
IF (X = 1)
    U_1 := F_1(H_min), U_2 := F_2(H_min)
ELSE
    U_1 := F_1(H_{f_1}), U_2 := F_2(H_{f_2})
END
```

resulting maximizing copula C_{F_1,F_2}. It is very interesting to observe that this copula appears quite pathological. It is highly margin-dependent and neither symmetric nor absolutely continuous. There is a significant singular component with total mass $p_\infty \approx 0.597$. In particular, this copula is typically not included in any financial toolbox, even though it should be because of its important meaning pointed out in the present article.

Conclusion
Using results from the coupling literature, it was shown how to compute explicitly the maximal probability for a joint default under given marginals. Moreover, this result was transferred into copula language, illustrating quite clearly how margin-dependent the problem is. Finally, a simple-to-implement simulation algorithm for the maximizing copula and an educational example were presented.

Acknowledgements The authors would like to thank Alfred Müller and an anonymous referee for valuable remarks.

References

1. den Hollander, W.T.F.: Probability Theory: Coupling. Lecture Notes Leiden University. Available at http://websites.math.leidenuniv.nl/probability/lecturenotes/CouplingLectures.pdf (2010)
2. Embrechts, P.: Copulas: a personal view. J. Risk Insur. **76**, 639–650 (2009)
3. Embrechts, P., Puccetti, G.: Bounds for functions of dependent risks. Financ. Stoch. **10**(3), 341–352 (2006)
4. Embrechts, P., Puccetti, G.: Bounds for functions of multivariate risks. J. Multivar. Anal. **97**(2), 526–547 (2006)
5. Embrechts, P., Höing, P.A., Puccetti, G.: Worst VaR scenarios. Insur. Math. Econ. **37**(1), 115–134 (2005)
6. Genest, C., Nešlehová, J.: A primer on copulas for count data. Astin Bull. **37**, 475–515 (2007)
7. Marshall, A.W.: Copulas, marginals and joint distributions. In: Rüschendorf, L., Schweizer, B., Taylor, M.D. (eds.) Distributions with Fixed Marginals and Related Topics, pp. 213–222. Institute of Mathematical Statistics, Hayward (1996)
8. Mikosch, T.: Copulas: tales and facts. Extremes **9**, 55–62 (2006)
9. Morini, M.: Mistakes in the market approach to correlation: a lesson for future stress-testing. In: Wehn, C.S., et al. (eds.) Rethinking Valuation and Pricing Models, pp. 331–359. Academic, Oxford (2012)
10. Puccetti, G., Rüschendorf, L.: Bounds for joint portfolios of dependent risks. Stat. Risk Model. **29**(2), 107–132 (2012)
11. Puccetti, G., Rüschendorf, L.: Computation of sharp bounds on the distribution of a function of dependent risks. J. Comput. Appl. Math. **236**(7), 1833–1840 (2012)
12. Puccetti, G., Rüschendorf, L.: Sharp bounds for sums of dependent risks. J. Appl. Probab. **50**(1), 42–53 (2013)

13. Puccetti, G., Rüschendorf, L.: Computation of sharp bounds on the expected value of a supermodular function of risks with given marginals. Working Paper (2013)
14. Schönbucher, P.J., Schubert, D.: Copula-dependent default risk in intensity models. Working Paper. Available at http://www.defaultrisk.com/pp_corr_22.htm (2001)
15. Sklar, M.: Fonctions de répartition à n dimensions et leurs marges. Publications de l'Institut de Statistique de L'Université de Paris **8**, 229–231 (1959)
16. Thorisson, H.: Coupling, Stationarity, and Regeneration. Springer, New York (2000)

Chapter 33
Optimization via Information Geometry

Luigi Malagò and Giovanni Pistone

33.1 Introduction

The present paper is based on the talk given by the second author on May 21, 2013, to the Seventh International Workshop on Simulation in Rimini. Some pieces of research that were announced in that talk have been subsequently published [19, 21, 22]. Here we give a general overview, references to latest published results, and a number of specific topics that have not been published elsewhere.

Let $(\Omega, \mathscr{F}, \mu)$ be a measure space, whose strictly positive probability densities form the algebraically open convex set $\mathscr{P}_>$. An *open statistical model* $(\mathscr{M}, \theta, B)$ is a parametrized subset of $\mathscr{P}_>$, that is, $\mathscr{M} \subset \mathscr{P}_>$ and $\theta \colon \mathscr{M} \to B$, where θ is a one-to-one mapping onto an open subset of a Banach space B. We assume in the following that Ω is endowed with a distance and $\mathscr{F}$ is its Borel σ-algebra.

If $f \colon \Omega \to \mathbb{R}$ is a bounded continuous function, the mapping $\mathscr{M} \ni p \mapsto \mathbb{E}_p[f]$ is a *Stochastic Relaxation* (SR) of f. The strict inequality $\mathbb{E}_p[f] < \sup_{\omega \in \Omega} f(\omega)$ holds for all $p \in \mathscr{M}$, unless f is constant. However, $\sup_{p \in \mathscr{M}} \mathbb{E}_p[f] = \sup_{\omega \in \Omega} f(\omega)$ if there exists a probability measure ν in the weak closure of $\mathscr{M} \cdot \mu$ whose support is contained in the set of maximizing points of f, that is to say

$$\nu \left\{ \omega \in \Omega \colon f(\omega) = \sup_{\omega \in \Omega} f(\omega) \right\} = 1, \quad \text{or} \quad \int f \, d\nu = \sup_{\omega \in \Omega} f(\omega).$$

L. Malagò
Dipartimento di Informatica, Università degli Studi di Milano, Via Comelico, 39/41, 20135 Milano, Italy
e-mail: malago@di.unimi.it

G. Pistone (✉)
de Castro Statistics, Collegio Carlo Alberto, Via Real Collegio 30, 10024 Moncalieri, Italy
e-mail: giovanni.pistone@carloalberto.org; giovanni.pistone@gmail.com

V.B. Melas et al. (eds.), *Topics in Statistical Simulation*, Springer Proceedings in Mathematics & Statistics 114, DOI 10.1007/978-1-4939-2104-1_33

Such a ν belongs to the border of $\mathcal{M} \cdot \mu$. For a discussion of the border issue for finite Ω, see [14]. Other relaxation methods have been considered, e.g., [4,25].

An *SR optimization method* is an algorithm producing a sequence $p_n \in \mathcal{M}$, $n \in \mathbb{N}$, which is expected to converge to the probability measure ν, so that $\lim_{n\to\infty} \mathbb{E}_{p_n}[f] = \sup_{\omega\in\Omega} f(\omega)$. Such algorithms are best studied in the framework of *Information Geometry* (IG), that is, the differential geometry of statistical models. See [3] for a general treatment of IG and [4,6,13,16–19] for applications to SR. All the quoted literature refers to the case where the model Banach space of the statistical manifold, i.e., the parameter space, is finite dimensional, $B = \mathbb{R}^d$. An infinite dimensional version of IG has been developed, see [22] for a recent presentation together with new results, and the references therein for a detailed bibliography. The nonparametric version is unavoidable in applications to evolution equations in Physics [21], and it is useful even when the sample space is finite [15].

33.2 Stochastic Relaxation on an Exponential Family

We recall some basic facts on exponential families, see [8].

1. The exponential family $q_\theta = \exp\left(\sum_{j=1}^d \theta_j T_j - \psi(\theta)\right) \cdot p$, $\mathbb{E}_p\left[T_j\right] = 0$, is a statistical model $\mathcal{M} = \{q_\theta\}$ with parametrization $q_\theta \mapsto \theta \in \mathbb{R}^d$.
2. $\psi(\theta) = \log\left(\mathbb{E}_p\left[e^{\theta\cdot T}\right]\right)$, $\theta \in \mathbb{R}^d$, is convex and lower semi-continuous.
3. ψ is analytic on the (nonempty) interior $\mathcal{U}$ of its proper domain.
4. $\nabla\psi(\theta) = \mathbb{E}_\theta[T]$, $T = (T_1, \ldots, T_d)$.
5. $\operatorname{Hess}\psi(\theta) = \operatorname{Var}_\theta(T)$.
6. $\mathcal{U} \ni \theta \mapsto \nabla\psi(\theta) = \eta \in \mathcal{N}$ is one-to-one, analytic, and monotone; $\mathcal{N}$ is the interior of the *marginal polytope*, i.e., the convex set generated by $\{T(\omega) \colon \omega \in \Omega\}$.
7. The gradient of the SR of f is
$$\nabla(\theta \mapsto \mathbb{E}_\theta[f]) = (\operatorname{Cov}_\theta(f, T_1), \ldots, \operatorname{Cov}_\theta(f, T_d)),$$
which suggests to take the least squares approximation of f on $\operatorname{Span}(T_1, \ldots, T_d)$ as direction of *steepest ascent*, see [18].
8. The representation of the gradient in the scalar product with respect to θ is called *natural gradient*, see [2,3,15].

Different methods can be employed to generate a maximizing sequence of densities p_n is a statistical model $\mathcal{M}$. A first example is given by Estimation of Distribution Algorithms (EDAs) [12], a large family of iterative algorithms where the parameters of a density are estimated after sampling and selection, in order to favor samples with larger values for f, see Example 1. Another approach is to evaluate the gradient of $\mathbb{E}_p[f]$ and follow the direction of the natural gradient over $\mathcal{M}$, as illustrated in Example 2.

Example 1 (EDA from [19]). An *Estimation of Distribution Algorithm* is an SR optimization algorithm based on sampling, selection, and estimation, see [12].

Input: N, M ▷ population size, selected population size
Input: $\mathcal{M} = \{p(x;\xi)\}$ ▷ parametric model
$t \leftarrow 0$
$\mathcal{P}^t = $ INITRANDOM() ▷ random initial population
repeat
$\mathcal{P}_s^t = $ SELECTION($\mathcal{P}^t, M$) ▷ select M samples
$\xi^{t+1} = $ ESTIMATION($\mathcal{P}_s^t, \mathcal{M}$) ▷ opt. model selection
$\mathcal{P}^{t+1} = $ SAMPLER(ξ^{t+1}, N) ▷ N samples
$t \leftarrow t + 1$
until STOPPINGCRITERIA()

Example 2 (SNGD from [19]). *Stochastic Natural Gradient Descent* [18] is an SR algorithm that requires the estimation of the gradient.

Input: N, λ ▷ population size, learning rate
Optional: M ▷ selected population size (default $M = N$)
$t \leftarrow 0$
$\theta^t \leftarrow (0, \ldots, 0)$ ▷ uniform distribution
$\mathcal{P}^t \leftarrow$ INITRANDOM() ▷ random initial population
repeat
$\mathcal{P}_s^t = $ SELECTION($\mathcal{P}^t, M$) ▷ opt. select M samples
$\hat{\nabla}\mathbb{E}[f] \leftarrow \widehat{\mathrm{Cov}}(f, T_i)_{i=1}^d$ ▷ empirical covariances
$\hat{I} \leftarrow [\widehat{\mathrm{Cov}}(T_i, T_j)]_{i,j=1}^d$ ▷ $\{T_i(x)\}$ may be learned
$\theta^{t+1} \leftarrow \theta^t - \lambda \hat{I}^{-1}\hat{\nabla}\mathbb{E}[f]$
$\mathcal{P}^{t+1} \leftarrow$ GIBBSSAMPLER(θ^{t+1}, N) ▷ N samples
$t \leftarrow t + 1$
until STOPPINGCRITERIA()

Finally, other algorithms are based on Bregman divergence. Example 3 illustrates the connection with the exponential family.

Example 3 (Binomial $B(n, p)$). On the finite sample space $\Omega = \{0, \ldots, n\}$ with $\mu(x) = \binom{n}{x}$, consider the exponential family $p(x;\theta) = \exp\left(\theta x - n\log\left(1 + \mathrm{e}^\theta\right)\right)$. With respect to the expectation parameter $\eta = n\mathrm{e}^\theta/(1 + \mathrm{e}^\theta) \in]0, n[$ we have $p(x;\eta) = (\eta/n)^x(1 - \eta/n)^{n-x}$, which is the standard presentation of the binomial density.

The standard presentation is defined for $\eta = 0, n$, where the exponential formula is not. In fact, the conjugate $\psi_*(\eta)$ of $\psi(\theta) = n\log\left(1 + \mathrm{e}^\theta\right)$ is

$$\psi_*(\eta) = \begin{cases} +\infty & \text{if } \eta < 0 \text{ or } \eta > n, \\ 0 & \text{if } \eta = 0, n, \\ \eta\log\left(\frac{\eta}{n-\eta}\right) - n\log\left(\frac{n}{n-\eta}\right) & \text{if } 0 < \eta < n. \end{cases}$$

We have

$$\log p(x;\eta) = \log\left(\frac{\eta}{n-\eta}\right)(x-\eta) + \psi_*(\eta), \quad \eta \in]0,n[$$
$$= \psi_*'(\eta)(x-\eta) + \psi_*(\eta) \leqslant \psi_*(x).$$

For $x \neq 0, n$, the sign of $\psi_*'(\eta)(x-\eta)$ is eventually negative as $\eta \to 0, n$, hence

$$\lim_{\eta\to 0,n} \log p(x;\eta) = \lim_{\eta\to 0,n} \psi_*'(\eta)(x-\eta) + \psi_*(\eta) = -\infty.$$

If $x = 0, n$, the sign of both $\psi_*'(\eta)(0-\eta)$ and $\psi_*'(\eta)(n-\eta)$ is eventually positive as $\eta \to 0$ and $\eta \to n$, respectively. The limit is bounded by $0 = \psi_*(x)$, for $x = 0, n$.

The argument above is actually general. It has been observed by [5] that the Bregman divergence $D_{\psi_*}(x\|\eta) = \psi_*(x) - \psi_*(\eta) - \psi_*'(\eta)(x-\eta) \geqslant 0$ provides an interesting form of the density as $p(x;\eta) = \mathrm{e}^{-D_{\psi_*}(x\|\eta)}\mathrm{e}^{\psi_*(x)} \propto \mathrm{e}^{-D_{\psi_*}(x\|\eta)}$.

33.3 Exponential Manifold

The set of positive probability densities $\mathscr{P}_>$ is a convex subset of $L^1(\mu)$. Given a $p \in \mathscr{P}_>$, every $q \in \mathscr{P}_>$ can be written as $q = \mathrm{e}^v \cdot p$ where $v = \log\left(\frac{q}{p}\right)$. Below we summarize, together with a few new details, results from [21,22] and the references therein, and the unpublished [24].

Definition 33.1 (Orlicz Φ-Space [11], [20, Chapter II], [23]). Define $\varphi(y) = \cosh y - 1$. The Orlicz Φ-space $L^\Phi(p)$ is the vector space of all random variables such that $\mathbb{E}_p[\Phi(\alpha u)]$ is finite for some $\alpha > 0$. Equivalently, it is the set of all random variables u whose Laplace transform under $p \cdot \mu$, $t \mapsto \hat{u}_p(t) = \mathbb{E}_p[\mathrm{e}^{tu}]$ is finite in a neighborhood of 0. We denote by $M^\Phi(p) \subset L^\Phi(p)$ the vector space of random variables whose Laplace transform is always finite.

Proposition 33.1 (Properties of the Φ-Space).

1. *The set $S_{\leqslant 1} = \{u \in L^\Phi(p) \colon \mathbb{E}_p[\Phi(u)] \leqslant 1\}$ is the closed unit ball of the complete norm*

$$\|u\|_p = \inf\left\{\rho > 0 \colon \mathbb{E}_p\left[\Phi\left(\frac{u}{\rho}\right)\right] \leqslant 1\right\}$$

on the Φ-space. For all $a \geqslant 1$ the continuous injections $L^\infty(\mu) \hookrightarrow L^\Phi(p) \hookrightarrow L^a(p)$ hold.

2. *$\|u\|_p = 1$ if either $\mathbb{E}_p[\Phi(u)] = 1$ or $\mathbb{E}_p[\Phi(u)] < 1$ and $\mathbb{E}_p\left[\Phi\left(\frac{u}{\rho}\right)\right] = \infty$ for $\rho > 1$. If $\|u\|_p > 1$, then $\|u\|_p \leqslant \mathbb{E}_p[\Phi(u)]$. In particular, $\lim_{\|u\|_p \to \infty} \mathbb{E}_p[\Phi(u)] = \infty$.*
3. *$M^\Phi(p)$ is a closed and separable subspace of $L^\Phi(p)$.*
4. *$L^\Phi(p) = L^\Phi(q)$ as Banach spaces if, and only if, $\int p^{1-\theta} q^\theta \, d\mu$ is finite on a neighborhood of $[0,1]$.*

Proof. 1. See [11], [20, Chapter II], [23].
2. The function $\mathbb{R}_\geqslant \ni \alpha \mapsto \hat{u}(t) = \mathbb{E}_p[\Phi(\alpha u)]$ is increasing, convex, lower semi-continuous. If for some $t_+ > 1$ the value $\hat{u}(t_+)$ is finite, we are in the first case and $\hat{u}(1) = 1$. Otherwise, we have $\hat{u}(1) \leqslant 1$. If $\|u\|_p > a > 1$, so that $\left\|\frac{a}{\|u\|_p} u\right\|_p > 1$, hence

$$1 < \mathbb{E}_p\left[\Phi\left(\frac{a}{\|u\|_p} u\right)\right] \leqslant \frac{a}{\|u\|_p} \mathbb{E}_p[\Phi(u)],$$

and $\|u\|_p < a\mathbb{E}_p[\Phi(u)]$, for all $a > 1$.
3. See [11], [20, Chapter II], [23].
4. See [9, 24].

Example 4 (Boolean State Space). In the case of a finite state space, the moment generating function is finite everywhere, but its computation can be challenging. We discuss in particular the Boolean case $\Omega = \{+1, -1\}^n$ with counting reference measure μ and uniform density $p(x) = 2^{-n}$, $x \in \Omega$. In this case there is a huge literature from statistical physics, e.g., [10, Ch. VII]. A generic real function on Ω—called pseudo-Boolean [7] in the combinatorial optimization literature—has the form $u(x) = \sum_{\alpha \in L} \hat{u}(\alpha) x^\alpha$, with $L = \{0,1\}^n$, $x^\alpha = \prod_{i=1}^n x_i^{\alpha_i}$, $\hat{u}(\alpha) = 2^{-n} \sum_{x \in \Omega} u(x) x^\alpha$.

As $\mathrm{e}^{ax} = \cosh(a) + \sinh(a)x$ if $x^2 = 1$ i.e., $x = \pm 1$, we have

$$\begin{aligned}
\mathrm{e}^{tu(x)} &= \exp\left(\sum_{\alpha \in \operatorname{Supp} \hat{u}} t\hat{u}(\alpha) x^\alpha\right) = \prod_{\alpha \in \operatorname{Supp} \hat{u}} \mathrm{e}^{t\hat{u}(\alpha) x^\alpha} \\
&= \prod_{\alpha \in \operatorname{Supp} \hat{u}} (\cosh(t\hat{u}(\alpha)) + \sinh(t\hat{u}(\alpha)) x^\alpha) \\
&= \sum_{B \subset \operatorname{Supp} \hat{u}} \prod_{\alpha \in B^c} \cosh(t\hat{u}(\alpha)) \prod_{\alpha \in B} \sinh(t\hat{u}(\alpha)) x^{\sum_{\alpha \in B} \alpha}.
\end{aligned}$$

The moment generating function of u under the uniform density p is

$$t \mapsto \sum_{B \in \mathscr{B}(\hat{u})} \prod_{\alpha \in B^c} \cosh(t\hat{u}(\alpha)) \prod_{\alpha \in B} \sinh(t\hat{u}(\alpha)),$$

where $\mathscr{B}(\hat{u})$ are those $B \subset \operatorname{Supp} \hat{u}$ such that $\sum_{\alpha \in B} \alpha = 0 \mod 2$. We have

$$\mathbb{E}_p[\Phi](tu) = \sum_{B \in \mathscr{B}_0(\hat{u})} \prod_{\alpha \in B^c} \cosh(t\hat{u}(\alpha)) \prod_{\alpha \in B} \sinh(t\hat{u}(\alpha)) - 1,$$

where $\mathscr{B}_0(\hat{u})$ are those $B \subset \operatorname{Supp} \hat{u}$ such that $\sum_{\alpha \in B} \alpha = 0 \mod 2$ and $\sum_{\alpha \in \operatorname{Supp} \hat{u}} \alpha = 0$.

If S is the $\{1, \dots, n\} \times \operatorname{Supp} \hat{u}$ matrix with elements α_i, we want to solve the system $Sb = 0 \mod 2$ to find all elements of $\mathscr{B}$; we add the equation $\sum b = 0 \mod 2$ to find $\mathscr{B}_0$. The simplest example is $u(x) = \sum_{i=1}^n c_i x_i$,

Example 5 (The Sphere is Not Smooth in General). We look for the moment generating function of the density

$$p(x) \propto (a + x)^{-\frac{3}{2}} \mathrm{e}^{-x}, \quad x > 0,$$

where a is a positive constant. From the incomplete gamma integral

$$\gamma - \frac{1}{2} x = \int_x^\infty s^{-\frac{1}{2}-1} \mathrm{e}^{-s} \, ds, \quad x > 0,$$

we have for $\theta, a > 0$,

$$\frac{d}{dx} \Gamma\left(-\frac{1}{2}, \theta(a + x)\right) = -\theta^{-\frac{1}{2}} \mathrm{e}^{-\theta a} (a + x)^{-\frac{3}{2}} \mathrm{e}^{-\theta x}.$$

We have, for $\theta \in \mathbb{R}$,

$$C(\theta, a) = \int_0^\infty (a + x)^{-\frac{3}{2}} \mathrm{e}^{-\theta x} \, dx = \begin{cases} \sqrt{\theta} \mathrm{e}^{\theta a} \Gamma\left(-\frac{1}{2}, \theta a\right) & \text{if } \theta > 0. \\ \frac{1}{2\sqrt{a}} & \text{if } \theta = 0, \\ +\infty & \text{if } \theta < 0. \end{cases}$$

or, $C(\theta, a) = \frac{1}{2} a^{-\frac{1}{2}} - \frac{\sqrt{\pi}\theta}{2} \mathrm{e}^{\theta a} R_{1/2,1}(\theta a)$ if $\theta \leqslant 1$, $+\infty$ otherwise, where $R_{1/2,1}$ is the survival function of the Gamma distribution with shape $1/2$ and scale 1.

The density p is obtained with $\theta = 1$,

$$p(x) = C(1, a)^{-1} (a + x)^{-\frac{3}{2}} \mathrm{e}^{-x} = \frac{(a + x)^{-\frac{3}{2}} \mathrm{e}^{-x}}{\mathrm{e}^a \Gamma\left(-\frac{1}{2}, a\right)}, \quad x > o,$$

and, for the random variable $u(x) = x$, the function

$$\alpha \mapsto \mathbb{E}_p[\Phi(\alpha u)] = \frac{1}{\mathrm{e}^a \Gamma\left(-\frac{1}{2}, a\right)} \int_0^\infty (a + x)^{-\frac{3}{2}} \frac{\mathrm{e}^{-(1-\alpha)x} + \mathrm{e}^{-(1+\alpha)x}}{2} \, dx - 1$$

$$= \frac{C(1-\alpha, a) + C(1+\alpha, a)}{2C(1,a)} - 1$$

is convex lower semi-continuous on $\alpha \in \mathbb{R}$, finite for $\alpha \in [-1, 1]$, infinite otherwise, hence not steep. Its value at $\alpha = 1$ is

$$\begin{aligned}\mathbb{E}_p[\Phi(u)] &= \frac{1}{e^a \Gamma\left(-\frac{1}{2}, a\right)} \int_0^\infty (a+x)^{-\frac{3}{2}} \frac{1+e^{-2x}}{2}\, dx - 1 \\ &= \frac{C(0,a) + C(2,a)}{2C(1,a)} - 1\end{aligned}$$

Example 6 (Normal Density). Let $p(x) = (2\pi)^{-1/2} e^{-(1/2)x^2}$. Consider a generic quadratic polynomial $u(x) = a + bx + \frac{1}{2}cx^2$. We have for $tc \neq 1$

$$t(a+bx+\frac{1}{2}cx^2) - \frac{1}{2}x^2 == -\frac{1}{2(1-tc)^{-1}}\left(x - \frac{tb}{1-tc}\right)^2 + \frac{1}{2}\frac{t^2b^2 - 2ta(1-tc)}{(1-tc)},$$

hence

$$\mathbb{E}_p\left[e^{tu}\right] = \begin{cases} +\infty & \text{if } tc \leqslant 1, \\ \sqrt{1-tc}\exp\left(\frac{1}{2}\frac{t^2b^2 - 2ta(1-tc)}{(1-tc)}\right) & \text{if } tc < 1. \end{cases}$$

If, and only if, $-1 < c < 1$, we have

$$\begin{aligned}\mathbb{E}_p[\Phi(u)] &= \frac{1}{2}\sqrt{1-c}\exp\left(\frac{1}{2}\frac{b^2 - a(1-c)}{(1-c)}\right) \\ &\quad + \frac{1}{2}\sqrt{1+c}\exp\left(\frac{1}{2}\frac{b^2 - a(1+c)}{(1+c)}\right) - 1.\end{aligned}$$

33.4 Vector Bundles

Vector bundles are constructed as sets of couples (p, v) with $p \in \mathscr{P}_>$ and v is some space of random variables such that $\mathbb{E}_p[v] = 0$. The tangent bundle is obtained when the vector space is $L_0^\Phi(p)$. The Hilbert bundle is defined as $H\mathscr{P}_> = \left\{(p, v) \colon p \in \mathscr{P}_>, v \in L_0^2(p)\right\}$. We refer to [21] and [15] were charts and affine connections on the Hilbert bundle are derived from the isometric transport

$$L_0^2(p) \ni u \mapsto \sqrt{\frac{p}{q}}u - \left(1 + \mathbb{E}_q\left[\sqrt{\frac{p}{q}}\right]\right)^{-1}\left(1 + \sqrt{\frac{p}{q}}\right)\mathbb{E}_q\left[\sqrt{\frac{p}{q}}u\right] \in L_0^2(q).$$

In turn, an isometric transport $U_p^q\colon L_0^2(p) \to L_0^2(q)$ can be used to compute the derivative of a vector field in the Hilbert bundle, for example the derivative of the gradient of a relaxed function.

The resulting second order structure is instrumental in computing the Hessian of the natural gradient of the SR function. This allows to design a second order approximation method, as it is suggested in [1] for general Riemannian manifolds, and applied to SR in [15]. A second order structure is also used to define the curvature of a statistical manifold and, possibly, to compute its geodesics, see [6] for applications to optimization.

References

1. Absil, P.A., Mahony, R., Sepulchre, R.: Optimization algorithms on matrix manifolds. Princeton University Press, Princeton (2008). With a foreword by Paul Van Dooren
2. Amari, S.I.: Natural gradient works efficiently in learning. Neural Comput. **10**(2), 251–276 (1998)
3. Amari, S., Nagaoka, H.: Methods of information geometry. American Mathematical Society, Providence (2000). Translated from the 1993 Japanese original by Daishi Harada
4. Arnold, L., Auger, A., Hansen, N., Ollivier, Y.: Information-Geometric Optimization Algorithms: A Unifying Picture via Invariance Principles (2011v1; 2013v2). ArXiv:1106.3708
5. Banerjee, A., Merugu, S., Dhillon, I.S., Ghosh, J.: Clustering with bregman divergences. J. Mach. Learn. Res. **6**, 1705–1749 (2005)
6. Bensadon, J.: Black-box optimization using geodesics in statistical manifolds. ArXiv:1309.7168
7. Boros, E., Hammer, P.L.: Pseudo-Boolean optimization. Discrete Appl. Math. **123**(1–3), 155–225 (2002). Workshop on Discrete Optimization, DO'99 (Piscataway, NJ) (2013 v1; 2013v2)
8. Brown, L.D.: Fundamentals of statistical exponential families with applications in statistical decision theory. No. 9 in IMS Lecture Notes. Monograph Series. Institute of Mathematical Statistics (1986)
9. Cena, A., Pistone, G.: Exponential statistical manifold. Ann. Inst. Stat. Math. **59**(1), 27–56 (2007)
10. Gallavotti, G.: Statistical Mechanics: A Short Treatise. Texts and Monographs in Physics. Springer, Berlin (1999)
11. Krasnosel'skii, M.A., Rutickii, Y.B.: Convex Functions and Orlicz Spaces. Noordhoff, Groningen (1961). Russian original: (1958) Fizmatgiz, Moskva
12. Larrañaga, P., Lozano, J.A. (eds.): Estimation of Distribution Algoritms. A New Tool for evolutionary Computation. Genetic Algorithms and Evolutionary Computation, vol. 2. Springer, New York (2001)
13. Malagò, L.: On the geometry of optimization based on the exponential family relaxation. Ph.D. thesis, Politecnico di Milano (2012)
14. Malagò, L., Pistone, G.: A note on the border of an exponential family. ArXiv:1012.0637v1 (2010)
15. Malagò, L., Pistone, G.: Combinatorial Optimization with Information Geometry: Newton Method. Entropy **16**(8), 4260–4289 (2014)
16. Malagò, L., Matteucci, M., Pistone, G.: Stochastic relaxation as a unifying approach in 0/1 programming. In: NIPS 2009 Workshop on Discrete Optimization in Machine Learning: Submodularity, Sparsity & Polyhedra (DISCML), Whistler, 11 Dec 2009
17. Malagò, L., Matteucci, M., Pistone, G.: Stochastic natural gradient descent by estimation of empirical covariances. In: Proceedings of IEEE CEC, pp. 949–956 (2011)

18. Malagò, L., Matteucci, M., Pistone, G.: Towards the geometry of estimation of distribution algorithms based on the exponential family. In: Proceedings of the 11th Workshop on Foundations of Genetic Algorithms, FOGA '11, pp. 230–242. ACM, New York (2011)
19. Malagò, L., Matteucci, M., Pistone, G.: Natural gradient, fitness modelling and model selection: a unifying perspective. In: Proceedings of IEEE CEC, pp. 486–493 (2013)
20. Musielak, J.: Orlicz Spaces and Modular Spaces. Lecture Notes in Mathematics, vol. 1034. Springer, Berlin (1983)
21. Pistone, G.: Examples of application of nonparametric information geometry to statistical physics. Entropy **15**(10), 4042–4065 (2013)
22. Pistone, G.: Nonparametric information geometry. In: Nielsen, F., Barbaresco, F. (eds.) Geometric Science of Information. LNCS, vol. 8085, pp. 5–36. Springer, Berlin/Heidelberg (2013). GSI 2013 Paris, August 28–30, 2013 Proceedings
23. Rao, M.M., Ren, Z.D.: Applications of Orlicz Spaces. Monographs and Textbooks in Pure and Applied Mathematics, vol. 250. Marcel Dekker, New York (2002)
24. Santacroce, M., Siri, P., Trivellato, B.: New results on mixture and exponential models by Orlicz spaces (2014, submitted)
25. Wierstra, D., Schaul, T., Peters, J., Schmidhuber, J.: Natural evolution strategies. In: Proceedings of IEEE CEC, pp. 3381–3387 (2008)

Chapter 34
Combined Nonparametric Tests for the Social Sciences

Marco Marozzi

34.1 Introduction

Non-normal data are common in social and psychological studies, as emphasized by Nanna and Sawilowsky [12] normality is the exception rather than the rule in applied research. Micceri [11] considered 440 data sets from psychological/social studies and concluded that none of them satisfied the normality assumptions, see also [2, 14, 15]. Moreover, social studies may have small sample sizes. These arguments are against parametric tests and in favor of nonparametric tests that are generally valid, robust, and powerful in situations where parametric tests are not [13]. In particular, in this chapter we consider permutation tests because they are particularly suitable for combined testing. Moreover, they do not even require random sampling, only exchangeability of observations between samples under the null hypothesis that the parent distributions are the same. It is important to emphasize that permutation testing is valid even when a non-random sample of n units is randomized into two groups to be compared. This circumstance is very common in social and biomedical studies, see [4].

Many nonparametric tests have been developed for comparing the distribution functions of two populations. These tests may be classified into four main classes: (1) tests for detecting mean/median differences, see, e.g., [5]; (2) tests for detecting variability differences, see, e.g., [7, 8]; (3) tests for jointly detecting mean/median and variability differences, see, e.g., [6]; (4) tests for detecting any differences between the distributions, see, e.g., [16].

The references listed above show that nonparametric combined tests have been very useful to address comparison problems of several types. Combined testing is

M. Marozzi (✉)
University of Calabria, Rende (CS), Italy
e-mail: marco.marozzi@unical.it

V.B. Melas et al. (eds.), *Topics in Statistical Simulation*, Springer Proceedings in Mathematics & Statistics 114, DOI 10.1007/978-1-4939-2104-1_34

an effective strategy because generally non-combined tests show good performance only for particular distributions. Since in many social studies there is no clear knowledge about the parent distribution, the problem of which test should be selected in practice arises. Our aim is to see whether nonparametric combined tests are useful also for detecting any differences (in means/medians, variability, shape) between distributions. We aim at proposing a test that even though was not the most powerful one for every distribution, it has good overall performance under every type of distribution, a combined test that inherits the good behavior shown by a certain number of single tests in particular situations.

Let $\mathbf{X}_i = (X_{i1}, \ldots, X_{in_i})$ be a random sample from a population with continuous distribution function $F_i(x)$, $i = 1, 2$, $n = n_1 + n_2$. Let $\mathbf{X} = (\mathbf{X}_1, \mathbf{X}_2)'$ be the combined sample and let $\mathbf{X}^* = (\mathbf{X}_1^*, \mathbf{X}_2^*)'$ be one of the $B = n!$ permutations of $\mathbf{X}$ with $\mathbf{X}_i^* = (X_{i1}^*, \ldots, X_{in_i}^*)$, $i = 1, 2$. Note that $F_i(x)$ is completely unknown, $i = 1, 2$. We would like to test the null hypothesis

$$H_0 : F_1(x) = F_2(x) \text{ for all } x \in (-\infty, \infty)$$

against the alternative hypothesis

$$H_1 : F_1(x) \neq F_2(x) \text{ for some } x \in (-\infty, \infty).$$

Traditional tests for the general two-sample problem are the Kolmogorov–Smirnov, Cramer Von Mises, and Anderson Darling tests. Zhang [16] proposed a unified approach that generates not only the traditional tests but also new nonparametric tests based on the likelihood ratio. We consider the Zhang tests that are analog to the traditional tests. The test statistics are

$$S_1 = -\frac{1}{n} \sum_{i=1}^{2} \sum_{j=1}^{n_i} \log\left(\frac{n_i}{j - 0.5} - 1\right) \log\left(\frac{n}{R_{ij} - 0.5} - 1\right) \tag{34.1}$$

which is the analog of the Cramer–Von Mises statistic and where R_{ij} denotes the rank of X_{ij}, $j = 1, \ldots, n_i$ in $\mathbf{X}$ in increasing order,

$$S_2 = \sum_{l=1}^{n} \sum_{i=1}^{2} n_i \frac{F_{il} \log F_{il} + (1 - F_{il}) \log(1 - F_{il})}{(l - 0.5)(n - l + 0.5)} \tag{34.2}$$

which is the analog of the Anderson–Darling statistic and where $F_{il} = \hat{F}_i(X_{(l)})$, $X(l)$ is the lth order statistic of the pooled sample, $l = 1, \ldots, n$, $\hat{F}_i$ is the empirical distribution function of the ith sample with correction at its discontinuous points so that $F_{il} = (j - 0.5)/n_i$ if $l = R_{ij}$ for some j or $F_{il} = j/n_i$ if $R_{ij} < l < R_{ij+1}$, with $R_{i0} = 1$ and $R_{in_i+1} = n + 1$,

$$S_3 = \max_{1 \leqslant l \leqslant n} \left[\sum_{i=1}^{2} n_i \left(F_{il} \log \frac{F_{il}}{F_l} + (1 - F_{il}) \log \frac{1 - F_{il}}{1 - F_l} \right) \right] \tag{34.3}$$

which is the analog of the Kolmogorov–Smirnov statistic and where $F_l = \hat{F}(X_{(l)})$, $\hat{F}$ is the empirical distribution function of the pooled sample with correction at its discontinuous points $X(l)$, $l = 1, \ldots, n$ so that $F_l = (l - 0.5)/n$. Note that large values of the statistics speak against the null hypothesis.

34.2 Combined Tests for the General Problem

A generic combined test statistic for the general two-sample problem is defined as $T_\psi = \psi(\mathbf{T})$ where ψ is a proper combining function, $\mathbf{T} = (T_1, \ldots, T_K)'$, $T_k = \frac{|S_k - E(S_k)|}{\sqrt{\mathrm{VAR}(S_k)}}$, S_k is a two-sided test statistic for the general two-sample problem whose large values speak against H_0, $E(S_k)$ and $\mathrm{VAR}(S_k)$ are, respectively, the mean and the variance of S_k, $k = 1, \ldots, K$, K is a natural number with $2 \leqslant K < \infty$. To be a proper combining function, ψ should satisfy some reasonable properties, see [13, pp. 123–124]. To keep things (relatively) simple, we do not follow the two-step procedure presented by Pesarin and Salmaso [13] based on p-value combination. Combinations of p-values are useful when a direct combining test is not easily available or difficult to justify. They are used also in different contexts than that considered here, like the multi-stage one [10]. Note that under H_0, $\mathbf{X}$ elements are exchangeable between samples and all permutations $\mathbf{X}^*$ are equally likely. Therefore permutation testing is appropriate. Let ${}_oT_\psi = T_\psi(\mathbf{X}_1, \mathbf{X}_2) = \psi({}_o\mathbf{T})$ denote the observed value of the combined test statistic where ${}_o\mathbf{T} = ({}_oT_1, \ldots, {}_oT_K)'$,

$${}_oT_k = T_k(\mathbf{X}_1, \mathbf{X}_2) = \frac{|{}_oS_k - E(S_k)|}{\sqrt{\mathrm{VAR}(S_k)}}$$

is the observed value of the standardized kth test statistic, ${}_oS_k = S_k(\mathbf{X}_1, \mathbf{X}_2)$ is the observed value of the non-standardized kth test statistic, $E(S_k) = \frac{1}{B}\sum_{b=1}^{B} {}_bS_k$, $\mathrm{VAR}(S_k) = \frac{1}{B}\sum_{b=1}^{B} ({}_bS_k - E(S_k))^2$, ${}_bS_k = S_k({}_b\mathbf{X}_1^*, {}_b\mathbf{X}_2^*)$ is the permutation value of S_k in the bth permutation ${}_b\mathbf{X}^* = ({}_b\mathbf{X}_1^*, {}_b\mathbf{X}_2^*)$ of $\mathbf{X}$, $b = 1, \ldots, B$. The p-value of the combined test T_ψ is the proportion of permutations of $\mathbf{X}$ which lead to values of the combined test statistic which are greater than or equal to ${}_oT_\psi$: $L_{T_\psi} = \frac{1}{B}\sum_{b=1}^{B} I({}_bT_\psi \geqslant {}_oT_\psi)$ where $I(.)$ denotes the indicator function and ${}_bT_\psi = T_\psi({}_b\mathbf{X}_1^*, {}_b\mathbf{X}_2^*) = \psi({}_b\mathbf{T})$ is the permutation value of T_ψ in ${}_b\mathbf{X}^*$, where ${}_b\mathbf{T} = ({}_bT_1, \ldots, {}_bT_K)'$ and

$${}_bT_k = T_k({}_b\mathbf{X}_1^*, {}_b\mathbf{X}_2^*) = \frac{|{}_bS_k - E(S_k)|}{\sqrt{\mathrm{VAR}(S_k)}}.$$

The optimal combining function is the one which corresponds to the uniformly most powerful combined test. Unfortunately, within the nonparametric framework, generally there exists the most powerful test only for particular parent distributions and for simple system of hypotheses. Birnbaum [1] showed that no method for combining independent tests is optimal in general and a fortiori this is also true when combining dependent tests. Note that T_k tests are very likely to be dependent because they test the same null hypothesis against the same alternative hypothesis and the corresponding test statistics are functions of the same data.

We consider two combining functions: the direct one and the one based on Mahalanobis distance which correspond to take as combined test statistic observed values, respectively

$$_oT_D = D(_o\mathbf{T}) = \sum_{k=1}^{K} {_oT_k}$$

and

$$_oT_M = M(_o\mathbf{T}) = {_o\mathbf{T}'}\mathbf{H}^{-1}{_o\mathbf{T}}$$

where $\mathbf{H} = [E(T_k T_h), k, h = 1, \ldots, K]$ is the correlation matrix between T_k, $k = 1, \ldots, K$, with $E(T_k T_h) = \frac{1}{B}\sum_{b=1}^{B} {_bT_k} \cdot {_bT_h}$.

In place of the direct combination, you might use the quadratic direct combination which corresponds to $_aT_{\text{QD}} = \text{QD}(_a\mathbf{T}) = \sum_{k=1}^{K} {_aT_k^2}$ with $a = o, b$, respectively, for the observed and permutation value of the combined test statistic. We do not consider T_{QD} because [13] shows that its power function is very close to that of T_M in all conditions. Note that T_{QD} and T_M are equivalent when $\mathbf{H}$ is the identity matrix or its elements are all 1. The maximum difference is in the intermediate situation between linearly independent T_ks and perfectly concordant T_ks. If $K = 2$, this happens when $E(T_k T_h) = \frac{1}{2}$. In general, different combining functions lead to different combined tests but the corresponding tests are asymptotically equivalent in the alternative [13]. It is impossible to find the best combination for any given testing problem without restrictions on the class of proper combining functions. Only locally optimal combinations can sometimes be obtained. If the practitioner finds the choice of the combining function too arbitrary, the combination procedure may be iterated, see [13, p. 133].

The null distribution of T_ψ can be obtained by computing T_ψ in all B permutations of $\mathbf{X}$. If T_ks were not affected by the order of elements within the two-samples (note that this can be assumed without much loss of generality) B reduces to the number of equally likely combinations of n_1 elements taken from n elements $C = n!/(n_1!n_2!)$. Although $C << B$, C increases very rapidly as n_1 and n_2 increase. Therefore in practice it might be computationally hard to compute the complete null distribution of T_ψ but we can rely upon an approximate but reasonably accurate estimate by considering a random sample of several thousands permutations of $\mathbf{X}$. See the next section for more details on the approximation error.

34.3 Comparison Study

In this section, we study size and power of some combined tests: T_{D12}, T_{D13}, and T_{D23} obtained by combining two tests via direct combination; T_{M12}, T_{M13}, and T_{M23} obtained by combining two tests via Mahalanobis combination; T_{D123} and T_{M123} obtained by combining all tests by direct and Mahalanobis combination, respectively. Note that [16] did not study the type-one error rate of his tests. This study is performed here.

It is very difficult to derive theoretical optimality properties for nonparametric tests with completely unknown distributions of the populations behind the samples. Therefore to study and compare type-one error rate and power of the tests we rely upon Monte Carlo simulation. We consider 10,000 Monte Carlo simulations. All tests are performed at the 0.05 nominal significance level. When a probability p is approximated by a proportion out of MC Monte Carlo replications, the error is of order $\sqrt{p(1-p)/\mathrm{MC}}$ when true test p-values are computed by considering all B permutations of $\mathbf{X}$. If this is not practical, the p-values are themselves estimated and the error is higher than before. We suggest to use 800 permutations when $\mathrm{MC} = 10{,}000$ that corresponds to an error of order $1.2\sqrt{p(1-p)/\mathrm{MC}}$. Therefore when a rejection probability close to 0.05 is approximated, the error is 0.00262. The error is maximum for $p = 0.5$ and it is 0.006. Note that we consider 1,000 permutations, that are as computational feasible as 800, to estimate $E(S_k), \mathrm{VAR}(S_k), E(T_k T_h)$ and the p-values of the tests (both non-combined and combined). Three situations are considered: (1) $N(0,1)$ vs $N(\mu,\sigma^2)$, (2) $N(\mu,\sigma^2)$ vs $G(r,1)$, (3) $U(0,1)$ vs $B(p,q)$, with various combinations of μ, σ, r, p, q. (n_1, n_2)=(10,10), (10,20), (20,20), (20,50), (50,50) are considered. A subset of the results are reported in Table 34.1 (email the author for receiving the whole set of results). Note that in situation 1 F_1 and F_2 may differ in location and/or scale but not in shape. In situation 2 shapes differ in all cases and we consider location and/or scale changes as well as no change in location nor in scale. In situation 3 F_1 is fixed whereas F_2 is arbitrary.

We noted that the tests maintain their sizes close to the nominal significance level. Analyzing power results, it is very interesting to note that there are cases where a combined test is more powerful than its components, for example this happens in situation 1 for the scale and for the location/scale alternatives. The S_1, S_2, and S_3 tests are positively dependent and informative on the null hypothesis. The combination assesses their dependence nonparametrically and can produce a synergism between the components so that the combined test is more powerful than the component tests. This synergism is not controllable because it is conditional on the data set and has been observed also by Marozzi [5]. In other cases like in situation 3, this does not happen but it is important to emphasize that this is not a drawback of the combination procedure, conversely it is generally expected that a combined test has an intermediate power with respect to its components because the less powerful component(s) contaminates the more powerful one(s) in that particular

Table 34.1 Some results of the size and power comparison study

	$N(0,1)$ vs $N(0,1)$					$N(2,1)$ vs $G(2,1)$				
n_1	10	10	20	20	50	10	10	20	20	50
n_2	10	20	20	50	50	10	20	20	50	50
S_1	0.052	0.047	0.053	0.053	0.049	0.079	0.057	0.115	0.105	0.321
S_2	0.053	0.048	0.052	0.054	0.049	0.078	0.062	0.113	0.121	0.330
S_3	0.025	0.047	0.043	0.052	0.047	0.052	0.078	0.103	0.149	0.300
T_{M12}	0.054	0.050	0.056	0.055	0.048	0.083	0.086	0.126	0.143	0.304
T_{D12}	0.052	0.047	0.053	0.054	0.049	0.078	0.060	0.114	0.113	0.326
T_{M13}	0.052	0.052	0.050	0.051	0.049	0.080	0.077	0.106	0.112	0.259
T_{D13}	0.052	0.049	0.053	0.054	0.048	0.086	0.075	0.123	0.135	0.332
T_{M23}	0.052	0.053	0.051	0.050	0.050	0.081	0.073	0.109	0.114	0.265
T_{D23}	0.052	0.050	0.054	0.053	0.048	0.087	0.078	0.124	0.144	0.338
T_{M123}	0.052	0.054	0.051	0.051	0.050	0.082	0.081	0.114	0.126	0.258
T_{D123}	0.052	0.049	0.053	0.055	0.048	0.085	0.071	0.122	0.130	0.336
	$N(0,1)$ vs $N(0.6,1)$					$N(2,2)$ vs $G(2,1)$				
n_1	10	10	20	20	50	10	10	20	20	50
n_2	10	20	20	50	50	10	20	20	50	50
S_1	0.236	0.289	0.427	0.563	0.793	0.066	0.084	0.076	0.172	0.236
S_2	0.235	0.287	0.418	0.559	0.785	0.065	0.082	0.078	0.178	0.260
S_3	0.137	0.231	0.315	0.465	0.690	0.038	0.077	0.070	0.177	0.305
T_{M12}	0.228	0.265	0.415	0.533	0.789	0.072	0.080	0.091	0.147	0.264
T_{D12}	0.234	0.288	0.423	0.560	0.789	0.066	0.083	0.078	0.175	0.249
T_{M13}	0.215	0.260	0.386	0.510	0.756	0.063	0.073	0.079	0.133	0.241
T_{D13}	0.227	0.275	0.402	0.545	0.770	0.067	0.083	0.082	0.182	0.283
T_{M23}	0.215	0.260	0.378	0.507	0.747	0.063	0.074	0.081	0.135	0.243
T_{D23}	0.224	0.272	0.397	0.542	0.765	0.067	0.082	0.082	0.185	0.295
T_{M123}	0.214	0.255	0.388	0.505	0.763	0.067	0.075	0.084	0.133	0.254
T_{D123}	0.229	0.281	0.411	0.555	0.779	0.067	0.083	0.081	0.182	0.278
	$N(0,1)$ vs $N(0,3)$					$N(4,6)$ vs $G(6,1)$				
n_1	10	10	20	20	50	10	10	20	20	50
n_2	10	20	20	50	50	10	20	20	50	50
S_1	0.081	0.041	0.235	0.197	0.783	0.384	0.530	0.710	0.889	0.990
S_2	0.085	0.054	0.253	0.257	0.799	0.387	0.535	0.716	0.892	0.990
S_3	0.053	0.092	0.188	0.293	0.638	0.214	0.419	0.545	0.777	0.961
T_{M12}	0.076	0.147	0.227	0.444	0.761	0.355	0.489	0.680	0.865	0.987
T_{D12}	0.082	0.048	0.245	0.228	0.792	0.386	0.534	0.714	0.891	0.990
T_{M13}	0.079	0.089	0.196	0.202	0.708	0.350	0.473	0.661	0.853	0.984
T_{D13}	0.091	0.072	0.239	0.269	0.747	0.361	0.492	0.679	0.859	0.987
T_{M23}	0.079	0.077	0.204	0.201	0.722	0.349	0.478	0.662	0.856	0.985
T_{D23}	0.093	0.081	0.247	0.301	0.756	0.362	0.494	0.681	0.860	0.987

(continued)

Table 34.1 (continued)

	$N(0,1)$ vs $N(0,3)$					$N(4,6)$ vs $G(6,1)$				
T_{M123}	0.076	0.105	0.190	0.308	0.701	0.339	0.462	0.645	0.844	0.983
T_{D123}	0.089	0.065	0.245	0.271	0.773	0.372	0.511	0.696	0.874	0.989
	$N(0,1)$ vs $N(0.6,2)$					$N(4,4)$ vs $G(6,1)$				
n_1	10	10	20	20	50	10	10	20	20	50
n_2	10	20	20	50	50	10	20	20	50	50
S_1	0.188	0.182	0.380	0.420	0.825	0.432	0.571	0.775	0.918	0.995
S_2	0.187	0.189	0.381	0.441	0.824	0.437	0.578	0.778	0.921	0.996
S_3	0.118	0.201	0.312	0.464	0.758	0.246	0.450	0.601	0.804	0.970
T_{M12}	0.172	0.189	0.351	0.421	0.797	0.403	0.528	0.746	0.895	0.993
T_{D12}	0.187	0.186	0.381	0.432	0.826	0.435	0.575	0.776	0.919	0.995
T_{M13}	0.176	0.201	0.339	0.407	0.765	0.398	0.515	0.731	0.886	0.993
T_{D13}	0.187	0.209	0.377	0.473	0.817	0.406	0.534	0.739	0.894	0.992
T_{M23}	0.172	0.197	0.338	0.407	0.762	0.400	0.523	0.732	0.889	0.993
T_{D23}	0.187	0.212	0.377	0.480	0.817	0.408	0.537	0.742	0.895	0.992
T_{M123}	0.171	0.193	0.328	0.397	0.752	0.385	0.501	0.717	0.875	0.992
T_{D123}	0.189	0.204	0.382	0.464	0.824	0.415	0.554	0.756	0.909	0.993

situation. In a different situation, the more powerful component(s) may become the less powerful one(s) and vice versa but the combined test is expected to have again an intermediate power. It is important to note that if the combining function leads to a convex acceptance region, then the power of the combined test cannot be less than the power of the least powerful component test, see [13]. This speaks in favor of the practical application of combined testing in particular to social studies when the hypothesis of normality is very often not satisfied. It is interesting to note that the direct combination generally produces more powerful tests than the Mahalanobis combination although the latter might seem to make a better use of the data. The results on the single tests are consistent with those of [16] that showed that the S_3 test is less powerful than the S_1 and S_2 tests. The S_3 test corresponds to the Kolmogorov–Smirnov test which is less powerful than the Anderson Darling and Cramer Von Mises tests, see, e.g., [9]. Among the combined tests, the test of choice is the T_{D123} test which always performs well even if it is not the most powerful test against all alternatives.

34.4 An Example of Social Experiment

In this section we analyze the data of a social experiment. The data reported at p. 68 of [3] were collected from a study comparing two teaching methods that were used to teach reading recovery in the fifth grade. The first method was a pullout program where 25 students were taken out of the classroom for half an hour a day, 4 days a week. The second method was a small group program where 25 students were taught in small groups for 45 min a day in the classroom, 4 days a week. After 4 weeks of the program, the students were assessed through a reading comprehension exam. We wish to test whether the teaching methods have no differential effect. We address the problem within the nonparametric framework because we are not comfortable to assume strict assumptions on the underlying distributions and [3] emphasized that the normal assumption is violated. Moreover, the groups to be compared are not genuine random samples since have been obtained through randomization of a non-random sample of students. Therefore a nonparametric test is not a proper method to analyze the data and we use the nonparametric tests studied before. We use them even if the variable (i.e., reading comprehension result) is discrete since ties are not present. The p-values of the tests have been estimated considering 1,000,000 permutations and are 0.00169 (S_1), 0.00183 (S_2), 0.00012 (S_3), 0.00175 (T_{D12}), 0.00167 (T_{M12}), 0.00039 (T_{D13}), 0.00079 (T_{M13}), 0.00040 (T_{D23}), 0.00058 (T_{M23}), 0.00063 (T_{D123}), 0.00078 (T_{M123}). All the tests find very strong evidence against the null hypothesis that the two teaching methods give the same results in reading comprehension.

Conclusion

The comparison study of the previous section, as many other ones, does not find the uniformly most powerful test for all the situations considered. This is not surprising, especially when addressing the general two-sample problem where the difference between F_1 and F_2 may be of any type: location, scale, kurtosis, skewness, and arbitrary mixtures of them. Marozzi [9] emphasizes that different tests are more powerful against different alternatives. The combination strategy may be effective because it aims at producing tests that inherit the best shown by component tests in very different situations. Although a combined test is not the most powerful test against all alternatives (such test does not exist for the general two-sample problem within the nonparametric framework i.e. without particular assumptions on F_1 and F_2), it is generally possible to find one that performs well in every situation as the T_{D123} test. This is a very useful tool for the practitioner that, as very often happens in social studies, faces a general two-sample with small sample size and problem without any clear idea on the distributions behind the samples or that is not comfortable to assume strict assumptions on the distributions.

References

1. Birnbaum, A.: Combining independent tests of significance. J. Am. Stat. Assoc. **49**, 559–575 (1954)
2. Blanca, M.J., Arnau, J., Lopez-Montiel, D., Bono, R., Bendayan, R.: Skewness and kurtosis in real data samples. Methodol. Eur. J. Res. Methods Behav. Soc. Sci. **9**(2), 78–84 (2013)

3. Corder, G.W., Foreman, D.I.: Nonparametric Statistics for Non-statisticians. Wiley, Hoboken (2009)
4. Ludbrook, J., Dudley, H.: Why permutation tests are superior to t and F tests in biomedical research. Am. Stat. **52**, 127–132 (1998)
5. Marozzi, M.: Multivariate tri-aspect non-parametric testing. J. Nonparametr. Stat. **19**, 269–282 (2007)
6. Marozzi, M.: Some notes on the location-scale Cucconi test. J. Nonparametr. Stat. **21**, 629–647 (2009)
7. Marozzi, M.: Levene type tests for the ratio of two scales. J. Stat. Comput. Simul. **81**, 815–826 (2011)
8. Marozzi, M.: A combined test for differences in scale based on the interquantile range. Stat. Pap. **53**, 61–72 (2012)
9. Marozzi, M.: Nonparametric simultaneous tests for location and scale testing: a comparison of several methods. Commun. Stat. Simul. C. **42**, 1298–1317 (2013)
10. Marozzi, M.: Adaptive choice of scale tests in flexible two-stage designs with applications in experimental ecology and clinical trials. J. Appl. Stat. **40**(4), 747–762 (2013)
11. Micceri, T.: The unicorn, the normal curve, and other improbable creatures. Psychol. Bull. **105**, 156–166 (1989)
12. Nanna, M.J., Sawilowsky, S.S.: Analysis of Likert scale data in disability and medical rehabilitation research. Psychol. Methods **3**, 55–67 (1998)
13. Pesarin, F., Salmaso, L.: Permutation Tests for Complex Data. Wiley, Chichester (2010)
14. Schmider, E., Ziegler, M., Danay, E., Beyer, L., Buhner, M.: Is it really robust? Reinvestigating the robustness of ANOVA against violations of the normal distribution assumption. Methodol. Eur. J. Res. Methods Behav. Soc. Sci. **6**(4), 147–151 (2001)
15. Wilcox, R.R., Keselman, H.J.: Using trimmed means to compare K measures corresponding to two independent groups. Multivar. Behav. Res. **36**(3), 421–444 (2010)
16. Zhang, J.: Powerful two-sample tests based on the likelihood ratio. Technometrics **48**, 95–103 (2006)

Chapter 35
The Use of the Scalar Monte Carlo Estimators for the Optimization of the Corresponding Vector Weight Algorithms

Ilya Medvedev

35.1 Introductory Information

The main object of study in this paper is the development and justification of the efficient weight Monte Carlo methods for estimating the linear functionals of the solution of the system of the integral equations of the second kind. Such equations describe many important processes in mathematical physics (especially in the theory of particle transfer).

Consider the following system of second-kind linear integral equations:

$$\varphi_i(x) = \sum_{j=1}^{m} \int_X k_{ij}(x, y)\varphi_j(y)dy + h_i(x) \tag{35.1}$$

or in the vector form $\Phi = \mathbf{K}\Phi + H$, where $H^{\mathrm{T}} = (h_1, \ldots, h_m)$,

$$\mathbf{K} \in [L_\infty \to L_\infty], \quad ||H||_{L_\infty} = \text{vrai} \sup_{i,x} |h_i(x)|,$$

and the integration is performed with respect to Lebesgue measure in the Euclidean $\varkappa$ space.

It is supposed that the spectral radius is $\lambda(\mathbf{K})$ is less than 1. In this case we have the following expansion of the solution to the Neumann series:

I. Medvedev (✉)
Institute of Computational Mathematics and Mathematical Geophysics SB RAS, Novosibirsk State University, Novosibirsk, Russia
e-mail: min@osmf.sscc.ru

V.B. Melas et al. (eds.), *Topics in Statistical Simulation*, Springer Proceedings in Mathematics & Statistics 114, DOI 10.1007/978-1-4939-2104-1_35

$$\Phi = \sum_{n=0}^{\infty} \mathbf{K}^n H. \tag{35.2}$$

The convergence of series (35.2) is sufficiently provided by the fulfillment of the inequality $||\mathbf{K}^{n_0}|| < 1$ for some $n_0 \geqslant 1$.

Consider a Markov chain $\{x_n\}, (n = 0, \dots, N)$ with some transition probability $p(x, y)$. The value $p(x) = 1 - \int\limits_X p(x, y)dy \geqslant 0$ is considered as the probability of breaking (stop) at the point x, N is a random number of the last state and $x_0 \equiv x$.

The standard Monte Carlo collision-based estimator is constructed for the value $\Phi(x)$ on the base following recursion

$$\boldsymbol{\xi}_x = H(x) + \delta_x Q(x, y)\boldsymbol{\xi}_y, \tag{35.3}$$

$$Q(x, y) = K(x, y)/p(x, y), \quad \Phi(x) = \mathrm{E}\boldsymbol{\xi}_x,$$

where $K(x, y)$ is the matrix of kernels $\{k_{ij}(x, y)\}, (i, j = 1, \dots, m)$ and δ_x is the chain nonbreak indicator function under the transition $x \to y$. Note that the relation $\Phi(x) = \mathrm{E}\boldsymbol{\xi}_x$ holds under the "unbiasedness conditions" [5]

$$p(x, y) > 0, \text{ if } \sum_{i,j=1}^{m} |k_{ij}(x, y)| > 0, \tag{35.4}$$

and under the additional condition $\lambda(\mathbf{K}_1) < 1$, where $\mathbf{K}_1$ is the operator obtained from $\mathbf{K}$ upon replacing the kernels by their absolute values.

The following equation for the covariation matrix $\Psi(x) = \mathrm{E}(\boldsymbol{\xi}_x\boldsymbol{\xi}_x^{\mathrm{T}})$ was presented in [3]

$$\Psi(x) = \chi(x) + \int\limits_X \frac{K(x, y)\Psi(y)K^{\mathrm{T}}(x, y)}{p(x, y)} dy, \tag{35.5}$$

or $\Psi = \chi + \mathbf{K}_p\Psi$, where $\chi = H\Phi^{\mathrm{T}} + \Phi H^{\mathrm{T}} - HH^{\mathrm{T}}$. This equation is considered in the space $\mathbf{L}_\infty$ of matrix-valued functions with the norm

$$||\Psi|| = \operatorname{vrai}\sup_{i,j,x} |\Psi_{i,j}(x)|.$$

By $\mathbf{K}_{p,1}$ we denote the operator obtained from $\mathbf{K}_p$ by replacing the kernels by their absolute values. It is supposed that $\mathbf{K}_{p,1} \in [\mathbf{L}_\infty \to \mathbf{L}_\infty]$

The following assertion was proved in [2] with the use of the method of recurrent "partial" averaging developed by the author [4].

Theorem 35.1. *If* $\lambda(\mathbf{K}_1) < 1$ *and* $\lambda(\mathbf{K}_{p,1}) < 1$, *then* $\Psi(x) = \mathrm{E}(\boldsymbol{\xi}_x \boldsymbol{\xi}_x^{\mathrm{T}})$ *is the solution to Eq.* (35.5) *and* $\Psi \in \mathbf{L}_\infty$.

It is not difficult to see that introducing an additional discrete coordinate with the values $1, 2, \ldots, m$ and the corresponding discrete integration measure with m "atoms" into the phase space, one can rewrite system (35.1) in the form of a single integral equation. This gives us the ability to construct scalar algorithms of the Monte Carlo method, including the simulation of "jumps" in the discrete coordinate. Replacing the argument x by (i, x) instead of (35.1) we get

$$\varphi(i, x) = \sum_{j=1}^{m} \int_X k((i, x), (j, y))\varphi(j, y)dy + h(i, x), \tag{35.6}$$

or $\varphi = K\varphi + h$. Here we have $k((i, x), (j, y)) = k_{ij}(x, y), h(i, x) = h_i(x)$ and $\varphi(i, x) = \varphi_i(x)$. The following Markov chain is constructed according to this representation:

$$(i_0, x_0), (i_1, x_1), \ldots, (i_N, x_N), \tag{35.7}$$

where $(i_0, x_0) = (i, x)$. The transition density for $(i, x) \to (j, y)$ is determined in chain (35.7) by the set of densities $p_{ij}(x, y) = p((i, x), (j, y))$ so that

$$P(i \to j|x) = p_{ij}(x) = \int_X p_{ij}(x, y)dy, \quad \sum_{j=1}^{m} p_{ij}(x) = q_i(x) \leqslant 1, \quad i, j = 1, \ldots, m.$$

The quantity $p_i(x) = 1 - q_i(x)$ here is the probability of breaking (in other words, stop) of the trajectory in its transition from the state (i, x); in the case of nonbreaking and the transition $i \to j$ the next phase state is distributed according to the conditional probability density $r_{ij}(x, y) = p_{ij}(x, y)/p_{ij}(x)$.

A collision-based estimator is uniquely determined for Eq. (35.6) by the recursion

$$\xi_{(i,x)} = h(i, x) + \delta_{(i,x)} q((i, x), (j, y))\xi_{(j,y)}, \tag{35.8}$$

where

$$q((i, x), (j, y)) = k((i, x), (j, y))/p((i, x), (j, y))$$

and $\delta_{(i,x)}$ is the nonbreaking indicator, i.e., $P(\delta_{(i,x)} = 1) = q_i(x)$, $P(\delta_{(i,x)} = 0) = 1 - q_i(x)$. The unbiasedness conditions here take the following form:

$$p((i, x), (j, y)) \neq 0, \text{ if } k((i, x), (j, y)) \neq 0 \ \forall (i, x), (j, y). \tag{35.9}$$

35.2 "Majorant" Methods of Study of Variance Boundedness

As was mentioned in Sect. 35.1, standard methods for checking the variance finiteness for a weighted estimator require a study of the spectral radius for the operator $||K_p||$ corresponding to the integral equation for the second moment of the weighted estimator [5]. A criterion for checking the finiteness of a weight estimator variance that is based on the construction of the appropriate majorant adjoint equation and use of the "partial-value" modelling was proposed in [4]. In this section we present a generalization of the criterion mentioned above and also its certain modifications for a weight scalar estimator of the solution of the system (35.6).

Let $\mathbf{t}' = (t_1', t_2') \in T = T_1 \times T_2$ be a set of two auxiliary values (possibly vector ones) chosen in order to implement a transition in a Markov chain. In the modified phase space $T \times X = \{(\mathbf{t}', x)\}$, one can write down the sub-stochastic kernel of system (35.6) in the form

$$k_{ij}((\mathbf{t}, x), (\mathbf{t}', y)) = \delta(y - y(x, \mathbf{t}'))k_{ij}^{(1)}(x, t_1')k_{ij}^{(2)}((x, t_1'), t_2'),$$

where $y(x, \mathbf{t}')$ is the function determining the new value of the standard Euclidean coordinates over x and the values of the auxiliary variables $\mathbf{t}'$. In addition, assume that $\forall x \in X$

$$\sum_{j=1}^{m} \int_{T_1} k_{ij}^{(1)}(x, t_1')dt_1' = 1 - \alpha_i(x) \leqslant 1. \tag{35.10}$$

Let the transitional densities have the form

$$p_{ij}^{(2)}((x, t_1'), t_2') \equiv k_{ij}^{(2)}((x, t_1'), t_2'), \qquad p_{ij}^{(1)}(x, t_1') = \frac{k_{ij}^{(1)}(x, t_1')u_j^{(1)}(x, t_1')}{[Ku](i, x)}, \tag{35.11}$$

where

$$u_j^{(1)}(x, t_1') = \int_{T_2} \int_{X} \delta(y - y(x, \mathbf{t}'))k_{ij}^{(2)}((x, t_1'), t_2')u_j(y)dy\,dt_2' =$$

$$\int_{T_2} k_{ij}^{(2)}((x, t_1'), t_2')u_j(y(x, \mathbf{t}'))dt_2'. \tag{35.12}$$

$$u = Ku + \hat{h}, \;\; \operatorname{supp} h \subseteq \operatorname{supp} \hat{h}, \;\; \frac{h}{\hat{h}} \leqslant C < \infty \;\; \forall x \in \operatorname{supp} \hat{h}, \tag{35.13}$$

Theorem 35.2 ([4]). *If all the functional elements of system* (35.6) *are nonnegative, then the variance of the collision-based estimator* ξ_x *is finite under the corresponding value simulation of the first auxiliary random variable* t_1' *(see* (35.11)*).*

Proof. The following equality can be verified by direct substitution:

$$[K_p u](i,x) = \sum_{j=1}^{m} \int_{T_1} \frac{\left(k_{ij}^{(1)}(x,t_1')\right)^2}{p_{ij}^{(1)}(x,t_1')} \left[\int_{T_2} \int_X k_{ij}^{(2)}((x,t_1'),t_2')\delta(y - y(x,\mathbf{t}'))u_j(y)dydt_2' \right] dt_1'$$

$$= \sum_{j=1}^{m} \int_{T_1} \frac{\left(k_{ij}^{(1)}(x,t_1')\right)^2 [Ku](i,x)}{k_{ij}^{(1)}(x,t_1')u_j^{(1)}(x,t_1')} \left[\int_{T_2} k_{ij}^{(2)}((x,t_1'),t_2')u_j(y(x,\mathbf{t}'))dt_2' \right] dt_1'$$

$$= [Ku](i,x) \sum_{j=1}^{m} \int_{T_1} k_1(x,t_1')_1 dt' = [Ku](i,x)(1 - \alpha_i(x))$$

$$= (u(i,x) - \hat{h}(i,x))(1 - \alpha_i(x))$$

This equality can be rewritten in the operator form $u = K_p u + \alpha(u - \hat{h}) + \hat{h}$.

The proof is then constructed on the basis of a step-by-step integration of the latter operator equation by the analogy with the proof of Theorem 4.3 from [4], formulated for the case of a single integral equation. □

For alternating sign $k_{ij}^{(1)}(x,\cdot), k_{ij}^{(2)}((x,\cdot),\cdot)$ and $h(i,x)$ the Theorem 35.2 is valid if we assume $u = K_1 u + |\hat{h}|$, and replace $k_{ij}^{(1)}(x,\cdot), k_{ij}^{(2)}((x,\cdot),\cdot)$ by $|k_{ij}^{(1)}(x,\cdot)|, |k_{ij}^{(2)}((x,\cdot),\cdot)|$ in the expressions (35.11),(35.12).

Note that the search for the solution u of the majorant Eq. (35.13) is practically equivalent to the search for the solution of the original problem. In this context, we propose to consider the approximate and probably more simple majorant Eq. (35.6)

$$g = \tilde{K}g + |\hat{h}|, \ \text{supp}\, h \subseteq \text{supp}\, \hat{h}, \ \frac{h}{\hat{h}} \leqslant C < \infty \ \forall x \in \text{supp}\, \hat{h}, \tag{35.14}$$

with nonnegative elements $\tilde{k}_{ij}^{(1)}(x,\cdot)$ instead of $k_{ij}^{(1)}(x,\cdot)$.

The conditions $|k_{ij}^{(1)}(x,\cdot)| \leqslant \tilde{k}_{ij}^{(1)}(x,\cdot)$, $\rho(\tilde{K}) < 1$ allow us to determine the partial value density

$$\tilde{p}_{ij}^{(1)}(x,t_1) = \frac{k_1^{(1)}(x,t_1')\, g_j^{(1)}(x,t_1')}{[\tilde{K}g](i,x)} \tag{35.15}$$

and to formulate the following result.

Theorem 35.3. *The variance of the collision-based estimator* ξ_x *is finite under partial value modelling of the first auxiliary random variable* t_1' *(see* (35.15)*).*

Theorem 35.3 can be proved similarly to the proof of Theorem 35.2.

Note that the verification of the inequalities $\lambda(K_p) < 1$ or $||K_p|| < 1$ for a scalar estimator of a vector solution is essentially easier than the verification of the corresponding inequality for a vector estimator. Nevertheless, due to the additional simulation of 'jumps' in the discrete coordinate, the variance of the scalar estimator on the average is greater than the variance of the corresponding vector estimator. For example, it was indicated in [2] that if the transition density does not depend on i, j, i.e.

$$p((i,x)(j,y)) = m^{-1}p(x,y), \quad \int_X p(x,y)dy \leqslant 1,$$

then the following relation holds: $\mathrm{D}\xi_{(i,x)} \geqslant \mathrm{D}\boldsymbol{\xi}_{x,i}$, because it is sufficiently clear that in this case we have

$$\boldsymbol{\xi}_{x,i} = \mathrm{E}\big(\xi_{(i,x)}|x_0,\ldots,x_N\big). \tag{35.16}$$

Taking the latter remark into account, we can formulate the following result.

Lemma 35.1. *If $\lambda(K_p) < 1$, then the variance of the vector estimator $\boldsymbol{\xi}_x$ with the transition density $p(x,y)$ is finite.*

Note that the assertion of Lemma 35.1 directly implies that if $\lambda(K_p) < 1$, then $\lambda(\mathbf{K}_{p,1}) < 1$.

35.3 Algorithms with Branching

It is known [1–3] that the variance of the weighed estimator $\mathrm{D}\boldsymbol{\xi}_x$ is finite if $\lambda(\mathbf{K}_{p,1}) < 1$. The estimation of the value $\lambda(\mathbf{K}_{p,1})$ for real problems requires a separate and laborious theoretical study. For example, using semiheuristic analytic calculations, numerical estimates, and integrating the resolvent, it was shown in [6] that the value $\lambda(\mathbf{K}_p)$) in problems of radiation transfer under polarization is close to the product of the similar spectral radius $\lambda(S_p)$ for an infinite medium (which can be calculated analytically) and the spectral radius of the scalar integral operator related to the "unit" matrix of scattering, which can be easily estimated. The ability to proceed to consideration of $\lambda(\mathbf{K}_p)$ is due to the majorant property of the first component of the Stokes vector. In particular, in the case of molecular scattering for a "physical" simulation, it has been obtained that $\lambda(\mathbf{K}_p) < 1$ for $p > 0.151$, where p is the lower bound of the absorption probability in the medium. One can essentially decrease the value of p by modification of the transfer process by substituting $\sigma \to \sigma_s$, $\sigma_c \to 0$ [1,5], where $\sigma = \sigma_s + \sigma_c$ is the total cross-section and σ_s and σ_c are scattering and absorption sections, respectively. The absorption is taken into account in this modification with the use of the corresponding weight factor, the following estimate is valid [6]

$$\lambda(\mathbf{K}_p) \leqslant \frac{1-p}{1+p}\lambda(S_p),$$

and $\lambda(\mathbf{K}_p) < 1$ for $p > 0.082$ in the case of molecular scattering.

Thus, for $p < 0.082$ the variance of the vector estimator can be infinitely large and the justification of the vector Monte Carlo algorithm applications remains open. In this case one may use a scalar weight estimator with "branching" of the trajectory.

Consider Eq. (35.6) with nonnegative elements $k((i,x),(j,y))$, $h(i,x)$ and define the collision-based estimator with branching [4]. To do that, introduce the integer-valued random variable $\nu = \nu((i,x),(j,y))$ (number of "branches") so that

$$P(\nu = [q]) = 1 + [q] - q, \quad P(\nu = 1 + [q]) = q - [q], \tag{35.17}$$

$q = q((i,x),(j,y))$. It is not difficult to check in this case that $\mathrm{E}\nu = q$ and the distribution (35.17) determines the minimal value of $\mathrm{D}\nu$ in the class of random integer-valued variables with the fixed value $\mathrm{E}\nu = q$ [4].

Hereafter we assume that $|q| < C < \infty$. Let the random variable $\zeta_{(i,x)}$ be determined by the recursion

$$\zeta_{(i,x)} = h(i,x) + \delta_{(i,x)} \sum_{n=1}^{\nu} \zeta^{(n)}_{(j,y)}, \tag{35.18}$$

where $\zeta^{(n)}_{(j,y)}$ are independent implementations of $\zeta_{(j,y)}$.

Lemma 35.2 ([5]). *If $\lambda(K_1) < 1$, then under the assumptions formulated above the relation $\mathrm{E}\zeta_{(i,x)} = \varphi(i,x)$ holds.*

Proof. Since all the elements in (35.18) are nonnegative, by Wald's identity we have

$$\mathrm{E}\sum_{n=1}^{\nu} \zeta^{(n)}_{(j,y)}, = \mathrm{E}\nu\mathrm{E}\zeta_{(j,y)}, .$$

This equality is also valid in the case $\mathrm{E}\zeta_{(j,y)} = +\infty$, because the nonnegativity of the elements of the problem implies

$$\mathrm{E}\sum_{n=1}^{\nu} \zeta^{(n)}_{(j,y)} = \mathrm{E}\mathrm{E}\left(\sum_{n=1}^{\nu} \zeta^{(n)}_{(j,y)} |\nu\right).$$

Therefore, the value $\mathrm{E}\zeta_{x_0}$ can be sequentially calculated by a recursion of form (35.8). □

In order to estimate the solution to original system (35.6) with alternating-sign elements $k((i,x),(j,y)), h(i,x)$ one should apply the substitution $q \to |q|$ in the expression (35.17) and use the random variable

$$\eta_{(i,x)} = h(i,x) + \delta_{(i,x)}\mathrm{sgn}(q)\sum_{n=1}^{\nu}\eta_{(j,y)}^{(n)},$$

where $\eta_{(j,y)}^{(n)}$ are independent implementations of $\eta_{(j,y)}$. The definition obviously implies $|\eta_{(i,x)}| \leqslant \zeta_{(i,x)}^{(1)}$, where $\zeta_{(i,x)}^{(1)}$ is the estimator of form (35.18) for the system (35.6) with the elements $|k((i,x)(j,y))|, |h(i,x)|$. Under the above assumptions, the value $\mathrm{E}\zeta_{(i,x)}^{(1)}$ is finite. Due to Lebesgue's theorem on dominated convergence, we have the equality $\mathrm{E}\eta_{(i,x)} = \varphi(i,x)$.

Theorem 35.4. *If the conditions of Lemma 35.2 hold, then the value* $\mathrm{E}\eta_{(i,x)}^2 < \infty$ *is determined by the Neumann series for Eq.* (35.6) *with the replacement of* $h(i,x)$ *by*

$$H(i,x) = h(i,x)\{2\varphi(i,x) - h(i,x)\} + \sum_{j=1}^{m}\int_X p((i,x),(j,y))\gamma\varphi^2(j,y)dy, \tag{35.19}$$

where $\gamma = (2q - 1 - [q])[q]$.

Proof. The proof follows from recurrent partial probabilistic averaging of the equality

$$\zeta_{(i,x)}^2 = h^2(i,x) + 2\delta_{(i,x)}h(i,x)\sum_{n=1}^{\nu}\zeta_{(j,y)}^{(n)} + 2\delta_{(i,x)}\sum_{n=1}^{\nu}\sum_{l=n+1}^{\nu}\zeta_{(j,y)}^{(n)}\zeta_{(j,y)}^{(l)} + \delta_{(i,x)}\sum_{n=1}^{\nu}\left(\zeta_{(j,y)}^{(n)}\right)^2.$$

□

Let us note that if we assume $h(i,x) \equiv 1$ in the system (35.6) then the value $\varphi_1(i,x) = \mathrm{E}\zeta_{(i,x)}^{(1)}$ coincides with the mean $\mathrm{E}\mu$ of the total number $\mu(i,x)$ of branches in the branching trajectory. Obviously, the value $\mathrm{E}\mu$ is linearly related to the average time T_b of simulation of a single branch trajectory. Assuming all the above, we can formulate the following lemma.

Lemma 35.3. *If* $\lambda(K_1) < 1$, *then the value* T_b *is bounded.*

Now we study the possibility of branching of trajectories for a vector estimator. Here and below we assume that the random number of "branches" $\nu(x,y)$ is nonnegative, bounded, and has some probability distribution. Define the random variable $\boldsymbol{\zeta}_x$ by the following recursion:

$$\boldsymbol{\zeta}_x = H(x) + \delta_x\frac{Q(x,y)}{\mathrm{E}\nu(x,y)}\sum_{n=1}^{\nu}\boldsymbol{\zeta}_y{}^{(n)}, \tag{35.20}$$

where $\boldsymbol{\zeta}_y{}^{(n)}$ are independent implementations of $\boldsymbol{\zeta}_x$. Repeating sequentially the calculations for $\boldsymbol{\zeta}_x$ as in the proof of Lemma 35.2, one can verify the following assertion.

Lemma 35.4. *If $\lambda(\mathbf{K}_1) < 1$, then under the assumptions presented above the relation* $\mathrm{E}\boldsymbol{\zeta}_x = \Phi(x)$ *holds.*

Theorem 35.5. *If all the components of system* (35.1) *are nonnegative, then the value* $\Psi(x) = \mathrm{E}(\boldsymbol{\zeta}_x \boldsymbol{\zeta}_x^{\mathrm{T}})$ *is determined by the Neumann series for the equation*

$$\Psi(x) = \tilde{\chi}(x) + \int_X \frac{K(x,y)\Psi(y)K^{\mathrm{T}}(x,y)}{\mathrm{E}\nu(x,y)p(x,y)}dy, \tag{35.21}$$

or $\Psi = \tilde{\chi} + \mathbf{K}_{\mathbf{p}}^{(1)}\Psi$, *where*

$$\tilde{\chi}(x) = \chi(x) + \int_X \frac{K(x,y)\mathrm{E}\big(\nu(x,y)(\nu(x,y)-1)\big)\Phi(y)\Phi^{\mathrm{T}}(y)K^{\mathrm{T}}(x,y)}{\big(\mathrm{E}\nu(x,y)\big)^2 p(x,y)}dy$$

Proof. The proof follows from recurrent partial probabilistic averaging of the equality

$$\boldsymbol{\zeta}_x\boldsymbol{\zeta}_x^{\mathrm{T}} = \Big(H(x) + \delta_x \frac{Q(x,y)}{\mathrm{E}\nu(x,y)} \sum_{n=1}^{\nu} \boldsymbol{\zeta}_y{}^{(n)}\Big)\Big(H^{\mathrm{T}}(x) + \delta_x\Big(\sum_{n=1}^{\nu} \boldsymbol{\zeta}_y{}^{(n)}\Big)^{\mathrm{T}} \frac{Q^{\mathrm{T}}(x,y)}{\mathrm{E}\nu(x,y)}\Big)$$

$$= H(x)H^{\mathrm{T}}(x) + \delta_x \frac{Q(x,y)}{\mathrm{E}\nu(x,y)} \sum_{n=1}^{\nu} \boldsymbol{\zeta}_y{}^{(n)} H^{\mathrm{T}}(x) + \delta_x H(x)\Big(\sum_{n=1}^{\nu} \boldsymbol{\zeta}_y{}^{(n)}\Big)^{\mathrm{T}} \frac{Q^{\mathrm{T}}(x,y)}{\mathrm{E}\nu(x,y)}$$

$$+\delta_x \frac{Q(x,y)}{\mathrm{E}\nu(x,y)} \sum_{n=1}^{\nu} \boldsymbol{\zeta}_y{}^{(n)} \Big(\sum_{n=1}^{\nu} \boldsymbol{\zeta}_y{}^{(n)}\Big)^{\mathrm{T}} \frac{Q^{\mathrm{T}}(x,y)}{\mathrm{E}\nu(x,y)}.$$ □

Note that operator $\mathbf{K}_{\mathbf{p}}^{(1)}$ in relation (35.21) differs from the corresponding operator $\mathbf{K}_{\mathbf{p}}$ from (35.5) in the presence of the additional factor $1/\mathrm{E}\nu(\cdot,\cdot)$ in the integrand. This fact gives us an ability to choose the corresponding distribution for the random number of branches $\nu(x,y)$ to decrease the variance $\mathrm{D}\boldsymbol{\zeta}_x$ or, which is more important, $\lambda(\mathbf{K}_{\mathbf{p}}^{(1)})$ in comparison with $\mathrm{D}\boldsymbol{\xi}_x$ or $\lambda(\mathbf{K}_p)$, respectively. In this case it is extremely important to study in advance that the mean of the total number of branches in bounded. This study can be simplified if we notice that for vector estimator with branching (35.20) one can construct the corresponding randomized scalar estimator with branching

$$\tilde{\zeta}_{(i,x)} = h(i,x) + \delta_{(i,x)} \frac{mk_{ij}(x,y)}{p(x,y)\mathrm{E}\nu(x,y)} \sum_{n=1}^{\nu} \tilde{\zeta}_{(j,y)}^{(n)}.$$

In this case the corresponding inequality (35.16) and analogues of Lemma 35.1 and its remark are valid for the estimators $\boldsymbol{\zeta}_{x,i}, \tilde{\zeta}_{(i,x)}$.

Acknowledgements The author is grateful to G.A. Mikhailov, the corresponding member of the Russian Academy of Sciences for useful advice. This work was supported by Russian Foundation of Basic Research (grants 12-01-00034a, 13-01-00441a, 13-01-00746a and 12-01-31328-mol a) and by the Leading Scientific Schools program, project no. NSh 5111.2014.1.

References

1. Marchuk, G.I., Mikhailov, G.A., Nazaraliev, M.A., Darbinjan, R.A., Kargin, B.A., Elepov, B.S.: The Monte Carlo Methods in Atmospheric Optics. Springer, Heidelberg (1980)
2. Medvedev, I.N., Mikhailov, G.A.: Probabilistic-algebraic algorithms of Monte Carlo methods. Russ. J. Numer. Anal. Math. Model. **26**(3), 323–336 (2011)
3. Mikhailov, G.A.: Optimization of Weighted Monte Carlo Methods. Springer, New York (1992)
4. Mikhailov, G.A., Medvedev, I.N.: The use of adjoint equations in Monte Carlo Methods. Omega Print, Novosibirsk (2009, in Russian)
5. Mikhailov, G.A., Voitishek, A.V.: Numerical Statistical Simulation: Monte Carlo Methods. Izdat. Tsentr Aka demiya, Moscow (2006, in Russian)
6. Mikhailov, G.A., Ukhinov, S.A., Chimaeva, A.S.: Variance of a standard vector Monte Carlo estimator in the theory of polarized radiative transfer. Comput. Math. Math. Phys. **46**(11), 2099–2113 (2006)

Chapter 36
Additive Cost Modelling in Clinical Trial

Guillaume Mijoule, Nathan Minois, Vladimir V. Anisimov, and Nicolas Savy

36.1 Introduction

In the framework of a clinical trial, an important and mandatory parameter of the clinical trial protocol is the Necessary Sample Size, the number n of patients to be recruited. A natural question is how long it takes to recruit these patients.

The use of Poisson process to describe the recruitment process is an accepted approach (Senn [6], Carter et al. [4]). However, the huge variability of the rates of the recruitment processes among centres were not taken into account. There were many investigations on this way and now we are able to claim that, to date, the easiest to handle and most relevant model is the Poisson-gamma model developed in [2] and further extended in [1, 5]. This model assumes that patients arrive at different centres according to randomly delayed Poisson processes where the rates are gamma-distributed.

In [3], authors introduce a more elaborated model in which the distinction is made between the screened (recruited) patients and the randomized patients who are patients satisfying the inclusion criteria (the other ones quit the trial). In what

G. Mijoule
University of Paris 10, Nanterre, France
e-mail: guillaume.mijoule@u-paris10.fr

N. Minois
INSERM Unit 1027, Toulouse, France
e-mail: nathan.minois@inserm.fr

V.V. Anisimov
Quintiles, Reading, UK
e-mail: Vladimir.Anisimov@quintiles.com

N. Savy (✉)
Toulouse Institute of Mathematics, Toulouse, France
e-mail: nicolas.savy@math.univ-toulouse.fr

V.B. Melas et al. (eds.), *Topics in Statistical Simulation*, Springer Proceedings in Mathematics & Statistics 114, DOI 10.1007/978-1-4939-2104-1_36

follow, $N^S(t)$ (resp. $N^R(t)$) denotes the number of screened (resp. randomized) patients at time t. The instant of interest is the first time denoted by τ when the process N^R attains n:

$$\tau = \left\{ \inf_{t \geqslant 0} : N^R(t) = n \right\}. \tag{36.1}$$

The paper aims to give the very first step of a model for multicentric clinical trial cost. The dynamic of the cost denoted $t \to C(t)$ is directed by the dynamic of the recruitment process we assume to be Poisson-Gamma. Given constants which are usually used to estimate the total cost of a trial, we introduce an additive cost model (defined in Sect. 36.2.2). This model allows us to compute parameters such that the expectation $\mathbb{E}[C(t)]$ for a given t or slightly more complicated but of paramount interest, $\mathbb{E}[C(\tau)]$. These parameters are really useful tools for the monitoring of a clinical trial.

The paper is organized as follows. Section 36.2 describes the Poisson-gamma model with screening failures, and introduces the cost model. Section 36.3 gives the main results regarding the expected cost of the trial, first focusing on the simpler non-Bayesian case. Section 36.4 applies those results in a simulation study.

36.2 An Empirical Bayesian Model for the Cost of Clinical Trials with Patients' Drop-Out

Consider a multicentric clinical trial where M centres are involved. In this section we describe the empirical Bayesian setting for modelling of patients' arrival and screening failure, and the associated cost model.

36.2.1 The Poisson-Gamma Model with Patients' Screening Failures

We assume that patients arrive at centres according to a Poisson-gamma process. The recruitment process in i-th centre is a Poisson process with rate λ_i where λ_i has a gamma distribution with parameters (α, β) and pdf $\varkappa e^{-\beta x} x^{\alpha-1} \mathbf{1}_{\{x>0\}}$ ($\varkappa$ is a normalizing constant). Processes in different centres are assumed independent. In papers [1, 2, 5] the validity of this model in the framework of clinical trials was intensively studied.

Now assume that a patient arriving in the i-th centre has a probability p_i of succeeding the screening process [3]. To account for variability of p_i among centres, we use again a Bayesian setting where we assume p_i are independent and distributed as a beta distribution of parameters (ψ_1, ψ_2), with pdf $\varkappa x^{\psi_1 - 1} x^{\psi_2 - 1} \mathbf{1}_{\{0<x<1\}}$ ($\varkappa$ is a normalizing constant).

36.2.2 The Cost Model

For centre i, we categorize the different costs of a clinical trial as follows:

- a fixed cost for a screened patient,
- a fixed cost for a randomized patient (on top of the screening cost),
- a time-depending cost for a randomized patient,
- a fixed cost for opening a centre,
- a time-depending cost for an active centre.

The model we propose is an additive cost model which expresses the total cost at time t of the i-th centre, denoted $C_i(t)$ by:

$$C_i(t) = J_i N_i^R(t) + K_i N_i^S(t) + \sum_{0 \leqslant T_n^i \leqslant t} g_i(t, T_n^i) + F_i + G_i t,$$

where J_i, K_i, F_i and G_i are constants (in general roughly known by the investigator of the centre). The time-depending cost for a randomized patient starts when a patient is included. It is thus natural that g_i is some function of both variables t and T_n^i, the randomization instant of n-th patient in i-th centre. We make the following hypotheses on the functions g_i:

- $g_i : \mathbb{R}_+ \times \mathbb{R}_+ \to \mathbb{R}$ is measurable,
- $g_i(t, s) = 0$ if $t < s$,
- $\forall t \geqslant 0$, $g_i(t, .)$ is continuous on $[0, t]$.

$N_i^R(t)$ (resp. $N_i^S(t)$) is the number of randomized (resp. screened) patients at time t in i-th centre. Notice that

$$\sum_{0 \leqslant T_n^i \leqslant t} g_i(t, T_n^i) = \int_0^t g_i(t, s) dN_i^R(s),$$

where the integral is to be understood as a Stieltjès one.

Finally, let $N^R(t) = \sum_{i=1}^M N_i^R(t)$ the total number of randomized patients at time t and $C = \sum_{i=1}^M C_i$ the total cost process. Patients' recruitment stops as soon as the process N^R reaches n. The recruitment time, τ, is defined by (36.1). Note that τ is a stopping time in the natural filtration of N^R.

In the following, we set

$$F = \sum_{i=1}^M F_i \quad \text{and} \quad G = \sum_{i=1}^M G_i.$$

Definition 36.1. The mean cost of the trial $\mathbb{E}\left[C(\tau)\right]$, denoted $\mathscr{C}$, is

$$\mathscr{C} = \mathbb{E}\left[\sum_{i=1}^{M}\left(J_i N_i^R(\tau) + K_i N_i^S(\tau)\right)\right] + \mathbb{E}\left[\sum_{i=1}^{M}\int_0^{\tau} g_i(\tau,s)dN_i^R(s)\right] + F + G\,\mathbb{E}\left[\tau\right].$$

36.3 Calculation of the Mean Cost

36.3.1 Non-random Recruitment Rates and Probabilities of Screening

We first investigate the simpler model where recruitment rates and probabilities of screening are known. In this case, patients arrive in centres according to standard homogenous Poisson processes. A simple conditional argument will give the general result in a Bayesian setting. Thus, we assume $(\lambda_i)_{1\leqslant i\leqslant M}$ and $(p_i)_{1\leqslant i\leqslant M}$ are known.

We have for any i the expansion

$$N_i^S = N_i^R + N_i^L, \tag{36.2}$$

where N_i^R is the aforementioned Poisson process of randomized patients with rate $p_i\lambda_i$ and N^L is an independent Poisson process with rate $(1-p_i)\lambda_i$, representing the number of screening failures in centre i over time. Finally, denote $\Lambda_1 = \sum_{i=1}^M p_i\lambda_i$.

Recall that, since N^R is a Poisson process with rate Λ_1, then, in the non-Bayesian setting, τ has a Gamma distribution with parameters (n, Λ_1). We let $p_{(n,\Lambda_1)}$ be the density of this distribution. The following lemma is the non-Bayesian version of our main theorem.

Lemma 36.1. *Assume* $(\lambda_i)_{1\leqslant i\leqslant M}$ *and* $(p_i)_{1\leqslant i\leqslant M}$ *are known. Let* $\Phi_1 = \sum_{i=1}^M J_i p_i\lambda_i$, $\Phi_2 = \sum_{i=1}^M K_i\lambda_i$ *and for any* $t > 0$,

$$G(t) = \sum_{i=1}^{M}\frac{p_i\lambda_i}{\Lambda_1}\,g_i(t,t) \quad \textit{and} \quad \tilde{G}(t) = \sum_{i=1}^{M}\frac{p_i\lambda_i}{\Lambda_1}\frac{1}{t}\int_0^t g_i(t,s)\,ds.$$

Then

$$\mathscr{C} = n\frac{\Phi_1+\Phi_2}{\Lambda_1} + \int_0^{+\infty} G(t)p_{(n,\Lambda_1)}(dt) + (n-1)\int_0^{+\infty}\tilde{G}(t)p_{(n,\Lambda_1)}(dt) + G\frac{n}{\Lambda_1} + F. \tag{36.3}$$

Proof. First, remark that all functions in (36.3) are positive and measurable, so the integrals are well defined. A standard result implies that $N_i^R(\tau)$ has a binomial distribution $\mathscr{B}\left(n, \frac{p_i\lambda_i}{\Lambda_1}\right)$. Thus,

$$\mathbb{E}\left[\sum_{i=1}^{M}(J_i + K_i)N_i^R(\tau)\right] = \sum_{i=1}^{M}(J_i + K_i)n\frac{p_i\lambda_i}{\Lambda_1} = \frac{n\Phi_1}{\Lambda_1} + \frac{n}{\Lambda_1}\sum_{i=1}^{M} K_i p_i \lambda_i.$$

Since N_i^L and τ are independent, and since $\mathbb{E}[\tau] = \frac{n}{\Lambda_1}$, by conditioning we get

$$\mathbb{E}\left[\sum_{i=1}^{M} K_i N_i^L(\tau)\right] = \sum_{i=1}^{M} K_i \mathbb{E}[(1-p_i)\lambda_i\tau] = \frac{n}{\Lambda_1}\sum_{i=1}^{M} K_i(1-p_i)\lambda_i.$$

Recalling (36.2), we obtain the first term in (36.3). It remains to show that

$$\mathbb{E}\left[\sum_{i=1}^{M}\int_0^{\tau} g_i(\tau,s)dN_i^R(s)\right] = \int_0^{+\infty} G(t)p_{(n,\Lambda_1)}(dt) + (n-1)\int_0^{+\infty}\tilde{G}(t)\, p_{(n,\Lambda_1)}(dt). \tag{36.4}$$

We have

$$\mathbb{E}\left[\int_0^{\tau} g_i(\tau,s)dN_i^R(s)\right] = \int_0^{+\infty}\mathbb{E}\left[\int_0^t g_i(t,s)dN_i^R(s)\,\middle|\,\tau = t\right]p_{(n,\Lambda_1)}(dt),$$

Assume first that for all $t > 0$, the restriction of $s \mapsto g_i(t,s)$ to $[0,t]$ is differentiable. Let $\partial_2 g_i(t,.)$ be this derivative. Also assume that $\forall t > 0, \sup_{0\leqslant s\leqslant t}|\partial_2 g_i(t,s)| < +\infty$. An integration by parts gives

$$\int_0^t g_i(t,s)dN^R(s) = g_i(t,t)N_i^R(t) - \int_0^t \partial_2 g_i(t,s)N_i^R(s)ds.$$

This leads

$$\mathbb{E}\left[\int_0^t g_i(t,s)dN_i^R(s)\,\middle|\,\tau = t\right]$$
$$= g_i(t,t)\mathbb{E}\left[N_i^R(t)|\,\tau = t\right] - \mathbb{E}\left[\int_0^t \partial_2 g_i(t,s)N_i^R(s)ds\,\middle|\,\tau = t\right].$$

Knowing $\{\tau = t\}$, $N_i^R(t)$ has a binomial distribution $\mathscr{B}\left(n, \frac{p_i\lambda_i}{\Lambda_1}\right)$. Moreover, given $\{\tau = t\}$, we can bound from above $|\partial_2 g_i(t,s)N_i^R(s)| \leqslant n \sup_{0\leqslant s\leqslant t}|\partial_2 g_i(t,s)|$, so that Fubini's theorem applies and

$$\mathbb{E}\left[\int_0^t g_i(t,s)dN_i^R(s)\,\middle|\,\tau = t\right] = n\, g_i(t,t)\frac{p_i\lambda_i}{\Lambda_1} - \int_0^t \partial_2 g_i(t,s)\mathbb{E}\left[N_i^R(s)\mid \tau = t\right]ds.$$

It remains to prove that $\forall s < t$,

$$\mathbb{E}\left[N_i^R(s) \mid \tau = t\right] = (n-1)\frac{p_i \lambda_i}{\Lambda_1}\frac{s}{t}. \tag{36.5}$$

Given $\{\tau = t\}$, there is a probability $\frac{p_i\lambda_i}{\Lambda_1}$ that N_i^R jumps at t. Knowing this event, the $N_i^R(t) - 1$ remaining jumps of N_i^R in $[0, t[$ are uniformly distributed. The same argument for the case where N_i^R does not jump at t implies

$$\mathbb{E}\left[N_i^R(s) \mid \tau = t\right] = \frac{p_i\lambda_i}{\Lambda_1}\mathbb{E}\left[N_i^R(t) - 1 \mid \tau = t\right]\frac{s}{t} + \left(1 - \frac{p_i\lambda_i}{\Lambda_1}\right)\mathbb{E}\left[N_i^R(t) \mid \tau = t\right]\frac{s}{t},$$

which leads to (36.5). Reintegrating by parts, we obtain (36.4).

Finally, the density of $\mathscr{C}^1([0,t])$ in $\mathscr{C}^0([0,t])$ for the uniform norm completes the proof in the case where $g_i(t,.)$ is only continuous on $[0,t]$, $\forall t > 0$.

36.3.2 Bayesian Setting: Random Recruitment Rates and Probabilities of Screening Success

Now, we assume the initial rates are distributed according to a prior gamma distribution and the probabilities of screening have a beta distribution. At some interim time t_1, assume i-th centre has screened n_i patients and randomized k_i patients. A Bayesian re-estimation shows that, given n_i and k_i, the posterior rate λ_i has a gamma distribution with parameters $(\alpha + n_i, \beta + t_1)$, and the probability of screening p_i has a beta distribution with parameters $(\psi_1 + k_i, \psi_2 + n_i - k_i)$ [3]. Our main theorem is a consequence of Lemma 36.1.

Theorem 36.1. *Let $\Phi_1 = \sum_{i=1}^M J_i p_i \lambda_i$, and $\Phi_2 = \sum_{i=1}^M K_i \lambda_i$. The mean cost reads*

$$\mathscr{C} = n\,\mathbb{E}\left[\frac{\Phi_1 + \Phi_2}{\Lambda_1}\right] + \int_0^{+\infty} e^{-t}\frac{t^{n-1}}{(n-1)!}\sum_{i=1}^M \mathbb{E}\left[\frac{p_i\lambda_i}{\Lambda_1} g_i(t/\Lambda_1, t/\Lambda_1)\right] dt$$
$$+ \int_0^{+\infty} e^{-t}\frac{t^{n-2}}{(n-2)!}\mathbb{E}\left[\int_0^{t/\Lambda_1}\sum_{i=1}^M g_i(t/\Lambda_1, s)p_i\lambda_i\, ds\right] dt + Gn\mathbb{E}\left[\frac{1}{\Lambda_1}\right] + F.$$

Proof. Conditioning by $(\lambda_1, \ldots, \lambda_M)$ and $(p_1, \ldots, p_M)$, we can make use of Lemma 36.1. The change of variable $x = t/\Lambda_1$ in the integrals leads to the result.

36.3.3 Mean Cost Variation When Closing a Centre

In this section, we calculate the mean cost variation when closing a particular centre. For the sake of notational simplicity, we make some assumptions, namely that each randomized patient yields a linear cost over time. This would be the case if, for instance, patients have to remain in observation until the trial ends. This means g_i is defined by $g_i(t,s) = L_i(t-s)\mathbf{1}_{\{t \geq s\}}$, where L_i is some positive constant. Moreover, we also assume the constants K_i, J_i and L_i do not depend on i and we write $K_i \equiv K$, $J_i \equiv J$ and $L_i \equiv L$.

Corollary 36.1. *Denote* $\Lambda_2 = \sum_{i=1}^{M}(1-p_i)\lambda_i$. *The closure of i-th centre implies a variation of the mean cost of the trial* $\Delta\mathscr{C}_i$ *given by:*

$$\Delta\mathscr{C}_i = \mathbb{E}\left[\frac{n\lambda_i K(p_i\Lambda_2 - (1-p_i)\Lambda_1) + \frac{n(n-1)}{2}\lambda_i p_i L + n\lambda_i p_i G - nG_i\Lambda_1}{\Lambda_1(\Lambda_1 - p_i\lambda_i)}\right] - F_i.$$

Proof. When closing centre i, the new mean cost is given in Theorem 36.1 by replacing Λ_1 by $\Lambda_1 - p_i\lambda_i$ and by summing over all indices except i. The proof is then a straightforward calculation.

36.4 Simulation Study

We apply the result of Corollary 36.1 in a simulation study. The parameters used in simulation scenario are $\alpha = 1.2$ and $\mu = \alpha/\beta = 0.2$ for the recruitment process and $\psi_1 = 3, \psi_2 = 1$ for the screening probability. In Fig. 36.1, we plot, for different sets of constants K and L, and for each centre, the variation in recruitment time and total cost when closing this centre.

When L/K is large, the mean cost is expected to be correlated to the recruitment time since most of the cost has linear increasing in time. In this case, closing a centre should never profitable. This is well shown by crosses in Fig. 36.1.

On the other hand, a small value of L/K means most of the cost is due to patients' screening cost; thus, closing centres with high probabilities of drop-out is expected be profitable. This is what we observe in simulations: for instance, the triangle and the circle in the bottom of Fig. 36.1 represent the centre with highest probability of drop-out.

Conclusion

The model described here is an additive model for the cost of a multicentric clinical trial. The process describing patients' arrival and drop-outs takes into

(continued)

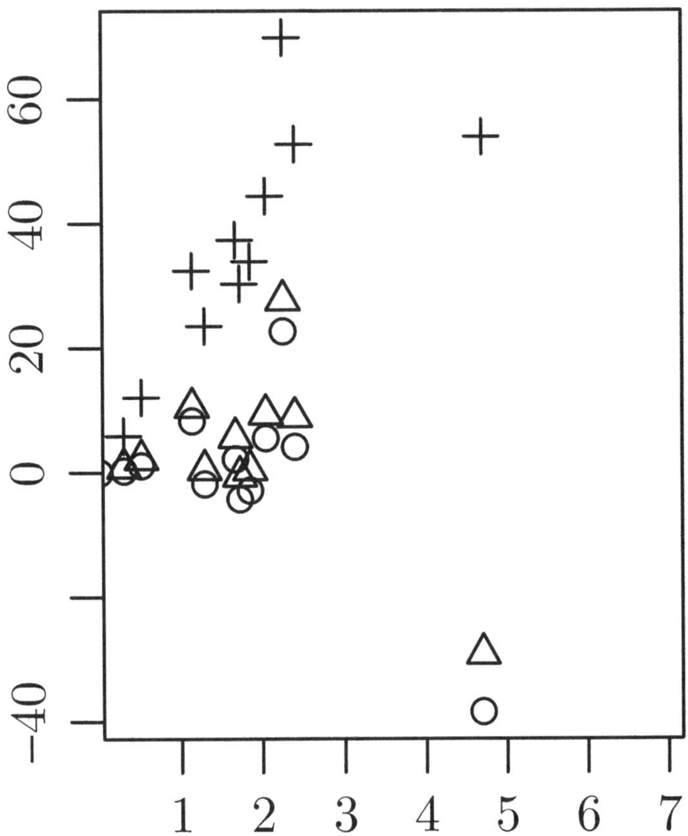

Fig. 36.1 Cost and recruitment time variation when closing a centre, for three different parameter sets. $G_i \equiv F_i \equiv 0$. $K_i \equiv 10$. *Crosses*: $L = 0.1$, *triangles*: $L = 0.01$, *circles*: $L = 0.001$

account the variability in the recruitment rates and in the probabilities of screening failures between centres. The expected cost of the trial is reachable. It yields an useful tool to determine whether closing a centre is profitable. We show its applicability in a simulation study. The main difficulty in practical applications will be the estimation of the constants describing the different costs.

Acknowledgements The authors thank Sandrine Andrieu, Valérie Lauwers-Cances and Stéphanie Savy for valuable discussions on this topic. This research has received the help from IRESP during the call for proposals launched in 2012 as a part of French "Cancer Plan 2009–2013".

References

1. Anisimov, V.V.: Statistical modeling of clinical trials (recruitment and randomization). Commun. Stat. Theory Methods **40**(19–20), 3684–3699 (2011)
2. Anisimov, V.V., Fedorov, V.V.: Modelling, prediction and adaptive adjustment of recruitment in multicentre trials. Stat. Med. **26**(27), 4958–4975 (2007)
3. Anisimov, V., Mijoule, G., Savy, N.: Statistical modelling of recruitment in multicentre clinical trials with patients' dropout. Stat. Med. (2014, in progress)
4. Carter, R.E.: Application of stochastic processes to participant recruitment in clinical trials. Control. Clin. Trials **25**(5), 429–436 (2004)
5. Mijoule, G., Savy, S., Savy, N.: Models for patients' recruitment in clinical trials and sensitivity analysis. Stat. Med. **31**(16), 1655–1674 (2012)
6. Senn, S.: Some controversies in planning and analysing multi-centre trials. Stat. Med. **17**, 1753–1765 (1998)

Chapter 37
Mathematical Problems of Statistical Simulation of the Polarized Radiation Transfer

Gennady A. Mikhailov, Anna S. Korda, and Sergey A. Ukhinov

37.1 Introduction

Light propagation can be treated as a random Markov chain of photon-substance collisions that lead to either photon scattering or photon absorption. In the Monte Carlo method, the trajectories of this chain are simulated on a computer and statistical estimates for the desired functionals are computed. The construction of random trajectories for a physical model of the process is known as direct simulation. No weights are used, and the variances of Monte Carlo estimates are always finite (see [1]). In the case of considered polarized radiation, a general matrix-weighted algorithms for solving systems of radiative transfer integral equations with allowance for polarization were constructed and preliminarily studied in [1,4].

This paper is devoted to additional researches of the variant of the matrix-weight algorithm based on direct simulation of "scalar" transfer process. Due to the fact that the appropriate statistical estimates can have the infinite variance, the method of "ℓ-fold polarization", in which recalculation of a Stokes vector on a "scalar" trajectory is carried out no more, than ℓ times, is offered deprived of this deficiency. Thus polarization is not exactly taken into account, but errors of required estimates can be quite small.

Also this paper examines the finiteness of the variance of corresponding standard vector Monte Carlo estimates, which is required for constructing the correct confidence intervals. To this end, in [4] is considered the system of integral equations defining the covariance matrix of a weighted vector estimate. Numerical estimates based on the iteration of the resolvent showed that the spectral radius of the corresponding matrix-integral operator is fairly close to the product of the

G.A. Mikhailov • A.S. Korda • S.A. Ukhinov (✉)
Institute of Computational Mathematics and Mathematical Geophysics SB RAS, Novosibirsk State University, 630090 Novosibirsk, Russia
e-mail: sausau@ngs.ru

V.B. Melas et al. (eds.), *Topics in Statistical Simulation*, Springer Proceedings in Mathematics & Statistics 114, DOI 10.1007/978-1-4939-2104-1_37

spectral radius for an infinite medium, which is calculated analytically, and the easy-to-estimate spectral radius of the scalar integral operator associated with an "identity" scattering matrix. In the purpose of enhancement of analytical study of this practically important factorization, in this paper is given obtained at [3] dual (to the one which is considered in [2]) representation of the mean square error of the estimates of considered functionals.

37.2 General Information

Various methods are available for describing the polarization properties of light. The most widespread and convenient method is that proposed by Stokes in 1852, who introduced four parameters $I,\ Q,\ U,\ V$ with the dimension of intensity, which determine the intensity, degree of polarization, polarization plane, and degree of ellipticity of radiation. In what follows, we consider the corresponding components of the Stokes vector function of light intensity:

$$\mathbf{I}(\mathbf{r},\omega) = \left(I_1(\mathbf{r},\omega), I_2(\mathbf{r},\omega), I_3(\mathbf{r},\omega), I_4(\mathbf{r},\omega)\right)^T.$$

The simplest "phenomenological" Markov model of polarized radiative transfer arises when the medium is assumed to be isotropic. The only difference from the standard scalar model is that the scattering phase function is replaced with a scattering matrix, which transforms the Stokes vector associated with a given "photon" at a scattering point (see, e.g., [1]).

We used the following notations: $x = (\mathbf{r},\omega)$ is a point of the phase space, $\mathbf{r}$ is a point of R^3 space, $\omega = (a,b,c)$ is a unit direction vector aligned with the run of the particle $(a^2 + b^2 + c^2 = 1)$; $\mu = (\omega,\omega')$ is the cosine of the scattering angle, φ is the azimuthal scattering angle, $r_{11}(\mu)$ is the scattering phase function, $\sigma(\mathbf{r})$ is the extinction coefficient, $q(\mathbf{r})$ is the probability of scattering, l is the free path, $p_\chi(l;\mathbf{r}',\omega)$ is the sub-stochastic distribution density of the free path l from the point $\mathbf{r}'$ in the direction ω: $p_\chi(l;\mathbf{r}',\omega) = \sigma(\mathbf{r}' + \omega l)\exp\left(-\tau_{\text{op}}(l;\mathbf{r}',\omega)\right),\quad l \leqslant l^*(\mathbf{r}',\omega)$; $\tau_{\text{op}}(l;\mathbf{r}',\omega) = \tau_{\text{op}}(\mathbf{r}',\mathbf{r}) = \int\limits_0^l \sigma(\mathbf{r}' + s\omega)ds$ is the optical length of the interval $[\mathbf{r}',\mathbf{r}' + l\omega = \mathbf{r}]$, and $l^*(\mathbf{r}',\omega)$ is the distance from the point $\mathbf{r}'$ in the direction ω up to the boundary of the medium, which may be assumed to be convex. Here, the trajectory can terminate since the particle escapes from the medium.

Let $F(x), H(x)$ be the column vectors of the functions $f_1(x),\ldots,f_4(x)$ and $h_1(x),\ldots,h_4(x)$, respectively, and

$$\Phi(x) = (\varphi_1(x),\varphi_2(x),\varphi_3(x),\varphi_4(x))^T = \sigma(\mathbf{r})\mathbf{I}(x)$$

is the vector density of collisions.

The system of integral equations describing radiative transfer with allowance for polarization has the following matrix kernel:

$$K(x', x) = \frac{q(\mathbf{r}')e^{-\tau_{op}(\mathbf{r}',\mathbf{r})}\sigma(\mathbf{r})P(\omega', \omega, \mathbf{r}')}{|\mathbf{r} - \mathbf{r}'|^2} \times \delta\left(\omega - \frac{\mathbf{r} - \mathbf{r}'}{|\mathbf{r} - \mathbf{r}'|}\right).$$

Thus, we have a vector-integral equation of transfer with allowance for polarization with respect to the vector function Φ:

$$\Phi(x) = \int_X K(x', x)\Phi(x')dx' + F(x), \quad \Phi = \mathbf{K}\Phi + F. \tag{37.1}$$

Let's call the operator $\mathbf{K}$ the matrix-integral transfer operator. Monte Carlo algorithms are based on a representation of the solution of Eq. (37.1) in the form of a Neumann series. Such a representation holds if the norm of the operator $\mathbf{K}$ (or of its power $\mathbf{K}^{n_0}$) is less than unity [1,4].

Linear functionals of the solution of the integral equation are usually estimated by applying Monte Carlo methods. In the case of a system of second-kind integral equations, the general Monte Carlo algorithm for estimating such functionals can be described as follows.

Suppose that we want to calculate the functional

$$\mathrm{I}_H = (\Phi, H) = \sum_{i=1}^{m} \int_X \varphi_i(x)h_i(x)dx = \sum_{n=0}^{\infty}(\mathbf{K}^n F, H).$$

Here, H is a vector function with absolutely bounded components; i.e., $H \in L_\infty$. A homogeneous Markov chain $\{x_n\}$ in the phase space X is defined by the probability density $\pi(x)$ of the initial state x_0, by the transition probability density $r(x', x)$ from x' to x, and by the probability $p(x')$ that the trajectory terminates in the transition from the state x'. The function $p(x', x) = r(x', x)[1 - p(x')])$ is called the transition density.

An auxiliary random vector $\mathbf{Q}$ of weights is defined by the formulas

$$\mathbf{Q}_0 = \frac{F(x_0)}{\pi(x_0)}, \quad \mathbf{Q}_n = [K(x_{n-1}, x_n)/p(x_{n-1}, x_n)]\mathbf{Q}_{n-1}, \quad Q_n^{(i)} = \sum_{j=1}^{4} Q_{n-1}^{(j)} \frac{k_{ij}(x_{n-1}, x_n)}{p(x_{n-1}, x_n)}.$$

By analogy with a single integral equation, it is shown (see [1, 2]) that $\mathrm{I}_H = (\Phi, H) = \mathsf{E}\zeta$, where

$$\zeta = \sum_{n=0}^{N} \mathbf{Q}_n^T H(x_n) = \sum_{n=0}^{N}\sum_{i=1}^{4} Q_n^{(i)} H_i(x_n). \tag{37.2}$$

Here, N is the random index of the last state of the chain. Relation (37.2) describes the Monte Carlo algorithm for estimating I_H. The substantiation of this relation essentially relies on the expansion of the solutions of equations in the

Neumann series (see [1]). Since the first component in (37.2) is nonnegative, it can be averaged term by term. The remaining components can be averaged because of the majorant property of the first component (see [1]).

Consider the Monte Carlo algorithm for computing the intensity and polarization of multiply scattered light. The simplest part in this problem is the transition probability density $r(x', x)$, which is defined by the kernel $k_{11}(x', x)$ corresponding to radiative transfer without allowance for polarization. Obviously, in the simulation of such process, the vector of "weights" after scattering has to be transformed by a matrix with the elements $k_{ij}(x', x)/k_{11}(x', x)$.

As was mentioned above, a light ray is characterized by the Stokes vector $\mathbf{I} = (I, Q, U, V)$. The unscattered solar light $\mathbf{I}_0$ is assumed to be natural; i.e., $\mathbf{I}_0 = (I_0, 0, 0, 0)^{\mathrm{T}}$.

After scattering, the Stokes vector $\mathbf{I}$ is transformed according to the formula

$$\mathbf{I}(\mathbf{r}, \omega) = P(\omega', \omega, \mathbf{r}) \cdot \mathbf{I}(\mathbf{r}, \omega'),$$

where $P(\omega', \omega, \mathbf{r}) = L(\pi - i_2) R(\omega', \omega, \mathbf{r}) L(-i_1)/2\pi$,

$$L(i) = \begin{pmatrix} 1 & 0 & 0 & 0 \\ 0 & \cos 2i & \sin 2i & 0 \\ 0 & -\sin 2i & \cos 2i & 0 \\ 0 & 0 & 0 & 1 \end{pmatrix}.$$

Here, i_1 is the angle between the plane ω', s and the scattering plane ω', ω; i_2 is the angle between the scattering plane ω', ω and the plane ω, s; and s is a vector of the local spherical system of coordinates [1].

For an anisotropic medium, all 16 components of the scattering matrix $R(\omega', \omega, \mathbf{r})$ are generally different. For an isotropic medium, the scattering matrix simplifies to

$$R(\omega', \omega, \mathbf{r}) = \begin{pmatrix} r_{11} & r_{12} & 0 & 0 \\ r_{21} & r_{22} & 0 & 0 \\ 0 & 0 & r_{33} & r_{34} \\ 0 & 0 & -r_{43} & r_{44} \end{pmatrix}, \quad r_{ij} \equiv r_{ij}(\mu, \mathbf{r}).$$

If the scattering particles are homogeneous spheres, then $r_{11} = r_{22}$, $r_{12} = r_{21}$, $r_{33} = r_{44}$, $r_{34} = r_{43}$. The matrix R is normalized so that $\int_{-1}^{1} r_{11}(\mu)\, d\mu = 1$.

New photon's direction ω after scattering is defined by the scattering angle θ and the azimuthal angle φ. The cosine μ of the angle θ is simulated according to the r_{11}, i.e., according to the scattering phase function. The angle $\varphi \in (0, 2\pi)$ is assumed to be isotropic and is equal to that between the planes ω', s and ω, ω' measured counterclockwise when viewed against the incident ray ω'. Thus, the azimuthal angle is equal to i_1. After the new direction was chosen, i_1 and i_2 can be found using spherical trigonometry formulas.

The procedure for updating the Stokes vector after scattering includes the formulas

$$\begin{aligned} I(\mathbf{r},\omega) &= r_{11} \cdot I(\mathbf{r},\omega') + r_{12} \cdot A, \\ Q(\mathbf{r},\omega) &= (r_{21} I(\mathbf{r},\omega') + A r_{22}) \cos 2i_2 - (r_{33} B - r_{34} V(\mathbf{r},\omega')) \sin 2i_2, \\ U(\mathbf{r},\omega) &= (r_{21} I(\mathbf{r},\omega') + A r_{22}) \sin 2i_2 + (r_{33} B - r_{34} V(\mathbf{r},\omega')) \cos 2i_2, \\ V(\mathbf{r},\omega) &= r_{43} B + r_{44} V(\mathbf{r},\omega'), \end{aligned} \tag{37.3}$$

where $A = Q(\mathbf{r},\omega')\cos 2i_1 - U(\mathbf{r},\omega')\sin 2i_1$, $B = Q(\mathbf{r},\omega')\sin 2i_1 + U(\mathbf{r},\omega')\cos 2i_1$.

37.3 Method of ℓ-Fold Polarization

The "scalar" integral equation $\varphi = K\varphi + f$ [1] corresponding to the base scalar model of radiation transfer can be written in the vector form:

$$\Phi_0 = \mathbf{K}_0 \Phi_0 + F_0,$$

where $\Phi_0 = (\varphi, 0, 0, 0)^{\mathrm{T}}$, $F_0 = (f, 0, 0, 0)^{\mathrm{T}}$ and $\mathbf{K}_0$ is matrix-integral operator corresponding to the diagonal scattering matrix: $R = \mathrm{diag}(r_{11}, r_{11}, r_{11}, r_{11})$.

After ℓ iterations of Eq. (37.1) beginning with Φ_0, we get such approximation to the solution Φ:

$$\Phi_\ell = \mathbf{K}^\ell \Phi_0 + \sum_{n=0}^{\ell-1} \mathbf{K}^n F = \sum_{n=0}^{\infty} \mathbf{K}^\ell \mathbf{K}_0^n F_0 + \sum_{n=0}^{\ell-1} \mathbf{K}^n F. \tag{37.4}$$

We designate the usage of formula (37.4) for the approximate computation as "the method of ℓ-fold polarization".

For constructing the corresponding estimate we should use instead of ζ from (37.2) the following random variable:

$$\zeta_\ell = \sum_{n=0}^{\infty} \delta_n q_n \tilde{Q}_n^{\mathrm{T}} H(x_{n+\ell}) + \sum_{n=0}^{\min(\ell-1, N)} Q_n^{\mathrm{T}} H(x_n).$$

Here q_n are scalar weights, i.e.

$$q_0 = \frac{f(x_0)}{\pi(x_0)}, \quad q_n = q_{n-1} \frac{k_{11}(x_{n-1}, x_n)}{p(x_{n-1}, x_n)},$$

and the vector weight $\tilde{Q}_n$ corresponding to ℓ-fold polarization is calculated by the formula

$$\delta_{n+\ell}\frac{K(x_{n+\ell-1},x_{n+\ell})}{p(x_{n+\ell-1},x_{n+\ell})}\cdot\ldots\cdot\delta_{n+1}\frac{K(x_n,x_{n+1})}{p(x_n,x_{n+1})}I_0,$$

where $I_0=(1,0,0,0)^T$; δ_n is an indicator that a trajectory doesn't terminate before the state x_n.

The point of special interest for the solution of atmospheric optics problems is the estimate of an influence of polarization on the intensity of radiation, i.e. the difference $\Delta_p(x)=\varphi_1(x)-\varphi_1^{(0)}(x)$, where $\varphi_1^{(0)}(x)$ corresponds to approximate scalar model.

Quantity $\Delta_p(x)$ is an error in intensity estimate $\varphi_1(x)$ caused by non-account of polarization. We denote the value $\Delta_p(x)$ produced by ℓ-fold polarization as $\Delta_p^{(\ell)}(x)$.

If a source of radiation is non-polarized, i.e. $F_0=(f,0,0,0)^{\mathrm{T}}$, then we have, due to (37.3), $\Delta_p^{(1)}(x)=0$. Hence, "in first approximation" for estimate of $\Delta_p(x)$ we should use value $\Delta_p^{(2)}(x)$, whose statistical estimate is easy to find from formulas (37.3).

Let's denote x_n,x_{n+1},x_{n+2} as x'',x',x and let I_D be an indicator of domain $D\subset X$. In case $F\equiv F_0$ and $H=(I_D,0,0,0)^{\mathrm{T}}$, which corresponds to the estimate of the integral $\int\limits_D\varphi_1(x)dx$ for non-polarized source, we have from (37.3):

$$q_n\tilde{Q}_n^{\mathrm{T}}=q_n\Delta'\Delta\mathbf{I}_0'(r_{11}r_{11}'+r_{21}(\mu')r_{12}(\mu)\cos 2i_2'\cos 2i_1$$
$$-r_{22}(\mu')r_{12}(\mu)\sin 2i_2'\sin 2i_1),$$

where Δ' and Δ are the indicators that the trajectory doesn't terminate in transition to points x' and x, respectively. Due to finiteness of weight multipliers the vector norm $\|\tilde{Q}\|$ of auxiliary weight is uniformly bounded and $\mathsf{D}\zeta_\ell<\infty$ if $\mathsf{D}\zeta_0<\infty$. The last inequality holds in case of direct simulation for basic scalar model and also when absorption or escape from medium are not simulated and instead are accounted by weight multipliers, which are equal to probabilities of these events.

37.4 Criterion for the Finiteness of the $\mathsf{E}\zeta^2$

Consider the space $\mathbf{L}_1$ of matrix functions $\Psi(x)$ with norm $||\Psi||=\int\limits_X\sum\limits_{i,j=1}^m|\Psi_{i,j}(x)|dx$ and define the linear functional

$$(\Psi,\Psi^*)=\int\limits_X \mathrm{tr}[\Psi(x)\Psi^{*\mathrm{T}}(x)]dx=\int\limits_X\sum_{i,j=1}^m\Psi_{i,j}(x)\Psi_{i,j}^*(x)dx,$$

$\Psi^*\in\mathbf{L}_\infty$, $\|\Psi^*\|_{\mathbf{L}_\infty}=\operatorname{vrai}\sup\limits_{i,x}\|\Psi_i^*(x)\|$ [3].

Define also a linear operator $\mathbf{K}_p$ by

$$[\mathbf{K}_p\Psi](x) = \int\limits_X \frac{K^{\mathrm{T}}(y,x)\Psi(y)K(y,x)}{p(y,x)}dy$$

and, according to [2], linear operator $\mathbf{K}_p^*$:

$$[\mathbf{K}_p^*\Psi^*](x) = \int\limits_X \frac{K(x,y)\Psi^*(y)K^{\mathrm{T}}(x,y)}{p(x,y)}dy.$$

Since $\mathrm{tr}(AB) = \mathrm{tr}(BA)$, then $\mathrm{tr}(\Psi K \Psi^{*\mathrm{T}} K^{\mathrm{T}}) = \mathrm{tr}(K^{\mathrm{T}}\Psi K \Psi^{*\mathrm{T}})$, and therefore $(\Psi, \mathbf{K}_p^*\Psi^*) = (\mathbf{K}_p\Psi, \Psi^*)$. Moreover, we have $|(\Psi, \Psi^*)| \leqslant ||\Psi||_{\mathbf{L}_1}||\Psi^*||_{\mathbf{L}_\infty}$.

Hence $||\mathbf{K}_p||_{\mathbf{L}_1} = ||\mathbf{K}_p^*||_{\mathbf{L}_\infty}$ and $\rho(\mathbf{K}_p) = \rho(\mathbf{K}_p^*)$.

The operator $\mathbf{K}_p$ leaves invariant the cone $\mathbf{L}_1^+ \subset \mathbf{L}_1$ of symmetric nonnegative definite matrix functions, because the transformation $K^{\mathrm{T}}\Psi K$ preserves the nonnegative definiteness of matrices Ψ. From here the following statement turns out [3].

Theorem 37.1. *Suppose that* $\rho(\mathbf{K}_p) < 1$, $FF^{\mathrm{T}}/\pi_0 \in \mathbf{L}_1$, $H \in L_\infty$. *Then*

$$\mathrm{E}\zeta^2 = (\Psi, H[2\Phi^* - H]^{\mathrm{T}}),$$

where $\Phi^* = \mathbf{K}^*\Phi^* + H$, $\Psi = \mathbf{K}_p\Psi + FF^{\mathrm{T}}/\pi_0$, *and* $\Psi \in \mathbf{L}_1^+$.

Note that in [2] dual presentation of $\mathrm{E}\zeta^2$ was constructed:

$$\mathrm{E}\zeta^2 = \int\limits_X \frac{F^{\mathrm{T}}(x)\Psi^*(x)F(x)}{\pi(x)}dx = (\frac{FF^{\mathrm{T}}}{\pi}, \Psi^*),$$

where $\Psi^* = H\Phi^{*\mathrm{T}} + \Phi^* H^{\mathrm{T}} - HH^{\mathrm{T}} + \mathbf{K}_p^*\Psi^*$.

In [4] the spectral radius $\rho(\mathbf{K}_p)$ of the operator $\mathbf{K}_p$ was estimated by resolvent iterations on the basis of the limit relation of the form:

$$\frac{F^T[\lambda\mathbf{I} - \mathbf{K}]^{-(m+1)}H}{F^T[\lambda\mathbf{I} - \mathbf{K}]^{-m}H} \to \frac{1}{\lambda - \rho(\mathbf{K})}, \quad \lambda > \rho(\mathbf{K}), \quad \mathbf{I} = \mathrm{diag}(1,1,1,1).$$

In order to improve the convergence of the algorithm in place of $H(x_n)H^{\mathrm{T}}(x_n)$ was taken $\psi^{(0)}$, i.e. the first eigenfunction of the operator S_p, which represents the realization of $\mathbf{K}_p$ for the case of the infinite medium where $F^{\mathrm{T}} = (1,0,0,0)$.

It occurs that even for optically thin layers the approximate equality $\rho(\mathbf{K}_p) \approx \rho(S_p)\rho(L_p)$ is valid, where L_p is a scalar integral operator with the kernel $k_{11}^2(x',x)/p(x',x)$.

In [4] it is shown more detailed than in [2], that value $\lambda_0 = \rho(S_p)$ is the solution of the system of equations:

$$c_{11} + c_{21}a_1 = \lambda_0$$
$$c_{12} + (c_{22} + c_{33})a_1 + c_{43}a_2 = 2\lambda_0 a_1$$
$$c_{34}a_1 + c_{44}a_2 = \lambda_0 a_2$$

where $c_{ij} = \int\limits_{-1}^{1} \frac{r_{ij}^2(\mu)}{p_2(\mu)} d\mu$ and $p_2(\mu)$ is simulated distribution density of $\mu = (\omega, \omega')$.

It was found that for the aerosol scattering the value λ_0 is majorated with the value $\lambda_m = 1.178$, corresponded to the molecular scattering. On the other hand, for the real atmosphere layers value $\rho(L_p)$ is small, therefore $\rho(\mathbf{K}_p^*) < 1$ and $D\zeta < +\infty$.

In [4] the results of calculations of the spectral radii of the operators $\mathbf{K}_p$ and L_p for the molecular and the aerosol scattering are presented. Obtained values of $\rho(\mathbf{K}_p)/\rho(L_p)$ statistically insignificant differ from the analytically found values $\rho(S_p)$ and are estimated with sufficient accuracy using even only the first iteration of the resolvent.

On the basis of the dual representation obtained in [3] new approximate estimate of the $\rho(\mathbf{K}_p)$ is constructed:

$$\rho(\mathbf{K}_p) \approx \tilde{\rho}(\mathbf{K}_p) = \frac{(\mathbf{K}_p \Psi_0, \mathbf{I})}{(\Psi_0, \mathbf{I})} \approx C\rho(\tilde{L}_p)\rho(S_p), \tag{37.5}$$

and a value C is not significantly different from 1. Here $\mathbf{I} = \mathrm{diag}(1, 1, 1, 1)$, $\Psi_0 = \tilde{\Psi}^* \tilde{\psi}(x)$, $\tilde{\Psi}^*$ is considered above eigenmatrix of the operator S_p and $\tilde{\psi}(x)$ is the main eigenfunction of the scalar operator $\tilde{L}_p$, which corresponds to the radiation model with the replacement of anisotropic scattering on an isotropic, i.e. with $r_{11} \equiv 1/2$ and $\tilde{p}_{11}(\mu) \equiv 1/2$.

This estimate (37.5) isn't contrary to the numerical results given in [4], because for corresponding flat layers with the isotropic scattering it was obtained that $\rho(\tilde{L}_p|\tau = 1) \approx 0.62$, $\rho(\tilde{L}_p|\tau = 2) \approx 0.78$, $\rho(\tilde{L}_p|\tau = 4) \approx 0.9$, and these values are sufficiently close to the values of $\rho(L_p)$ from [4]. The estimate (37.5) can be recommended for practical use taking into account that for the optically thick media the substitution of the essentially anisotropic scattering with the isotropic scattering slightly increases the value $\rho(L_p)$.

Also the value $\rho_0(\tilde{L}_p|\tau = 10) \approx 0.974$ was obtained. Hence, we have for the flat layer with the optical thickness 10 and the molecular scattering: $\rho(\mathbf{K}_p) \approx 0.974 \times 1.178 = 1.15$, and with the aerosol scattering $\rho(\mathbf{K}_p) \approx 0.974 \times 1.02077 = 0.994$.

Acknowledgements This work was supported by the Russian Foundation for Basic Research (13-01-00441, 13-01-00746, 12-01-00034), and by MIP SB RAS (A-47, A-52).

References

1. Marchuk, G.I., Mikhailov, G.A., Nazaraliev, M.A., et al.: Monte Carlo Methods in Atmospheric Optics. Nauka, Novosibirsk (1976); Springer, Heidelberg (1980)
2. Mikhailov, G.A.: Optimization of Weighted Monte Carlo Methods. Nauka, Moscow (1987); Springer, Heidelberg (1992)
3. Mikhailov, G.A., Ukhinov, S.A.: Dual representation of the mean square of the Monte Carlo vector estimator. Doklady Math. **83**(3), 386–388 (2011)
4. Mikhailov, G.A., Ukhinov, S.A., Chimaeva, A.S.: Variance of a standard vector Monte Carlo estimate in the theory of polarized radiative transfer. Comput. Math. Math. Phys. **46**(11), 2006–2019 (2006)

Chapter 38
Using a Generalized Δ^2-Distribution for Constructing Exact D-Optimal Designs

Trifon I. Missov and Sergey M. Ermakov

38.1 Introduction

The construction of optimal designs according to a specified criterion is an optimization problem. The majority of relevant algorithms are based on generic methods in which the objective function is constructed in accordance with the chosen criterion. Focusing on the D-criterion, we will take advantage of the structure of the information matrix, whose determinant is to be maximized.

In this article we search for a D-optimal design of predefined size n. Namely, in a region X we specify a linear regression model with m linearly independent functions, and we look for an exact optimal design with n points ($n > m$), i.e., we have an experiment with n trials. We propose a procedure that is based on the properties of the information matrix, whose determinant is to be maximized. First, we normalize the determinant to a p.d.f. and develop a procedure for simulating random vectors from the resulting *generalized* Δ^2*-distribution* with parameters n and m (see [11]). The latter can be viewed as a natural extension of the Ermakov–Zolotoukhin Δ^2-distribution [6] that has a single parameter $m = n$. We simulate vectors from the generalized Δ^2-distribution and choose the sample modes as

T.I. Missov (✉)
Max Planck Institute for Demographic Research, Konrad-Zuse-Str. 1, 18057 Rostock, Germany

University of Rostock, Ulmenstr. 69, 18057 Rostock, Germany
e-mail: missov@demogr.mpg.de

S.M. Ermakov
Faculty of Mathematics and Mechanics, Department of Statistical Simulation, Saint Petersburg State University, 28 Universitetsky prospekt, 198504 Peterhof, Saint Petersburg, Russia
e-mail: sergej.ermakov@pobox.spbu.ru

V.B. Melas et al. (eds.), *Topics in Statistical Simulation*, Springer Proceedings in Mathematics & Statistics 114, DOI 10.1007/978-1-4939-2104-1_38

a starting generation of points for further optimization, which we perform by differential evolution (DE) [14]. The latter proved to find efficiently global optima for a number of complex objective functions [13].

38.2 Background

Consider $\varphi_1, \ldots, \varphi_m$ to be m linearly independent in a region X, $\dim X = s$, functions, continuous in a topology, in which X is compact. With no loss of generality we can treat $\varphi_1, \ldots, \varphi_m$ as an orthonormal system in $L^2(X, \mu)$, where μ is a σ-finite measure on X. Assume that at each point $x \in X$ a random variable Y_x is defined in such a way that $\mathrm{E}\, Y_x = \theta^T \varphi(x)$, where $\varphi(x) = (\varphi_1(x), \ldots, \varphi_m(x))^T$ is an $m \times 1$ vector of $L^2(X, \mu)$-orthonormal functions and $\theta = (\theta_1, \ldots, \theta_m)^T$ is an $m \times 1$ vector of unknown real parameters. We assume in addition that $\mathrm{Var}\, Y_x = \sigma^2$, $\mathrm{Cov}(Y_{x_1}, Y_{x_2}) = 0$ for $x, x_1, x_2 \in X$, $x_1 \neq x_2$. Denote by $D_n = (x_1, \ldots, x_n)$ a *discrete* design containing n points. The corresponding $n \times m$ design matrix is denoted by $X_n = \left|\left|\varphi_i(x_j)\right|\right|_{i=1,j=1}^{m,n}$. An *exact design* refers to the measure

$$\xi_N = \begin{pmatrix} x_{r_1} & x_{r_2} & \ldots & x_{r_N} \\ \frac{r_1}{n} & \frac{r_2}{n} & \ldots & \frac{r_N}{n} \end{pmatrix}, \tag{38.1}$$

where $x_{r_1} \neq x_{r_2} \neq \cdots \neq x_{r_N} \in D_n$, $N \leqslant n$, and r_i is the absolute frequency of x_{r_i} in D_n, $i = 1, \ldots, N$. The corresponding *information matrix* of ξ_N is given by

$$M(\xi_N) = \sum_{i=1}^{N} \varphi(y_i)\, \varphi^T(y_i)\, \frac{r_i}{n}. \tag{38.2}$$

An exact D-optimal design is a discrete measure (38.1) that maximizes $\det M(\xi_N)$. Its construction is based on numerical approximation procedures, most of which are based on Fedorov's sequential algorithm [7]. The associated difficulties concern convergence, choice of weights, and computational load (especially inverting the $m \times m$ information matrix at each step to assess the variance of the least squares estimate of the expected response, as well as the optimization procedure for the latter itself). These issues are addressed in a series of subsequent works (see, e.g., [1,5,15]), which offer solutions in special cases.

38.3 Simulation of the Generalized Δ^2-Distribution

The generalized Δ^2-distribution has a p.d.f. $\Delta^2_{n,m}$, proportional to the determinant of the information matrix of an n-point design in an m-parameter regression model

$$\Delta^2_{n,m}(Q) = \frac{(n-m)!}{n!} \det \left\| \sum_{i=1}^{n} \varphi_k(x_i)\varphi_l(x_i) \right\|_{k,l=1}^{m}, \tag{38.3}$$

where $Q = (x_1, \ldots, x_n)$ and $\varphi_1, \ldots, \varphi_m$, as previously, is an orthonormal system in $L^2(X, \mu)$. The orthonormality assumption is made primarily for simulation reasons. It is not restrictive in any way, as a linearly independent system can be easily orthogonalized by a Gram–Schmidt process, which does not alter the determinant in (38.3). The simulation procedure for the generalized Δ^2 distribution is based on the algorithm presented in [3, 4, 10]: we represent $\Delta^2_{n,m}(Q)$ as a product of conditional densities, which we iteratively simulate. The form of the conditional densities is given by the following:

Theorem 1. *Suppose $X = [0,1]^s$, μ is the Lebesgue measure, and $\varphi_1, \ldots, \varphi_m$ is an orthonormal system of functions in $L^2(X, \mu)$. For $x_1, \ldots, x_n \in X$ and $k = 1, \ldots, n-1$ denote*

$$p^{(n-k)}(x_1, \ldots, x_{n-k}) = \frac{(n-m)!}{n!} \int_X \det \left\| \sum_{i=1}^{n} \varphi_k(x_i)\varphi_l(x_i) \right\|_{k,l=1}^{m} dx_{n-k+1} \ldots dx_n, \tag{38.4}$$

$p^{(n)}(x_1, \ldots, x_n) = \Delta^2_{n,m}(x_1, \ldots, x_n)$ for $k = n$. Then the $(n-k)$-th conditional density $p_{n-k}(x_{n-k}|x_1, \ldots, x_{n-k+1})$ of $\Delta^2_{n,m}$ is given by

$$p_{n-k}(x_{n-k}\,|\,x_1, \ldots, x_{n-k+1}) = \frac{\sum\limits_{l=a_k}^{b_k} V_k^{m-l} \sum\limits_{\substack{1 \leqslant i_1 < \cdots < i_l \leqslant n-k \\ 1 \leqslant j_1 < \cdots < j_l \leqslant m}} \left(\det \left\| \varphi_{j_p}(x_{i_q}) \right\|_{p,q=1}^{l}\right)^2}{\sum\limits_{l=a_{k+1}}^{b_{k+1}} V_{k+1}^{m-l} \sum\limits_{\substack{1 \leqslant i_1 < \cdots < i_l \leqslant n-k-1 \\ 1 \leqslant j_1 < \cdots < j_l \leqslant m}} \left(\det \left\| \varphi_{j_p}(x_{i_q}) \right\|_{p,q=1}^{l}\right)^2}, \tag{38.5}$$

where $a_k = \max\{0, m-k\}$, $b_k = \min\{m, n-k\}$, $V_k^{m-l} = k!/(k-m+l)!$, and

$$\left(\det \left\| \varphi_{j_p}(x_{i_q}) \right\|_{p,q=1}^{l}\right)^2 = m \qquad \text{for } l = 0. \tag{38.6}$$

Proof. We will take advantage of

$$\det \left\| \sum_{i=1}^{n} \varphi_k(x_i)\varphi_l(x_i) \right\|_{k,l=1}^{m} = \sum_{1 \leqslant i_1 < \cdots < i_m \leqslant n} \left(\det \left\| \varphi_k(x_{i_j}) \right\|_{k,j=1}^{m}\right)^2, \tag{38.7}$$

(see Ermakov [2, p. 228]) and prove the theorem by induction. By integrating $\Delta^2_{n,m}(Q)$ with respect to x_n, we get

$$\frac{n!}{(n-m)!}p_{n-1}(x_1,\ldots,x_{n-1}) = \sum_{1\leqslant i_1<\cdots<i_m\leqslant n-1} \left(\det\left|\left|\varphi_p(x_{i_q})\right|\right|_{p,q=1}^{m}\right)^2 + \\ + m \sum_{\substack{1\leqslant i_1<\cdots<i_{m-1}\leqslant n-1 \\ 1\leqslant j_1<\cdots<j_{m-1}\leqslant m}} \left(\det\left|\left|\varphi_{j_p}(x_{i_q})\right|\right|_{p,q=1}^{m-1}\right)^2, \quad (38.8)$$

which proves the basis ($k = 1$). To perform the inductive step, suppose the statement holds for $k = r$, $1 \leqslant r < n-1$. We will prove that it holds for $k = r+1$, too. With no loss of generality let $m < n-m$. Then $p_{n-r}(x_1,\ldots,x_{n-r})$ has a different functional form for $1 \leqslant r < m$, for $m \leqslant r < n-m$, and for $n-m \leqslant r < n$. That is why we will perform the inductive step in each of the three cases separately. If $1 \leqslant r < m$, then $a_r = m-r$, $b_r = m$, and we have

$$\frac{n!}{(n-m)!}p_{n-r-1}(x_1,\ldots,x_{n-r-1}) = \frac{n!}{(n-m)!}\int_X p_{n-r}(x_1,\ldots,x_{n-r})\mu(dx_{n-r}) \\ = \sum_{l=m-r-1}^{m} C_{r+1}^{m-l} \sum_{\substack{1\leqslant i_1<\cdots<i_l\leqslant n-r-1 \\ 1\leqslant j_1<\cdots<j_l\leqslant m}} \left(\det\left|\left|\varphi_{j_p}(x_{i_q})\right|\right|_{p,q=1}^{l}\right)^2.$$

If $m \leqslant r < n-m$, then $a_r = 0$, $b_r = m$, and we have

$$\frac{n!}{(n-m)!}p_{n-r-1}(x_1,\ldots,x_{n-r-1}) = \frac{n!}{(n-m)!}\int_X p_{n-r}(x_1,\ldots,x_{n-r})\mu(dx_{n-r}) \\ = \sum_{l=0}^{m} C_{r+1}^{m-l} \sum_{\substack{1\leqslant i_1<\cdots<i_l\leqslant n-r-1 \\ 1\leqslant j_1<\cdots<j_l\leqslant m}} \left(\det\left|\left|\varphi_{j_p}(x_{i_q})\right|\right|_{p,q=1}^{l}\right)^2.$$

Finally, when $n-m \leqslant r < n$, $p_{n-r-1}(x_1,\ldots,x_{n-r-1})$ can be represented as a linear combination of determinants (of order $n-r-1$ and lower). In this case $a_r = 0$, $b_r = n-r$, and we have

$$\frac{n!}{(n-m)!}p_{n-r-1}(x_1,\ldots,x_{n-r-1}) = \frac{n!}{(n-m)!}\int_X p_{n-r}(x_1,\ldots,x_{n-r})\mu(dx_{n-r}) \\ = \sum_{l=0}^{n-r-1} C_{r+1}^{m-l-1} \sum_{\substack{1\leqslant i_1<\cdots<i_l\leqslant n-r-1 \\ 1\leqslant j_1<\cdots<j_l\leqslant m}} \left(\det\left|\left|\varphi_{j_p}(x_{i_q})\right|\right|_{p,q=1}^{l}\right)^2,$$

which completes the proof of the inductive step and the theorem. □

38.4 Examples: Exact D-Optimal Designs for Univariate and Bivariate Polynomial Regression

In this section we construct exact D-optimal designs for polynomial regression in $[0, 1]^s$, $s = 1, 2$ by implementing the three-step procedure presented in Sect. 38.1: we simulate random vectors from the $\Delta^2_{n,m}$, then pick up a subsample that leads to the N largest values of $\Delta^2_{n,m}(Q)$ $(N < n)$, and, finally, use DE with a starting generation from the previous step to allocate the global maximum.

38.4.1 *D*-Optimal Designs for Univariate Polynomial Regression

For a polynomial regression in [0, 1] the D-optimal design Q_D is concentrated at the two endpoints of the interval and in the roots of $\varphi_i'(x)$, $i = 3, \ldots, m$ (see, e.g., [9]). If $n = km$, $k \in \mathbb{Z}$, the exact D-optimal designs contain all these m points with equal weights k/n. If $n = km + p$, $p \in \mathbb{Z}$, $0 < p < m$, then $m - p$ points from Q_D are represented k times, while each of the remaining p points is represented $k + 1$ times. It is difficult to find a unifying pattern in the order by which x_i are sequentially added to the exact D-optimal design. For example, for $m = 3$ and $n = 4$ in almost all (989 of the 1,000 runs) of the three-step procedure the final exact D-optimal design was concentrated in $(0, 0.5, 0.5, 1)$, i.e. when we add a fourth point, it should be located in 0.5. However, for $m = 3$ and $n = 5$ the resulting exact D-optimal design was either $(0, 0, 0.5, 0.5, 1)$ or $(0, 0.5, 0.5, 1, 1)$ with almost the same number of occurrences (483 vs 517). We observed the same structure for $m = 3$ and higher $n = 3k + 1$ (0.5 comes first) or $n = 3k + 2$ (no distinct pattern whether 0 or 1 comes first). For $m = 4$ (the D-optimal design is concentrated in 0, 0.28, 0.72, and 1), and $n = 4k + 2$ the internal points are "added" first (in 99.2 % of the cases), but in a different order (looking at results for $n = 4k + 1$), followed by the endpoints (again in a different order). We observed the same principle for higher m, too, which might be indicating that the exact D-optimal design for certain n is not unique, i.e. $\Delta^2_{n,m}$ is not unimodal. This is in line with the theoretical findings of Gaffke and Krafft [8] for univariate quadratic regression.

38.4.2 *D*-Optimal Designs for Bivariate Polynomial Regression

Consider a polynomial regression in $[0, 1]^2$. Then for

$$\varphi_1 = 1, \qquad \varphi_2 = \sqrt{3}\,(2x - 1), \qquad \varphi_3 = \sqrt{3}\,(2y - 1) \tag{38.9}$$

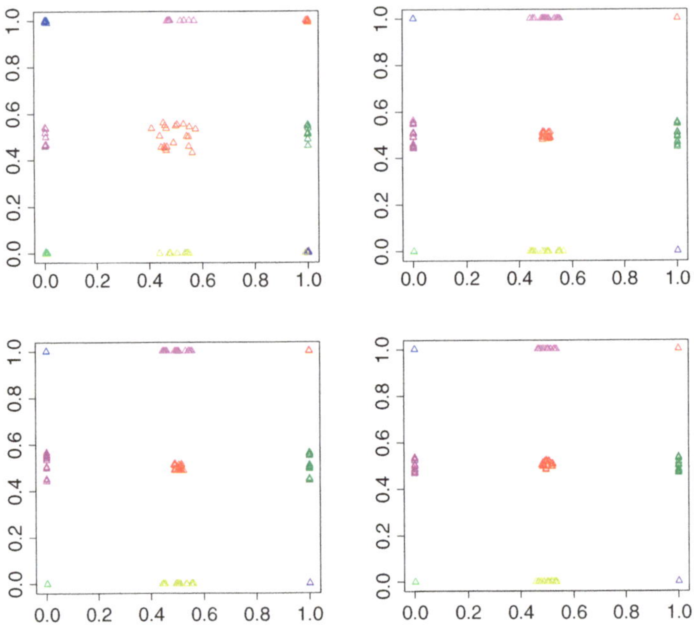

Fig. 38.1 D-optimal designs for quadratic regression in $[0,1]^2$: $m = 6$; $n = 7$ (*top left panel*), $n = 10$ (*top right panel*), $n = 13$ (*bottom left panel*), $n = 16$ (*bottom right panel*). Each design contains the four vertices, points on the border of the square, and exactly one internal point, which is located approximately at $(0.5, 0.5)$

the exact D-optimal design is allocated on the vertices of the square $[0,1]^2$ with varying weights. The pattern of "consecutive addition" is similar to the univariate case: if $n = km + p$, $m - p$ vertices appear k times and the other p vertices $k+1$ times. If we consider a quadratic regression, i.e. add

$$\varphi_4 = \sqrt{5}\,(6x^2 - 6x + 1), \qquad \varphi_5 = \sqrt{5}\,(6y^2 - 6y + 1), \qquad \varphi_6 = 12xy - 6x - 6y + 3$$

to (38.9), then the exact D-optimal design is no longer concentrated only on the vertices but also on the borders of the square (see Fig. 38.1), approximately halfway between the vertices. The latter is to be expected as 0.5 (x or y) is the root of the derivative of $\varphi_4, \varphi_5, \varphi_6$. The exact D-optimal design in this special case contains exactly one internal point, which is in line with Podkorytov's theoretical finding [12]. In the case of cubic regression ($m = 6$, see Fig. 38.2) the points on the borders correspond to the roots (0.28 and 0.72) of the derivatives of the cubic terms (see Appendix B), and there are three internal points. Note that the exact D-optimal design is *not* unique in all of the above cases.

If we continue further, the exact D-optimal design in $[0,1]^3$ for linear regression ($m = 4$) is located in 4 of the 8 vertices of the cube (with no unique solution), for quadratic regression ($m = 9$) in the 8 vertices and a point, which may lie on one of the sides or be internal, but in any case its coordinates equal the roots of φ_i', etc.

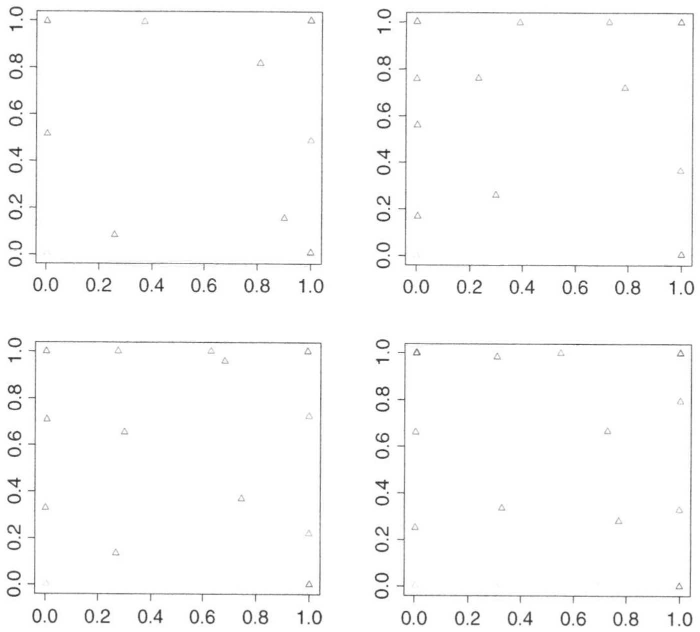

Fig. 38.2 D-optimal designs for cubic regression in $[0,1]^2$: $m = 10$; $n = 11$ (*top left panel*), $n = 14$ (*top right panel*), $n = 16$ (*bottom left panel*), $n = 17$ (*bottom right panel*). Each design contains the four vertices, points on the border of the square, and three internal points

Conclusion

We propose a general procedure for constructing exact D-optimal designs of predefined size n by taking advantage of the form of the associated information matrix, whose determinant is to be maximized. We normalize the latter to a pdf and simulate vectors from the resulting generalized Δ^2-distribution. We take a subset that delivers the highest N values of $\Delta^2_{n,m}(Q)$ and run a DE algorithm to allocate precisely the mode of $\Delta^2_{n,m}$. We illustrate this three-step procedure by constructing D-optimal designs for polynomial regression in $[0,1]^s$ ($s = 1, 2$). Affine transformations of the study region will not influence the procedure as only two things have to be adjusted: the set of orthonormal functions and the normalizing constant of $\Delta^2_{n,m}(Q)$, which has to take into account the s-measure of the region. In regions with a more complex structure orthonormality could be dropped and the simulation algorithm for $\Delta^2_{n,m}$ could be adjusted accordingly [11]. The three-step algorithm we present in this paper can be attractive in the sense that it is general and applicable to regression problems of any difficulty, in the presence of good software solutions for DE (available in almost all

(continued)

widely used statistical packages) and $\Delta^2_{n,m}$-simulation. Further research must focus on the performance of our procedure (and, in particular, DE as a global optimization algorithm) in regions of more complex structure and less trivial systems of linearly independent functions.

Acknowledgements We express our gratitude to Todor A. Angelov for his assistance in efficient programming.

References

1. Atwood, C.L.: Sequences converging to D-optimal designs of experiments. Ann. Stat. **1**(2), 342–352 (1973)
2. Ermakov, S.M.: Die Monte-Carlo Methode und verwandte Fragen. Deutsch, Wissenschaft (1975)
3. Ermakov, S.M., Missov, T.I.: Simulation of the Δ^2-distribution. Vestnik St. Petersburg University: Mathematics **4**(38), 53–60 (2005)
4. Ermakov, S.M., Missov, T.I.: On resampling-type methods in regression analysis. In: Chirkov, M.K. (ed.) Proceedings of the Smirnov Scientific Research Institute of Mathematics and Mechanics, pp. 27–36, Publishing House of the St. Petersburg State University, Saint Petersburg (2005)
5. Ermakov, S.M. Zhigljavsky, A.A.: Mathematical Theory of Optimal Design. Nauka, Moscow (1987)
6. Ermakov, S.M, Zolotukhin, V.G.: Polynomial approximations and the Monte Carlo Method. Probab. Theory Appl. **5**, 428–431 (1960)
7. Fedorov, V.V.: Theory of Optimal Experiments. Academic, New York (1972)
8. Gaffke, N., Krafft, O.: Exact D-optimum designs for quadratic regression. J. R. Stat. Soc. B **44**(3), 394–397 (1982)
9. Karlin, S., Studden, W.J.: Optimal experimental designs. Ann. Math. Stat. **37**(4), 783–815 (1966)
10. Missov, T.I.: Integral evaluation using the Δ^2-distribution. Simulation and illustration. Monte Carlo Methods Appl. **13**(2), 219–225 (2007)
11. Missov, T.I., Ermakov, S.M.: On importance sampling in the problem of global optimization. Monte Carlo Methods Appl. **15**(2), 135–144 (2009)
12. Podkorytov, A.N.: On the properties of D-optimal designs for quadratic regression. Vestnik LGU **2**(7), 163–166 (1975)
13. Price, K.V., Storn, R., Lampinen, J.: Differential Evolution: A Practical Approach to Global Optimization. Springer, Berlin/Heidelberg (2005)
14. Storn, R., Price, K.: Differential evolution—a simple and efficient heuristic for global optimization over continuous spaces. J. Global Optim. **11**, 341–359 (1997)
15. Tsay, J.-Y.: On the sequential construction of D-optimal designs. J. Am. Stat. Assoc. **71**(355), 671–674 (1976)

Chapter 39
Sample Size in Approximate Sequential Designs Under Several Violations of Prerequisites

Karl Moder

39.1 Introduction

In classical statistical analysis data are collected at the beginning and a specific test is applied. Based on some knowledge about α, β resp. power, an expected difference δ and about the variances in the underlying populations an appropriate sample size can be determined in advance. In sequential designs data are gathered and tested step by step. As soon as the test comes to a decision the procedure stops. So the sample size is a random variable. Only an upper limit for it can be calculated in advance. This upper limit of the sample size as well as the mean sample size is affected by the type of the distribution and the variances in the populations. The effects of these influences are presented here.

Several variants of sequential designs exist. This paper refers mainly to the triangular design [8], but also group sequential designs developed by [4, 5] and [3] are examined.

In triangular designs [8] with continuous monitoring the boundary values are on a straight line in the score scale for each boundary. So two regression lines which intersect each other define a continuation region. As in all sequential procedures recruitment of data must be stopped as soon as the continuation region is left. A maximum sample size can be calculated based on these regression parameters (a, b) by means of $n = \pm\frac{a\,(1+\sqrt{2\,b^2+1})}{b}$.

In this paper we restrict ourselves to testing hypotheses about means of normal distributions. The procedure in a two-sample situation corresponding to the t-test is as follows:

K. Moder (✉)
Institute of Applied Statistics and Computing, University of Natural Resources and Life Sciences, Vienna, Austria
e-mail: karl.moder@boku.ac.at

V.B. Melas et al. (eds.), *Topics in Statistical Simulation*, Springer Proceedings in Mathematics & Statistics 114, DOI 10.1007/978-1-4939-2104-1_39

Table 39.1 Sequential analysis for the data of Rasch [6] according to Whitehead [8]

nr	sex	height	v	z	ll.	ul.
1	female	165			-4.030	4.030
2	male	179	0.250	-1.000	-4.101	3.815
3	female	168	0.347	-1.385	-4.129	3.732
4	male	180	0.514	-1.971	-4.176	3.589
5	female	168	0.620	-2.408	-4.207	3.498
6	male	188	0.851	-2.791	-4.273	3.300
7	female	173	1.028	-3.100	-4.323	3.148
8	male	174	1.326	-3.283	-4.408	2.893
9	female	167	1.440	-3.753	-4.441	2.795
10	male	185	1.583	-4.282	-4.482	2.673
11	female	169	1.711	-4.729	-4.518	2.563

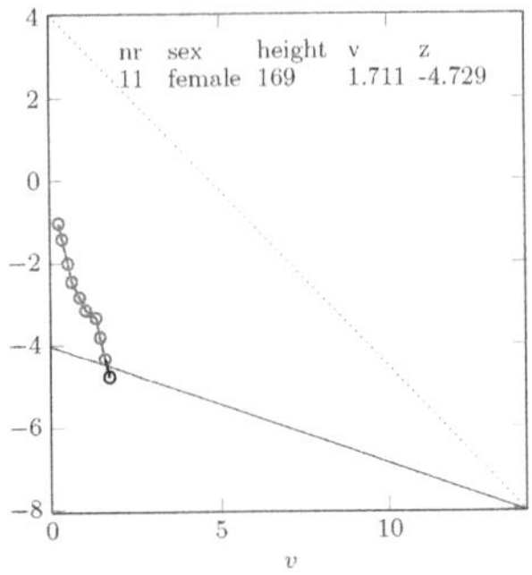

- Formulate H_0:
  $H_0 : \theta = \theta_0 \qquad H_A : \theta = \theta_1 < \theta_0 \quad \text{or} \quad \theta = \theta_2 > \theta = \theta_0$
- Define α, β:
- Calculate θ:
  $\theta = \frac{\mu_1 - \mu_2}{\sigma}$
- Calculate regression parameters:
  $a = (1 + u_{1-\beta}/u_{1-\alpha} \log(1/(2\alpha))/\theta$
  $b = \theta/[2(1 + u_{1-\beta}/u_{1-\alpha})]$
- Calculate the test statistic:
  $S_n^2 = \frac{1}{n_1+n_2}\left\{\sum_{i=1}^{n_1} x_{1i}^2 + \sum_{i=1}^{n_2} x_{2i}^2 - \left(\sum_{i=1}^{n_1} x_{1i} + \sum_{i=1}^{n_2} x_{2i}\right)^2 / (n_1 + n_2)\right\}$
  $z_n = \frac{n_1 n_2}{n_1+n_2} \frac{\bar{x}_1 - \bar{x}_2}{S_n}, \qquad v_n = \frac{n_1 n_2}{n_1+n_2} - \frac{z_n^2}{2(n_1+n_2)}$
- Continue sampling as long as z_n lies within the continuation region; otherwise accept/reject H_0:

continuation region	
$-a + 3bv_n < z_n < a + bv_n$	$\theta > \theta_0$
$a + bv_n < z_n < -a + 3bv_n$	$\theta < \theta_0$

The procedure is illustrated by means of a study on body height of female and male students [6]. $H_0 : \mu_1 = \mu_2, \quad \alpha = 0.05 \quad H_A : \mu_1 < \mu_2, \quad \beta = 0.05$

$\mu_1 = 170, \mu_2 = 178, \sigma = 7$

$\theta_1 = \frac{\mu_1 - \mu_2}{\sigma} = \frac{-8}{7} = -1.143$

$a = (1 + \frac{u(0.95)}{u(0.95)}) \log(\frac{1}{0.1}) \frac{1}{\theta_1} = -4.0295$

$b = \frac{-1.143}{2(1+\frac{u(0.95)}{u(0.95)}} = -0.2857$

Intersection of regression: $v_{\max}$=14.103, $z_{\max} = -8.059$

max. number of observ.: $n = 29$ (t-test: 17)

The results of the depicted procedure are shown in Table 39.1 and in the accompanying graphics. Entering the eleventh person, z_n falls below the lower limit (ll.) of the continuation region. So we have to reject our null hypothesis and stop

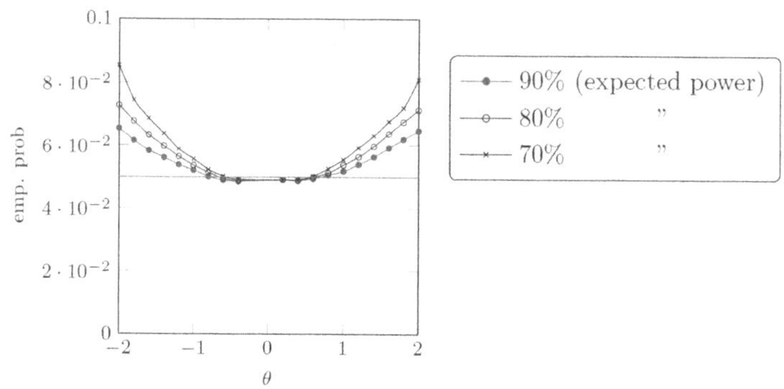

Fig. 39.1 Empirical type I error rate for the triangular test for different presumed values of θ but a true θ of 0 ($\alpha = 0.05, \sigma_1 = \sigma_2 = 1$)

the procedure. Although the maximum sample size (29) for the sequential analysis is higher than that of the t-test (17) only eleven observations are needed to make a decision.

In a simulation study the effects of missing prerequisites for this kind of test were evaluated.

39.2 Simulation Results

Several situations with respect to heterogeneity of variances, skewness, and kurtosis were examined by means of a simulation study. The standard deviation in the first distribution was fixed to 1, whereas that of the second distribution varied between 1, 2, and 3. Based on the Fleishman system [1] skewed distributions were generated ($\gamma_1 = -3, -1, 0, 1, 3$). For kurtosis γ_2 was set to $-15, 0, 5, 15$.

A Fortran code was developed to evaluate triangular designs. Each simulation run is based on 1 million analyses. SAS [7] procedures for the methods of [3, 4] were used. Here only 10,000 analyses were performed per simulation run because of time reasons.

39.2.1 Type I Error Rate

Sequential designs are based on reasonable assumptions about parameters of the underlying distributions. In contrast to fixed sample designs it is necessary to plan the experiment. So some knowledge about relevant differences between distributions is necessary.

Figure 39.1 shows empirical type I error rates if the presumed standardized difference (θ) between distributions varies between -2 to -0.4 and 0.4 to $+2$ in steps of 0.2. Values close to zero cause a and n to grow to infinity and are not considered.

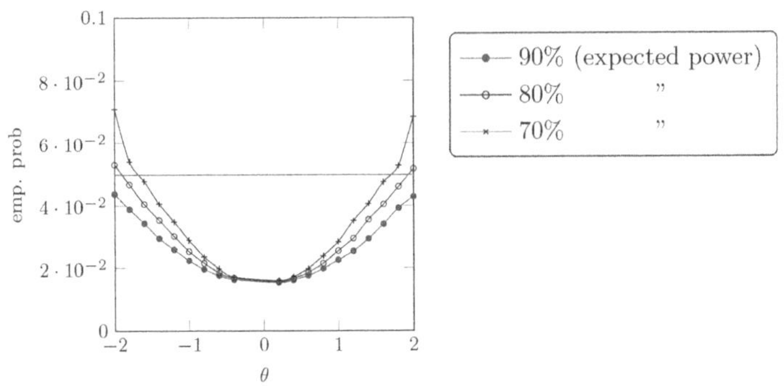

Fig. 39.2 Empirical type I error rate for the triangular test for different presumed values of θ but a true θ of 0 ($\alpha = 0.05, \sigma_1 = 1, \sigma_2 = 2$)

For high levels of presumed θs the predefined α value is exceeded. In this situation only small sample sizes are needed to reject or accept the null hypothesis. All regression parameters are based on the quantiles of the normal distribution. As these approximations work poorly for small sample sizes the α-value is not kept. Especially in cases when the assumed power is low this effect is intensified.

In Fig. 39.2 the situation is similar to that of Fig. 39.1, but variances are inhomogeneous.

Inhomogeneous variances decrease type I error rate in situations where θ is small resp. n is high. But for high θs (n small) α is exceeded and the influence of small sample sizes is more pronounced as with homogeneous variances. The situation gets worse if heterogeneity becomes more extreme.

39.2.2 Power

The following figures refer to the situation that the expected θ-value corresponds to the true θ. In this situation the empirical power should correspond to the predefined power level.

If variances are homogeneous, then the empirical power corresponds to the expected power as long as θ is small (which leads to large sample sizes) and variances are homogeneous (Fig. 39.3, picture on the left). As soon as variances are inhomogeneous the empirical power decreases dramatically (Fig. 39.3, picture on the right). If the standard deviation in the second population is twice as high as in the first, the power decreases from 90 to ~60 %. If the standard deviation of the second population is three times as high as in the first, the power decreases from 90 to ~20 %.

Table 39.2 shows sample sizes associated with the situations depicted in Fig. 39.3.

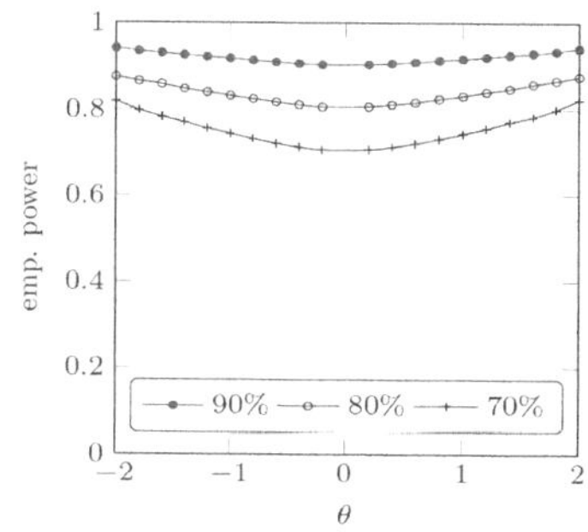

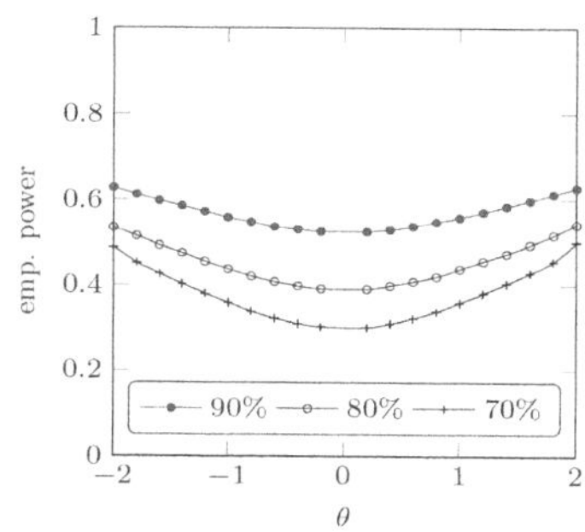

Fig. 39.3 Empirical power in comparison with theoretical power (90, 80, 70 %) for different levels of θ and homogeneous variances (*left picture*) as well as inhomogeneous variances ($\sigma_1 = 1$, $\sigma_2 = 2$) (*right picture*)

The sample size of the t-test is greater than the mean sample size of the triangular test (but smaller than the maximum sample size), no matter if variances are homogeneous or not. In case of inhomogeneous variances sample size for the triangular design depends on the standard deviation of the population which is used for calculation. If the higher one is used, the mean observed (seq.obs2) and maximum sample size (seq.max2) is smaller than in the situation where the smaller one is used (seq.obs1, seq.max1).

39.2.3 Skewness, Kurtosis

Based on the Fleishman system [1] distributions with different levels of kurtosis were generated. All sequential designs mentioned above were examined. No remarkable influences were found in regard to type I error rate and power.

In the case of skewed distributions the influence on power and type I error rate is high if the skewness differs in the distributions to be compared. As can be seen from Fig. 39.4 type I error rate raises up to 30 % in an one sided test situation if the distributions are skewed to a different level ($\gamma_{11} = -3$, $\gamma_{12} = 3$). The power is close to 1 for high θs. If variances are inhomogeneous too, it may happen, that sometimes type I error rate (no difference between means exist) exceeds the empirical power for situations where real differences exist even if the calculated power is 90 %. So in such situations the triangular test is completely inappropriate.

Similar results—with partly more pronounced deviations from expected alpha and power levels—can be found for group sequential designs [3, 5, 9] which are not shown here but were simulated based on equal assumptions about distributions.

Table 39.2 Observed and maximum sample sizes for the triangular design in comparison with the sample sizes for the t-test at different levels of θ if variances are homogeneous ($\sigma_1 = \sigma_2 = 1$) (Table on the left) or inhomogeneous ($\sigma_1 = 1, \sigma_2 = 2$) (Table on the right)

$\sigma_1 = \sigma_2 = 1$					$\sigma_1 = 1, \sigma_2 = 2$						
		Sample size					Sample size				
$1 - \beta$	θ	seq.obs	seq.max	t-test	$1 - \beta$	θ	seq.obs1	seq.obs2	seq.max1	seq.max2	t-test
90 %	2.0	7.8	14	12	90 %	2.0	9.4	7.6	14	29	24
	1.2	17.3	40	26		1.2	24.0	16.8	40	81	62
	0.4	132.6	364	216		0.4	207.4	134.0	364	729	538
80 %	2.0	6.6	10	10	80 %	2.0	7.2	6.6	10	21	18
	1.2	14.2	30	20		1.2	17.4	14.5	30	58	46
	0.4	106.9	264	156		0.4	148.9	116.1	264	526	390
70 %	2.0	5.7	8	8	70 %	2.0	5.8	5.6	8	16	14
	1.2	11.7	22	16		0.8	28.4	25.9	50	100	76
	0.4	86.4	200	120		0.4	111.1	99.9	200	400	296

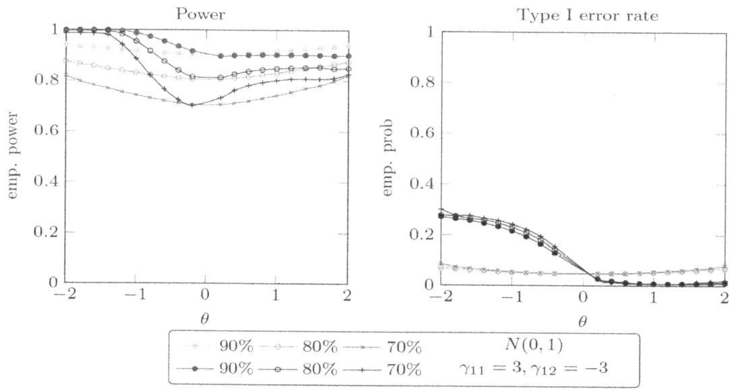

Fig. 39.4 Type I error rate and power for the triangular design if the distributions are skewed ($\gamma_{11} = -3, \gamma_{12} = 3$) and variances are homogeneous ($\sigma_1 = \sigma_2 = 1$)

Conclusions

Tests for triangular designs as well as group sequential designs are based on normal approximations. If differences between populations are high, then the calculated sample sizes are small. In the follow this normal approximation is very bad and type I error rate exceeds the predefined α-level, even if all other prerequisites are met.

Inhomogeneous variances show a high impact on power and type I error rate. Kurtosis does hardly affect power and type I error rate, whereas different skewness in populations may lead to a completely useless test.

Mean sample sizes are always smaller than with the ordinary non-sequential approach although the maximum necessary sample size to get a decision is higher.

Type one error rate and power in sequential designs depend to a very high degree on sample size and on prerequisites. So the use of these designs should be restricted to situations where prerequisites are met and sample size is not to small.

References

1. Fleishman, A.I.: A method for simulating non-normal distributions. Psychometrika **43**, 521–532 (1978)
2. Kittelson, J.M., Emerson, S.S.: A unifying family of group sequential test designs. Biometrics **55**, 874–882 (1999)
3. O'Brien, P.C., Fleming, T.R.: A multiple testing procedure for clinical trials. Biometrics **35**, 549–556 (1979)

4. Pocock, S.J.: Group sequential methods in the design and analysis of clinical trials. Biometrika **64**, 191–199 (1977)
5. Pocock, S.J.: Interim analyses for randomized clinical trials: the group sequential approach. Biometrics **38**, 153–162 (1982)
6. Rasch, D., Pilz, J., Verdooren, R., Gebhardt, A.: Optimal Experimental Design with R. Chapman & Hall/CRC, Boca Raton (2011)
7. SAS Institute Inc.: SAS/Stat 9.2 User's Guide. SAS Institute, Cary (2008)
8. Whitehead, J.: The Design and Analysis of Sequential Clinical Trials, 2nd edn. Ellis Horwood, NewYork (1997)
9. Whitehead, J., Stratton, I.: Group sequential clinical trials with triangular continuation regions. Biometrics **39**, 227–236 (1983)

Chapter 40
Numerical Stochastic Models of Meteorological Processes and Fields

Vasily Ogorodnikov, Nina Kargapolova, and Olga Sereseva

Numerical stochastic models of scalar and vector time-series, spatial and spatial-time random fields based on real data are widely used for solution of different problems in science and technology. As examples it is possible to refer to problems in atmospheric optics related to solar radiation scattering in clouds [13], to oceanologic problems related to rhythmic of oceanologic processes [2] and analysis of undulating surface (especially when freak waves appear) [14]. In statistical meteorology such models are used for study of extreme events (such as long-term frosts or drought), sudden drops of meteorological parameters or their unfavorable combinations [4, 10], for study of meteorological parameters' dynamic influence to natural and technical objects and processes, for prediction of forest fires and so on. Such models are also used in financial Mathematics and for telecommunication net' construction.

In this paper several approaches to modeling of random meteorological processes and fields with respect to spatial and time-specificity are considered. All models are based on long-term real data, obtained on 47 weather stations in Novosibirsk region, Perm and Astrakhan.

V. Ogorodnikov (✉) • N. Kargapolova
Institute of Computational Mathematics and Mathematical Geophysics SB RAS, prospect Akademika Lavrentjeva, 6, Novosibirsk 630090, Russia

Novosibirsk State Univesity, Pirogova, 2, Novosibirsk 630090, Russia
e-mail: ova@osmf.sscc.ru; nkargapolova@gmail.com

O. Sereseva
Institute of Computational Mathematics and Mathematical Geophysics SB RAS, prospect Akademika Lavrentjeva, 6, Novosibirsk 630090, Russia
e-mail: seresseva@mail.ru

V.B. Melas et al. (eds.), *Topics in Statistical Simulation*, Springer Proceedings in Mathematics & Statistics 114, DOI 10.1007/978-1-4939-2104-1_40

40.1 Modeling of Meteorological Indicator Time-Series with Daily Periodicity on Basis of Markov Chains

Since many of meteorological processes possess daily periodicity, for their simulation it is necessary to use algorithms which allow taking this daily rhythm into account. One of the approaches to simulation is based on special type of inhomogeneous Markov chains, studied in [5]. It is shown that for binary Markov chain $\xi_t, t = 0, 1, 2, \ldots$ with transition probability matrix that is a periodic time-function, following proposition holds.

Proposition. *One dimensional distribution of* ξ_t *is oscillating time-function. Limit distribution is a periodic time-function. Asymptotically process* ξ_t *is periodically correlated. Constant value series distribution, its first moment and variance are also periodic time-functions.*

Obtained analytical formulas, describing distribution, correlation structure and other characteristics of ξ_t, let analyze simulated process only on basis of estimated by real data characteristics of Markov chain.

Example. Let $\chi_t, t = 0, 1, 2, \ldots$ be air temperature, measured every 12 h (at midnight and noon) during a month. Indicator process $I(\chi_t)$ is defined as

$$I(\chi_t) = \begin{cases} 1, & \chi_t \geqslant c, \\ 0, & \chi_t < c, \end{cases}$$

where c (°C) is given level. Using $I(\chi_t)$ it is possible to estimate initial distribution and transition matrix (as time-function of period 2) of Markov chainξ_t. Table 40.1 gives probabilities that air temperature is higher than c (°C) at last measurement in month, obtained from real data and from model. Third column contains values of mean-square deviation arising from real data estimation (data for 32 years was used).

Table 40.1 Probabilities of c (°C)-level exceedance (Astrakhan, December)

c°C	$P(I(\chi_t) = 1)$	σ_T	$P(\xi_t = 1)$
−15	1.0000	0.0000	0.9975
−10	0.9063	0.0515	0.9667
−5	0.8438	0.0642	0.8976
−3	0.8125	0.0690	0.8291
−1	0.6875	0.0819	0.6291
0	0.5625	0.0877	0.5216
1	0.3438	0.0840	0.3782
3	0.2188	0.0731	0.2102
5	0.0938	0.0515	0.1052

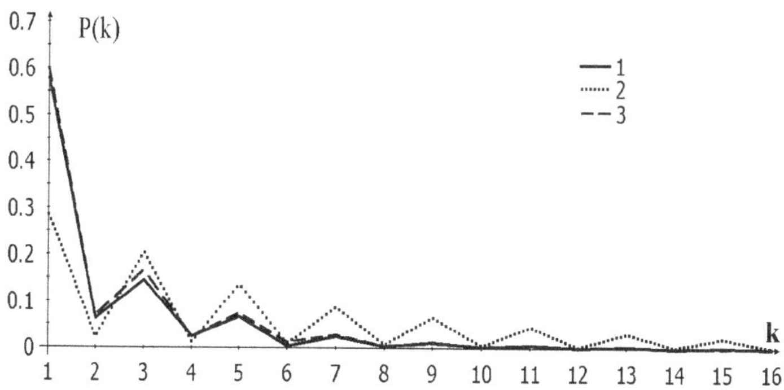

Fig. 40.1 Distribution of $(1,1)^T$-series length, $c^1 = 20°\text{C}$, $c^2 = 2\ \text{m/s}$. *Curve 1*: real data, *curve 2*: 1st order Markov chain, *curve 3*: 2nd order Markov chain

Similar approach can be used for analysis of complex of meteorological processes. Let $\mathbf{x}_t = \left(e_t^1, e_t^2\right)^T$ be a real data vector process: e_t^1—value of first meteorological process at moment t, e_t^2—value of second process at the same moment; c^1 and c^2 are corresponding given levels. Let's define indicator process

$$I(\mathbf{x}_t) = \begin{cases} (1,1)^T, & e_t^1 \geq \tilde{n}^1 \text{ and } e_t^2 \geq c^2, \\ (1,0)^T, & e_t^1 \geq \tilde{n}^1 \text{ and } e_t^2 < c^2, \\ (0,1)^T, & e_t^1 < \tilde{n}^1 \text{ and } e_t^2 \geq c^2, \\ (0,0)^T, & e_t^1 < \tilde{n}^1 \text{ and } e_t^2 < c^2. \end{cases} \tag{40.1}$$

Inhomogeneous vector Markov chain $\mathbf{X}_t$ with time-periodic transition probability matrix can be used as a model of process (40.1) It should be noted that order or chain can be varied. Period of transition matrix, as a time-function, is equal to number of measurement throughout a day. Initial distributions $P(\mathbf{X}_0 = (i,j)^T)$, $i, j \in \{0, 1\}$ of chain $\mathbf{X}_t$ and transition matrix are estimated by real data.

Figure 40.1 shows probabilities that value of studied process is equal to $(1,1)^T$ during k measurements, if e_t^1—air temperature, e_t^2—wind speed modulus. Probability for e_t^1—air temperature and e_t^2—relative humidity is given in Fig. 40.2. Estimations were made on basis of real data, obtained in July in Perm with 2 measurements per day. 100,000 model samples were used.

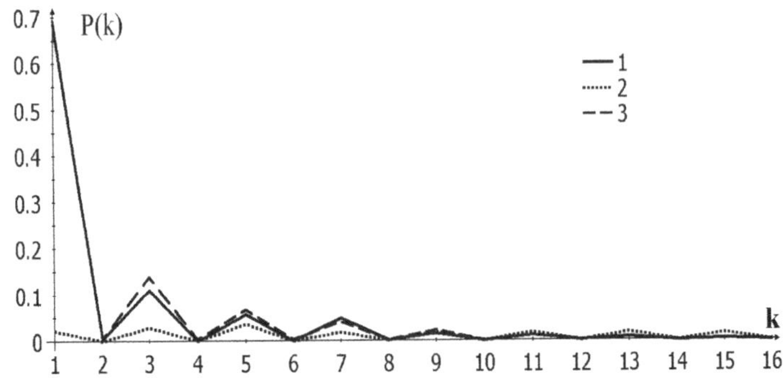

Fig. 40.2 Distribution of $(1,1)^T$-series length, $c^1 = 10°\text{C}$, $c^2 = 40\,\%$. *Curve 1*: real data, *curve 2*: 1st order Markov chain, *curve 3*: 2nd order Markov chain

40.2 Numerical Stochastic Model or Spatial and Spatial-Time Fields of Daily Sums of Liquid Precipitation

Spatial field $\{\eta_{ik}\}$ of daily sums of liquid precipitation on regular grid $\{x_{ik}\}$, $i = 1,\dots,n$, $k = 1,\dots,m$ can be represented in the form

$$\eta_{ik} = \omega_{ik}\chi_{ik}, \tag{40.2}$$

where $\{\omega_{ik}\}$—field of precipitation's indicators, that takes on a value of 0 or 1 with probabilities $P(\omega_{ik} = 1) = p_{ik}$ and $P(\omega_{ik} = 0) = 1 - p_{ik} = q_{ik}$, respectively, correlation matrix $S = \{s_{ij,kl}\}$, $i,k = 1,\dots,n$, $j,l = 1,\dots,m$; $\{\chi_{ik}\}$—conditional field of daily sums of precipitation, if there are precipitation, with one-dimensional conditional distribution $F_{ik}(x)$ and correlation matrix $Q = \{q_{ij,kl}\}$. For a field $\{\eta_{ik}\}$ probabilities $P(\eta_{ik} \geqslant 0.1) = p_{ik}$ and $P(\eta_{ik} = 0) = 1 - p_{ik}$ are equal to probabilities $P(\omega_{ik} = 1)$ and $P(\omega_{ik} = 0)$. Field $\{\omega_{ik}\}$ can be constructed as threshold-transformation of every element of Gaussian field $\{\xi_{ik}\}$ [1,4,6–12,15]:

$$\omega_{ik} = \begin{cases} 1, & \xi_{ik} \leqslant c_{ik} \\ 0, & \xi_{ik} > c_{ik} \end{cases}.$$

Here value of c_{ik} can be found from equation

$$p_{ik} = P(\xi_{ik} \leqslant c_{ik}) = \frac{1}{\sqrt{2\pi}} \int_{-\infty}^{c_{ik}} e^{-\frac{1}{2}u^2} du$$

when p_{ik} is given. Correlation matrixes $G = \{g_{ij,kl}\}$ of Gaussian field $\{\xi_{ik}\}$ and $S = \{s_{ij,kl}\}$ are connected by the relation

$$s_{ij,kl} = \frac{p_{ij}q_{kl} + p_{kl}q_{ij} - 2\left(T(c_{ij}, a_{ij,kl}) + T(c_{kl}, a_{kl,ij})\right)}{2\sqrt{p_{ij}q_{ij}\,p_{kl}q_{kl}}},$$

$$T(c_{ij}, a_{ij,kl}) = \frac{1}{2\pi}\int\limits_0^{a_{ij,kl}} e^{-\frac{c_{ij}^2(1+u^2)}{2}}\frac{du}{1+u^2}, \quad a_{ij,kl} = \sqrt{\frac{1-g_{ij,kl}}{1+g_{ij,kl}}}, \tag{40.3}$$

where $T(c,a)$—Owen's function.

To construct the field $\{\chi_{ik}\}$ method of inverse distribution function [9] can be used, and χ_{ik} can be computed as

$$\chi_{ik} = F_{ik}^{-1}\left(\Phi(\zeta_{ik})\right),$$

where ζ_{ik}—elements of Gaussian field $\{\zeta_{ik}\}$ with correlation matrix $H = \{h_{ij,kl}\}$ and

$$h_{ij,kl} = f(q_{ij,kl}). \tag{40.4}$$

It is important to note that with given $s_{ij,kl}$, $q_{ij,kl}$, $P(\omega_{ik} = 1)$ and $F_{ik}(x)$ solutions of (40.3) and (40.4) may not exist [9]. In some cases solutions exist, but matrixes G and H are not correlation matrixes. In such cases problem may be solved only approximately.

For model (40.2) it is necessary to estimate probabilities p_{ik}, q_{ik}, conditional one-dimensional distribution $F_{ik}(x)$ and correlation matrixes S and Q. Analysis of real precipitation fields in Novosibirsk region shows that many statistical characteristics are weakly dependent on position of weather station [3]. Some characteristics are slightly inhomogeneous, but in this paper suppose that field is homogeneous and consider probabilities $p_{ik} = p$ for field $\{\omega_{ik}\}$ and one-dimensional distribution function of field $\{\chi_{ik}\}$ independent from space coordinates. Corresponding estimations were done for the entire area on basis of real data obtained on all stations. For approximation of empirical distribution function by $F(x), x \in [0.1, \infty)$ a method, proposed by Marchenko in [8], was used. This method is based on combination of approximation with cubic splines and Weibull's distribution.

Due to suggested homogeneity all correlation matrixes have block Toeplitz structure, if grid is regular. So they can be approximated by:

$$\begin{aligned} r(x_s, y_s; x_h, y_h) &= r(x_s - x_h, y_s - y_h) = r(x, y) \\ &= \exp(-[ax^2 + bxy + cy^2]^\theta), \end{aligned}$$

where parameters a, b, c and θ are chosen to minimize mean-square difference between real and approximated functions [1]. Correlation function's isolines of simulated field are ellipses. Their major axes are codirectional with typical wind direction in considered area.

Homogeneous Gaussian fields with matrix correlation function (or block Toeplitz covariation matrix) were simulated according to the method of conditional distributions [7,9,10].

For estimation of ultimate water reserves on given area over given time, for study of precipitation spatial-time anomalies and other applications spatial-time models are intended. Spatial-time field can be considered as a sequence of spatial fields, where temporal and spatial-time correlation dependences are defined by real data. In trivial case of spatial-homogeneous and time-stationary field correlation function is a direct product of spatial and temporal correlation functions that are corresponding to direct product of spatial and temporal correlation matrixes. Simulation methods for Gaussian fields with such correlation structure are well known [10], and precipitation field can be constructed as above.

Verification of spatial fields'model, if information about field is given only in several points (weather stations), is more complicated problem in comparison with analogous problem for time-series. For example, if we'd like to compare some characteristics, estimated by model-made and real data, several problems appear. First of all, estimations based on real data have huge statistical error. This error is caused by length of time-interval, when physical conditions are unchanged and process may be considered as time-stationery. At the same time many characteristics require data-interpolation from station to arbitrary point of considered area. This interpolation also influences on accuracy of estimation.

Probability of event "total amount of precipitation on several stations is greater than given level c" was used for verification in this paper. Real data allow to estimate this probability without problems. But the model gives values only in grid nodes, so for estimation of probabilities it is necessary to interpolate data from node nearest to station. The less step of grid is, the less systematic inaccuracy associated with interpolation is. This inaccuracy can be studied and even excluded (if all stations are situated in grid nodes). But inaccuracy associated with assumed homogeneity of field can't be excluded or reduced. So accumulated error shows either assumption is acceptable or not. Six weather stations ($\nu = 6$), nearest to nodes of regular grid 30×25, were chosen. Figure 40.3 shows probabilities

$$P(A1(c)) = \left(\sum_{i=1}^{\nu} P(\eta_i \geqslant c)\right) \Big/ \nu$$
$$P(A2(c) = \left(\sum_{i=1}^{\nu-1} \sum_{j=i+1}^{\nu} P(\eta_i \geqslant c,\, \eta_j \geqslant c)\right) \Big/ (\nu(\nu-1)/2),$$
$$P(A6(c) = P(\eta_1 \geqslant c, \ldots, \eta_\nu \geqslant c)$$

calculated on real and model-made data for different levels c. Since model was done with assumption of homogeneity, probabilities $P(A1(c))$, estimated on real data, were averaged over chosen stations probabilities $P(A2(c))$ were averaged over mandatory station-pairs combination. It should be noted, that real data estimations have essential statistical error because of rather small amount of data. Model-made data may be used as additional information for further investigations.

Finally we use considered model for study of extreme rainfall regime. Essential characteristic, that is widely used for estimation of water reserves in given area, is

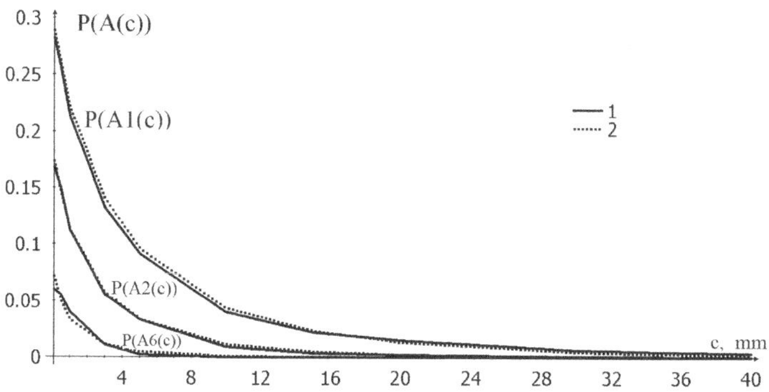

Fig. 40.3 Probabilities $P(A1(c))$, $P(A2(c))$ and $P(A6(c))$, estimated on real and model data. *Curve 1*: real data, *curve 2*: model data

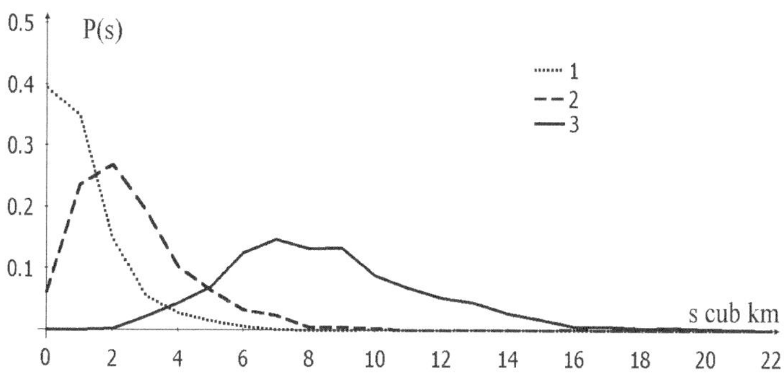

Fig. 40.4 Model-made distribution densities of total precipitation amount in considered area. *Curve 1*: in 5 days, *curve 2*: in 10 days, *curve 3*: in 30 days

distribution of total precipitation amount. Figure 40.4 shows distribution densities of total precipitation amount in considered area in 5, 10 and 30 days.

In this paper all estimations are done for homogeneous spatial and spatial-time fields of daily sums of liquid precipitation. For modeling of heterogeneous fields it is necessary to have information about the field as a function of space coordinates. It is easy to simulate inhomogeneous field if heterogeneity appears only in one-point characteristics (e.g. in probabilities of non-zero precipitation in one given point). For such simulation it is necessary to have corresponding data on stations and to interpolate it to grid nodes. One of the approaches to simulation of heterogeneous in correlations spatial-time fields of precipitation in given in [12] and is based on simulation of joint series on stations with due regard to their cross correlation and with following stochastic interpolation of field values on stations to grid nodes. Models of conditional Gaussian fields with given values in selected points may be used for stochastic interpolation. Approximate methods for simulation of conditional non-Gaussian fields are based on method of inverse distribution function and on algorithms for simulation of conditional Gaussian field. Appropriate modification of these algorithms may be used for simulation of precipitation fields [11].

It should be noted that in some cases (for example, when grid is condensed or when it is necessary to estimate ultimate water reserves in given area over a long period with spatial-time model) modeling can be exceedingly time-consuming. In these cases it is useful to make calculations on multiprocessor computers, and proposed models are good parallelizable.

Acknowledgements This work was supported by the Russian Foundation for Basic Research (grants 12-01-00727-a, 12-05-00169-a).

References

1. Anisimova, A.V.: Numerical modeling of liquid precipitation's indicator random fields. In: Proceedings of young scientists' conference, pp. 3–15. Novosibirsk (1997) (in Russian)
2. Dragan, Y.P., Rozhkov, V.A., Yavorskiy, I.N.: Methods of probabilistic analysis of oceanologic processes rhythmic. Gidrometeoizdat, Leningrad (1987) (in Russian)
3. Drobyshev, A.D., Marchenko, A.S., Ogorodnikov, V.A., Chizhikov, V.D.: Statistical structure of time series for daily sums of liquid precipitations in the plane part of Novosibirsk region. In: Proc. West Sib. Research Inst, 86, pp. 44–66. Goskomgidromet (1989) (in Russian)
4. Evstafieva, A.I., Khlebnikova, E.I., Ogorodnikov, V.A.: Numerical stochastic models for complexes of time series of weather elements. Russ. J. Numer Anal. Math. Model. **20**(6), 535–548 (2005)
5. Kargapolova, N.A., Ogorodnikov, V.A.: Inhomogeneous Markov chains with periodic matrices of transition probabilities and their application to simulation of meteorologocal processes. Russ. J. Numer Anal. Math. Model. **27**(3), 213–228 (2012)
6. Kleiber, W., Katz, R.W., Rajagopalan, B.: Daily spatiotemporal precipitation simulation using latent and transformed Gaussian processes. Water Resour. Res. **48**, W01523 (2012). doi:10.1029/2011WR011105
7. Marple, S.L.: Digital Spectral Analysis with Applications. Prentice-Hall, Englewood Cliffs (1987) (in Russian)
8. Marchenko, A.S.: Approximation of empirical probability distribution for daily sums of liquid precipitation. In: Proc. West Sib. Research Inst, 86, pp. 66–74. Goskomgidromet (1989) (in Russian)
9. Mikhailov, G.A., Voitishek, A.V.: Numerical Statistical Modelling. Monte Carlo Methods. Akademia, Moscow (2006) (in Russian)
10. Ogorodnikov, V.A., Prigarin, S.M.: Numerical Modelling of Random Processes and Fields: Algorithms and Applications. VSP, Utrecht (1996)
11. Ogorodnikov, V.A., Kargapolova, N.A., Sereseva, O.V.: Numerical stochastic model of spatial fields of daily sums of liquid precipitation. Russ. J. Numer Anal. Math. Model. **28**(2), 187–2008 (2013)
12. Ogorodnikov, V.A., Sereseva, O.V.: Numerical stochastic model of spatial fields of daily sums of liquid precipitation. In: Proceedings of the International Workshop, Applied Methods of Statistical Analysis. Simulations and Statistical Inference, pp. 221–226. Novosibirsk (2013)
13. Prigarin, S.M., Marshak, A.: Numerical model of broken clouds adapted to results of observations. Atmos. Oceanic Opt. **18**(3), 236–242 (2005)

14. Prigarin, S.M., Litvenko, K.V.: Numerical simulation of sea surface and extreme ocean waves with stochastic spectral models. In: Proceedings of the International Workshop, Applied Methods of Statistical Analysis. Simulations and Statistical Inference, pp. 394–402. Novosibirsk (2013)
15. Ukhinova, O.S., Ogorodnikov, V.A.: Stochastic models of spatial-time fields of precipitation sums. In: Proceedings of the 6th St. Peterburg Workshop on simulation, pp. 193–197. St. Peterburg (2009)

Chapter 41
Comparison of Resampling Techniques for the Non-causality Hypothesis

Angeliki Papana, Catherine Kyrtsou, Dimitris Kugiumtzis, and Cees G.H. Diks

41.1 Introduction

Resampling techniques are utilized for the construction of the empirical null distribution of a test statistic, when the asymptotic distribution cannot be established. We are concerned with the inter-dependence structure of multivariate time series. The generated resampled time series have to capture statistical properties of the original time series but also satisfy the corresponding null hypothesis, H_0, of no inter-dependence between two time series in the presence of the other variables [7]. Bootstrapping, first introduced in [1], aims at estimating the properties of a test statistic when sampling from an approximating distribution. For time series, bootstraps must be carried out in a way that they suitably capture the dependence

A. Papana (✉)
University of Macedonia, 156 Egnatias street, 54006 Thessaloniki, Greece
e-mail: angeliki.papana@gmail.com

C. Kyrtsou
University of Macedonia, Thessaloniki, Greece

University of Strasbourg, BETA, Strasbourg, France

University of Paris 10, Economix, ISC-Paris, Ile-de-France
e-mail: ckyrtsou@uom.gr

D. Kugiumtzis
Department of Electrical and Computer Engineering, Aristotle University of Thessaloniki, 54124 Thessaloniki, Greece
e-mail: dkugiu@auth.gr

C.G.H. Diks
Center for Nonlinear Dynamics in Economics and Finance (CeNDEF), University of Amsterdam, Roetersstraat 11, 1018 WB, Amsterdam, The Netherlands
e-mail: C.G.H.Diks@uva.nl

V.B. Melas et al. (eds.), *Topics in Statistical Simulation*, Springer Proceedings in Mathematics & Statistics 114, DOI 10.1007/978-1-4939-2104-1_41

structure of the data generation process consistent with the H_0, and be otherwise random, e.g. [6, 11, 12]. Similarly, randomization methods preserve the dependence structure consistent with H_0 when randomly shuffling the time series [3, 13, 15].

The transfer entropy (TE) is a non-parametric measure that quantifies the amount of directed transfer of between two random processes [14]. The TE is a non-symmetrical measure defined in terms of transition probabilities that provides information about the direction of the dependencies. The TE from a process X to another process Y is the amount of additional information (reduction of uncertainty) about the future values of Y provided by knowing past values of X and Y instead of past values of Y alone. The advantages of TE are that it is model-free, makes no assumptions about the distribution of the data and is effective in case of non-linear signals. The partial transfer entropy (PTE) is a multivariate extension of the TE [9, 16].

Resampling techniques are utilized for the H_0 of non-causality, i.e. no causal effects from one variable (driver) to another one (response), in the presence of the remaining observed variables (confounding variables). A suitable statistic, sensitive to the inter-dependence of the time series, is the PTE. The causality test is actually a significance test for the PTE and in the absence of asymptotic distribution for the PTE, resampling is required.

The appropriateness of six resampling schemes for the null hypothesis H_0 of non-causality is examined. Specifically, we combine two resampling methods: (1) the time shifted surrogates [13] and (2) the stationary bootstrap [11], with three independence settings of the time series adapted for the non-causality test: (a) resampling only the time series of the driving variable, (b) resampling separately the driving and the response time series, and (c) resampling separately the driving and the response time series, while destroying the dependence of the future and past of the response variable. The properties of the test for the six resampling schemes, i.e. the empirical distribution of PTE, the size and power of the test, are assessed in a simulation study.

The structure of the paper is as follows. In Sect. 41.2, the PTE is briefly discussed and in Sect. 41.3 the resampling methods and the independence settings are presented. In Sect. 41.4, the resampling schemes are evaluated with means of simulations on different coupled and uncoupled multivariate systems. The conclusions are drawn in Sect. 41.5.

41.2 PTE

The PTE is a multivariate information measure [9, 16], introduced as an extension of the bivariate measure of transfer entropy (TE) [14]. The TE quantifies the amount of information explained in a response variable Y at h time steps ahead from the state of a driving variable X accounting for the concurrent state of Y. Let $\{x_t, y_t\}$, $t = 1, \dots, n$ be the observed time series of two variables, and $\mathbf{x}_t = (x_t, x_{t-\tau}, \dots, x_{t-(m-1)\tau})'$ and $\mathbf{y}_t = (y_t, y_{t-\tau}, \dots, y_{t-(m-1)\tau})'$ the reconstructed state space vectors, where m is the embedding dimension and τ is the time delay. The TE from X to Y is the conditional mutual information $I(y_{t+h}; \mathbf{x}_t | \mathbf{y}_t)$:

$$\begin{aligned} \mathrm{TE}_{X \to Y} &= I(y_{t+h}; \mathbf{x}_t | \mathbf{y}_t) = \sum p(y_{t+h}, \mathbf{x}_t, \mathbf{y}_t) \log \frac{p(y_{t+h} | \mathbf{x}_t, \mathbf{y}_t)}{p(y_{t+h} | \mathbf{y}_t)} \\ &= H(\mathbf{x}_t, \mathbf{y}_t) - H(y_{t+h}, \mathbf{x}_t, \mathbf{y}_t) + H(y_{t+h}, \mathbf{y}_t) - H(\mathbf{y}_t), \end{aligned} \tag{41.1}$$

where TE is given either based on probability distributions ($p(x)$ is the probability mass function of the discretized variable x) or entropy terms ($H(\mathbf{x}) = -\int f(\mathbf{x}) \log f(\mathbf{x}) d\mathbf{x}$ is the differential entropy of the vector variable $\mathbf{x}$ with probability density function $f(\mathbf{x})$).

The PTE accounts for the direct coupling of X to Y conditioning on the confounding variables of a multivariate system, collectively denoted Z. It is given by

$$\begin{aligned} \mathrm{PTE}_{X \to Y | Z} &= I(y_{t+h}; \mathbf{x}_t | \mathbf{y}_t, \mathbf{z}_t) \\ &= H(\mathbf{x}_t, \mathbf{y}_t, \mathbf{z}_t) - H(y_{t+h}, \mathbf{x}_t, \mathbf{y}_t, \mathbf{z}_t) + H(y_{t+h}, \mathbf{y}_t, \mathbf{z}_t) - H(\mathbf{y}_t, \mathbf{z}_t). \end{aligned} \tag{41.2}$$

The estimation of the TE and PTE relies upon the estimation of the probability density functions. Different types of estimators exist, such as histogram-based (e.g. by discretizing the state space to equidistant intervals at each axis), kernel-based and using correlation sums. Here, we use the nearest neighbor estimator [5], which is proved to be effective especially for high-dimensional data [17].

Theoretically, the PTE (and TE) should be zero in the case of no causal effects. However, a bias can be present due to various reasons, e.g. the estimation method for the entropies and subsequently densities, the selection of the embedding parameters, the finite sample size and the noise level as well [8]. In order to determine whether a PTE value indicates a weak coupling or whether it is not statistically significant, a resampling method should be used to assess its statistical significance.

41.3 Resampling Techniques

We first present the two resampling methods of time shifted surrogates and stationary bootstrap, and then the three independence settings. Time shifted surrogates preserve the dynamics of a time series $\{x_1, \ldots, x_n\}$, while the couplings are destroyed [13]. They are formed by cyclically time shifting the components of $\{x_1, \ldots, x_n\}$. To formulate them from a time series with length n, an integer d is randomly chosen ($d < n$) and the d first values of the time series are moved to the end: $\{x_t^*\} = \{x_{d+1}, \ldots, x_n, x_1, \ldots, x_d\}$. For testing $X \to Y$ in a bivariate time series, the pair $\{x_t^*, y_t\}$ is consistent with the H_0 of non-causality.

The stationary bootstrap has been proposed for the calculation of standard errors and the construction of confidence intervals for a statistic based on weakly dependent stationary observations [11]. The bootstrap series are generated by resampling blocks of random size, where the length of each block has a geometric distribution. For a fixed probability p, block lengths L_i are generated with probability $p(L_i = k) = (1-p)^{(k-1)}$ for $k = 1, 2, \ldots$. The starting time points of the blocks I_i are drawn from the discrete uniform distribution on $\{1, \ldots, n-k\}$. A bootstrap time series $\{x_t^*\}$ is formed by first starting with a random block as defined above $B_{I_1,L_1} = \{x_{I_1}, x_{I_1+1}, \ldots, x_{I_1+L_1-1}\}$, and blocks are added until length n is reached.

Three independence settings are considered for both resampling methods, all consistent with the H_0 of non-causality from X to Y conditioned on Z. The first setting, denoted A, is to resample only the time series of the driving variable X. This is the standard approach for surrogate test for the significance of causality measures [2, 10, 13]. The intrinsic dynamics of the variable X is preserved in the resampled time series $\{x_t^*\}$ but the coupling between X^* and Y is destroyed, so that H_0 is fulfilled and $\text{PTE}_{X^* \to Y|Z} = 0$. The variables X and Y as well as X and Z are independent, however the pair of variables (Y, Z) preserves its interdependence.

The second setting, denoted B, suggests to randomize both the driving variable X and the response Y, i.e. resampled time series $\{x_t^*\}$ and $\{y_t^*\}$ are generated. Again, the intrinsic dynamics of both X and Y are preserved but the coupling between them is destroyed, H_0 is fulfilled and $\text{PTE}_{X^* \to Y^*|Z} = 0$. In this case, independence holds for all variable pairs (X, Y), (Y, Z) and (X, Z). However, there is still no complete independence between all arguments in the definition of PTE, as y_{t+h} preserves by construction of $\{y_t^*\}$ its dependence on $\mathbf{y}_t$.

Finally, we consider the third setting of complete independence of all variables involved in the definition of PTE, denoted C, i.e. in addition to the resampling of X and Y, also y_{t+h} is resampled separately. Thus all terms in PTE, i.e. y_{t+h}, $\mathbf{x}_t$, $\mathbf{y}_t$ and $\mathbf{z}_t$ are independent, and H_0 is again fulfilled.

Combining the two resampling methods (time shifted surrogates and stationary bootstraps) and the three independence settings (A, B and C), six resampling schemes are formulated that are utilized to test the null hypothesis of no causal effects among the variables of multivariate systems.

41.4 Simulation Study

In the simulation study we apply the significance test for the PTE with the six resampling schemes to multiple realizations of different simulation systems. Specifically, we estimate the PTE from 100 realizations per simulation system. For each realization and each resampling scheme, $M = 100$ resampled time series are generated. Let us denote q_0 the PTE value from one realization of a system and $q_1, q_2, \ldots, q_M$ the PTE values from the resampled time series for this particular realization and for a specific resampling scheme. The rejection of H_0 of no causal effects is decided by the rank ordering of the PTE values computed on the original time series, q_0, and the resampled time series, $q_1, q_2, \ldots, q_M$. For the one-sided test, if r_0 is the rank of q_0 when ranking the list $q_0, q_1, \ldots, q_M$ in ascending order, the

p-value of the test is $1-(r_0-0.326)/(M+1+0.348)$, by applying the correction in [19].

The simulation systems that have been used in this study are the following:

1. Three coupled Hénon maps, with nonlinear couplings ($X_1 \rightarrow X_2$, $X_2 \rightarrow X_3$) (System 5 in [10]) with equal coupling strengths c for $X_1 \rightarrow X_2$ and $X_2 \rightarrow X_3$. We set $c = 0$ (uncoupled case), $c = 0.3$ and $c = 0.5$ (strong coupling). The Hénon map is a well-known discrete-time dynamical system that exhibits chaotic behavior [4].
2. A vector autoregressive process of 4 variables and order 5, VAR(5), with linear couplings ($X_1 \rightarrow X_3$, $X_2 \rightarrow X_1$, $X_2 \rightarrow X_3$, $X_4 \rightarrow X_2$) (Eq. (12) in [18]).
3. Five coupled Hénon maps, with nonlinear couplings ($X_1 \rightarrow X_2$, $X_2 \rightarrow X_3$, $X_3 \rightarrow X_4$, $X_4 \rightarrow X_5$) defined similarly to system 1. We consider again equal coupling strengths c, and set $c = 0$ (uncoupled case), $c = 0.2$ and $c = 0.4$ (strong coupling).

We consider time series lengths $n = 512$ and 2048. To estimate the PTE, we set the embedding dimension m to appropriate values for each system, i.e. $m = 2$ for system 1 and 3, $m = 5$ for system 2, the delay time $\tau = 1$ and the time step ahead $h = 1$ (as defined in [14]). The number of nearest neighbors for the estimation of the probability distributions is 10.

In terms of presentation, we focus on the sensitivity of the PTE (percentage of rejection of H_0 when there is true direct causality), as well as the specificity of the PTE (percentage of no rejection of H_0 when there is no direct causality), at the significance level $\alpha = 0.05$. The notation $X_2 \rightarrow X_1|Z$ denotes the Granger causality from X_2 to X_1, accounting for the presence of confounding variables $Z = X_3, \ldots$. For brevity, we use the notation $X_2 \rightarrow X_1$ instead of $X_2 \rightarrow X_1|Z$, implying the conditioning on the confounding variables. The notation of Granger causality for other pairs of variables is analogous.

System 1. The mean PTE values are negatively biased in the uncoupled case ($c = 0$). For $c = 0.3$ and $c = 0.5$, they are much larger when direct couplings exist ($X_1 \rightarrow X_2$, $X_2 \rightarrow X_3$) than the rest of the directions, and increase with n. Regarding the indirect coupling $X_1 \rightarrow X_3$, PTE increases with n for $c = 0.5$ (mean $PTE_{X_1 \rightarrow X_3} = -0.0002$ for $n = 512$ and 0.0075 for $n = 2048$).

We evaluate how the null distribution of the PTE from the six resampling schemes differs with respect to the original PTE values. For $c = 0$, all the resampling schemes correctly indicate the absence of couplings; the percentages of significant PTE values vary from 0 to 12 % (see Table 41.1). Considering $c = 0.3$, the schemes B and C indicate correctly the couplings, while scheme A indicates the spurious one $X_2 \rightarrow X_1$ and the indirect one $X_1 \rightarrow X_3$. The percentage of erroneously rejected H_0 for non-existing or indirect couplings tends to increase with c and the time series length for all resampling schemes, the most robust being 1C and 2C.

It turns out that when the resampled time series become more independent, the percentage of spurious couplings decreases. The most independent resampling schemes 1C and 2C give smallest rejection rate since the null distribution for the test

Table 41.1 Percentage of significant PTE values for system 1 for $n = 512/2048$, for the six resampling schemes

$c = 0$	$X_1 \rightarrow X_2$	$X_2 \rightarrow X_1$	$X_2 \rightarrow X_3$	$X_3 \rightarrow X_2$	$X_1 \rightarrow X_3$	$X_3 \rightarrow X_1$
1A	2/3	3/9	3/5	3/5	6/4	4/10
1B	4/1	5/12	4/5	3/4	4/6	5/10
1C	1/0	0	2/0	0	1/0	1/0
2A	2/1	3/7	3/5	1/2	6/5	3/12
2B	2/0	1	1/0	1/0	4/1	1/3
2C	0	0	0	0	2/0	0
$c = 0.3$	$\mathbf{X_1 \rightarrow X_2}$	$X_2 \rightarrow X_1$	$\mathbf{X_2 \rightarrow X_3}$	$X_3 \rightarrow X_2$	$X_1 \rightarrow X_3$	$X_3 \rightarrow X_1$
1A	**100**	11/30	**100**	14/13	15/36	5/4
1B	**100**	9/31	**100**	3/2	8/7	3/4
1C	**100**	3/0	**87/100**	0	1/0	0
2A	**100**	9/26	**100**	8/11	9/27	4/3
2B	**100**	3/11	**100**	1/0	2	2/0
2C	**100**	2/0	**100**	0	0	0
$c = 0.5$	$\mathbf{X_1 \rightarrow X_2}$	$X_2 \rightarrow X_1$	$\mathbf{X_2 \rightarrow X_3}$	$X_3 \rightarrow X_2$	$X_1 \rightarrow X_3$	$X_3 \rightarrow X_1$
1A	**100**	8/32	**100**	11/14	32/95	8
1B	**100**	2/25	**100**	3/0	6/68	8/1
1C	**100**	0	**100**	0	2/32	0
2A	**100**	4/24	**100**	9/11	23/93	6/4
2B	**100**	1/9	**100**	1/0	4/57	2/1
2C	**100**	0	**100**	0	1/33	0

The directions of true couplings are highlighted. A single number is displayed when the same percentage corresponds to both n

is more spread and displaced to the right as the resampling changes from the least independent scheme (scheme A) to the most independent one (C) (see Fig. 41.1).

System 2. The mean PTE values from 100 realizations of the second system for the directions of the true couplings are larger than for the other directions and increase with n, with the exception of $X_2 \rightarrow X_3$ that is at a lower level and does not increase with n (see Table 41.2). Concerning the uncoupled directions, the mean PTE values vary from 0.0013 to 0.0095 for both n and they decrease with n (the three largest mean PTE values across all non-direct couplings are reported in Table 41.2).

The true couplings $X_2 \rightarrow X_1$, $X_1 \rightarrow X_3$, $X_4 \rightarrow X_2$ are well established by the significance test (see Table 41.2), while no spurious causalities are identified (percentage of significant PTE vary from 0 to 6 % at the uncoupled directions). The weak coupling $X_2 \rightarrow X_3$ is detected only by the scheme A, with a power of the test increasing with n. We note again that the surrogate/bootstrap PTE values increase as the resampled time series become more independent (see Fig. 41.2).

System 3. No couplings are noted in the uncoupled case ($c = 0$) for system 3; the percentage of significant PTE values range from 0 to 11 % for all the resampling schemes and both time series lengths. The PTE is also effective for $c = 0.2$ (see Table 41.3). As resampled time series become more independent, a loss in the power

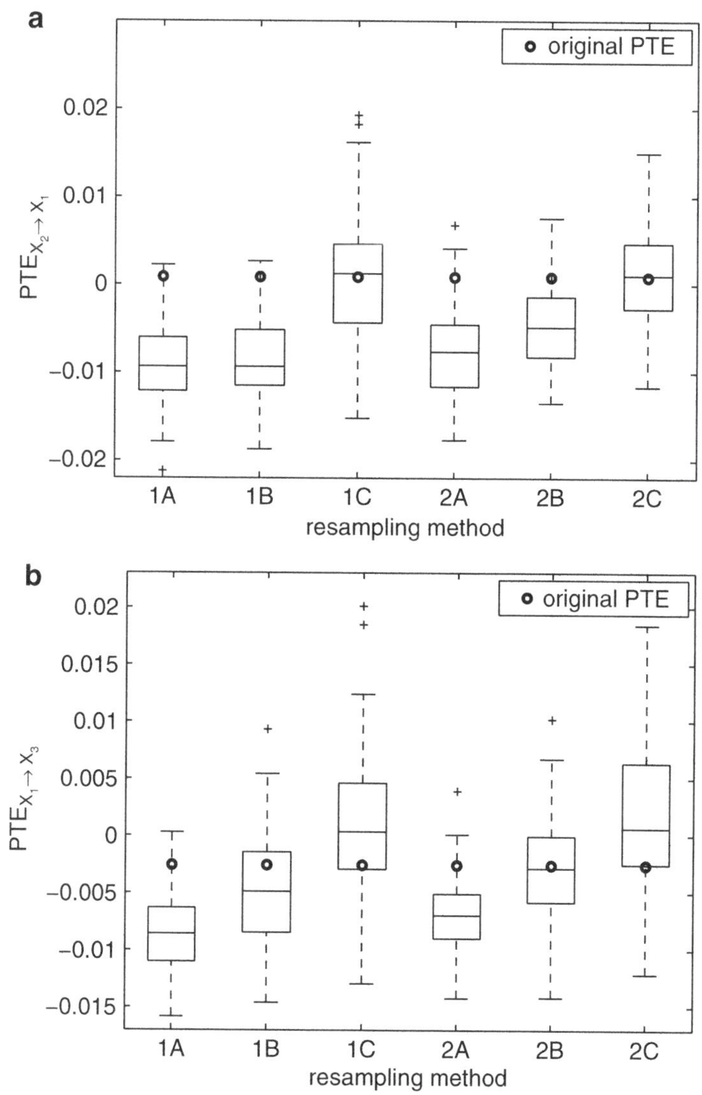

Fig. 41.1 Boxplots of surrogate/bootstrap PTE values and original PTE value from one realization of system 1 for $c = 0.3$ and $n = 2048$, for the directions (**a**) $X_2 \rightarrow X_1$ and (**b**) $X_1 \rightarrow X_3$

of the test is observed for $n = 512$. For the strong coupling strength $c = 0.4$, indirect and spurious couplings are observed for $n = 2048$ based on the resampling scheme A, e.g. we obtained for scheme 1A: 49 % for $X_1 \rightarrow X_3$, 60 % for $X_2 \rightarrow X_4$, 64 % for $X_3 \rightarrow X_5$, 19 % for $X_2 \rightarrow X_1$, 18 % for $X_3 \rightarrow X_2$, 22 % for $X_4 \rightarrow X_3$ and 27 % for $X_5 \rightarrow X_4$. Similar results are observed for scheme 2A. Scheme B indicates the spurious coupling $X_2 \rightarrow X_4$, but at a lower percentage than scheme A. Only the true couplings are indicated using the resampling methods C (see Table 41.3).

Table 41.2 Mean PTE values and percentage of significant PTE values from 100 realizations of system 2

mean PTE	$\mathbf{X_2 \rightarrow X_1}$	$\mathbf{X_1 \rightarrow X_3}$	$\mathbf{X_4 \rightarrow X_2}$	$\mathbf{X_2 \rightarrow X_3}$	$X_2 \rightarrow X_4$	$X_3 \rightarrow X_4$	$X_1 \rightarrow X_4$
$n = 512$	**0.0920**	**0.0772**	**0.0998**	**0.0060**	0.0095	0.0071	0.0067
$n = 2048$	**0.1221**	**0.0965**	**0.1355**	**0.0059**	0.0061	0.0042	0.0034
$n = 512/2048$	$\mathbf{X_2 \rightarrow X_1}$	$\mathbf{X_1 \rightarrow X_3}$	$\mathbf{X_4 \rightarrow X_2}$	$\mathbf{X_2 \rightarrow X_3}$	$X_2 \rightarrow X_4$	$X_3 \rightarrow X_4$	$X_1 \rightarrow X_4$
1A	**100**	**100**	**100**	**22/66**	3	1/0	1/0
1B	**100**	**99/ 100**	**100**	**0**	3/1	1/0	0
1C	**100**	**100**	**100**	**4/6**	0	0	0
2A	**100**	**100**	**100**	**14/60**	1/3	1/0	1/0
2B	**100**	**100**	**100**	**0**	3/4	2/0	1/0
2C	**100**	**100**	**100**	**5/14**	0	0	0

The format of the table is as for Table 41.1

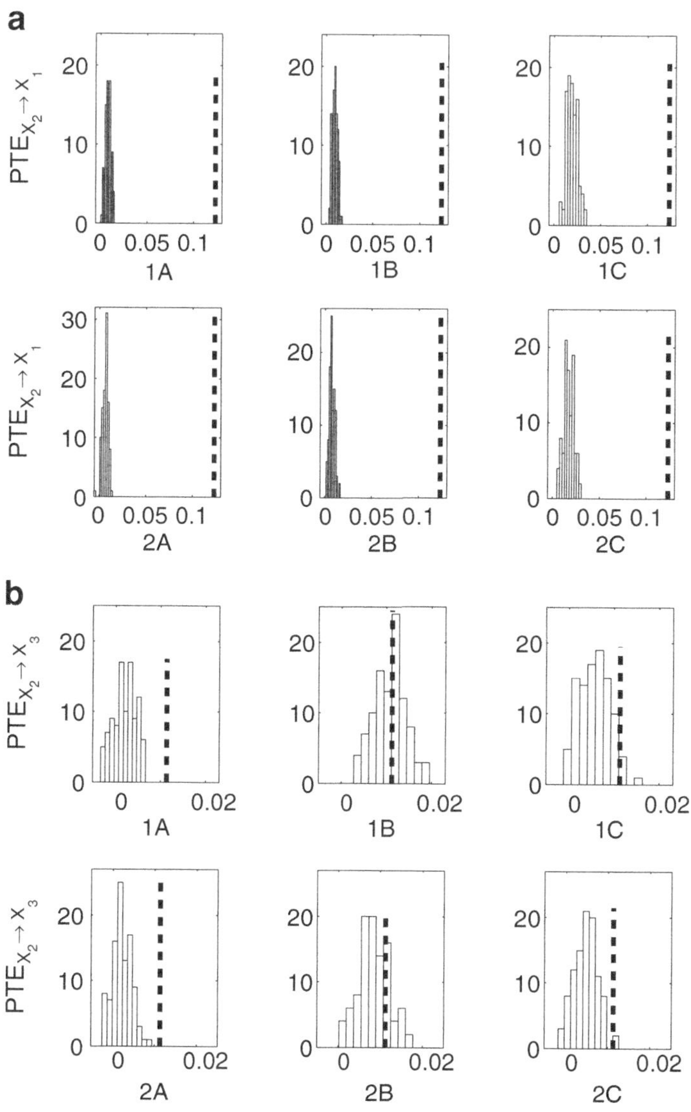

Fig. 41.2 Distribution of surrogate/bootstrap PTE values and original PTE value (*vertical dotted line*) from one realization of system 2 with $n = 2048$, for the directions (**a**) $X_2 \rightarrow X_1$ and (**b**) $X_2 \rightarrow X_3$

Table 41.3 Percentage of significant PTE values from 100 realizations of system 3 for $n = 512/2048$, for the true couplings, an indirect coupling ($X_2 \to X_4$) and an uncoupled case ($X_5 \to X_4$)

$c = 0.2$	$\mathbf{X_1 \to X_2}$	$\mathbf{X_2 \to X_3}$	$\mathbf{X_3 \to X_4}$	$\mathbf{X_4 \to X_5}$	$X_2 \to X_4$	$X_5 \to X_4$
1A	**54/100**	**58/100**	**63/100**	**54/100**	8/5	10/7
1B	**54/100**	**57/100**	**59/100**	**52/100**	6/2	9/5
1C	**31/100**	**22/98**	**18/99**	**16/99**	1/0	0
2A	**53/100**	**62/100**	**67/100**	**59/100**	8/7	14/9
2B	**48/100**	**60/100**	**64/100**	**57/100**	7/0	11/1
2C	**29/100**	**29/100**	**28/100**	**26/99**	1/0	2/0
$c = 0.4$	$\mathbf{X_1 \to X_2}$	$\mathbf{X_2 \to X_3}$	$\mathbf{X_3 \to X_4}$	$\mathbf{X_4 \to X_5}$	$X_2 \to X_4$	$X_5 \to X_4$
1A	**100**	**100**	**98/100**	**100**	15/60	18/27
1B	**100**	**100**	**99/100**	**99/ 100**	6/21	3/6
1C	**100**	**83/100**	**86/ 100**	**84/100**	1	1/0
2A	**100**	**100**	**100**	**100**	21/65	22
2B	**100**	**100**	**100**	**100**	9/25	10/8
2C	**96/100**	**96/100**	**96/100**	**96/100**	1	2/1

41.5 Discussion

The importance of assessing the correct statistical significance for the PTE has been explored in a simulation study. Concerning the resampled time series, by definition, the mutual information of X and Y conditioned on Z should be in theory zero, i.e. $I(Y; X|Z) = 0$. The formulation of more independent resampled data (schemes B and C) compared to the standard technique (scheme A) seems to improve the bias of the test statistic and helps prevent false indications of couplings in the case of the nonlinear coupled systems. The size and the power of the test are improved, especially if strong couplings exist. However, when the couplings are linear, scheme A seems to be more efficient in identifying weak couplings.

It turns out that when the PTE is estimated for an increasing level of randomness in the resampled time series, the estimated PTE values also increase, while the distribution of PTE from the resampled time series gets wider and less spurious couplings are thus detected. This higher specificity comes at the cost of lower sensitivity, and vice versa. Thus, none of the six resampling schemes turns out to be optimal, but it becomes clear that the significance test for the PTE gets more conservative as resampling is more random.

The aforementioned resampling schemes can be utilized for any test statistic in order to examine the null hypothesis of no causal effects. Since the efficiency of a causality measure is determined in terms of the corresponding resampling technique that is used, the usefulness of each of the examined resampling schemes will be further investigated for different causality measures.

Acknowledgements The research project is implemented within the framework of the Action "Supporting Postdoctoral Researchers" of the Operational Program "Education and Lifelong Learning" (Action's Beneficiary: General Secretariat for Research and Technology), and is co-financed by the European Social Fund (ESF) and the Greek State.

References

1. Efron, B.: Bootstrap methods: another look at the jackknife. Ann. Stat. **7**(1), 1–26 (1979)
2. Faes, L., Porta, A., Nollo, G.: Mutual nonlinear prediction as a tool to evaluate coupling strength and directionality in bivariate time series: comparison among different strategies based on k nearest neighbors. Phys. Rev. E **78**, 026,201 (2008)
3. Faes, L., Porta, A., Nollo, G.: Testing frequency-domain causality in multivariate time series. IEEE Trans. Biomed. Eng. **57**(8), 1897–1906 (2010)
4. Hénon, M.: A two-dimensional mapping with a strange attractor. Commun. Math. Phys. **50**(1), 69–77 (1976)
5. Kraskov, A., Stögbauer, H., Grassberger, P.: Estimating mutual information. Phys. Rev. E **69**(6), 066,138 (2004)
6. Künsch, H.: The jackknife and the bootstrap for general stationary observations. Ann. Stat. **17**, 1217–1241 (1989)
7. Lahiri, S.: Resampling Methods for Dependent Data. Springer, New York (2003)
8. Papana, A., Kugiumtzis, D., Larsson, P.: Reducing the bias of causality measures. Phys. Rev. E **83**, 036207 (2011)
9. Papana, A., Kugiumtzis, D., Larsson, P.: Detection of direct causal effects and application to electroencephalogram analysis. Int. J. Bifurcat. Chaos **22**(9), 1250222 (2012)
10. Papana, A., Kyrtsou, C., Kugiumtzis, D., Diks, C.: Simulation study of direct causality measures in multivariate time series. Entropy **15**(7), 2635–2661 (2013)
11. Politis, D., Romano, J.: The stationary bootstrap. J. Am. Stat. Assoc. **89**(428), 1303–1313 (1994)
12. Politis, D., Romano, J., Lai, T.: Bootstrap confidence bands for spectra and cross-spectra. IEEE Trans. Sig. Proc. **40**, 1206–1215 (1992)
13. Quian Quiroga, R., Kraskov, A., Kreuz, T., Grassberger, P.: Performance of different synchronization measures in real data: a case study on electroencephalographic signals. Phys. Rev. E **65**, 041903 (2002)
14. Schreiber, T.: Measuring information transfer. Phys. Rev. Lett. **85**(2), 461–464 (2000)
15. Theiler, J., Eubank, S., Longtin, A., Galdrikian, B., Farmer, J.: Testing for nonlinearity in time series: the method of surrogate data. Phys. D **58**, 77–94 (1992)
16. Vakorin, V., Krakovska, O., McIntosh, A.: Confounding effects of indirect connections on causality estimation. J. Neurosci. Methods **184**, 152–160 (2009)
17. Vlachos, I., Kugiumtzis, D.: Non-uniform state space reconstruction and coupling detection. Phys. Rev. E **82**, 016207 (2010)
18. Winterhalder, M., Schelter, B., Hesse, W., Schwab, K., Leistritz, L., Klan, D., Bauer, R., Timmer, J., Witte, H.: Comparison of linear signal processing techniques to infer directed interactions in multivariate neural systems. Signal Proc. **85**(11), 2137–2160 (2005)
19. Yu, G., Huang, C.: A distribution free plotting position. Stoch. Env. Res. Risk Assess. **15**(6), 462–476 (2001)

Chapter 42
A Review of Multilevel Modeling: Some Methodological Issues and Advances

Giulia Roli and Paola Monari

42.1 Introduction

Multilevel modeling is a recently new class of statistical methods, firstly introduced in 1987 by Goldstein and later by Bryk and Raudenbush [2] and Hox [17]. This approach for data analysis is a generalization of linear and generalized linear regressions, when the structure of data is nested, i.e. base level units are grouped into higher level units involving their own variability and dependencies among the related observations. The nested or *multilevel* structure of data is a common phenomenon, especially in behavioral and social sciences, where the study of the relationship between individuals and society is of crucial importance and, thus, the dependence of data becomes a focal interest of the research. Moreover, the hierarchy of data can be a nuisance generated by the sampling design, such as in the multi-stage sampling, which is frequently employed in the traditional surveys to reduce the costs of data collection. Whatever the dependence arises from, it is "neither accidental nor ignorable" [13]. Indeed, the risks of drawing wrong conclusions are high if the clustering of the data is disregarded [38].

Mainly thanks to the wide range of applicability and the great increase of statistical softwares [8], in the last decades multilevel modeling has enjoyed an explosion of published papers and books in both methodological and application field. Its popularity is well reflected by the raise in the published books up to now: 6 books published in the period from 1972 to 1992, 6 in 1994 to 1998, 14 in 1999 to 2003, 25 in 2004 to 2008 [37].

The first usage of multilevel modeling refers to educational research aiming at estimating the effect of school enrollment on students' achievement. To date,

G. Roli • P. Monari (✉)
Dipartimento di Scienze Statistiche, Università di Bologna, Bologna, Italy
e-mail: g.roli@unibo.it; paola.monari@unibo.it

V.B. Melas et al. (eds.), *Topics in Statistical Simulation*, Springer Proceedings in Mathematics & Statistics 114, DOI 10.1007/978-1-4939-2104-1_42

several applications have been made in virtually all the disciplines, such as medical, economic, genetic, demographic researches. Multilevel modeling appears in the statistical literature under different but equivalent terms: *multi-level linear models* in sociological research and in health sciences [13], *mixed-effects models* and *random-effects models* in biometric applications [9, 21], *random-coefficient regression models* in the econometric literature [34], *covariance components models* in the more generic statistical literature [7, 25]. Raudenbush and Bryk [33] adopted the term *hierarchical linear models* as a more general way to refer to the structural feature of data in several applications, as firstly introduced by Lindley and Smith in 1972.

Currently, there is a need to not only develop the research on multilevel approach for the analysis of complex data, but also to have instructions to properly address the usage. This work aims at summarizing methodological aspects related to multilevel models, illustrating good-practices, advantages, and limits by reviewing applications in various fields, such as socio-economic, educational, health, and medical sciences. To date, only few reviews are available to report practices of multilevel applications, mainly restricted to education field (see, e.g., [5, 35]). We further focus our attention on the latest advances of multilevel modeling towards, for example, the inclusion of latent variables, such as multilevel structural equation and latent class models [36], and the increasing use of the Bayesian inference approach [10].

The paper develops by firstly introducing the research aims and the modeling framework of multilevel regression modeling, as well as the basic assumptions usually invoked (Sect. 42.2). In particular, some practical issues concerning the model specification are considered in Sect. 42.2.1. Different aspects of parameter estimation, according to softwares usually employed, are showed in Sect. 42.3. Latest advances in multilevel modeling are listed in Sect. 42.4.

42.2 Modeling Framework

The starting point for the employment of multilevel modeling is represented by a hierachical structure of data. In general, we refer to a data hierarchy as consisting units grouped at different levels [38]. In the simplest cases, a two-level hierarchy of data is considered with units nested in clusters, each one with its own variability, with the potential to be easily extended to higher levels. The most typical cases lie in units clustered into organizations, such as teachers in schools, pupils in classes, families in neighborhoods, employees in firms, children in families, animals in litters, patients in doctors or hospitals, respondents in interviewers. Other examples consist in longitudinal or panel data with multiple measures nested into single individuals. Such structures are typically strong hierarchies because there is much more variation among level-2 units (subjects) than among level-1 units (measurement). Consequently, most of the books on multilevel analysis deal with this kind of nested data separately by referring to *repeated measures models*.

Units at the more disaggregated level are usually named level-1 units, but other terms are used, such as micro-level, micro- secondary or elementary units. Aggregation units are called level-2 units, but also macro-level, macro- or primary units or, simply, clusters.

The first consequence of a hierarchical structure of data is the dependency of level-1 units within level-2 units. As a result, the use of standard regression analysis is improper. A common procedure with two-level data is to aggregate the micro-level data to the macro-level, e.g. averaging by macro-units, with serious risks of errors, such as "shift of meaning" and "ecological fallacy." Moreover, aggregation neglects the original data structure and cannot examine the potential cross-level interaction effects [38]. Conversely, multilevel analysis approach address to the problem of handling hierarchical data by representing each level by its own sub-model and, thus, fulfilling several research purposes:

- improving the estimation of the level-1 effects under investigation (i.e., all the available information at both levels are efficiently used in order to exploit both the group features and the relations existing in the overall sample);
- evaluating of the cross-level effects (e.g., how variables measured at one level affect relations occurring at another);
- decomposing of the variance–covariance components among levels;
- generalizing standard methods (e.g., ordinary regression techniques represent a special case, in which there is only one level).

To introduce the modeling, we consider a general framework for a multilevel generalized linear regression of two-level clustered data. In particular, let us consider an outcome Y whose distribution is from the exponential family with mean μ and denote with y_{ij} the outcome value for each level-1 unit i (with $i = 1, \ldots, N_j$) in level-2 unit j (with $j = 1, \ldots, J$). Let us further consider a set of level-1 covariates, i.e. information related to level-1 units, denoted by x_{ijk} (with $k = 1, \ldots, K$). By referring to the typical multi-stage structure of multilevel modeling, at level one we have

$$y_{ij} = \mu_{ij} + \varepsilon_{ij}$$

$$g(\mu_{ij}) = \eta_{ij} = \alpha_j + \sum_{k=1}^{K} \beta_{jk} x_{ijk} \tag{42.1}$$

where $g(\cdot)$ is the differentiable monotonic link function relating the outcome with the linear predictor η_{ij}, α_j and β_{jk} are the random intercepts and coefficients across level-2 units, respectively. A special case is multilevel linear regression where $g(\cdot)$ is the identity function and the normal distribution of level-1 residuals ε_{ij} with null mean is assumed. Several submodels can be considered by setting only some parameters as random across level-2 units. For instance, in a random-intercept model only the parameters α_j are assumed to vary and β_k are fixed across clusters.

If no level-1 covariates are included in the model specification, we refer to a one-way ANOVA with random effects or fully unconditional model.

At level two the random intercepts and coefficients are regressed on Q level-2 covariates, that is some kinds of available information on clusters denoted by z_{qj} (with $q = 1, \ldots, Q$)

$$\alpha_j = \psi_{00} + \sum_{q=1}^{Q} \psi_{0q} z_{qj} + u_{0j}$$

$$\beta_{jk} = \psi_{k0} + \sum_{q=1}^{Q} \psi_{kq} z_{qj} + u_{jk}, \; \forall \; k \tag{42.2}$$

where ψ_s are the level-2 parameters (intercepts and coefficients) and u's the level-2 residuals. A typical assumption for the level-2 residuals is the multivariate normal distribution with null means and a variance covariance structure, which can be both simplified, e.g. by assuming null covariances, or complicated, e.g. by considering different matrixes across clusters. Level-1 and level-2 residuals are assumed to be independent.

A combined version of model equations is often considered by embedding regressions 42.2 into model 42.1

$$g(\mu_{ij}) = \psi_{00} + \sum_{q=1}^{Q} \psi_{0q} z_{qj} + u_{0j} + \sum_{k=1}^{K} \left(\psi_{k0} x_{ijk} + \sum_{q=1}^{Q} \psi_{kq} z_{qj} x_{ijk} + u_{kq} x_{ijk} \right) \tag{42.3}$$

This version is required by some statistical softwares to compute estimates of the crucial parameters and allows to show the cross-level interaction of level-1 and level-2 covariates implicitly accounted by the model.

42.2.1 Some Issues on Model Development and Specification

A first issue concerning the model specification arises from sufficient sample sizes to yield accurate estimates. Several authors (see, e.g., [1, 26]) showed that the regression coefficients, their standard errors, and variance components are unbiased independently of the sample size. Conversely, the number of level-2 units is crucial, as the standard errors of the level-2 variances are under-estimated when the number of level-2 units is lower than 100. They suggest a number of 50 level-2 units as sufficient basing on the results from a simulation study yielding an acceptable noncoverage rate of about 7.3 %.

In addition to the sample sizes at the separate levels, the size of the intraclass correlation (ICC), i.e. the proportion of variance in the outcome due to the level-2 units, also may affect the accuracy of the estimates ([12]). In the linear multilevel model, the ICC can be estimated by specifying an empty model at both levels which allows to decompose the variance of the outcome Y into the variance of level-1 units σ_e (i.e., the variance of ε_{ij}) and that of groups σ_u (i.e., the variance of level-2 residuals u_{0j})

$$y_{ij} = \psi_{00} + u_{0j} + \varepsilon_{ij}$$

Using this model, the ICC is defined as $\frac{u_{0j}}{u_{0j}+\varepsilon_{ij}}$. Several simulation research (see, e.g., [26, 29]) show that ICC had no effect on the non-coverage rates, even if it decreases as ICC increases. However, an ICC not lower than 0.15 is generally recommended.

Another topic is related to the centering or not centering solutions regarding both level-1 and level-2 predictors. An important debate has been carried out about this issue involving several authors (see, e.g., [20, 32]). Paccagnella in 2006 ([31]) reviewed the essential issues and the main conclusions that can be drawn. In particular, the scaling is introduced to measure contextual effects in a relative way, as well as addressing to collinearity problems, and it is particularly useful in social applications. Two methods are usually employed: grand mean and group mean centering of model variables. Grand mean centering of a covariate X_k, $(x_{ijk} - \overline{x}_k)$, is just a reparametrization of the model and does not cause problems. The criticisms concern group-mean centering $(x_{ijk} - \overline{x}_{jk})$ which leads to not equivalent models. Conversely, non centering technique can be a consequence of poor quality of the group mean or by considering that mean is not the only available variable for measuring contextual effects. Several authors showed that the decision of centering depends on whether the model has been specified and on the purposes of the analysis. In particular, centering can be the solution if there is the aim of distinguishing level-2 effects from level-1 characteristics, in case of problems of collinearity and if faster convergences are needed. Conversely, not centering is adopted when the research interests are on individual effects, to deal with more parsimonious models and to yield more intuitive interpretations of the parameter estimates.

42.3 Parameter Estimation and Softwares

To estimate the parameters involved in a multilevel model several methods can be employed. In the simplest case of linear multilevel modeling and under a frequentist perspective, Maximum-Likelihood (ML) and Restricted Maximum-Likelihood (REML) estimation techniques are usually employed by using many different algorithms: EM algorithm [6]; Newton–Raphson algorithm [24], implemented in the procedure PROC MIXED of SAS; Fisher scoring algorithm [25] in the software VARCL; IGLS algorithm [11] in the software MLwiN; mix of EM and Fisher scoring algorithms in the software HLM. ML estimation yields unbiased estimates of the fixed effects but biased estimates of the variance components. REML estimates are further unbiased for the variance components thanks to the removing of the fixed parameters from the likelihood function.

In multilevel generalized linear models parameter estimation is more complex, as involving approximations to maximum likelihood through Gauss–Hermite quadrature, adaptive Gauss–Hermite quadrature, Monte Carlo integration, or LaplaceŠs method. The most frequently used methods are based on a first- or second-order Taylor series expansion around an estimate of the fixed and random portions of the model. This is referred to as Penalized Quasi-Likelihood (PQL) estimation. In Marginal Quasi-Likelihood (MQL) estimation, the Taylor series is expanded around the fixed part of the model. These methods are implemented in PROC GLIMMIX and NLMIXED of SAS, GLAMM of STATA, MLwiN and HLM softwares.

Bootstrapping can be further employed especially to deal with the bias in the variance estimates and standard errors in both parametric [28] and nonparametric [3] versions. This method is implemented in MLwiN software.

Bayesian estimation through Monte Carlo Markov Chain (MCMC) methods and Gibbs sampler can be further employed by using, e.g., WinBUGS or BUGS softwares, ensuring accurate estimates also in small databases (see also the next section).

42.4 Some Advances

In the last decades several advances of multilevel techniques towards more complex situations have been made. First, together with a hierarchical structure of data, the presence of some variables of interests unmeasured directly but only by a set of items or fallible instruments is considered. In such cases, statistical literature refers to *multilevel models with latent variables*, or *multilevel regression and structural equation models* ([19, 30, 33].

Second, *hierarchical Bayesian models* can be considered as a natural completion of the hierarchical structure of modeling involved by multilevel approach. Under this Bayesian perspective, all parameters are viewed as random and, as a consequence, (hyper-) prior distributions need to be specified. Empirical Bayes [27], Semi Bayes [14], Fully Bayesian and Bayes Empirical Bayes approaches [23] can be adopted, yielding less biased estimates, more robust and more conservative tests, also in small and sparse datasets and with a lower number of level-2 units [15, 16, 33, 39].

Third, spatial and spatio-temporal analysis is a particular multilevel model, when the clusters are geographical areas and/or occasions [22]. In such cases, the variance–covariance parameters are more complex than those introduced before. For instance, in spatial analysis areal proximities can be considered to better explain the spatial distribution of the outcomes.

Then, when the response variable is more than one we refer to *multivariate multilevel models* with several applications in medical and social researches [18].

Finally, commonly in biological, agricultural, environmental, and medical applications for continuous repeated measurements, there are situations where the linearity assumption no more holds. In such cases, *nonlinear multilevel models* need to be considered [4].

References

1. Browne, W.J., Draper, D.: Implementation and performance issues in the Bayesian and likelihood fitting of multilevel models. Comput. Stat. **15**, 391–420 (2000)
2. Bryk, A.S., Raudenbush, S.W.: Hierarchical Linear Models in Social and Behavioral Research: Applications and Data Analysis Methods, 1st edn. Sage Publications, Newbury Park (1992)
3. Carpenter, J.: Test inversion bootstrap confidence intervals. Methodol. R. Stat. Soc. **61**, 159–172 (1999)
4. Davidian, M., Giltinan, D.M.: Nonlinear models for repeated measurements: an overview and update. J. Agric. Biol. Environ. Stat. **8**, 387–419 (2003)
5. Dedrick, R.F., Ferron, J.M., Hess, M.R., Hogarty, K.Y., Kromrey, J.D., Lang, T.R., Niles, J., Lee, R.: Multilevel modeling: a review of methodological issues and applications. Rev. Educ. Res. **79**(1), 69–102 (2009)
6. Dempster, A.P., Laird, N.M., Rubin, D.B.: Maximum likelihood from incomplete data via the EM algorithm. Methodol. R. Stat. Soc. **39**(1), 1–38 (1977)
7. Dempster, A.P., Rubin, D.B., Tsutakawa, R.K.: Estimation in covariance components models. Am. Stat. Assoc. **76**(374), 341–353 (1981)
8. de Leeuw J., Kreft I.G.G.: Software for Multilevel Analysis. Department of Statistics Papers, Department of Statistics, UCLA, Los Angeles (1999)
9. Elston, R.C., Grizzle, J.E.: Estimation of time-response curves and their confidence bands. Biometrics **18**(2), 148–159 (1962)
10. Gelman, A., Carlin, J.B., Stern, H.S., Rubin, D.B.: Bayesian data analysis, 2nd edn. Chapman & Hall, New York (2003)
11. Goldstein, H.: Multilevel mixed linear model analysis using iterative generalized least squares. Biometrika **78**, 43–56 (1986)
12. Goldstein, H.: Multilevel Models in Education and Social Research. Charles Griffin & Co, London; Oxford University Press, New York (1987)
13. Goldstein, H.: Multilevel Statistical Models. Multilevel Models. Institute of Education, London (1999)
14. Greenland, S.: A semi-bayes approach to the analysis of correlated multiple associations, with an application to an occipational cancermortality study. Stat. Med. **11**, 219–230 (1992)
15. Greenland, S.: Bayesian perspectives for epidemiological research: I. Foundations and basic methods. Int. J. Epidemiol. **35**, 765–775 (2006)
16. Greenland, S.: Bayesian perspectives for epidemiological research: II. Regression analysis. Int. J. Epidemiol. **36**, 195–202 (2007)
17. Hox, J.J.: Applied Multilevel Analysis. TT-Publikaties, Amsterdam (1995)
18. Hox, J.J.: Multilevel Analysis. Techniques and Applications. Lawrence Erlbaum Associates, Mahwah (2002)
19. Jöreskog, K.G.: A general method for estimating a linear structural equation system. In: Goldberger, A., Duncan, O. (eds.) Structural Equation Models in the Social Sciences, pp. 85–112. Seminar Press, New York (1973)
20. Kreft, I.G.G., de Leeuw, J., Aiken, L.: The effect of different forms of centering in hierarchical linear models. Multivar. Behav. Res. **30**(1), 1–21 (1995)
21. Laird, N.M., Ware, J.H.: Random-effects models for longitudinal data. Biometrics **38**(4), 963–974 (1982)

22. Lawson, A.B.: Bayesian Disease Mapping. CRC press, New York (2009)
23. Lindley, D.V., Smith, A.F.M.: Bayes estimates for the linear model (with discussion). Methodol. R. Stat. Soc. **34**, 1–41 (1972)
24. Lindstrom, M.L., Bates, D.M.: Newton–Raphson and EM algorithms for linear mixed-effects models for repeated-measures data. J. Am. Stat. Assoc. **83**(404), 1014–1021 (1988)
25. Longford, N.: A fast scoring algorithm for maximum liklihood estimation in unbalanced models with nested random effects. Biometrika **74**(4), 817–827 (1987)
26. Maas, C.J.M., Hox, J.J.: Sufficient sample sizes for multilevel modeling. Methodol. Eur. **1**, 85–91 (2005)
27. Maritz, J., Lwin, T. Empirical Bayes Methods. Chapman and Hall/CRC, London (1989)
28. Meijer, E., Busing, F.M.T.A., Van der Leeden, R.: Estimating bootstrap confidence intervals for two-level models. In: Hox, J.J., De Leeuw, E.D. (eds.) Assumptions, Robustness, and Estimation Methods in Multivariate Modeling, pp. 35–47. TT Publicaties, Amsterdam (1998)
29. Moineddin, R., Matheson, F.I., Glazier, R.H.: A simulation study of sample size for multilevel logistic regression models. BMC Med. Res. Methodol. **7**(34), 1–10 (2007)
30. Muthén, B.: Latent variable modeling with longitudinal and multilevel data. In: Raftery, A. (ed.) Sociological Methodology, pp. 453–480. Blackwell Publishers, Boston (1997)
31. Paccagnella, O.: Centering or not centering in multilevel models? the role of the group mean and the assessment of group effects. Eval. Rev. **30**(1), 66–85 (2006)
32. Raudenbush, S.W.: "Centering" predictors in multilevel analysis: choices and consequences. Multilevel Model. Newslett. **1**(2), 10–12 (1989)
33. Raudenbush, S.W., Bryk, A.S.: Hierarchical Linear Models: Applications and Data Analysis Methods, 2nd edn. Sage, Newbury Park (2002)
34. Rosenberg, B.: Linear regression with randomly dispersed parameters. Biometrika **60**, 61–75 (1973)
35. Schreiber, J.B., Griffin, B.W.: Review of multilevel modeling and multilevel studies in the journal of educational research (1992–2002). J. Educ. Res. **98**, 24–33 (2004)
36. Skrondal, A., Rabe-Hesketh, S.: Generalized Latent Variable Modeling: Multilevel, Longitudinal and Structural Equation Models. Chapman & Hall/CRC, Boca Raton (2004)
37. Skrondal, A., Rabe-Hesketh, S.: Multilevel Modelling, vol. 1–4. Sage, London (2010)
38. Snijders, T.A.B., Bosker, R.J.: Multilevel Analysis. An Introduction to Basic and Advanced Multilevel Modeling. Sage, London (1999)
39. Stegmueller, D.: How many countries to you need for multilevel modeling? a comparison of frequentist and Bayesian approaches. Am. J. Polit. Sci. **57**, 748–761 (2013)

Chapter 43
On a Generalization of the Modified Gravity Model

Diana Santalova

43.1 Introduction

Many models have been suggested for passenger and migration flows estimation, and special attention has been paid to the gravity models and their modifications [2–4, 6, 7]. One of the modifications (hereinafter referred to as *modified gravity model*) was considered by Andronov and Santalova in [1, 8].

The modified gravity model is a nonlinear regression model for passenger correspondences estimation between pairs of spatial points depending on distance between them, population at every such point, and various predictors. The model for a correspondence between points i and l can be written as

$$Y_{i,l} = \frac{(h_i h_l)^\nu}{(d_{i,l})^\tau} exp(a + (c_{(i)} + c_{(l)})\alpha + g_{(i,l)}\beta + Z_{i,l}), \qquad (43.1)$$

where a, $\alpha = (\alpha_1, \alpha_2, \ldots, \alpha_m)^T$ and $\beta = (\beta_1, \beta_2, \ldots, \beta_m)^T$ are unknown regression parameters, ν and τ are unknown shape parameters, $c_{(i)} = (c_{i,1}, c_{i,2}, \ldots, c_{i,m})$ and $g_{(i,l)} = (c_{i,1}c_{l,1}, \ldots, c_{i,m}c_{l,m})$ are known m-vector-rows, $\{Z_{i,l}\}$ are i.i.d. random variables with zero mean and unknown variance σ^2. The matrix of $Y_{i,l}$ is called the *correspondence matrix*. Correspondence matrix is required for any transport model as input information.

Unknown parameters of the model (43.1) and correspondences are estimated using *aggregated data*, i.e. total numbers of passenger departures (for simplicity *departures*) at every point in a considered time interval. The departures Y_i at a point i are presented as a sum of correspondences over other points l:

D. Santalova (✉)
Tartu University, 2 J.Liivi street, Tartu 50409, Estonia
e-mail: diana.santalova@ut.ee

V.B. Melas et al. (eds.), *Topics in Statistical Simulation*, Springer Proceedings in Mathematics & Statistics 114, DOI 10.1007/978-1-4939-2104-1_43

$$Y_i = \sum_{\substack{l=1 \\ i \neq l}}^{n} Y_{i,l} = \sum_{\substack{l=1 \\ i \neq l}}^{n} \frac{(h_i h_l)^{\nu}}{(d_{i,l})^{\tau}} exp(a + (c_{(i)} + c_{(l)})\alpha + g_{(i,l)}\beta + Z_{i,l}). \quad (43.2)$$

The model under the assumption that error term $Z_{i,l}$ is normally distributed random variable has been estimated in [1], and statistical properties of the estimates have been verified in [8].

Another assumption of the model is that estimated correspondences $Y^*_{i,l}$ are symmetric. It means that estimated passenger departures Y_i^* at every point i must be equal to the estimated passenger arrivals at the same point. One must be warned that we deal with the so-called *raw* correspondences, i.e. without expanding them into *commuting*, *duty* or *leisure* trips. Since raw correspondences cannot be symmetric and departures do not equal to arrivals, the model assumption of symmetry is quite often violated.

Besides, the present approach for collecting statistical data about raw correspondences (developed and used by EUROSTAT) allows to fix only the data about passengers *embarked* and *disembarked* at a point, including international transit traffic. Due to this nonconformity, using the model (43.1) for correspondences estimation can cause loss of the estimation accuracy (which can be observed as increasing the mean square error, for example).

One of the approaches to diminish the mean square error in such a situation might be a "brute force method", which can be implemented in estimation of two vectors of parameters. The first vector is estimated from the total numbers of departed (embarked) passengers $\{Y_i^E\}$, and the second one from total numbers of arrived (disembarked) passengers $\{Y_i^D\}$. This approach was tested for railway passenger correspondences estimation between the EU member states, and the mean square error was diminished up to 43 % compared with ordinary estimation procedure [9].

Another approach is based on a hypothesis that non-symmetry of the correspondences can violate the assumption about normality of error distribution. So, we intend to generalize the model with normally distributed errors to a model with skew-normal error distribution, and to propose a method of its estimation.

The paper is organized as follows. In the next section two generalizations of the model (43.1) are suggested. In Sect. 43.3 the correctness of the suggested generalizations is investigated. Experimental results are presented in the end of the paper. Finally further research directions are discussed.

43.2 Generalized Models

Let us denote an estimate of $Y_{i,l}$ as $Y^*_{i,l}$ and make the following assumptions:

- $Y^*_{i,l} > 0$ for $i \neq l$,
- $Y^*_{i,i} = 0$,
- $Y^*_{i,l} = Y^*_{l,i}$.

We suggest two generalizations of the model (43.1). First of them, say *M-Model*, contains multiplicative error term, and the second one, *A-Model*—an additive error term. For both models the error term $Z_{i,l}$ is distributed by skew-normal law. Its density function in the univariate case is (see, for example, [5])

$$\frac{2}{\sigma}\varphi\Big(\frac{x-\mu}{\sigma}\Big)\,\Phi\Big(\lambda\frac{x-\mu}{\sigma}\Big), \qquad x \in \mathfrak{R}, \tag{43.3}$$

where φ and Φ denote the standard normal density and distribution functions, respectively. So, the error term $Z_{i,l} \sim \mathsf{SN}\left(\mu, \sigma^2, \lambda\right)$, where μ is the location, $\sigma > 0$ is the scale and $\lambda \in \mathfrak{R}$ is the shape parameter, respectively. When $\lambda = 0$, we return to the normal distribution $\mathbb{N}(\mu, \sigma^2)$; otherwise, the distribution is positively or negatively asymmetric, in agreement with the sign of λ. The following parameter δ is related to the shape parameter via the relationship:

$$\delta = \frac{\sigma\lambda}{\sqrt{1+\sigma^2\lambda^2}}, \qquad \delta \in (-1, 1)\,. \tag{43.4}$$

In further derivations the first moments of univariate skew-normally distributed variable are needed, which are as follows (see [5]):

$$\begin{aligned} \mathsf{E}(Z_{i,l}) &= \mu + \sqrt{\frac{2}{\pi}}\sigma\delta = \mu + \sqrt{\frac{2}{\pi}}\frac{\sigma^2\lambda}{\sqrt{1+\sigma^2\lambda^2}}, \\ \mathsf{D}(Z_{i,l}) &= \sigma^2\left(1 - \frac{2}{\pi}\delta^2\right) = \sigma^2\left(1 - \frac{2}{\pi}\frac{\sigma^2\lambda^2}{1+\sigma^2\lambda^2}\right). \end{aligned} \tag{43.5}$$

43.2.1 M-Model

The multiplicative model for correspondences is given in the following way:

$$Y_{i,l} = \frac{(h_i h_l)^{\nu}}{(d_{i,l})^{\tau}} \exp(a + (c_{(i)} + c_{(l)})\alpha + g_{(i,l)}\beta) Z_{i,l}, \tag{43.6}$$

where the error term $Z_{i,l}$ is distributed according to skew-normal distribution with *non-zero mean* due to the model's structure. The total number of departed passengers Y_i at every point i is the sum of the relevant correspondences over other points:

$$Y_i = \sum_{\substack{l=1 \\ i \neq l}}^{n} Y_{i,l} = \sum_{\substack{l=1 \\ i \neq l}}^{n} \frac{(h_i h_l)^{\nu}}{(d_{i,l})^{\tau}} \exp(a + (c_{(i)} + c_{(l)})\alpha + g_{(i,l)}\beta) Z_{i,l}. \tag{43.7}$$

The expectation and variance of a correspondence $Y_{i,l}$, $i \neq l$, straightforwardly are

$$\mathsf{E}(Y_{i,l}) = \frac{(h_i h_l)^{\nu}}{(d_{i,l})^{\tau}} \exp(a + (c_{(i)} + c_{(l)})\alpha + g_{(i,l)}\beta)\mathsf{E}(Z_{i,l}), \tag{43.8}$$

$$\mathsf{D}(Y_{i,l}) = \frac{(h_i h_l)^{2\nu}}{(d_{i,l})^{2\tau}} \exp\left(2(a + (c_{(i)} + c_{(l)})\alpha + g_{(i,l)}\beta)\right)\mathsf{D}(Z_{i,l}), \tag{43.9}$$

where $\mathsf{E}(Z_{i,l})$ and $\mathsf{D}(Z_{i,l})$ are as (43.5).

43.2.2 A-Model

For the model with the additive error term let us rewrite the model (43.1) as

$$Y_{i,l} = \frac{(h_i h_l)^{\nu}}{(d_{i,l})^{\tau}} \exp(a + (c_{(i)} + c_{(l)})\alpha + g_{(i,l)}\beta) \exp(Z_{i,l}), \tag{43.10}$$

where the error term $Z_{i,l}$ is distributed according to skew-normal distribution with *zero mean*, when $\mu = -\sqrt{\frac{2}{\pi}}\delta\sigma$, unknown variance σ^2 and shape parameter $\lambda \in \Re$. As for (43.6), the total number of departed passengers Y_i at every point i is:

$$Y_i = \sum_{\substack{l=1 \\ i \neq l}}^{n} Y_{i,l} = \sum_{\substack{l=1 \\ i \neq l}}^{n} \frac{(h_i h_l)^{\nu}}{(d_{i,l})^{\tau}} \exp(a + (c_{(i)} + c_{(l)})\alpha + g_{(i,l)}\beta) \exp(Z_{i,l}). \tag{43.11}$$

Like in Sect. 43.2.1, the expectation and variance for a correspondence $Y_{i,l}$, $i \neq l$, are

$$\mathsf{E}(Y_{i,l}) = \frac{(h_i h_l)^{\nu}}{(d_{i,l})^{\tau}} \exp\big(a + (c_{(i)} + c_{(l)})\alpha + g_{(i,l)}\beta\big)\mathsf{E}(\exp(Z_{i,l})), \tag{43.12}$$

$$\mathsf{D}(Y_{i,l}) = \frac{(h_i h_l)^{2\nu}}{(d_{i,l})^{2\tau}} \exp\big(2(a + (c_{(i)} + c_{(l)})\alpha + g_{(i,l)}\beta)\big)\mathsf{D}(\exp(Z_{i,l})). \tag{43.13}$$

To obtain the closed form of expectation and variance of $Y_{i,l}$ one must derive expectation and variance for $\exp(Z_{i,l})$. They are stated in the following Lemma.

Lemma 43.1. *The approximated expected value of* $\exp(Z_{i,l})$ *is*

$$\mathsf{E}(\exp(Z_{i,l})) \approx 1 + \frac{\sigma^2}{2}\left(1 - \frac{2}{\pi}\delta^2\right), \tag{43.14}$$

and approximated variance is

$$\mathsf{D}(\exp(Z_{i,l})) \approx \sigma^2\left(1 - \frac{2}{\pi}\delta^2\right) + \frac{\sigma^2}{2} + \sqrt{\frac{2}{\pi}}\left(\frac{4}{\pi} - 1\right)\delta^3. \tag{43.15}$$

Proof. From the expansion into a Taylor series we get

$$\mathsf{E}(\exp(Z_{i,l})) = \mathsf{E}\left(1 + Z_{i,l} + \frac{1}{2}Z_{i,l}^2 + \ldots\right) \approx 1 + \mathsf{E}(Z_{i,l}) + \frac{1}{2}\mathsf{E}(Z_{i,l}^2),$$

where $\mathsf{E}(Z_{i,l}) = 0$ by assumption (43.10). Then

$$\mathsf{E}(\exp(Z_{i,l})) \approx 1 + \frac{1}{2}\mathsf{E}(Z_{i,l}^2) \approx 1 + \frac{1}{2}\mathsf{D}(Z_{i,l}),$$

where substituting $\mathsf{D}(Z_{i,l})$ as (43.5) gives us expression (43.14).

For proving expression (43.15) let us use expansion into Taylor series as well:

$$\mathsf{D}(\exp(Z_{i,l})) = \mathsf{D}\left(1 + Z_{i,l} + \frac{1}{2}Z_{i,l}^2 + \ldots\right) \approx \mathsf{D}(Z_{i,l} + \frac{1}{2}Z_{i,l}^2) =$$

$$\mathsf{D}(Z_{i,l}) + \frac{1}{4}\mathsf{D}(Z_{i,l}^2) + \mathrm{Cov}(Z_{i,l}, Z_{i,l}^2), \tag{43.16}$$

where $\mathsf{D}(Z_{i,l})$ is taken from (43.5), and covariance $\mathrm{Cov}(Z_{i,l}, Z_{i,l}^2)$ is derived as follows:

$$\mathrm{Cov}(Z_{i,l}, Z_{i,l}^2) = \mathsf{E}\big((Z_{i,l} - \mathsf{E}(Z_{i,l}))(Z_{i,l}^2 - \mathsf{E}(Z_{i,l}^2))\big) = \mathsf{E}(Z_{i,l}^3) = \sqrt{\frac{2}{\pi}}\left(\frac{4}{\pi} - 1\right)\delta^3, \tag{43.17}$$

since $\mathsf{E}(Z_{i,l}) = 0$, and the third moment is given in [5]. For $\mathsf{D}(Z_{i,l}^2)$ take into account that here $Z_{i,l}^2 \sim \chi_1^2$ (see, for example, [5]). So, if $Z_{i,l} \sim \mathsf{SN}\left(0, \sigma^2, \delta\right)$ and $V_{i,l} \sim \mathsf{SN}\,(0, 1, \delta)$, then $Z_{i,l} = \sigma V_{i,l}$ and $Z_{i,l}^2 = \sigma^2 V_{i,l}^2$, where $V_{i,l}^2 \sim \chi_1^2$. It is known that $\mathsf{D}(Z_{i,l}^2) = \sigma^2\mathsf{D}(V_{i,l}^2) = 2\sigma^2$. So, assembling all stated relations above gives us the expression (43.15). □

Next Subsection contains derivations for getting estimates which are common for both models.

43.2.3 Estimation Method

For both models, the expected value and variance of a departure are written as sums of expectations or variances of relevant correspondences:

$$\mathsf{E}(Y_i) = \sum_{l=1}^{n} \mathsf{E}(Y_{i,l}), \quad \mathsf{D}(Y_i) = \sum_{l=1}^{n} \mathsf{D}(Y_{i,l}), \tag{43.18}$$

where $\mathsf{E}(Y_{i,l})$ and $\mathsf{D}(Y_{i,l})$ are calculated by formulas (43.8) and (43.9) for M-Model, and by formulas (43.12–43.15) for A-Model, respectively.

The covariance $W_{i,l}$ between the departures Y_i and Y_l, if $i \neq l$, for both models is the following:

$$W_{i,l} = \mathrm{Cov}(Y_i, Y_l) = \mathsf{D}(Y_{i,l}), \tag{43.19}$$

where $\mathsf{D}(Y_{i,l})$ is calculated by (43.9) for M-Model and by (43.13), (43.15) for A-Model.

It is assumed that departures $\{Y_i\}$ in both models are calculated from many correspondences, that implies weak dependency and normal distribution. The Log-Likelihood function for the sample $Y = (Y_1, Y_2, \ldots, Y_n)$ for both models is

$$l\,(a, \alpha, \beta, \nu, \tau, \mu, \sigma, \delta) = -\frac{1}{2} ln\,(|\mathsf{W}|) - \frac{1}{2}\,(Y - \mathsf{E}(Y))^T\,\mathsf{W}^{-1}\,(Y - \mathsf{E}(Y)), \tag{43.20}$$

with the only difference that the parameter μ is not estimated in A-Model. The values of $\mathsf{E}(Y) = (\mathsf{E}(Y_1), \mathsf{E}(Y_2), \ldots, \mathsf{E}(Y_n))$ and the matrix W are calculated using (43.18) and (43.19).

43.3 Case Study

In this section the efficiency of the estimation method of the M-Model and A-Model will be evaluated empirically using principles of simulation modeling. The corresponding procedure has been developed and tested in [8]. Briefly, the models and the vector of true parameters $\Theta_0 = (a, \alpha, \beta, \nu, \tau, \mu, \sigma, \lambda)$ are fixed. The correspondences $Y_{i,l}$ for every pair of points (i, l) are generated by (43.6) for M-Model, and by (43.10) for A-Model. The departures from the points are calculated by (43.7) for M-Model, and by (43.11) for A-Model. Further a maximum likelihood estimate of the vector of parameters is obtained from (43.20). These steps are repeated $k_{\max}$ times, which gives us a set of estimates.

For unbiasedness study the average of the $k_{\max}$ obtained estimates has been used, $\bar{\Theta}_U = \frac{1}{k_{\max}} \sum_{k=1}^{k_{\max}} \hat{\Theta}^k$. An estimator of the parameter $\hat{\Theta}$ is unbiased if $\mathsf{E}\hat{\Theta} = \Theta$. According to the law of large numbers, $\bar{\Theta}_U \xrightarrow{P} \mathsf{E}\hat{\Theta}$. Thus, if this limit equals Θ, then the considered estimate is *unbiased*.

For analysis of consistency the sequence $\hat{\Theta}^k_C = \{\tilde{\Theta}^k\}$ of moving averages $\tilde{\Theta}^k = \frac{1}{k} \sum_{j=1}^{k} \hat{\Theta}^j$ has been considered. A sequence of estimates $\hat{\Theta}^k_C$ is *consistent* if $\hat{\Theta}^k_C \xrightarrow{P} \Theta$. It can be interpreted empirically as $\hat{\Theta}^k_C \xrightarrow[k \to k_{\max}]{} \Theta$, providing that the number of samples $k_{\max}$ is large.

In simulations statistical data from 75 cities are used. Variables of interest $\{Y_i\}$ are the total inland road passenger departures from cities, in thousands of passengers. The following qualitative covariates are chosen for the experiments:

c_1 – attractiveness of a city as an industrial center,
c_2 – significance of a city as a cultural/social center.

The values of these covariates are determined by experts, besides c_1 and c_2 take values from the set of gradations $S = \{0, 1, 2\}$, where gradation 0 means low level of significance or attractiveness, 1—middle level and 2—high level.

The total departures for M-Model and A-Model were generated from the above mentioned statistical data and the following vector of true parameters:

$$\Theta_0 = (a, \alpha_1, \alpha_2, \beta_1, \beta_2, \nu, \tau, \mu, \sigma, \lambda) = (2, 1, 1, 1, 1, 1, 2, 2, 1, 2)\,. \quad (43.21)$$

The results of estimation are stated below.

Looking back to our previous experience in δ (43.4) estimation, one must say that the covariance matrix W (43.19) often was singular, and several values of the Log-Likelihood function were unbounded. In addition, several estimates of δ exceeded its feasible range, which corresponds to the boundary values of the index of skewness $\gamma_1 \in (-0.99527, 0.99527)$ (see [5]). In such cases the corresponding values of the parameter λ were undefined [10].

Now we estimate the shape parameter λ directly. The average estimated values $\bar{\Theta} = (1.96, 1.00, 0.98, 1.00, 1.02, 1.00, 2.01, 1.61, 1.07, 2.05)$ of the vector Θ_0 are obtained from 4,000 estimates. One can see that the average estimate of location parameter μ is biased. The average optimal value of the Log-Likelihood function is 500, and the corresponding value of mean square error is 2.55×10^7.

The A-Model seems to be more natural than the M-Model due to the additive error term, besides the A-Model is free of location parameter μ. The vector of model parameters Θ_0 is as (43.21), where location parameter μ is preset to zero. The average estimated values $\bar{\Theta} = (2.34, 1.01, 1.02, 0.98, 0.97, 1.01, 1.97, 1.44, 2.17)$ of the vector Θ_0 are obtained from 2250 estimates. One can see that the average estimates of the parameters a, σ and λ are biased. The average optimal value of the Log-Likelihood function is 586, and the corresponding value of mean square error is 5.01×10^7.

Concerning biasedness of certain parameters estimates for both models, one can conclude that a more accurate optimization procedure is needed. Besides, the number of parameters to be estimated is large. One must also take into account that the expressions of the expectation and variance for the A-Model are approximated. Further we intend to analyze the models w.r.t. different values of the parameter λ.

Acknowledgements The research was supported by the European Union through the European Social Fund (Mobilitas grant No GMTMS280MJ) and Estonian Science Foundation (grant No ETF9127).

References

1. Andronov, A., Santalova, D.: Statistical estimates for modified gravity model by aggregated data. Commu. Stat. Simul. Comput. **41**(2), 730–745 (2012)
2. Erlander, S., Stewart, N.F.: The Gravity Model in Transportation Analysis: Theory and Extensions. VSP, Utrecht (1990)
3. Grosche, T., Rothlauf, F., Heinzl, A.: Gravity models for airline passenger volume estimation. J. Air Trans. Manag. **13**(4), 175–183 (2007)
4. Karemera, D., Oguledo, V.I., Davis, B.: A gravity model analysis of international migration to North America. Appl. Econ. **32**(13), 1745–1755 (2000)
5. Kotz, S., Nadarajah, S.: Multivariate t Distributions and their Applications. Cambridge University Press, Cambridge (2004)
6. Lewera, J.J., Van den Berg, H.: A gravity model of immigration. Econ. Lett. **99**(1), 164–167 (2008)
7. Otruzar, J., Willumsen, L.G.: Modeling Transport, 4th edn. Wiley, New York (2011)
8. Santalova, D.: Experimental analysis of statistical properties of parameter estimation of modified gravity model. Autom. Control Comput. Sci. **47**(2), 99–106 (2013)
9. Santalova, D.: Non-symmetrical transport flows estimation using the modified gravity model. In: Mariagiulia M. (eds.) 7th International Workshop on Simulation, pp. 312–313. Department of Statistical Sciences, Unit of Rimini, University of Bologna, Bologna, May 21–25 (2013). (Dipartimento di Scienze Statistiche "Paolo Fortunati", Alma Mater Studiorum Universita di Bologna, Serie Ricerche)
10. Santalova, D.: Migration flows estimation using the modified gravity model. In: Christos H.S. (eds.) 15th Applied Stochastic Models and Data Analysis International Conference, 25–28 June 2013, Mataró (Barcelona), International Society for the Advancement of Science and Technology, pp. 186–186 (2013)

Chapter 44
Bivariate Lorenz Curves Based on the Sarmanov–Lee Distribution

José María Sarabia and Vanesa Jordá

44.1 Introduction

The interest of academics in assessing country levels of well-being has shifted from an evaluation based solely on economic aspects to a more comprehensive conception of such a process, which has an intrinsic multidimensional nature. The different works in [3, 4, 9, 14, 21, 25] and [26] move in this direction. Different approaches have been proposed in the literature to measure inequality in well-being, being the most satisfactory one the consideration of multidimensional inequality measures since this methodology takes into account inequality within each dimension and the degree of association among them. However, as in the unidimensional case, these measures only offer overall conclusions about the evolution of the distribution of well-being, thus other statistical tools are needed to analyze the evolution of inequality in different parts of the distribution. In this context, the extension of the univariate Lorenz curve to higher dimensions is not an obvious task. The three existing definitions were proposed by Taguchi [22, 23], Arnold [1] and Koshevoy and Mosler [11], who introduced the concepts of Lorenz zonoid and Gini zonoid index.

The main contributions of this paper are the following. Using the definition proposed by Arnold [1], closed expressions for the bivariate Lorenz curve are given. To do that, we use a type of models based on the class of bivariate distributions with given marginals described by Sarmanov and Lee [13, 20]. This model presents several advantages. In particular, the expression of the bivariate Lorenz curve can be easily interpreted as a convex linear combination of products of classical and generalized Lorenz curves. We obtain a closed expression of the bivariate

J.M. Sarabia (✉) • V. Jordá
Department of Economics, University of Cantabria, Avda. de los Castros s/n, 39005 Santander, Cantabria, Spain
e-mail: sarabiaj@unican.es; jordav@unican.es

V.B. Melas et al. (eds.), *Topics in Statistical Simulation*, Springer Proceedings in Mathematics & Statistics 114, DOI 10.1007/978-1-4939-2104-1_44

Gini index [2] in terms of the classical and the generalized Gini indices of the marginal distributions. We prove that this index can be decomposed in two factors, corresponding to the equality within and between variables.

The contents of the paper are as follows. In Sect. 44.2 we present preliminary results including the definition of univariate Lorenz and concentration curves and a short review about the three different definitions of bivariate Lorenz curves proposed in the literature. An explicit expression for the Arnold's bivariate Lorenz curve is also given. In Sect. 44.3 we introduce the bivariate Sarmanov–Lee Lorez curve, also obtaining a simple closed expression for this curve and for its corresponding bivariate Gini index, according to the Arnold's definition (see [2]). A decomposition of this index in two factors is given. A bivariate Pareto Lorenz curve based on the FGM family is presented in 44.4. Other aspects are briefly discussed in Sect. 44.5.

44.2 Preliminary Results

44.2.1 Univariate Lorenz and Concentration Curves

Let $\mathscr{L}$ denote the class of univariate distributions functions with positive finite expectations and $\mathscr{L}_+$ denote the class of all distributions in with $F(0) = 0$ corresponding to non-negative random variables. We use the following definition proposed in [6].

Definition 44.1. The Lorenz curve L of a random variable X with cumulative distribution function $F \in \mathscr{L}$ is

$$L(u;F) = \frac{\int_0^u F^{-1}(y)dy}{\int_0^1 F^{-1}(y)dy} = \frac{\int_0^u F^{-1}(y)dy}{E[X]}, \quad 0 \leqslant u \leqslant 1, \tag{44.1}$$

where $F^{-1}(y) = \sup\{x : F(x) \leqslant y\}$ if $0 \leqslant y < 1$ and $F^{-1}(y) = \sup\{x : F(x) < 1\}$ if $y = 1$.

Now, we present the concept of concentration curve introduced by Arnold [8]. Let $g(x)$ be a continuous function of x such that its first derivative exists and $g(x) \geqslant 0$. If the mean $E_F[g(X)]$ exits, then one can define

$$L_g(y;F) = \frac{\int_0^x g(x)dF(x)}{E_F[g(X)]},$$

where $y = g(x)$ and $f(x)$ and $F(x)$ are, respectively, the probability density function (PDF) and the cumulative distribution function (CDF) of the random variable X. The implicit relation between $L_g(g(x); F)$ and $F(x)$ is called the concentration curve of the function $g(X)$. The concentration curve also admits an implicit representation.

44.2.2 Arnold's Definition of Bivariate Lorenz Curve

The first definition of a bivariate Lorenz curve was proposed by Taguchi [22–24]. Unfortunately, this definition was not symmetric and its extension to higher dimensions did not look simple. Other definition of a multidimensional Lorenz curve was initially proposed by Koshevoy [10], who identified a suitable definition. Thereafter, further results were obtained by Koshevoy and Mosler [11, 12, 16]. These authors introduced the so-called Lorenz zonoid, which is a convex American football-subset of the three-dimensional unit cube that includes the points $(0, 0, 0)$ and $(1, 1, 1)$. While the extension to higher dimensions is fairly direct, the computation of these formulas in parametric income distributions is not straightforward.

The following definition was proposed by Arnold (see [1, 2]) and it is a quite natural extension of (44.1) to higher dimensions. Let $\boldsymbol{X} = (X_1, X_2)^\top$ be a bivariate random variable with bivariate probability distribution function F_{12} on R_+^2 having finite second and positive first moments. We denote by F_i, $i = 1, 2$ the marginal CDFs corresponding to X_i, $i = 1, 2$, respectively.

Definition 44.2. The Lorenz surface of F_{12} is the graph of the function,

$$L(u_1, u_2; F_{12}) = \frac{\int_0^{s_1}\int_0^{s_2} x_1 x_2 dF_{12}(x_1, x_2)}{\int_0^{\infty}\int_0^{\infty} x_1 x_2 dF_{12}(x_1, x_2)}, \tag{44.2}$$

where

$$u_1 = \int_0^{s_1} dF_1(x_1), \quad u_2 = \int_0^{s_2} dF_2(x_2), \quad 0 \leqslant u, v \leqslant 1.$$

The two-attribute Gini–Arnold index $GA(F_{12})$ is defined as

$$GA(F_{12}) = 4\int_0^1\int_0^1 [u_1 u_2 - L(u_1, u_2; F_{12})] du_1 du_2, \tag{44.3}$$

where the egalitarian surface is given by $L_0(u_1, u_2; F_0) = u_1 u_2$. Since the previous definition has not been explored in detail in the literature, we highlight some of its properties. If F_{12} is a product distribution function, then

$$L(u_1, u_2; F_{12}) = L(u_1; F_1) L(u_2; F_2),$$

which is just the product of the marginal Lorenz curves. If we denote by F_a the one-point distribution at $a \in R^2_+$, that is, the egalitarian distribution at a, then the egalitarian distribution has bivariate Lorenz curve $L(u_1, u_2; F_a) = u_1 u_2$. In the case of a product distribution, the two-attribute Gini–Arnold defined in (44.3) can be written as

$$1 - GA(F_{12}) = [1 - G(F_1)][1 - G(F_2)].$$

Our results are based on an explicit expression of the bivariate Lorenz curve defined in (44.2), which admits the following representation.

Theorem 44.1. *The bivariate Lorenz curve can be written in the explicit form,*

$$L(u_1, u_2; F_{12}) = \frac{1}{E[X_1 X_2]} \int_0^{u_1} \int_0^{u_2} A(x_1, x_2) dx_1 dx_2, \quad 0 \leqslant u_1, u_2 \leqslant 1,$$

where

$$A(x_1, x_2) = \frac{F_1^{-1}(x_1) F_2^{-1}(x_2) f_{12}(F_1^{-1}(x_1), F_2^{-1}(x_2))}{f_1(F_1^{-1}(x_1)) f_2(F_2^{-1}(x_2))}. \tag{44.4}$$

Proof. The proof is direct making the change of variable $(u_1, u_2) = (F_1(x_1), F_2(x_2))$ in (44.2). □

44.3 The Bivariate Sarmanov–Lee Lorenz Curve

In this section, we introduce the so-called bivariate Sarmanov–Lee Lorenz curve. As a previous step, we present the bivariate Sarmanov–Lee distribution.

Let $\mathbf{X} = (X_1, X_2)^\top$ be a bivariate Sarmanov–Lee (SL) distribution with joint PDF,

$$f(x_1, x_2) = f_1(x_1) f_2(x_2) \{1 + w\varphi_1(x_1)\varphi_2(x_2)\}, \tag{44.5}$$

where $f_1(x_1)$ and $f_2(x_2)$ are the univariate PDF marginals, $\varphi_i(t)$, $i = 1, 2$ are bounded nonconstant functions such that

$$\int_{-\infty}^{\infty} \varphi_i(t) f_i(t) dt = 0, \quad i = 1, 2,$$

and w is a real number which satisfies the condition $1 + w\varphi_1(x_1)\varphi_2(x_2) \geqslant 0$ for all x_1 and x_2. We denote $\mu_i = E[X_i] = \int_{-\infty}^{\infty} t f_i(t) dt$, $i = 1, 2$, $\sigma_i^2 = \mathrm{var}[X_i] = \int_{-\infty}^{\infty} (t - \mu_i)^2 f_i(t) dt$, $i = 1, 2$ and $\nu_i = E[X_i \varphi_i(X_i)] = \int_{-\infty}^{\infty} t\varphi_i(t) f_i(t) dt$, $i = 1, 2$. Properties of this family have been explored in [13]. Moments and regressions of this family can be easily obtained.

Note that (44.5) and its associated copula have two components: a first component corresponding to the marginal distributions and the second component which defines the structure of dependence, given by the parameter w and the functions $\varphi_i(u)$, $i = 1, 2$. These two components will be translated to the structure of the associated bivariate Lorenz curve, and the corresponding bivariate Gini index. In relation with other families with given marginals, the Sarmanov–Lee copula has several advantages: its joint PDF and CDF are quite simple; the covariance structure in general is not limited and its different probabilistic features can be obtained in an explicit form. On the other hand, the SL distribution includes several relevant special cases, including the classical Farlie–Gumbel–Morgenstern (FGM) distribution and the variations proposed in [7] and [5].

44.3.1 Main Result

The bivariate SL Lorenz curve is obtained using (44.5) in Eq. (44.2).

Theorem 44.2. *Let* $\mathbf{X} = (X_1, X_2)^\top$ *be a bivariate Sarmanov–Lee distribution with joint PDF (44.5) characterized by non-negative marginals satisfying* $E[X_1] < \infty$, $E[X_2] < \infty$ *and* $E[X_1 X_2] < \infty$. *Then, the bivariate Lorenz curve is given by*

$$L_{\mathrm{SL}}(u_1, u_2; F_{12}) = \pi L(u_1; F_1) L(u_2; F_2) + (1 - \pi) L_{g_1}(u_1; F_1) L_{g_2}(u_2; F_2), \tag{44.6}$$

where

$$\pi = \frac{\mu_1 \mu_2}{E[X_1 X_2]} = \frac{\mu_1 \mu_2}{\mu_1 \mu_2 + w \nu_1 \nu_2},$$

and $L(u_i; F_i)$, $i = 1, 2$ *are the Lorenz curves of the marginal distributions* X_i, $i = 1, 2$ *respectively, and* $L_{g_i}(u_i; F_i)$, $i = 1, 2$, *represent the concentration curves of the random variables* $g_i(X_i) = X_i \varphi_i(X_i)$, $i = 1, 2$, *respectively.*

Proof. The function (44.4) for the Sarmanov–Lee distribution can be written of the form

$$A_{\mathrm{SL}}(x_1, x_2) = F_1^{-1}(x_1) F_2^{-1}(x_2) \left\{1 + w \varphi_1(F_1^{-1}(x_1)) \varphi_2(F_2^{-1}(x_2))\right\},$$

and integrating in the domain $(0, u) \times (0, v)$ we obtain

$$\mu_1 \mu_2 L(u_1; F_1) L(u_2; F_2) + w E_{F_1}[g_1(X_1)] E_{F_2}[g_2(X_2)] L_{g_1}(u_1; F_1) L_{g_2}(u_2; F_2),$$

and after normalized we obtain (44.6). □

The interpretation of (44.6) is quite direct: the bivariate Lorenz curve can be expressed as a convex linear combination of two components: (a) a first component corresponding to the product of the marginal Lorenz curves (marginal component) and a second component corresponding to the product of the concentration Lorenz curves (structure dependence component).

44.3.2 Bivariate Gini index

The following result provides a convenient expression of the two-attribute bivariate Gini defined in (44.3), which permits a simple decomposition of the equality in two factors. The first component represents the equality within variables and the second factor represents the equality between variables.

Theorem 44.3. *Let* $\mathbf{X} = (X_1, X_2)^\top$ *be a bivariate Sarmanov–Lee distribution with bivariate Lorenz curve* $L(u, v; F_{12})$. *The two-attribute bivariate Gini index defined in (44.3) is given by*

$$1 - G(F_{12}) = \pi[1 - G(F_1)] \cdot [1 - G(F_2)] + (1 - \pi)[1 - G_{g_1}(F_1)] \cdot [1 - G_{g_2}(F_2)],$$

where $G(F_i)$, $i = 1, 2$ *are the Gini indices of the marginal Lorenz curves, and* $G_{g_i}(F_i)$, $i = 1, 2$ *represent the concentration indices of the concentration Lorenz curves* $L_{g_i}(u_i, F_i)$, $i = 1, 2$.

Proof. The proof is direct using expression (44.6) and taking into account that $G(F_{12}) = 1 - 4\int_0^1 \int_0^1 L(u, v; F_{12}) du dv$. □

Then, the overall equality (OE) given by $1 - G(F_{12})$ can be decomposed in two factors,

$$OE = EW + EB,$$

where

$$\begin{aligned} OE &= 1 - G(F_{12}), \\ EW &= \pi[1 - G(F_1)][1 - G(F_2)], \\ EB &= (1 - \pi)[1 - G_{g_1}(F_1)][1 - G_{g_2}(F_2)]. \end{aligned}$$

EW represents the equality within variables and the second factor, EB, represents the equality between variables which includes the structure of dependence of the underlying bivariate income distribution through the functions g_i, $i = 1, 2$. Note that the decomposition is well defined since $0 \leqslant OE \leqslant 1$ and $0 \leqslant EW \leqslant 1$ and hence $0 \leqslant EB \leqslant 1$.

44.4 Bivariate Pareto Lorenz Curve Based on the FGM Family

In this section, we present an example of bivariate Lorenz curve. Let $\mathbf{X} = (X_1, X_2)^\top$ be a bivariate FGM with classical Pareto marginals and joint PDF,

$$f_{12}(x_1, x_2; \alpha, \sigma) = f_1(x_1) f_2(x_2)\{1 + w[1 - 2F_1(x_1)][1 - 2F_2(x_2)]\}, \tag{44.7}$$

where

$$F_i(x_i) = 1 - \left(\frac{x}{\sigma_i}\right)^{-\alpha_i}, \quad x_i \geqslant \sigma_i, \quad i = 1, 2,$$

$$f_i(x_i) = \frac{\alpha_i}{\sigma_i}\left(\frac{x}{\sigma_i}\right)^{-\alpha_i - 1}, \quad x_i \geqslant \sigma_i, \quad i = 1, 2,$$

are the CDF and the PDF of the classical Pareto distributions, respectively [1], with $\alpha_i > 1, \sigma_i > 0, i = 1, 2, -1 \leqslant w \leqslant 1$ and $\varphi_i(x_i) = 1 - 2F_i(x_i), i = 1, 2$ in (44.5).

Using (44.6) with $g_i(x_i) = x_i[1 - 2F_i(x_i)]$, $i = 1, 2$ and after some computations, the bivariate Lorenz curve associated with (44.7) is

$$L_{\text{FGM}}(u_1, u_2; F_{12}) = \pi_w L(u_1; \alpha_1) L(u_2; \alpha_2) + (1 - \pi_w) L_{g_1}(u_1; \alpha_1) L_{g_2}(u_2; \alpha_2),$$

where the Lorenz and the concentration curves are given, respectively, by

$$L(u_i; \alpha_i) = 1 - (1 - u_i)^{1 - 1/\alpha_i}, \quad 0 \leqslant u \leqslant 1, \quad i = 1, 2,$$

$$L_{g_i}(u_i; \alpha_i) = 1 - (1 - u_i)^{1 - 1/\alpha_i}[1 + 2(\alpha_i - 1)u_i], \quad 0 \leqslant u \leqslant 1, \quad i = 1, 2,$$

and,

$$\pi_w = \frac{(2\alpha_1 - 1)(2\alpha_2 - 1)}{(2\alpha_1 - 1)(2\alpha_2 - 1) + w}.$$

The bivariate Gini index is given by (using (44.3))

$$G(\alpha_1, \alpha_2) = \frac{(3\alpha_1 - 1)(3\alpha_2 - 1)(2\alpha_1 + 2\alpha_2 - 3) + [h(\alpha_1, \alpha_2)]w}{(3\alpha_1 - 1)(3\alpha_2 - 1)[(1 - 2\alpha_1)(1 - 2\alpha_2) + w]},$$

where

$$h(\alpha_1, \alpha_2) = -3 - 4\alpha_1^2(\alpha_2 - 1)^2 + (5 - 4\alpha_2)\alpha_2 + \alpha_1(5 + \alpha_2(8\alpha_2 - 7)).$$

44.5 Extensions and Additional Properties

Other alternative families of bivariate Lorenz curves can be constructed, including models based on conditional specification [18], models with marginals specified in terms of univariate Lorenz curves (see [17, 19]) and models based on mixtures of distributions, which allow us to incorporate heterogeneity factors in the inequality analysis. Furthermore, the concepts developed in this chapter can be extended to dimensions higher than two.

Moreover, some stochastic orderings related with the Lorenz curves can be defined. We denote by $\mathscr{L}_+^k$ the set of all k-dimensional nonnegative random vectors $\mathbf{X}$ and $\mathbf{Y}$ with finite marginal expectations, that is $E[X_i] \in R_{++}$ and $E[Y_i] \in R_{++}$. Let $\mathbf{X}, \mathbf{Y} \in \mathscr{L}_+^k$, and we define the following order (see [15]): $\mathbf{X} \preceq_L \mathbf{Y}$ if $L(u; F_{\mathbf{X}}) \geq L(u; F_{\mathbf{Y}})$.

Theorem 44.4. *Let* $\mathbf{X}, \mathbf{Y} \in \mathscr{L}_+^2$ *with the same Sarmanov–Lee copula. Then, if* $X_i \preceq_L Y_i$, *and* $X_i \preceq_{Lg_i} Y_i$ $i = 1, 2$, *then* $\mathbf{X} \preceq_L \mathbf{Y}$.

Proof. The proof is direct based on the expression of the bivariate Sarmanov–Lee Lorenz curve defined in (44.6). □

Acknowledgements The authors thank to Ministerio de Economía y Competitividad (project ECO2010-15455) and Ministerio de Educación (FPU AP-2010-4907) for partial support.

References

1. Arnold, B.C.: Pareto Distributions. International Co-operative Publishing House, Fairland (1983)
2. Arnold, B.C.: Majorization and the Lorenz durve. In: Lecture Notes in Statistics, vol. 43. Springer, New York (1987)
3. Atkinson, A.B.: Multidimensional deprivation: contrasting social welfare and counting approaches. J. Econ. Inequal. **1**, 51–65 (2003)
4. Atkinson, A.B., Bourguignon, F.: The comparison of multi-dimensioned distributions of economic status. Rev. Econ. Stud. **49**, 183–201 (1982)
5. Bairamov, I., Kotz, S.: On a new family of positive quadrant dependent bivariate distributions. Int. Math. J. **3**, 1247–1254 (2003)
6. Gastwirth, J.L.: A general definition of the Lorenz curve. Econometrica **39**, 1037–1039 (1971)
7. Huang, J.S., Kotz, S.: Modifications of the Farlie–Gumbel–Morgenstern distributions a tough hill to climb. Metrika **49**, 135–145 (1999)
8. Kakwani, N.C.: Applications of Lorenz curves in economic analysis. Econometrica **45**, 719–728 (1977)
9. Kolm, S.C.: Multidimensional equalitarianisms. Q. J. Econ. **91**, 1–13 (1977)
10. Koshevoy, G.: Multivariate Lorenz majorization. Soc. Choice Welf. **12**, 93–102 (1995)
11. Koshevoy, G., Mosler, K.: The Lorenz zonoid of a multivariate distribution. J. Am. Stat. Assoc. **91**, 873–882 (1996)
12. Koshevoy, G., Mosler, K.: Multivariate gini indices. J. Multivar. Anal. **60**, 252–276 (1997)
13. Lee, M.L.T.: Properties of the Sarmanov family of bivariate distributions. Commun. Stat. A-Theory **25**, 1207–1222 (1996)
14. Maasoumi, E.: The measurement and decomposition of multi-dimensional inequality. Econometrica **54**, 991–997 (1986)
15. Marshall, A.W., Olkin, I., Arnold, B.C.: Inequalities: Theory of Majorization and its Applications, 2nd edn. Springer, New York (2011)
16. Mosler, K.: Multivariate dispersion, central regions and depth: the lift zonoid approach. In: Lecture Notes Statistics, vol. 165. Springer, Berlin (2002)
17. Sarabia, J.M.: Parametric Lorenz curves: models and applications. In: Chotikapanich, D. (ed.) Modeling Income Distributions and Lorenz Curves. Series: Economic Studies in Inequality, Social Exclusion and Well-being, vol. 4, pp. 167–190. Springer, New York (2008)

18. Sarabia, J.M., Castillo, E., Pascual, M., Sarabia, M.: Bivariate income distributions with lognormal conditionals. J. Econ. Inequal. **5**, 371–383 (2007)
19. Sarabia, J.M., Castillo, E., Slottje, D.: An ordered family of Lorenz curves. J. Econ. **91**, 43–60 (1999)
20. Sarmanov, O.V.: Generalized normal correlation and two-dimensional frechet classes. Doklady Sov. Math. **168**, 596–599 (1966)
21. Slottje, D.J.: Relative price changes and inequality in the size distribution of various components. J. Bus. Econ. Stat. **5**, 19–26 (1987)
22. Taguchi, T.: On the two-dimensional concentration surface and extensions of concentration coefficient and Pareto distribution to the two-dimensional case-I. Ann. Inst. Stat. Math. **24**, 355–382 (1972)
23. Taguchi, T.: On the two-dimensional concentration surface and extensions of concentration coefficient and Pareto distribution to the two-dimensional case-II. Ann. Inst. Stat. Math. **24**, 599–619 (1972)
24. Taguchi, T.: On the structure of multivariate concentration—some relationships among the concentration surface and two variate mean difference and regressions. Comput. Stat. Data Anal. **6**, 307–334 (1988)
25. Tsui, K.Y.: Multidimensional generalizations of the relative and absolute inequality indices: the Atkinson–Kolm–Sen approach. J. Econ. Theory **67**, 251–265 (1995)
26. Tsui, K.Y.: Multidimensional inequality and multidimensional generalized entropy measures: an axiomatic derivation. Soc. Choice Welf. **16**, 145–157 (1999)

Chapter 45
Models with Cross-Effect of Survival Functions in the Analysis of Patients with Multiple Myeloma

Mariia Semenova, Ekaterina Chimitova, Oleg Rukavitsyn, and Alexander Bitukov

45.1 Introduction

Accelerated life models are used more and more often in oncology and hematology studies for estimation of the effect of explanatory variables on lifetime distribution and for estimation of the survival function under given covariate values, see [6].

The most popular and most widely applied survival regression model is the proportional hazards model (called also the Cox model) introduced by Sir David Cox. The popularity of this model is based on the fact that there are simple semiparametric estimation procedures which can be used when the form of the survival distribution function is not specified, see [4]. The survival functions for different values of covariates according to the Cox proportional hazard (PH) model do not intersect. However, in practice this condition often does not hold. Then, we need to apply some more complicated models which allow decreasing, increasing, or nonmonotonic behavior of the ratio of hazard rate functions.

Following [1] and [2], we illustrate possible applications of the Hsieh model (see [5]) and the simple cross-effect model, which are particularly useful for the analysis of survival data with one crossing point.

M. Semenova • E. Chimitova (✉)
Novosibirsk State Technical University, Novosibirsk, Russia
e-mail: vedernikova.m.a@gmail.com; ekaterina.chimitova@gmail.com

O. Rukavitsyn • A. Bitukov
The Hematology Center, Main Military Clinical Hospital named after N.N.Burdenko, Moscow, Russia
e-mail: 82465@bk.ru

V.B. Melas et al. (eds.), *Topics in Statistical Simulation*, Springer Proceedings in Mathematics & Statistics 114, DOI 10.1007/978-1-4939-2104-1_45

45.2 Parametric Models

Suppose that each individual in a population has a lifetime T_x under a vector of covariates $x = (x_1, x_2, \ldots, x_m)^T$. Let us denote by $S_x(t) = P(T_x \geqslant t) = 1 - F_x(t)$ the survival function and by $\lambda_x(t)$ and $\Lambda_x(t)$ the hazard rate function and the cumulative hazard rate function of T_x, respectively.

In survival analysis, lifetimes are usually right censored. The observed data usually are of the form $(t_1, \delta_1), \ldots, (t_n, \delta_n)$, where $\delta_i = 1$ if t_i is an observed complete lifetime, while $\delta_i = 0$ if t_i is a censoring time, which simply means that the lifetime of the i-th individual is greater than t_i.

45.2.1 *Proportional Hazards Model*

The cumulative hazard rate for the Cox proportional hazards model is given by

$$\Lambda_x(t; \beta, \theta) = \exp\left(\beta^T \cdot x\right) \Lambda_0(t; \theta), \tag{45.1}$$

where β is the vector of unknown regression parameters, $\Lambda_0(t;\theta)$ is the baseline cumulative hazard rate function, which belongs to a specified class of hazard rate functions.

This model implies that the ratio of hazard rates under different values of covariate x_2 and x_1 is constant over time:

$$\frac{\lambda_{x_2}(t)}{\lambda_{x_1}(t)} = \frac{\exp\left(\beta^T \cdot x_2\right)}{\exp\left(\beta^T \cdot x_1\right)} = \text{const.} \tag{45.2}$$

However, this model is rather restrictive and is not applicable when the ratios of hazard rates are not constant in time. There may be an interaction between covariates and time, in which case hazards are not proportional.

45.2.2 *Hsieh Model*

According to the idea of Hsieh, one possible way to obtain a nonmonotonic behavior of ratios of hazard rates is to take a power function of the baseline cumulative hazard function. Namely, Hsieh proposed the model given by

$$\Lambda_x(t; \beta, \gamma, \theta) = \exp\left(\beta^T \cdot x\right) \{\Lambda_0(t;\theta)\}^{\exp\left(\gamma^T \cdot x\right)}. \tag{45.3}$$

The parameters β and γ are m-dimensional. It is a generalization of the proportional hazards model taking the power $\exp(\gamma^T x)$ of $\Lambda_0(t;\theta)$ instead of the power 1. It is easy to show that the Hsieh model implies that the hazard ratio between different fixed covariates is increasing from 0 to ∞ or decreasing from ∞ to 0. The model of Hsieh does not contain interesting alternatives to crossing: the hazard rates under different constant covariates cross for any values of the parameters β and $\gamma \neq 0$ [5]. This model implies that the ratio

$$\frac{\lambda_{x_2}(t)}{\lambda_{x_1}(t)} = \{\Lambda_0(t;\theta)\}^{\exp(\gamma^T \cdot x)-1}$$

is monotone, $\frac{\lambda_{x_2}(0)}{\lambda_{x_1}(0)} = 0$ and $\frac{\lambda_{x_2}(\infty)}{\lambda_{x_1}(\infty)} = \infty$ or vice versa, so there exists the point $t_0 : \frac{\lambda_{x_2}(t_0)}{\lambda_{x_1}(t_0)} = 1$. If $\gamma = 0$, then the hazard rates coincide [2].

45.2.3 Simple Cross-Effect Model

A more versatile model including not only crossing but also going away of hazard rates is the simple cross-effect model [2] given by

$$\Lambda_x(t;\beta,\gamma,\theta) = \left(1 + \exp\left((\beta+\gamma)^T \cdot x\right) \Lambda_0(t;\theta)\right)^{\exp\left(-\gamma^T \cdot x\right)} - 1. \qquad (45.4)$$

The ratio

$$\frac{\lambda_{x_2}(t)}{\lambda_{x_1}(t)} = \exp\left(\beta^T \cdot x\right) \left(1 + \exp\left((\beta+\gamma)^T \cdot x\right) \Lambda_0(t;\theta)\right)^{\exp\left(-\gamma^T \cdot x\right)-1}$$

is monotone, $\frac{\lambda_{x_2}(0)}{\lambda_{x_1}(0)} = \exp\left(\beta^T \cdot x\right)$, $\frac{\lambda_{x_2}(\infty)}{\lambda_{x_1}(\infty)} = \infty$, if $\gamma < 0$, and $\frac{\lambda_{x_2}(\infty)}{\lambda_{x_1}(\infty)} = 0$, if $\gamma > 0$. So, the hazard rates may cross or go away but cannot converge or approach (in sense of the ratio at the point t).

To test the goodness-of-fit of the described models to an observed data, it is possible to use the approach based on the residuals $R_i = \Lambda_x\left(t_i;\hat{\beta},\hat{\gamma},\hat{\theta}\right), i = 1,\ldots,n$, which should fit closely to the standard exponential distribution if the model is indeed "correct." Testing the hypothesis H_0 whether the samples of observed residuals belong to a particular distribution can be carried out by means of Kolmogorov, Cramer-von Mises-Smirnov, and Anderson–Darling tests and using the maximum likelihood estimates of unknown parameters [3].

45.3 Analysis of Patients with Multiple Myeloma

This investigation of patients with multiple myeloma was carried out in the Hematology Center, in the Main Military Clinical Hospital named after N.N.Burdenko. The purpose of the investigation is to compare the response time to the treatment in two groups of patients. The difference in these groups is in the fact that the first group received chemotherapy with Bortezomibe, which is marketed as Velcade by Millennium Pharmaceuticals.

45.3.1 Data Description

The data, presented in Table 45.1, include observations of 60 patients, 4 of which were censored. Patients in the study were randomly assigned to one of two treatment groups: chemotherapy without Bortezomibe ($x_1 = 0$) or chemotherapy together with Bortezomibe ($x_1 = 1$).

In addition to treatment, several factors were also observed: type of response (the value $x_2 = 1$ corresponds to the general response, $x_2 = 0$ corresponds to the progression of the disease), sex ($x_3 = 1$ means that the patient is male, $x_3 = 0$ means that the patient is female), and age in years (x_4). Table 45.1 also gives the response times in months (t) and the censoring indicator δ. So, the number of patients fallen into different groups is given in Table 45.2.

There are 38 observations in the first group and 22 observations in the second one. It should be noted that 4 observations are independent randomly censored observations. Moreover, we will take into account the age of patients as a covariate in the survival models.

Table 45.1 The data of patients with multiply myeloma

t	δ	x_1	x_2	x_3	x_4	t	δ	x_1	x_2	x_3	x_4	t	δ	x_1	x_2	x_3	x_4
61	1	1	1	1	64	62	1	1	0	1	75	7	1	0	1	0	66
50	1	1	0	1	81	3	1	1	1	1	64	2	1	0	1	0	60
2	1	1	1	0	71	26	1	1	1	1	61	262	0	0	1	1	68
36	1	1	0	1	69	22	1	1	0	1	72	81	1	0	1	0	81
14	1	1	0	0	74	46	1	1	0	1	59	33	1	0	0	0	79
27	1	1	0	1	83	3	1	1	1	1	77	215	1	0	0	1	65
1	1	1	1	0	46	16	1	1	1	1	66	57	1	0	0	0	85
4	1	1	1	1	80	10	1	1	0	0	46	17	1	0	0	1	89
27	1	1	0	0	58	25	1	1	1	0	55	26	1	0	1	0	75
115	0	1	0	1	50	6	1	1	1	1	48	7	1	0	0	1	47
13	1	1	0	1	85	5	1	1	1	0	51	30	1	0	0	1	75
2	1	1	1	1	56	30	1	1	0	1	81	2	1	0	0	1	66
3	1	1	1	1	57	25	1	1	0	1	58	26	1	0	1	1	76
25	1	1	1	1	71	39	1	1	0	1	77	20	0	0	1	1	37
4	1	1	1	1	64	6	1	1	1	1	65	5	0	0	0	0	57
62	1	1	0	0	57	83	1	1	0	0	69	8	1	0	1	0	73
9	1	1	1	0	71	24	1	1	1	1	52	127	1	0	0	0	79
10	1	1	0	0	56	3	1	1	1	1	49	149	1	0	0	1	87
7	1	1	1	1	55	7	1	0	1	1	61	10	1	0	1	1	65
54	1	1	0	1	75	2	1	0	1	1	45	8	1	0	1	1	61

Table 45.2 The plan of experiment

x_2	x_3	$x_1 = 0$	$x_1 = 1$	$\sum$
0	0	4	6	10
0	1	6	12	18
1	0	5	5	10
1	1	7	15	22
$\sum$		22	38	60

Table 45.3 AIC for the Hsieh and SCE models

Model	AIC
Exponential Hsieh	489.45
Exponential SCE	487.41
Weibull Hsieh	491.37
Weibull SCE	485.62
Gamma Hsieh	490.02
Gamma SCE	482.59
Lognormal Hsieh	479.51
Lognormal SCE	478.19

45.3.2 Simulation Results

First of all, we estimated survival functions for patients in two groups of treatment using nonparametric Kaplan–Meier estimates since the sample is censored. The estimates of survival functions intersect once. By this reason the proportional hazards model can be inappropriate for these data (proportional hazard assumption was not hold). We compared the Hsieh models and the simple cross-effect (SCE) models with different baseline distributions by the Akaike information criterion ($AIC = 2k - 2\log L$, where k is the number of estimated parameters and L is the maximized likelihood function). The obtained values for the considered models are given in Table 45.3.

The minimal AIC value is equal to 478.19 for the lognormal SCE model, thus we propose using the SCE model with lognormal baseline distribution for relating the distribution of response time to the scheme of chemotherapy and other factors. In this case, the baseline hazard rate function has the following form:

$$\Lambda_0\left(t;\theta\right) = -\log\left(\frac{1}{2} - \frac{1}{2\sqrt{\pi}}\Gamma\left(\frac{1}{2\theta_2}\log^2\left(t/\theta_1\right), \frac{1}{2}\right)\right).$$

In Table 45.4, there are maximum likelihood estimates of the model parameters $\theta = (\theta_1, \theta_2)$, $\beta = (\beta_1, \beta_2, \beta_3, \beta_4)^T$ and $\gamma = (\gamma_1, \gamma_2, \gamma_3, \gamma_4)^T$ and the p-values of the Wald test for testing insignificance of parameters.

As can be seen from Table 45.4, the parameters for the first and second covariates in the model, namely type of chemotherapy and type of response, are significant (p-values are less than 0.05). The obtained statistics of Kolmogorov, Cramer-von Mises-Smirnov, and Anderson–Darling tests are $S_k = 0.50$, $S_{\omega^2} = 0.034$ and

Table 45.4 Estimates of parameters of the lognormal SCE model

Model parameters	MLEs of parameters	p-value of the Wald test
θ_1	90.81	
θ_2	1.21	
β_1, γ_1	0.60, −0.46	0.03
β_2, γ_2	4.36, 2.17	0.001
β_3, γ_3	−0.25, 0.19	0.18
β_4, γ_4	0.65, −0.007	0.63

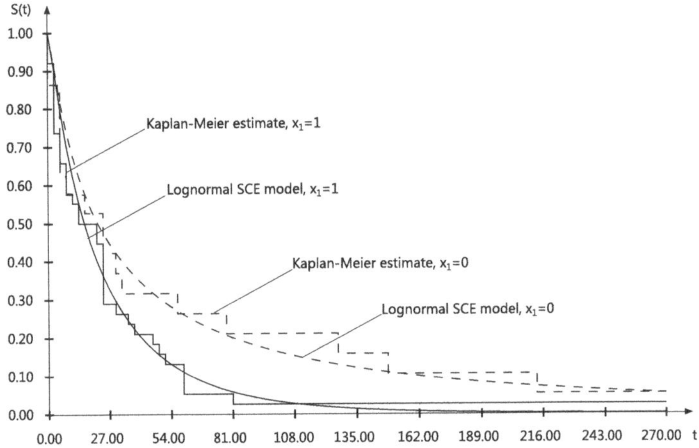

Fig. 45.1 The Kaplan–Meier estimates and corresponding survival functions of the SCE model

$S_{\Omega^2} = 0.25$, and the corresponding p-values are equal to 0.92, 0.89, and 0.96, respectively. So, the goodness-of-fit hypothesis of the lognormal SCE model is not rejected.

In Fig. 45.1, the nonparametric Kaplan–Meier estimates and the corresponding survival functions of the SCE model are presented.

So, it is possible to conclude that the response in the group of patients, treated with Bortezomibe, was achieved significantly faster than in the control group, in which patients were taken chemotherapy without Bortezomibe. Moreover, in the case of general response (which combines such cases as complete, partial, minimal response and stabilization) the lifetime till response is significantly less than in the case of progression of the disease.

Acknowledgements This research has been supported by the Russian Ministry of Education and Science as part of the state task No 2014/138 (project No 1689).

References

1. Bagdonavicius, V., Nikulin, M.: Accelerated Life Models. Chapman and Hall/CRC, Boca Raton (2002)
2. Bagdonavicus, V., Levuliene, R., Nikulin, M.: Modeling and testing of presence of hazard rates crossing under censoring. Comm. Stat. Sim. Comp. **41**, 980–991 (2012)
3. Balakrishnan, N., Chimitova, E., Galanova, N., Vedernikova, M.: Testing goodness-of-fit of parametric AFT and PH models with residuals. Comm. Stat. Sim. Comp. **42**, 1352–1367 (2013)
4. Cox, D.R.: Regression models and life tables (with discussion). J. R. Stat. Soc. Ser. B **34**, 187–220 (1972)
5. Hsieh, F.: On heteroscedastic hazards regression models: theory and application. J. R. Stat. Soc. Ser. B **63**, 63–79 (2001)
6. Klein, J.P., Moeschberger, M.L.: Survival Analysis. Springer, New York (1997)

Chapter 46
Monte Carlo Method for Partial Differential Equations

Alexander Sipin

46.1 Introduction

The unbiased estimators for solutions $u(x)$ of boundary value problems for PDEs are usually constructed on trajectories of Markov processes in domain $\mathscr{D}$ in R^n or on the boundary $\partial\mathscr{D}$. The transition function $P(x, dy)$ of such processes $\{x_i\}_{i=0}^{\infty}$ is usually the kernel of integral equation

$$u(x) = \int_Q u(y)P(x, dy) + F(x), x \in Q \tag{46.1}$$

in the space $M(Q)$ of the measurable bounded functions on the compact Q. Here, $Q = \overline{\mathscr{D}}$ or $Q = \partial\mathscr{D}$, the function $F(x)$ is defined by boundary conditions and right part of differential equation.

Let K be integral operator in the Eq. (46.1). If $\|K\| < 1$, we may use von-Neumann–Ulam scheme [3] to construct the unbiased estimators for $u(x)$. In case of $\|K\| = 1$ and $F(x) \geqslant 0$ for any bounded solution $u(x)$ of Eq. (46.1) we have representation [1, 4]:

$$u(x) = \sum_{i=0}^{\infty} K^i F(x) + K^{\infty}u(x), x \in Q, \tag{46.2}$$

where

$$K^{\infty}u(x) = \lim_{i\to\infty} K^i u(x). \tag{46.3}$$

A. Sipin (✉)
Vologda State Pedagogical University, S.Orlov 6, Vologda, Russia
e-mail: cac1909@mail.ru

V.B. Melas et al. (eds.), *Topics in Statistical Simulation*, Springer Proceedings in Mathematics & Statistics 114, DOI 10.1007/978-1-4939-2104-1_46

Really, due to (46.1), we have inequality

$$\sum_{i=0}^{n} K^i F(x) = u(x) - K^{n+1}u(x) \leqslant 2\|u\|.$$

So, the row (46.2) converges and the limit (46.3) exists.

The function $v(x) = K^\infty u(x)$ is the solution of the equation $v(x) = Kv(x)$. It is called invariant function for $P(x, dy)$.

The function

$$GF(x) = \sum_{i=0}^{\infty} K^i F(x)$$

is the potential of the function $F(x)$.

The function $u(x)$ is called excessive one for $P(x, dy)$, if the inequality $u(x) \geqslant Ku(x)$ is true. Due to $K^\infty GF(x) = 0$, the excessive function $u(x)$ have a unique decomposition to sum of potential and invariant functions.

Let Δ be an absorbing state, $\tau = \inf\{n | x_n = \Delta\}$, $u(\Delta) = 0$ and $\rho(x, \Gamma)$ is the distance between x and a set Γ. The Markov chain $\{x_i\}_{i=0}^{\infty}$ in the domain $\mathscr{D}$ with transition function $P(x, dy)$ must have additional properties :

1. $P_x(\tau = \infty) > 0$,
2. $P_x(x_i \to x_\infty, x_\infty \in \partial\mathscr{D} | \tau = \infty) = 1$ or
3. $P_x(\rho(x_i, \partial\mathscr{D}) \to 0 | \tau = \infty) = 1$,
4. $E_x \min(\tau, \tau_\varepsilon) < \infty$ for $\tau_\varepsilon = \inf(i : \rho(x_i, \partial\mathscr{D}) < \varepsilon)$.

which permit us to obtain unbiased and ε-biased estimators for $u(x)$.

Using invariant and excessive functions for $P(x, dy)$, we can construct simple conditions, which yield properties (1–4). Markov chains, usually used in Monte Carlo algorithms for boundary value problems [3], satisfy these conditions. These results are applied to the "walk in hemispheres" [2] and the "walk on cylinders" [5] processes.

46.2 Some Properties of the Markov Chain

Here we investigate some properties of the Markov chain $\{x_i\}_{i=0}^{\infty}$, which starts from the point $x_0 = x$.

Let $\{\mathfrak{A}_i\}_{i=0}^{\infty}$ be an increasing family of σ-algebras, $\{x_i\}$ be $\mathfrak{A}_i$—measurable, χ_i be the indicator of the set $\{\tau > i\}$. For any bounded excessive solution $u(x)$ of the Eq. (46.1) we define *standard sequence of unbiased estimators* as

$$\eta_i = \sum_{j=0}^{i-1} F(x_j)\chi_j + \chi_i u(x_i). \qquad (46.4)$$

It is evident that $\{\eta_i, \mathfrak{A}_i\}_{i=0}^{\infty}$ is uniformly integrated martingale and

$$E_x \chi_i u(x_i) = K^i u(x).$$

Hence, $\eta_\infty = \lim \eta_i$ exist P_x a.s. and $E_x \eta_\infty = u(x)$. The row $\sum_{j=0}^{\infty} F(x_j)\chi_j$ converges P_x a.s. by B.Levy's Theorem. We have also $K^\infty u(x) = E_x \lim \chi_i u(x_i)$ by Lebesgue's Theorem.

Theorem 46.1. *1. If for any $x \in Q$ probability $P_x(\tau < \infty) = 1$, then any bounded excessive function is potential.*

2. If the function $u(x) \equiv 1$ is potential, then for any $x \in Q$ probability $P_x(\tau < \infty) = 1$.

3. If nonnegative invariant function $v(x)$ exists and $v(x) > 0$, then $P_x(\tau = \infty) > 0$.

4. Let $F(x)$ be continuous function, $F(x) \geqslant 0$ and $GF(x) < \infty$ for all $x \in Q$. Let $\Gamma = \{x \in Q | F(x) = 0\}$. If $P_x(\tau = \infty) > 0$, then

$$P_x(\lim \rho(x_i, \Gamma) = 0 | \tau = \infty) = 1.$$

5. Let $F_i(x)$ be continuous function, $F_i(x) \geqslant 0$ and $GF_i(x) < \infty$ for all $x \in Q$ and $i = 1, 2$. Let $\Gamma_i = \{x \in Q | F_i(x) = 0\}$ for $i = 1, 2$ and $P_x(\tau = \infty) > 0$, then

$$P_x(\lim \rho(x_i, \Gamma_1 \cap \Gamma_2) = 0 | \tau = \infty) = 1.$$

6. Let $v(x)$ be continuous invariant function, $Kv^2(x)$ is continuous and

$$\Gamma = \{x \in Q | v^2(x) = Kv^2(x)\}.$$

If $P_x(\tau = \infty) > 0$, then

$$P_x(\lim \rho(x_i, \Gamma) = 0 | \tau = \infty) = 1.$$

7. Let $v(x) \geqslant 0$ and $-v(x)$ be continuous excessive function or $v(x) \leqslant 0$ and $v(x)$ is continuous excessive function, $Kv^2(x)$ is continuous and

$$\Gamma = \{x \in Q | v^2(x) = Kv^2(x)\}.$$

If $P_x(\tau = \infty) > 0$, then

$$P_x(\lim \rho(x_i, \Gamma) = 0 | \tau = \infty) = 1.$$

8. In last two cases 6 and 7 for $x \in \Gamma$ function $v(y) = const = v(x)$ P_x a.s. and if $v(x) \neq 0$, then $P(x, Q) = 1$.

Proof. 1. The function $u_1(x) \equiv 1$ is excessive and

$$K^{\infty}u_1(x) = E_x \lim \chi_i = P_x(\tau = \infty) = 0.$$

For any bounded excessive function $u(x)$ we have

$$|K^{\infty}u(x)| \leqslant \limsup K^i |u(x)| \leqslant \|u\| K^{\infty}u_1(x) = 0.$$

2. If $u(x) \equiv 1 = Gg(x)$, then

$$g(x) = P(x, \Delta),\ K^i g(x) = P_x(\tau = i)$$

and

$$P_x(\tau < \infty) = \sum_{i=0}^{\infty} P_x(\tau = i) = \sum_{i=0}^{\infty} K^i g(x) = 1.$$

3. Let χ_∞ be an indicator of the set $\{\tau = \infty\}$ then

$$0 < v(x) = K^{\infty}v(x) = E_x \lim \chi_i v(x_i) \leqslant E_x \chi_\infty \sup v(x) = P_x(\tau_1 = \infty) \sup v(x).$$

Hence, $P_x(\tau_1 = \infty) > 0$.
4. Now P_x a.s. $\lim F(x_i)\chi_i = 0$ and $\forall i (\chi_i = 1)$ in the set $\{\tau = \infty\}$, hence P_x a.s. $\lim F(x_i) = 0$ in this set. Let the sequence $\{\rho(x_{i_k}, \Gamma)\}_{k=0}^{\infty}$ converges. The set Q is compact, so the sequence $\{x_{i_k}\}_{k=0}^{\infty}$ have a converging subsequence.Without the loss of generality we may suppose that $\{x_{i_k}\}_{k=0}^{\infty}$ converges to a point $\tilde{x}$. We have $0 = \lim F(x_{i_k}) = F(\tilde{x})$ by the continuity of $F(x)$. So, $\tilde{x} \in \Gamma$ and $\rho(\tilde{x}, \Gamma) = 0$. The set Γ is closed and $\lim \rho(x, \Gamma)$ is continuous function of x, hence $\lim \rho(x_{i_k}, \Gamma) = \rho(\tilde{x}, \Gamma) = 0$. The sequence $\{\rho(x_i, \Gamma)\}_{k=0}^{\infty}$ is bounded and it has a converging subsequence, so $\lim \rho(x_i, \Gamma) = 0$.
5. The function $F(x) = F_1(x) + F_2(x)$ have a finite potential and $\Gamma = \Gamma_1 \cap \Gamma_2$, so item 4 yields item 5.
6. For a function $v(x)$ we have inequality $F(x) = Kv^2(x) - v^2(x) \geqslant 0$. Hence, $-v^2(x)$ is excessive function. Now item 6 is a corollary of item 4.
7. Really,

$$F(x) = Kv^2(x) - v^2(x) = \int_Q v^2(y)P(x, dy) - v^2(x)$$

$$\geqslant 2v(x)Kv(x) - 2v^2(x)P(x, Q) + v^2(x)P(x, Q) - v^2(x) \geqslant v^2(x)(1 - P(x, Q)) \geqslant 0.$$

Hence, $-v^2(x)$ is excessive function. Now item 7 is a corollary of item 4.
8. This item is evident.

□

As a rule, the sequence $\{x_i\}_{i=0}^{\infty}$ converges a.s. Particulary, the following Theorem is valid.

Theorem 46.2. *1. Let there exist bounded excessive functions $w_m(x)$, $m = 1, \ldots, n$, such as for coordinate function $v_m(x)$, sum $w_m(x) + v_m(x)$ or difference $w_m(x) - v_m(x)$ is excessive function than the Markov chain $\{x_i\}_{i=0}^{\infty}$ converges on the set $\{\tau = \infty\}$ a.s.*

2. Let there exist constants w_m, $m = 1, \ldots, n$, such as for coordinate function $v_m(x)$, sum $w_m + v_m(x)$ or difference $w_m - v_m(x)$ is excessive function than the Markov chain $\{x_i\}_{i=0}^{\infty}$ converges on the set $\{\tau = \infty\}$ a.s.

3. Let coordinate function $v_m(x)$ or $-v_m(x)$ $m = 1, \ldots, n$ is excessive function, or $v_m(x)$ is invariant function, than the Markov chain $\{x_i\}_{i=0}^{\infty}$ converges on the set $\{\tau = \infty\}$ a.s.

Proof. (1). Let $h_m(x) = w_m(x) - v_m(x)$ be an excessive function. *Standard sequence of unbiased estimators* (46.4) for $h_m(x)$ and sequence $\{\chi_i h_m(x_i)\}_{i=0}^{\infty}$ converges a.s. *Standard sequence of unbiased estimators* (46.4) for $w_m(x)$ and sequence $\{\chi_i w_m(x_i)\}_{i=0}^{\infty}$ converges a.s. So, for $m = 1, \ldots, n$ sequence $\{\chi_i v_m(x_i)\}_{i=0}^{\infty}$ converge a.s. Similary, we can prove the assertions (2) and (3) □

46.3 Statistical Estimators

The standard sequence of unbiased estimators (46.4) is not realizable, since it contains the function $F(x)$ and $u(x)$ with unknown values. To obtain realized estimator one applies or an unbiased estimators, or a ε—biased estimators of these functions. We describe the appropriate procedure, following [3].

We assume that the Markov chain $\{x_i\}_{i=0}^{\infty}$ satisfy following condition

(a) The sequence $\{x_i\}_{i=0}^{\infty}$ a.s. converges on the set $\{\tau = \infty\}$ to a point $x_\infty \in \partial Q$.

For $\delta > 0$ we define $\tau_1 = \inf\{i \,|\, \rho(x_i, \partial Q) < \delta\}$ and $\tau_\delta = \min(\tau_1, \tau)$. For a Markov chain satisfying (a) the value of τ_δ is a.s. finite.

The sequence of unbiased estimators $\{\xi_i\}_{i=0}^{\infty}$ for solution $u(x)$ of the problem (46.1) is called *admissible*, if there exists a sequence σ -algebras $\{\mathfrak{B}_i\}_{i=0}^{\infty}$ such that $\mathfrak{A}_i \subseteq \mathfrak{B}_i$ and $\mathfrak{B}_i \subseteq \mathfrak{B}_{i+1}$, and ξ_i is given by $\xi_i = \zeta_i + \chi_i u(x_i)$, where ζ_i is $\mathfrak{B}_i$-measurable. The sequence of *admissible* estimators define random variable ξ_δ by equality $\xi_\delta = \zeta_{\tau_\delta} + \chi u(x^*_{\tau_\delta})$, where χ—indicator of event $\{\tau > \tau_1\}$, $x^*_{\tau_\delta}$—point of the boundary and $\rho(x^*_{\tau_\delta}, x_{\tau_\delta}) \leqslant \delta$.

Properties of this estimator were obtained in the following theorem.

Theorem 46.3 ([3], theorem 2.3.2). *If an admissible sequence of estimators $\{\xi_i\}_{i=0}^{\infty}$ is a square integrable martingale for the filtration $\{\mathfrak{B}_i\}_{i=0}^{\infty}$, defined above, then a random variable ξ_δ is $\varepsilon(\delta)$-biased estimator of $u(x)$ ($\varepsilon(\delta)$—modulus of continuity of $u(x)$), and its variance is a bounded function of a parameter δ.*

The following lemma gives a condition of the square integrability for the standard sequence of estimators.

Lemma 46.1. *If the Eq. (46.1) has bounded solutions for $F(x)$ and $|F(x)|$, then the standard sequence of unbiased estimators is a square integrable martingale relative to the filtration $\{\mathfrak{A}_i\}_{i=0}^{\infty}$.*

Proof. The potential $GF^2(x) \leqslant \|F\|G|F(x)| < \infty$. Therefore, for the standard sequence of estimators we have an inequality

$$E_x\eta_i^2 \leqslant 2E_x\left(\sum_{j=0}^{\infty}\chi_j|F(x_j)|\right)^2 + 2\|u\|^2$$

$$= 2E_x\sum_{j=0}^{\infty}\chi_j F^2(x_j) + 4E_x\left(\sum_{j=0}^{\infty}\chi_j|F(x_j)| \sum_{m=j+1}^{\infty}\chi_m|F(x_m)|\right) + 2\|u\|^2$$

$$= 2GF^2(x) + 4E_x\sum_{j=0}^{\infty}\chi_j|F(x_j)|E_x\left(\chi_j\sum_{m=j+1}^{\infty}\chi_m|F(x_m)| \mid \mathfrak{A}_j\right) + 2\|u\|^2$$

$$= 2GF^2(x) + 4E_x\sum_{j=0}^{\infty}\chi_j|F(x_j)|(G|F|(x_j) - |F(x_j)|) + 2\|u\|^2$$

$$\leqslant 2GF^2(x) + 4\|(G|F| - |F|)\|G|F|(x) + 2\|u\|^2.$$

□

The function $F(x)$ is presented usually in the form $F(x) = h(x)Ef(Y)$, where the random variable Y has a distribution that depends on x, the function $f(y)$ is the right-hand side of the differential equation, or the value of its solutions at the boundary. Let $\{y_j\}_{j=0}^{\infty}$ be a sequence of random variables such that

$$F(x_i) = h(x_i)E(f(y_i) \mid \mathfrak{A}_i) \tag{46.5}$$

a.s. and $\mathfrak{B}_i$ is a minimal σ—algebra generated by $\mathfrak{A}_i$ and the sequence $\{y_j\}_{j=0}^{i}$, then the sequence of unbiased estimators

$$\xi_i = \sum_{j=0}^{i-1} h(x_j)f(y_j)\chi_j + \chi_i u(x_i) \tag{46.6}$$

is *admissible*. Such estimators are be traditionally called estimators by *collisions*. If the Eq. (46.1) has a bounded solution $\tilde{u}(x)$ for the $F(x) = h(x)$, then by

Lemma 46.1 it follows that the sequence of unbiased estimators

$$\tilde{\xi}_i = \sum_{j=0}^{i-1} h(x_j)\chi_j + \chi_i \tilde{u}(x_i) \tag{46.7}$$

is a square integrable martingale for filtration $\{\mathfrak{A}_i\}_{i=0}^{\infty}$. This implies the lemma.

Lemma 46.2. *Let the Eq. (46.1) has a bounded solution for the $F(x) = h(x)$. If $f(x)$ is bounded function and $F(x)$ have the form (46.5) then the Eq. (46.1) also has a bounded solution . The sequence of unbiased estimators (46.6) is square integrable martingale for filtration $\{\mathfrak{B}_i\}_{i=0}^{\infty}$.*

Note that the condition of Lemma 46.2 is valid if the corresponding boundary value problems have bounded solutions of the required smoothness for the constant right-hand side $f(x)$ (the zero boundary condition) and zero right-hand side of the differential equation (with constant boundary condition). In the last case, the function $h(x) = P(x, \Gamma_0)$, where Γ_0—a lot of points on the boundary ∂Q, included in the support of $P(x, dy)$. In this case, $h(x)$ has a sense of probability of trajectory absorbtion on the boundary ∂Q. In this case, as a vector of y_i, usually, use the following point of the trajectory, that is defined as $y_i = x_{i+1}$. As a result we have the estimator by *absorption*.

To obtain an exact upper bound of the expectation $E_x \tau_\delta$ commonly used renewal theorem. Using the estimators (46.7), we easily obtain that the expectation is finite.

Lemma 46.3. *Let the Eq. (46.1) has bounded solutions for the $F(x) = h(x)$ and there is a constant $c(\delta)$, such that the inequality $h(x) \geqslant c(\delta) > 0$ fulfilled for $x \in Q$ and $\rho(x, \partial Q) \geqslant \delta$. Then we have inequality $E_x \tau_\delta \leqslant Gh(x)/c(\delta)$.*

46.4 Some Applications

Now we investigate Random Walks on Hemispheres processes [2], which applies for solving various boundary problems for Laplace operator. Here we discuss only one of them.

Let some plane Π divide a domain $\mathscr{D} \subset R^3$ into two sub-domains $\mathscr{D}_+$ and $\mathscr{D}_-$. Let $u(x)$ be a harmonic function in $\mathscr{D}_+$ and $\mathscr{D}_-$. We suppose that $u(x)$ is continuous one in $\overline{\mathscr{D}}$ and $u(x), x \in \partial\mathscr{D}$ is known.

We denote the normal to plane by ν. It has a direction from $\mathscr{D}_-$ to $\mathscr{D}_+$. Let ν be an orth of the first coordinate axis. Hence, $v_1(x) = 0$ is the plane equation. Let $\lambda = \text{const} > 0$ and $\lambda \neq 1$ the equation

$$\lambda \frac{\partial u}{\partial \nu_+} = \frac{\partial u}{\partial \nu_-}, \tag{46.8}$$

connects the normal derivations for $u(x), x \in \Pi$. Described boundary condition defines unique function which is harmonic one in $\mathscr{D}_+$ and $\mathscr{D}_-$.

Now we define a transition function for Random Walks on Hemispheres. For $x \in \mathscr{D}_+$ ($x \in \mathscr{D}_-$) we define $S(x)$ as a maximal hemisphere which satisfies the next conditions

- $S(x) \subset \overline{\mathscr{D}}_+ \quad (S(x) \subset \overline{\mathscr{D}}_-)$,
- The plane part of hemisphere lies on the plain Π,
- the center of the hemisphere $x_0 \in \mathscr{D}$,
- x lies in the direction ν ($-\nu$) from x_0.

Here $R(x)$ is the radius of the hemisphere $S(x)$. If such hemisphere does not exist, then $S(x) \subset \overline{\mathscr{D}}_+ \quad (S(x) \subset \overline{\mathscr{D}}_-)$ is a maximal sphere with center x.

From Green formula we have

$$u(x) = -\int_{S(x)} \frac{\partial G(x,y)}{\partial \nu_y} u(y) d_y S.$$

Here $\partial G(x,y)/\partial \nu_y$ is normal derivation of Green function $G(x,y)$. Hence, the transition function of Markov chain is

$$P(x,dy) = -\frac{\partial G(x,y)}{\partial \nu_y} d_y S.$$

If $x \in \Pi$, then $S(x) \subset \overline{\mathscr{D}}$ is a maximal sphere with center x and $R(x)$ is its radius. Let $S_+ = S(x) \cap \overline{\mathscr{D}}_+$ and $S_- = S(x) \cap \overline{\mathscr{D}}_-$. Green formula and condition (46.8) give an integral equations for $u(x)$

$$u(x) = \frac{\lambda}{1+\lambda} \frac{1}{2\pi R^2} \int_{S_+} u(y)\, d_y S + \frac{1}{1+\lambda} \frac{1}{2\pi R^2} \int_{S_-} u(y)\, d_y S. \tag{46.9}$$

Hence, for $x \in \Pi$ transition function $P(x,dy)$ is the mix of two uniform distributions on S_+ and S_- with probabilities $\lambda/(1+\lambda)$ and $1/(1+\lambda)$, respectively.

Evidently, coordinate functions $v_2(x)$ and $v_3(x)$ are invariant ones for the kernel $P(x,dy)$. On the plain $v_1(x) = 0$ for $\lambda < 1$ an inequality

$$\int_{S(x)} P(x,dy) v_1(y) = \frac{\lambda}{1+\lambda} \frac{1}{2\pi R^2} \int_{S_+} v_1(y)\, d_y S + \frac{1}{1+\lambda} \frac{1}{2\pi R^2} \int_{S_-} v_1(y)\, d_y S =$$

$$= \frac{\lambda - 1}{1+\lambda} \frac{1}{2\pi R^2} \int_{S_+} v_1(y)\, d_y S < 0$$

is valid, so coordinate function $v_1(x)$ is excessive one. For $\lambda > 1$ the function $-v_1(x)$ is excessive one.

Due to Theorem 46.2 the *Random Walks on Hemispheres* converges to some random point x_∞. The function $v_2(x)$ is invariant one for $P(x, dy)$. Due to item 6 of the Theorem 46.1 $x_\infty \in \Gamma$ for $\Gamma = \{x \in \overline{\mathscr{D}} | v_2^2(x) = K v_2^2(x)\}$. Note that $x \in \Gamma$ is equivalent to the fact that $v_2(Y) = v_2(x)$ a.s. if Y have a distribution $P(x, dy) = \delta(y - x)dy$, what fulfills only for $x \in \partial\mathscr{D}$.

Acknowledgements Work is supported by RFBR grant 14-01-00271a.

References

1. Doob, J.L.: Discrete potential theory and boundaries. J. Math. Mech. **8**(3), 433–458 (1959)
2. Ermakov, S.M., Sipin, A.S.: The "walk in hemispheres" process and its applications to solving boundary value problems. Vestnik St.Petersburg University: Mathematics **42**(3), 155–163 (2009)
3. Ermakov, S.M., Nekrutkin, V.V., Sipin A.S.: Random Processes for Classical Equations of Mathematical Physics. Kluwer, Boston (1989)
4. Meyer, P.A.: Probability and Potentials. Blaisddell Publishing, Waldham, Massachusetts, Toronto/London (1966)
5. Sipin, A.S., Bogdanov, I.I. : "Random walk on cylinders" for heat equation. In: Tchirkov, M.K. (eds.) Mathematical Models. Theory and Applications VVM, Issue 13, pp. 25–36. Sc. Res. Inst. Chem., St.Petersburg University (2012) (in Russion)

Chapter 47
The Calculation of Effective Electro-Physical Parameters for a Multiscale Isotropic Medium

Olga Soboleva and Ekaterina Kurochkina

47.1 Governing Equations and Approximation of a Medium

Wave propagation in complex inhomogeneous media is an urgent problem in many fields of research. In electromagnetics, these problems arise in such applications as estimation of soil water content, well logging methods, etc. In order to compute the electromagnetic fields in an arbitrary medium, one must numerically solve Maxwell's equations. The large-scale variations of coefficients as compared with wavelength are taken into account in these models with the help of some boundary conditions. The numerical solution of the problem with variations of parameters on all the scales requires high computational costs. The small-scale heterogeneities are taken into account by the effective parameters. In this case, equations are found on the scales that can be numerically resolved.

It has been experimentally shown that the irregularity of electric conductivity, permeability, porosity, density abruptly increases as the scale of measurement decreases. The spatial positions of the small-scale heterogeneities are very seldom exactly known. It is customary to assume the parameters with the small-scale variations to be random fields characterized by the joint probability distribution functions. In this case, the solution of the effective equations must be close to the ensemble-averaged solution of the initial problem. For such problems, a well-known procedure of the subgrid modeling [3] is often used. To apply the subgrid modeling method, we need a "scale regular" medium. It has been experimentally shown

O. Soboleva (✉)
Institute of Computational Mathematics and Mathematical Geophysics SB RAS, pr. Lavrentieva 6, Novosibirsk 630090, Russia
e-mail: olgasob@gmail.com

E. Kurochkina
Institute of Thermophysics SB RAS, Kutateladze st. 1, Novosibirsk 630090, Russia
e-mail: Kurochkina@itp.nsc.ru

V.B. Melas et al. (eds.), *Topics in Statistical Simulation*, Springer Proceedings in Mathematics & Statistics 114, DOI 10.1007/978-1-4939-2104-1_47

that many natural media are "scale regular" in the sense that their parameters, for example, permeability, porosity, density, electric conductivity can be approximated by fractals and multiplicative cascades [1,5]. The effective coefficients in the quasi-steady Maxwell's equations for a multiscale isotropic medium are described in [6]. In the present paper, the correlated fields of electric conductivity and permittivity are approximated by a multiplicative continuous cascade. We obtain formulas of effective coefficients for Maxwell's equations in the frequency domain when the following condition $\sigma(\mathbf{x})/(\omega\varepsilon(\mathbf{x})) < 1$ is satisfied. Usually, this condition is valid for ε and σ in moist soil at high frequencies.

The Maxwell's equations in the time-harmonic form with an impressed current source $\mathbf{F}$ in a 3D-medium are given by

$$\begin{aligned} \mathrm{rot}\mathbf{H}(\mathbf{x}) &= (-i\omega\varepsilon(\mathbf{x}) + \sigma(\mathbf{x}))\,\mathbf{E}(\mathbf{x}) + \mathbf{F}, \\ \mathrm{rot}\mathbf{E} &= i\omega\mu\mathbf{H}, \end{aligned} \tag{47.1}$$

where $\mathbf{E}$ and $\mathbf{H}$ are the vectors of electric and magnetic field strengths, respectively; μ is the magnetic permeability; $\mathbf{x}$ is the vector of spatial coordinates. The magnetic permeability is assumed to be equal to the magnetic permeability of vacuum. At infinity, the radiation conditions must be satisfied. The wavelength is assumed to be large as compared with the maximum scale of heterogeneities of the medium L.

For the approximation of the coefficients $\sigma(\mathbf{x})$, $\varepsilon(\mathbf{x})$, we use the approach described in [7]. Let, for example, the field of permittivity be known. This means that the field is measured on a small scale l_0 at each point $\mathbf{x}$, $\varepsilon(\mathbf{x})_{l_0} = \varepsilon(\mathbf{x})$. To pass to a coarser scale grid, it is not sufficient to smooth the field $\varepsilon(\mathbf{x})_{l_0}$ on a scale l, $l > l_0$. The field thus smoothed is not a physical parameter that can describe the physical process, governed by Eq. (47.1), on the scales (l, L). This is due to the fact that the fluctuations of permittivity on the scale interval (l_0, l) correlate with the fluctuations of the electric field strength $\mathbf{E}$ induced by the permittivity. To find a permittivity that could describe an ensemble-averaged physical process on the scales (l, L), system (47.1) will be used. Following [4], consider a dimensionless field ψ, which is equal to the ratio of two fields obtained by smoothing the field $\varepsilon(\mathbf{x})_{l_0}$ on two different scales l', l. Let $\varepsilon(\mathbf{x})_l$ denote the parameter $\varepsilon(\mathbf{x})_{l_0}$ smoothed on the scale l. Then $\psi(\mathbf{x}, l, l') = \varepsilon(\mathbf{x})_{l'}/\varepsilon(\mathbf{x})_l, \quad l' < l$. Expanding the field ψ into a power series in $l - l'$ and retaining first order terms of the series, at $l' \to l$, we obtain the equation:

$$\frac{\partial \ln \varepsilon(\mathbf{x})_l}{\partial \ln l} = \chi(\mathbf{x}, l), \tag{47.2}$$

where $\chi(\mathbf{x}, l') = (\partial\psi(\mathbf{x}, l', l'y)/\partial y)\,|_{y=1}$. The solution of Eq. (47.2) is

$$\varepsilon(\mathbf{x})_{l_0} = \varepsilon_0 \exp\left(-\int_{l_0}^{L} \chi(\mathbf{x}, l_1)\frac{dl_1}{l_1}\right), \tag{47.3}$$

where ε_0 is a constant. The field χ determines the statistical properties of the permittivity. According to the limit theorem for sums of independent random variables: if the variance of $\chi(\mathbf{x}, l)$ is finite, the integral in (47.3) tends to a field with a normal distribution as the ratio L/l_0 increases. If the variance of $\chi(\mathbf{x}, l)$ is infinite and there exists a non-degenerate limit of the integral in (47.3), the integral tends to a field with a stable distribution. In this paper, it is assumed that the field $\chi(\mathbf{x}, l)$ is isotropic with a normal distribution and a statistically homogeneous correlation function:

$$\Phi^{\chi\chi}\left(\mathbf{x}, \mathbf{y}, l, l'\right) = \Phi^{\chi\chi}(|\mathbf{x}-\mathbf{y}|, l, l')\delta\left(\ln l - \ln l'\right). \tag{47.4}$$

It follows from (47.4) that the fluctuations of $\chi(\mathbf{x}, l)$ on different scales do not correlate. This assumption is standard in the scaling models [4]. This is due to the fact that the statistical dependence is small if the scales of fluctuations are different. To derive subgrid formulas to calculate effective coefficients, this assumption may be ignored. However, this assumption is important for the numerical simulation of the field ε. For a scale invariant medium, for any positive K, we have

$$\Phi^{\chi\chi}(|\mathbf{x}-\mathbf{y}|, l, l') = \Phi^{\chi\chi}(K|\mathbf{x}-\mathbf{y}|, Kl, Kl').$$

In a scale invariant medium, the correlation function does not depend on the scale at $\mathbf{x} = \mathbf{y}$, and the following estimation is obtained for $l_0 < l_\eta < r < L$[7]:

$$< \varepsilon(\mathbf{x})_{l_0}\varepsilon(\mathbf{x}+\mathbf{r})_{l_0} >\sim C\,(r/L)^{-\Phi_0^{\chi\chi}}, \tag{47.5}$$

where $C = \varepsilon_0^2\,(L/l_0)^{-2<\chi>}\,e^{-\Phi_0^{\chi\chi}\gamma/2}$, γ is the Euler constant. Here the angle brackets denote ensemble averaging. If for any l the equality $< \varepsilon(\mathbf{x})_l >= \varepsilon_0$ is valid, then it follows from (47.3), (47.4) that $\Phi_0^{\chi\chi} = 2 < \chi >$. As the minimum scale l_0 tends to zero, the permittivity field described in (47.3) becomes a multifractal and we obtain an irregular field on a Cantor-type set to be nonzero.

The conductivity coefficient $\sigma(\mathbf{x})$ is constructed by analogy with the permittivity coefficient:

$$\sigma(\mathbf{x})_{l_0} = \sigma_0 \exp\left(-\int_{l_0}^{L} \varphi(\mathbf{x}, l_1)\frac{dl_1}{l_1}\right). \tag{47.6}$$

The function $\varphi(\mathbf{x}, l)$ is assumed to have the normal distribution and to be delta-correlated in the logarithm of the scale. The correlation between the permittivity and conductivity fields is determined by the correlation of the fields $\chi(\mathbf{x}, l')$ and $\varphi(\mathbf{x}, l')$:

$$\Phi^{\varphi\chi}\left(\mathbf{x}, \mathbf{y}, l, l'\right) = \Phi^{\varphi\chi}(|\mathbf{x}-\mathbf{y}|, l, l')\delta\left(\ln l - \ln l'\right). \tag{47.7}$$

47.2 Subgrid Model

The electric conductivity and permittivity functions $\sigma(\mathbf{x})_{l_0}$, $\varepsilon(\mathbf{x})_{l_0}$ are divided into two components with respect to the scale l. The large-scale (ongrid) components $\sigma(\mathbf{x}, l)$, $\varepsilon(\mathbf{x}, l)$ are obtained, respectively, by statistical averaging over all $\varphi(\mathbf{x}, l_1)$ and $\chi(\mathbf{x}, l_1)$ with $l_0 < l_1 < l$, $l - l_0 = dl$, where dl is small. The small-scale (subgrid) components are equal to $\sigma'(\mathbf{x}) = \sigma(\mathbf{x})_{l_0} - \sigma(\mathbf{x}, l)$, $\varepsilon'(\mathbf{x}) = \varepsilon(\mathbf{x})_{l_0} - \varepsilon(\mathbf{x}, l)$:

$$\varepsilon(\mathbf{x}, l) = \varepsilon_0 \exp\left[-\int_l^L \chi(\mathbf{x}, l_1)\frac{dl_1}{l_1}\right]\left\langle \exp\left[-\int_{l_0}^{l} \chi(\mathbf{x}, l_1)\frac{dl_1}{l_1}\right]\right\rangle$$

$$\varepsilon'(\mathbf{x}) = \varepsilon(\mathbf{x}, l)\left[\frac{1}{\left\langle \exp\left[-\int\limits_{l_0}^{l} \chi(\mathbf{x}, l_1)\frac{dl_1}{l_1}\right]\right\rangle} \exp\left[-\int\limits_{l_0}^{l} \chi(\mathbf{x}, l_1)\frac{dl_1}{l_1}\right] - 1\right]. \quad (47.8)$$

The coefficients $\sigma(\mathbf{x}, l)$, $\sigma'(\mathbf{x})$ are calculated in the same way. The large-scale (ongrid) components of the electric and magnetic field strengths $\mathbf{E}(\mathbf{x}, l)$, $\mathbf{H}(\mathbf{x}, l)$ are obtained by averaging the solutions to system (47.1), in which the large-scale components of the conductivity $\sigma(\mathbf{x}, l)$ and the permittivity $\varepsilon(\mathbf{x}, l)$ are fixed and the small components $\sigma'(\mathbf{x})$, $\varepsilon'(\mathbf{x})$ are random variables. The subgrid components of the electric and magnetic field strengths are equal to $\mathbf{H}'(\mathbf{x}) = \mathbf{H}(\mathbf{x}) - \mathbf{H}(\mathbf{x}, l)$, $\mathbf{E}'(\mathbf{x}) = \mathbf{E}(\mathbf{x}) - \mathbf{E}(\mathbf{x}, l)$. Substituting the relations for $\mathbf{E}(\mathbf{x})$, $\mathbf{H}(\mathbf{x})$, and $\sigma(\mathbf{x})$, $\varepsilon(\mathbf{x})$ into system (47.1) and averaging over small-scale components, we have

$$\begin{aligned} \mathrm{rot}\mathbf{H}(\mathbf{x}, l) &= (-i\omega\varepsilon(\mathbf{x}, l) + \sigma(\mathbf{x}, l))\mathbf{E}(\mathbf{x}, l) + \left\langle\left(\sigma' - i\omega\varepsilon'\right)\mathbf{E}'\right\rangle + \mathbf{F}, \\ \mathrm{rot}\mathbf{E}(\mathbf{x}, l) &= \mu i\omega\mathbf{H}(\mathbf{x}, l). \end{aligned} \quad (47.9)$$

The subgrid term $\langle(-i\omega\varepsilon' + \sigma')\mathbf{E}'\rangle$ in system (47.9) is unknown. The form of this term in (47.9) determines a subgrid model. The subgrid term is estimated using perturbation theory. Subtracting system (47.9) from system (47.1) and taking into account only the first order terms, we obtain the subgrid equations:

$$\begin{aligned} \mathrm{rot}\mathbf{H}'(\mathbf{x}) &= (\sigma(\mathbf{x}, l) - i\omega\varepsilon(\mathbf{x}, l))\mathbf{E}'(\mathbf{x}) + \left(\sigma'(\mathbf{x}) - i\omega\varepsilon'(\mathbf{x})\right)\mathbf{E}(\mathbf{x}, l), \\ \mathrm{rot}\mathbf{E}'(\mathbf{x}) &= \mu i\omega\mathbf{H}'(\mathbf{x}). \end{aligned} \quad (47.10)$$

The variable $\mathbf{E}(\mathbf{x}, l)$ on the right-hand side of (47.10) is assumed to be known. Using "frozen-coefficients" method, as a first approximation we can write down the solution of system (47.10) for the components of the electric field strength:

$$E'_\alpha \approx \Omega \int_{-\infty}^{\infty} \frac{e^{ikr}}{r} \left(-i\omega\varepsilon'\left(\mathbf{x}'\right) + \sigma'\left(\mathbf{x}'\right)\right) E_\alpha\left(\mathbf{x}', l\right) d\mathbf{x}' \tag{47.11}$$

$$+\Omega_1 \int_{-\infty}^{\infty} \frac{\partial}{\partial x_\alpha} \frac{\partial}{\partial x_\beta} \frac{e^{ikr}}{r} \left(-i\omega\varepsilon'\left(\mathbf{x}'\right) + \sigma'\left(\mathbf{x}'\right)\right) E_\beta\left(\mathbf{x}', l\right) d\mathbf{x}',$$

where $\Omega = i\omega\mu/(4\pi)$, $\Omega_1 = 1/(4\pi(-i\omega\varepsilon(\mathbf{x}, l) + \sigma(\mathbf{x}, l)))$, $r = |\mathbf{x} - \mathbf{x}'|$, $k^2 = \omega\mu(\omega\varepsilon(\mathbf{x}, l) + i\sigma(\mathbf{x}, l))$. Here the summation of repeated indices is implied. We take the square root such that $Re\,k > 0,\ Im\,k > 0$. Using (47.11), the subgrid term can be written down as

$$\left\langle \left(\sigma'(\mathbf{x}) - i\omega\varepsilon'(\mathbf{x})\right) E'_\alpha(\mathbf{x}) \right\rangle$$

$$\approx \Omega \int_{-\infty}^{\infty} \frac{e^{ikr}}{r} \left\langle \left(\sigma'(\mathbf{x}) - i\omega\varepsilon'(\mathbf{x})\right)\left(-i\omega\varepsilon'\left(\mathbf{x}'\right) + \sigma'\left(\mathbf{x}'\right)\right) \right\rangle E_\alpha\left(\mathbf{x}', l\right) d\mathbf{x}' \tag{47.12}$$

$$+\Omega_1 \int_{-\infty}^{\infty} \frac{\partial}{\partial x'_\alpha} \frac{\partial}{\partial x'_\beta} \frac{1}{r} e^{ikr} \left\langle \left(\sigma'(\mathbf{x}) - i\omega\varepsilon'(\mathbf{x})\right)\left(\sigma'\left(\mathbf{x}'\right) - i\omega\varepsilon'\left(\mathbf{x}'\right)\right) \right\rangle E_\beta\left(\mathbf{x}', l\right) d\mathbf{x}'.$$

The wavelength is assumed to be large as compared with the maximum scale of heterogeneities of the medium L, $l < L$. Following [6], for $\omega\mu L^2|(i\omega\varepsilon(\mathbf{x}, l) + \sigma(\mathbf{x}, l))| \ll 1$, we obtain estimation of the subgrid term in (47.9)

$$\left\langle -i\omega\varepsilon'(\mathbf{x}) E'_\alpha(\mathbf{x}) \right\rangle + \left\langle \sigma'(\mathbf{x}) E'_\alpha(\mathbf{x}) \right\rangle \approx \frac{1}{3} \Phi_0^{\chi\chi} i\omega\varepsilon(\mathbf{x}, l) E_\alpha(\mathbf{x}, l) \frac{dl}{l}$$

$$-\left(\frac{2}{3}\Phi_0^{\chi\varphi} - \frac{1}{3}\Phi_0^{\chi\chi}\right) \frac{dl}{l} \sigma(\mathbf{x}, l) E_\alpha(\mathbf{x}, l). \tag{47.13}$$

Substituting (47.13) into (47.9), we have:

$$\mathrm{rot}\mathbf{H}(\mathbf{x}, l) = \left[\sigma_{l0} \exp\left[-\int_l^L \varphi(\mathbf{x}, l_1) \frac{dl_1}{l_1}\right] - i\omega\varepsilon_{l0} \exp\left[-\int_l^L \chi(\mathbf{x}, l_1) \frac{dl_1}{l_1}\right]\right] \mathbf{E}(\mathbf{x}, l)$$

$$\mathrm{rot}\mathbf{E}(\mathbf{x}, l) = i\omega\mu\mathbf{H}(\mathbf{x}, l), \tag{47.14}$$

$$\varepsilon_{l0} = \left(1 - \frac{\Phi_0^{\chi\chi}}{3} \frac{dl}{l}\right)\left[1 + \left(\frac{\Phi_0^{\chi\chi}}{2} - \langle\chi\rangle\right) \frac{dl}{l}\right] \varepsilon_0, \tag{47.15}$$

$$\sigma_{l0} = \left[1 - \left(\frac{2}{3}\Phi_0^{\chi\varphi} - \frac{1}{3}\Phi_0^{\chi\chi}\right) \frac{dl}{l}\right]\left[1 + \left(\frac{\Phi_0^{\varphi\varphi}}{2} - \langle\varphi\rangle\right) \frac{dl}{l}\right] \sigma_0.$$

As $dl \to 0$ in (47.15), we obtain the equation

$$\frac{d\ln\varepsilon_{0l}}{d\ln l} = \frac{1}{6}\Phi_0^{\chi\chi} - \langle\chi\rangle,$$
$$\frac{d\ln\sigma_{0l}}{d\ln l} = -\frac{2}{3}\Phi_0^{\chi\varphi} + \frac{1}{3}\Phi_0^{\chi\chi} + \frac{1}{2}\Phi_0^{\varphi\varphi} - \langle\varphi\rangle. \qquad (47.16)$$

For the scale-invariant medium the effective equations have the following simple form:

$$\mathrm{rot}\mathbf{H}(\mathbf{x},l) = -i\omega\left(\frac{l}{L}\right)^{\alpha}\varepsilon_l(\mathbf{x})\mathbf{E}(\mathbf{x},l) + \left(\frac{l}{L}\right)^{\beta}\sigma_l(\mathbf{x})\mathbf{E}(\mathbf{x},l) + \mathbf{F},$$
$$\mathrm{rot}\mathbf{E}(\mathbf{x},l) = i\omega\mu\mathbf{H}(\mathbf{x},l), \qquad (47.17)$$

where $\alpha = \langle\chi\rangle - \Phi_0^{\chi\chi}/6$, $\beta = \langle\varphi\rangle + \frac{2}{3}\Phi_0^{\chi\varphi} - \frac{1}{3}\Phi_0^{\chi\chi} - \frac{1}{2}\Phi_0^{\varphi\varphi}$.

47.3 Numerical Simulation

The following numerical problem was solved in order to verify the formulas obtained above. Equations (47.1) are solved in a cube with edge L_0. The following dimensionless variables are used: $\hat{\mathbf{x}} = \mathbf{x}/L_0$, $\hat{\sigma} = \sigma/\sigma_0$, $\hat{\mathbf{H}} = \mathbf{H}/H_0$, $\hat{\mathbf{E}} = L_0\sigma_0/(k_1H_0)\mathbf{E}$, $k_1 = L_0\sqrt{\sigma_0\mu\omega}$, $k = k_1\sqrt{\hat{\sigma} - i\varkappa\hat{\varepsilon}}$, $\varkappa = \omega\varepsilon_0/\sigma_0$. In the calculation, the parameter $\varkappa$ is equal to 5. This corresponds to $\sigma_0/(\omega\varepsilon_0) = 0.2$. Thus, the problem is solved in a unit cube, with $\sigma_0 = 1$, $\varepsilon_0 = 1$, $k_1 = 4\sqrt{2}$. To satisfy the radiation conditions at infinity, the perfectly matched layers are used. The current source $F_{\hat{x}_1} = 0$, $F_{\hat{x}_2} = 0$, $F_{\hat{x}_3} = 0.5\exp(-q^2(\hat{x}_3 - 0.2)^2)$, $q = 60$ is located at the point $(0, 0, 0.2)$. In the domain $0.3 \leqslant \hat{x}_i < 1.3$, the conductivity and the permittivity are simulated by multiplicative cascades. The integrals in (47.3), (47.6) are approximated by finite difference formulas. A $256\times256\times256$ grid is used for the spatial variables in the domain $0.3 \leqslant \hat{x}_i < 1.3$. In these formulas, it is convenient to pass to a logarithm to base 2:

$$\sigma(\hat{\mathbf{x}})_{l_0} \approx 2^{-\sum\limits_{i=-8}^{0}\varphi(\hat{\mathbf{x}},\tau_i)\Delta\tau}, \quad \varepsilon(\hat{\mathbf{x}})_{l_0} \approx 2^{-\sum\limits_{i=-8}^{0}\chi(\hat{\mathbf{x}},\tau_i)\Delta\tau}, \qquad (47.18)$$

where $\langle\sigma(\hat{\mathbf{x}})_{l_0}\rangle = 1$, $\langle\varepsilon(\hat{\mathbf{x}})_{l_0}\rangle = 1$, $l = 2^{\tau}$, $\Delta\tau$ is the τ grid-size. In our calculations, $\Delta\tau$ is taken to be one. For random fields φ, χ, the following formulas are used for each τ_i:

$$\varphi(\hat{\mathbf{x}},\tau_i) = \sqrt{\frac{\Phi_0^{\varphi\varphi}}{\ln 2}}\zeta_1 + \frac{\Phi_0^{\varphi\varphi}}{2}, \quad \chi(\hat{\mathbf{x}},\tau_i) = \sqrt{\frac{\Phi_0^{\chi\chi}}{\ln 2}}(\rho\zeta_1 + \rho_1\zeta_2) + \frac{\Phi_0^{\chi\chi}}{2},$$

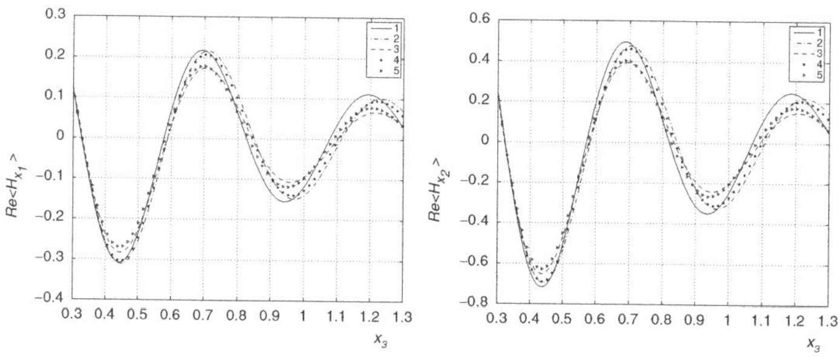

Fig. 47.1 Real parts of E_{x_1} , H_{x_2} obtained by: *1*—system (47.1) at $\sigma = 1, \varepsilon = 1$; *2*—effective system at $\rho = 1$; *3*—effective system with $\rho = -1$; *4*—numerical method at $\rho = 1$; *5*—numerical method at $\rho = -1$

where $\rho_1 = \sqrt{1-\rho^2}$, $\Phi_0^{\varphi\chi} = \rho\sqrt{\Phi_0^{\chi\chi}\Phi_0^{\varphi\varphi}}$, $-1 \leqslant \rho \leqslant 1$, $\zeta_1(\hat{\mathbf{x}}, \tau_i)$, $\zeta_2(\hat{\mathbf{x}}, \tau_i)$ are independent Gaussian random fields with unit variance, zero mean, and the following correlation function

$$\left\langle \zeta_1(\hat{\mathbf{x}}, \tau_i)\zeta_1(\hat{\mathbf{y}}, \tau_j)\right\rangle = \left\langle \zeta_2(\hat{\mathbf{x}}, \tau_i)\zeta_2(\hat{\mathbf{y}}, \tau_j)\right\rangle = \exp\left[-(\mathbf{x}-\mathbf{y})^2 / 2^{2\tau_i}\delta_{ij}\right].$$

The coefficients $\Phi_0^{\varphi\varphi} = 2\langle\varphi\rangle$, $\Phi_0^{\chi\chi} = 2\langle\chi\rangle$ are constants. To numerically simulate the Gaussian fields ζ_1, ζ_2, we use the algorithm from [9]. The delta-correlation in the scale logarithm means that the fields φ, χ are independently generated for each scale l_i. The number of terms in (47.18) is chosen so that probabilistic averaging can be replaced by volume averaging on the largest fluctuation scale. The smallest fluctuations scale is chosen in such a way as to approximate (47.1) by a difference scheme with a good accuracy on all the scales. We use a method based on a finite difference scheme proposed in [8] and a decomposition method from [2]. The fields in the exponents of (47.18) are generated as the sum of two scales: $i = -5, -4$. The minimum scale is $l_0 = 1/32$, the maximum scale is $L = 1/16$. The characteristics of the electric and magnetic field strengths are calculated on the scales (l_0, L). At each x_3, these fields are averaged over the planes (x_1, x_2). Then these fields are additionally averaged over the Gibbs ensemble. Equations (47.1) are solved 48 times. The fields thus obtained are compared with the solution to effective Eq. (47.17). In the calculations we use: $\Phi_0^{\varphi\varphi} = \Phi_0^{\chi\chi} = 0.4$, $<\varphi>=<\chi>= 0.2$, $\Phi_0^{\varphi\chi} = \rho\sqrt{\Phi_0^{\varphi\varphi}\Phi_0^{\chi\chi}}/\ln 2$, $\rho = 1$ or $\rho = -1$. Figure 47.1 shows a comparison between the mean fields obtained by the numerical method described above, the effective fields obtained by Eq. (47.17) and by the fields obtained by Eq. (47.1) with the coefficients $\sigma =< \sigma(\mathbf{x}) >= 1, \varepsilon =< \varepsilon(\mathbf{x}) >= 1$ (curve 1). Although curves 4, 5 in Fig. 47.1 small differ in magnitude from curve 1, curves 4, 5 decay faster than curve 1 for one wavelength. Such deviations will have an influence over a distance containing many wavelengths.

References

1. Bekele, A., Hudnall, H.W., Daigle, J.J., Prudente, A., Wolcott, M.: Scale dependent variability of soil electrical conductivity by indirect measures of soil properties. J. Terramech. **42**, 339–351 (2005)
2. Davydycheva, S., Drushkin, V., Habashy, T.: An efficient finite-difference scheme for electromalnetic logging in 3D anisotropic inhomogeneous media. Geophysics **68**, 1525–1530 (2003)
3. Hoffman, J.: Dynamic subgrid modelling for time dependent convection–diffusion–reaction equations with fractal solutions. Int. J. Numer. Methods Fluids **40**, 583–592 (2002)
4. Kolmogorov, A.N.: A refinement of previous hypotheses concerning the local structure of turbulence in a viscous incompressible fluid at high Reynolds number. J. Fluid Mech. **13**, 82–85 (1962)
5. Krylov, S.S., Lyubchich, V.A.: The apparent resistivity scaling and fractal structure of an iron formation. Izv. Phys. Solid Earth **38**, 1006–10012 (2002)
6. Kurochkina, E.P., Soboleva, O.N.: Effective coefficients of quasi-steady Maxwell's equations with multiscale isotropic random conductivity. Phys. A **390**, 231–244 (2011)
7. Kuz'min, G.A., Soboleva, O.N.: Subgrid modeling of filtration in porous self-similar media. J. Appl. Mech. Tech. Phys. **43**, 583–592 (2002)
8. Lebedev, V.I.: Difference analogies of orthogonal decjmpositions of basic differetial operators and some boundary value problems. J. Comput. Math. Math. Phys. **3**, 449–465 (1964) (in Russia)
9. Ogorodnikov, V.A., Prigarin, S.M.: Numerical Modeling of Random Processes and Fields: Algorithms and Applications, pp. 45–56. VSP, Utrecht (1996)

Chapter 48
An Approximate Solution of the Travelling Salesman Problem Based on the Metropolis Simulation with Annealing

Tatiana M. Tovstik

48.1 Introduction

The traveling salesman problem (TSP) is an NP-hard problem in combinatorial optimization [1]. The classic TSP is to look for the closed shortest possible path that passes through N given points and visits each point exactly one time. The first book about this problem was published in 1832 in Germany. Then in 1930 Karl Menger gave a mathematical formulation of the problem. In 1985 Lawler et al. [2] provided a comprehensive survey of all major research results until that date. In 1985 Hopfield and Tank [3] proposed to minimize energy by using neuron nets, in 1987 Durbin and Uillshoy [4] used the elastic net method to find the sub-optimal solution. In 1991 Reinelt published the Library $TSPLIB$ with the standard TSP [5] of various complexity. Now this Library is continuously supplemented on the Internet. In 1992 Laporte [6] published an overview of the exact and approximate TSP algorithms, and possible applications of the TSP. Among the exact algorithms he indicated the integer linear programming formulations [7], the related branch-and-bound algorithms, various shortest spanning bound and related algorithms [8, 9]. Finding the global extremum is possible by the Monte-Carlo Method in combination with the power method, which was first mentioned in the monograph [9] by Ermakov and was described in detail in [10].

The dynamic simulation method by Metropolis [11] is used for a variety of optimization problems. This method with annealing allows to find the global minimum of a functional. The length of the path is considered to be the functional in the TSP. The different heuristic algorithms differ from each other by the choice

T.M. Tovstik (✉)
St.Petersburg State University, St.Petersburg, Russia
e-mail: Peter.Tovstik@mail.ru

V.B. Melas et al. (eds.), *Topics in Statistical Simulation*, Springer Proceedings in Mathematics & Statistics 114, DOI 10.1007/978-1-4939-2104-1_48

of the initial and following approximations. In [12] two different mutation strategies for generating next solutions are proposed. The two-change (or the swap) operator and its modifications are used in [13–15].

In this work the symmetric TSP is studied, and the investigations from the paper [16] are continued. Normalized path length is used as criterion. Construction of the initial approximation is discussed. The following approximations are obtained by using the Metropolis method with annealing. This method is also applied to separate parts of the path. The choice and the change of an annealing coefficient is discussed. Special features of the proposed algorithm involve removing of self-sections and visual monitoring of intermediate results. The presented examples show that the path received with this algorithm is close to the optimal one.

48.2 The Algorithm of the Initial Approximation Construction

It is well known [11] that when using the Metropolis method to the problems with the large N it is important to start from the good initial configuration.

By using linear transformations initial points can be reflected into a unit square. We shall consider three variants of initial configurations.

Variant 1. The polar co-ordinates r, φ with the origin in the square center are introduced. In the initial approximation all points are numbered according the growth of the angle φ. This numbering is fixed as the initial configuration.

Variant 2. The numbering of points is executed according to the motion of rays as it is shown in the left side of Fig. 48.1. The rays rotate around the points with the Cartesian co-ordinates (0, 0), (0.25, 0.25), and (0.25, −0.25).

Fig. 48.1 The initial numbering of 5,000 random uniformly distributed points (*left*), and the final path close to optimal one with $\gamma = 0.759$ (*right*)

Variant 3. The points in a unit square with the Cartesian co-ordinates (x, y) are numbered according to the growth of co-ordinates x, and if for some points the coordinates x coincide then the numbering of these points is executed according to the growth of co-ordinates y.

These heuristic variants are established during the numerical experiments. We use the variant 1 for a comparatively small number of points ($N \leqslant 3000$) (see Examples 2 and 3). The Variant 2 is more convenient for a larger number of points with distribution close to the random uniform one (see Example 1). The Variant 3 can be used if some points lie on the x-lines. And it can be applied for the electronic plates design. In the case of large N it is convenient to use the parallel calculations optimizing the separate parts of the path (see Example 4).

48.3 The Metropolis Method with Annealing

To obtain the following approximations we simulate the Markovian process by the Metropolis method with *annealing*.

Let X be the finite set, and its elements $x \in X$ be named *configurations*. Let us introduce the real *energy function* $H(x)$ equal to the *path length*. The Metropolis method allows to find the configuration x with $H(x)$ close to minimum. Here the path length (or the energy function) is used as a criterion. In some papers [12] the average path length is used as a criterion. The last criterion is not acceptable for us because we study a sequence of paths and not a set of random paths.

In the TSP co-ordinates of N points are given, and the configuration (or the path) x is defined by the order of the passage of points and does not depend on the passage direction (the problem is *symmetric*). As the path is closed the choice of starting point is not important. If for the kth approximation $x = x^{(k)}$, then

$$x^{(k)} = \left(j_1^{(k)} \to j_2^{(k)} \to \cdots \to j_N^{(k)} \to j_{N+1}^{(k)} = j_1^{(k)}\right), \tag{48.1}$$

where $j_1^{(k)}$ is the starting point number and $j_i^{(k)}$ is the number of the ith point in the passage. For $k = 0$ the relation (48.1) defines the initial approximation.

The energy function $H(x)$ for the configuration $x = x^{(k)}$ is equal

$$H(x) = H(x^{(k)}) = \sum_{i=1}^{N} r_i^{(k)}, \qquad r_i^{(k)} = r(j_i^{(k)}, j_{i+1}^{(k)}), \tag{48.2}$$

where $r(j_i^{(k)}, j_m^{(k)})$ are the distances between the points.

We fulfill the transition from the kth approximation to the $(k + 1)$th one by the so called *two-change*, operator, which consists of the following. First we choose at random two numbers of the path (48.1). Let them be $j_i^{(k)}$ and $j_m^{(k)}$ $(m - i > 2)$. Then we construct the *test configuration* y in which compared with (48.1) the part of path between the points $j_{i+1}^{(k)}$ and $j_m^{(k)}$ is passed in the opposite order

$$y = \left(j_1^{(k)} \to \ldots \to j_{i+1}^{(k)} \to j_m^{(k)} \to j_{m-1}^{(k)} \to \ldots \to j_{i+1}^{(k)} \to j_{m+1}^{(k)} \to j_{m+2}^{(k)} \to \ldots \to j_{N+1}^{(k)}\right). \tag{48.3}$$

We call by the *neighbor set* $\delta(x)$ the set of paths, which can be obtained from the given path x by a single two-change. The set $\delta(x)$ consists of $N(N-3)/2$ elements, and $x \notin \delta(x)$.

The test path y is accepted as the next approximation (namely $x^{(k+1)} = y$) with the probability 1 if $H(y) \leqslant H(x^{(k)})$, and with the probability P_y $(0 < P_y < 1)$ if $H(y) > H(x^{(k)})$. In the opposite case $x^{(k+1)} = x^{(k)}$. Here

$$P_y = \mathbf{P}(x^{(k+1)} = y \mid \Delta H > 0) = \exp(-\beta \Delta H), \quad \Delta H = H(y) - H(x^{(k)}). \tag{48.4}$$

This way it is possible to leave a local minimum, and the process is called *an annealing*, with the *annealing coefficient* β [1]. From relations (48.2)–(48.4) it follows

$$\Delta H = r(j_{i-1}^{(k)}, j_m^{(k)}) + r(j_i^{(k)}, j_{m+1}^{(k)}) - r(j_{i-1}^{(k)}, j_i^{(k)}) - r(j_m^{(k)}, j_{m+1}^{(k)}). \tag{48.5}$$

To propose the way of calculation and changing of β we first consider study the case of a random uniform distribution of points within a unit square. We put $\beta = \beta_0/\sigma(\Delta H)$, where $\sigma(\Delta H)$ is a root-mean-square of the random value ΔH. To understand the connection between β_0 and P_y, which here is random, we give Table 48.1 in which the dependency $P_0(\beta_0)$ is presented. Here P_0 is the expectation of P_y

$$P_0 = \mathbf{E} P_y = \mathbf{E}\left(\exp\left(-\beta_0 \frac{\Delta H}{\sigma(\Delta H)}\right) \middle| \, \Delta H > 0\right). \tag{48.6}$$

The values P_0 are calculated by the Monte-Carlo method.

We re-write ΔH in the form

$$\Delta H = \eta_1 + \eta_2 - \xi_1 - \xi_2, \qquad \Delta H > 0 \tag{48.7}$$

where $\eta_1, \eta_2, \xi_1, \xi_2$ are the corresponding distances in (48.5). The following approximate relation is valid [16]

$$\sigma^2(\Delta H | \Delta H > 0) \approx 0.75(\sigma^2(\xi_1) + \sigma^2(\eta_1)), \tag{48.8}$$

Table 48.1 The dependence $P_0(\beta_0)$

β_0	4	5	6	7	8	9	10	11	12
P_0	0.195	0.166	0.144	0.128	0.116	0.106	0.097	0.090	0.084

which gives the annealing coefficient

$$\beta = \frac{\beta_1}{\sqrt{\sigma^2(\xi_1) + \sigma^2(\eta_1)}}, \quad \beta_1 \approx \frac{\beta_0}{\sqrt{0.75}}. \tag{48.9}$$

These results are obtained for the random uniformly distributed points. We generalize the relation (48.9) for the case of non-random points and take into account that the right side of (48.9) depends on the number of the iteration k.

For the path k we find

$$\hat{r}^{(k)} = \frac{1}{N}\sum_i r_i^{(k)}, \quad (\hat{\sigma}^{(k)})^2 = \frac{1}{N}\sum_i (r_i^{(k)})^2 - (\hat{r}^{(k)})^2, \tag{48.10}$$

and we propose to use the following expression for β

$$\beta = \frac{\beta_*}{\sqrt{(\hat{\sigma}^{(k)})^2 + (\hat{\sigma}^{(k+1)})^2}} \tag{48.11}$$

in which the dispersions $\sigma^2(\xi_1)$ and $\sigma^2(\eta_1)$ are changed by $(\hat{\sigma}^{(k)})^2$ and $(\hat{\sigma}^{(k+1)})^2$, and the value $(\hat{\sigma}^{(k+1)})^2$ is calculated for the test path y. The value β_* is to be chosen. Based on the numerical experiments in the following examples we take $4 \leqslant \beta_* \leqslant 7$.

Finally instead of (48.4) we recommend to use the relation

$$P = \exp\left(-\beta_* \frac{\Delta H}{\sqrt{(\hat{\sigma}^{(k)})^2 + (\hat{\sigma}^{(k+1)})^2}}\right). \tag{48.12}$$

48.4 The Energy Minimization in the Separate Parts of Path

In case of a large number of points N for minimizing the time of calculation is convenient to seek the energy minimum successively in the *separate parts of path*.

Let *raster* Γ be the sequence of points

$$j_i^{(k)} \to j_{i+1}^{(k)} \to \dots j_{m-1}^{(k)} \to j_m^{(k)}, \tag{48.13}$$

$L(\Gamma) = m - i + 1$ be the raster length, then

$$H(\Gamma) = \sum_{s=i}^{m-1} r_s^{(k)} \tag{48.14}$$

is the raster energy.

We minimize the raster energy by the successive two-changes, but the raster ends $j_i^{(k)}$ and $j_m^{(k)}$ are fixed and not included in the two-change process. The rasters move along the entire path, and the successive rasters overlap.

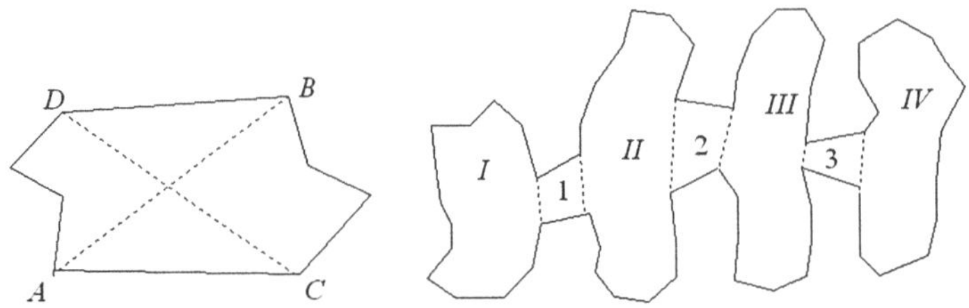

Fig. 48.2 The self-sections moving off (*left*); the subgroups gathering (*right*)

To find rasters which energy can be minimized we use the visual control (or monitoring) of intermediate results.

The self-sections moving off always leads to the path decrease. In Fig. 48.2 (left) the old path ABCDA and the new path ACBDA are shown.

For an initial approximation of Variant 3 (see the Example 4 and Fig. 48.2 (right)). We divide points into some separate subgroups (in the Example 4 we take 4 subgroups I, II, III, VI). We seek the path close to the optimal one for each subgroup. Then we merge subgroups using the neighboring points, so that in the quadrangles $1, 2, 3$ there are not points, and change the points numbering (the similar algorithm is described in [6]). Then we use rasters near the points of contact. At last we use the two-change operator for the entire set of points.

48.5 The Normalized Path Length

We introduce the *normalized path length* γ by relation

$$\gamma = H/\sqrt{NA}, \tag{48.15}$$

where A is the area occupied by points. Inequality $0.655 < \gamma < 0.92$ is fulfilled [17] for the random uniform distribution of points, and supposedly

$$\gamma \approx 0.749. \tag{48.16}$$

To estimate the quality of approximation we compare the obtained normalized path length with the value (48.16).

48.6 Examples

To find the path close to optimal one we use the two-change, the minimization in the separate parts of path. We delete the self-crossing of path that is the partial case of the two-change, and always leads to the energy decrease. We use rasters

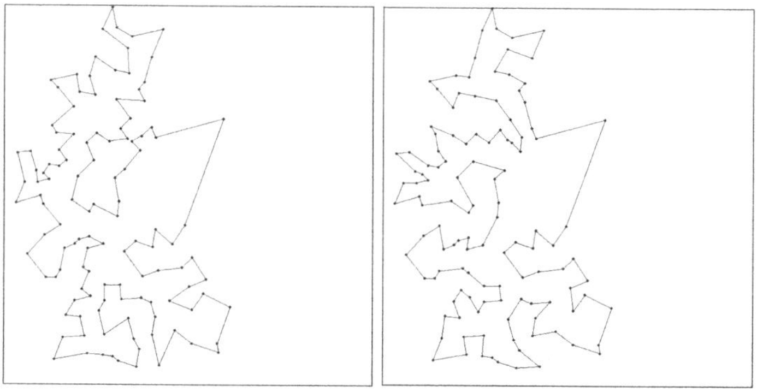

Fig. 48.3 The "optimal" path by Groetschel with $N = 120$. $H_{GR} = 1666.5$ (*left*); our path with $H_* = 1645.5$ (*right*)

with $L \leqslant 15$ to delete the part of self-crossings. At the every step we calculate the annealing coefficient taking β_* in (48.12) in the limits $4 \leqslant \beta_* \leqslant 7$. The visual control (or monitoring) allows us to find the parts of path which can be made shorter by using the rasters. The FORTRAN (Developer Studio) is used.

Example 1. We take as initial data the $N = 5,000$ random uniformly distributed points. The initial approximation with $H = 151.6$, $\gamma = 4.793$ in the form of Variant 2 in Sect. 48.2 is shown in the left side of Fig. 48.1. The final path with $H_* = 53.7$, $\gamma_* = 0.759$ is shown on the right side of Fig. 48.1. The value γ_* is close to the optimal value (48.16) (by star we mark our results).

Example 2. For this and following examples data are taken from Internet library TSPLIB. The example $GR120$ contains $N = 120$ points. The "optimal" path with $H_{Gr} = H_{opt} = 1666.5$ and $\gamma_{GR} = 0.749$, shown on the left side of Fig. 48.3, was obtained by Groetschel in 1977. Later Ermakov and Leora [10] obtained the better result with $H_{EL} = 1654.8$, $\gamma_{EL} = 0.744$. Our result with $H_*(120) = 1645.6$, $\gamma_* = 0.740$ is presented on the right side of Fig. 48.3.

Example 3. This example with $N = 666$ is named as $GR666$ in the library $TSPLIB$. The "optimal" path with $H_{opt} = 3952.5$ is given there (see left side of Fig. 48.4). Our result with $H_* = 3240.5$, $\gamma_* = 0.498$ is shown on the right side of Fig. 48.4. The value $\gamma_* = 0.498$ is far from the value (48.16) because the points distribution is far from uniform (there are the points of concentration).

Example 4. The initial data $XQC2175$ from $TSPLIB$ contain $N = 2175$ points. The optimal variant is absent in $TSPLIB$. As initial approximation Variant 3 from Sect. 48.2 is taken, and it coincides with the initial data. Two ways of calculations are used. In the first of them the data are divided into four groups: (1–550), (551–1115), (1116–1614), (1615–2175). For each group the suboptimal path is found, and then these paths are merged into the entire path with $H_0 = 7198.8$ and $\gamma_0 = 0.647$.

Fig. 48.4 The "optimal" path of TSPLIB with $N = 666$, $H_{\text{opt}} = 3952.5$ (*left*); our path with $H_{*} = 3240.5$ (*right*)

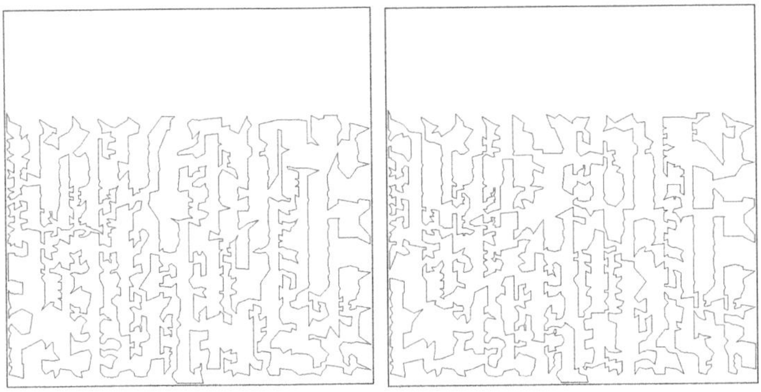

Fig. 48.5 Our path obtained by the initial approximation of the Variant 3 with $N = 2175$, $H_{*1} = 7135.4$, $\gamma_1 = 0.641$ (*left*); our path obtained by the initial approximation of the Variant 1 with $H_{*2} = 7188.9$, $\gamma_2 = 0.646$ (*right*)

The following optimization of the entire path gives $H_{*1} = 7135.4$ and $\gamma_1 = 0.641$. The obtained path is shown on the left side of Fig. 48.5.

In the second way the initial data are taken according to Variant 1. After optimization we get $H_{*2} = 7188.9$ and $\gamma_2 = 0.646$ (see the right side of Fig. 48.5). As we see the Variant 3 gives the better result than the Variant 1.

Conclusions

An approximate algorithm for a symmetric TSP solution is proposed. The Metropolis simulation with annealing is used. The algorithm features consist in using of the good initial approximation, choice of the annealing coefficient,

(continued)

minimization of the separate parts of path and deleting of the path cross-sections. The algorithm effectiveness is confirmed in the above examples. In some cases obtained results are better than ones from the TSPLIB. Although this algorithm is based on the two-change operation, but in detail it essentially differs from heuristic algorithms used by the other authors.

Acknowledgements The work is supported by Russian Foundation of Basic Researches (grant 11.01.00769-a).

References

1. Winkler, G.: Image Analysis, Random Fields and Dynamic Momte Carlo Methods. Springer, Berlin (1995)
2. Lawler, E,L., Lestra, J.K., Rinnooy Kan, A.H.G., Shmous, D.B.: The Traveling Salesman Problem. A Guided Tourof Combinatorial Optimization. Willey, Chichester (1985)
3. Hopfield, J.J., Tank, D.W.: Neural computation of decisions in optimization problems. Biol. Cybern. **52**, 141–152 (1985)
4. Durbin, R., Willshaw, D.: An analogue approach to the traveling salesman problem using an elastic net method. Nature **326**, 689 (1987)
5. Reinelt, G.: TSPLIB–A traveling salesman problem library. ORSA J. Comput. **3**(4), 376–384 (1991)
6. Laporte, G.: The traveiing salesman problem: an overview of exact and approximate algorithms. Eur. J. Oper. Res. **59**, 231–247 (1992)
7. Dantzig, G.B., Jonson, S.M.: On a linear-programming combinatorial approach to the traveling-salesman problem. **7**, 58–66 (1959)
8. Carpaneto, G., Martello, S., Toth, P.: Algorithms and codes for the assignment problem. In: Simeone, B., Toth, P., Gallo, G., Maffioli, F., Pallottino, S. (eds.) FOR-TRAN Codes for Network Optimization. Annals of Operations Research, vol. 13, pp. 193–223 (1988)
9. Ermakov, S.M.: Monte Carlo Methods and Close Problems. Nauka, Moscow (1975)
10. Ermakov, S.M., Leopa, S.N.: Optimization by metropolis method. In: Mathematical Models. Theory and Applications, vol. 11, pp. 72–83. St.Petersburg University (2010)
11. Metropolis, N., Rosenbluth, A.W., Rosenbluth, M.N., Teller, A.H., Teller, E.J.: Equations of state calculations by fast computing machines. J. Chem. Phys. **21**, 1087–1092 (1953)
12. Bayram, H., Sahin R.: A new simulated annealing approach for traveling salesman problem. Math. Comput. Approaches **18**(3) (2013)
13. Lin, S., Kernigan, B.W.: An effective heuristic algorithm for the travelling- salesman problem. Oper. Res. **21**, 498 (1973)
14. Geng, H., Chen Z., Wang, W., Shi, D., Zhao, K.: Solving the traveling salesman problem based on adaptive simulated annealing algorithm. Appl. Soft. Comput. **11**(4), 3680–3689 (2011)
15. Wang, W., Tian, D., Li, Y.: An improved simulated annealing algorithm for traveling salesman problem. In: Proceedings of the 2012 Information Technology and Software Engineering, vol. 211, pp. 525–532. Springer, Berlin (2013)
16. Tovstik, T.M., Zhukova, E.V.: Algorithm of the approximate solution of the traveling salesman problem. Vestnik St.Petersburg Univ. Ser. **1**(1), 101–109 (2013)
17. Beardwood, J., Haton, J.H., Hammersley, J.M.: The shortest path through many points. Proc. Cambridge Philos. Soc. **55**, 299–327 (1959)

Chapter 49
The Supertrack Approach as a Classical Monte Carlo Scheme

Egor Tsvetkov

49.1 Introduction

In [1], linear functionals on the solutions of Boltzmann equations are called Boltzmann tallies. Monte Carlo methods based on the Neumann–Ulam scheme can be used to estimate Boltzmann tallies. However, not all real-world calculations can be represented in that form. In particular, physical quantities that depend on collective effects of particles cannot be described with Boltzmann tallies.

The energy deposited by a particle in a sensitive volume (the so-called pulse height tally) is the classic example of non-Boltzmann tally. The second example is a device that detects coincidences. Such devices found their application, for example, in positron emission tomography and gamma-ray astronomy. In neither of these two cases can the detector function be represented as a sum

$$q(S) = q(x_0) + q(x_1) + \cdots + q(x_k),$$

where S is the trajectory, x_i are the coordinates of the particles before the collisions, $i = 1, 2, \ldots, k$, $q(S)$ is the detector response on trajectory S, and $q(x_i)$ is the detector response on the single collision event.

Many approaches that use variance reduction techniques to estimate non-Boltzmann tallies have been suggested. Lappa [9] proposed non-imitating methods to estimate any moment of a Boltzmann tally. Uchaikin [13–15] and Lappa [10] considered a special class of tallies in which the functionals were represented as

E. Tsvetkov (✉)
Moscow Institute of Physics and Technology, 141700, Institutskiy sidestreet, Dolgoprudny, Russia
e-mail: tsvetkov_egor@mail.ru

V.B. Melas et al. (eds.), *Topics in Statistical Simulation*, Springer Proceedings in Mathematics & Statistics 114, DOI 10.1007/978-1-4939-2104-1_49

the sum of the effects of collisions and the effects of free runs. Borisov and Panin [2] proposed an approach to estimate a pulse-height tally with a variance reduction technique that includes the so-called contributions.

Booth introduced the most general variance reduction technique to estimate non-Boltzmann tallies [1]. This technique is a set of rules of the main idea which considers a branching trajectory as an indivisible collection of tracks. Only physical reactions can create new trajectory branches. Variance reduction techniques cannot create new branches because a whole collection of tracks is split when the splitting technique is used. This collection of tracks is called a supertrack. The supertrack approach can be easily accessed with the famous MCNP code. Numerical experiments showed that the supertrack approach is about five times faster than the analog Monte Carlo [6].

Although the supertrack approach has clear physical interpretation, the strict theoretical substantiation of it has not been proposed yet. The goal of the present article is to build a strict mathematical basis for the supertrack approach. The novelty of this paper consists in deriving the supertrack approach from the general Monte Carlo scheme, more exactly, we deduce the rules how to sample trajectories and calculate their statistical weights. The rules proved to be identical to those that were formulated in [1]. This can be treated as substantiation of the supertrack approach.

For each technique we prove the unbiasedness and explain why this technique is more efficient than the analog Monte Carlo. We operate with the variance but the exact measure of efficiency of Monte Carlo algorithms is the figure of merit (FOM, [3]). FOM requires the numerical experiments that are not conducted in the present work (see [6]).

In this article, we depart from describing the probability space on the set of all branching trajectories. The state of a particle is, for us, described by three values $x = (\mathbf{x}, \mathbf{u}, E)$, where $\mathbf{x}$ is the position of the particle, $\mathbf{u}$ is the direction of the velocity of the particle, and E is the generalized energy of the particle. The generalized energy of a particle contains information about the type of particle as well as its physical energy. The set of all x is the phase space X. We assume X to be a rectangle in $\mathbb{R}^7$, so a σ-algebra exists on X. Hereinafter, the exact nature of X is inessential.

49.2 Probability Space

The way how to build the probability space $(\mathbb{S}, \mathscr{H}, P)$ on the set of all branching trajectories $\mathbb{S}$ has been described in the literature, e.g. [4, 5, 8, 11]. Below, we describe the branching trajectory in a special way and perform an algebraic transformations of the probability density function $p(S)$ that induces the measure P.

49.2.1 The Set of Branching Trajectories

In physical terms, the branching trajectory is a tree (as in graph theory) with coordinates in the phase space X that are assigned to each tree node. The set of all trees is enumerable; therefore, we can use a natural value n to encode the structure of the tree.

We enumerate the nodes in a tree using an enumeration by generations. The enumeration is performed in the following order. The root of the tree at which the primary particle originates is counted as node 0. The node at which the primary particle encounters its first collision is counted as node 1. After that all nodes of the next generation are enumerated; the order of enumeration within a generation is not important. The last node is counted as k_n.

We assign to every node of the tree the coordinates of the particle immediately before the collision represented by that node. The coordinates are denoted by $x_1, x_2, \ldots, x_{k_n}$. The phase coordinates of the primary particle are x_0.

Therefore, the branching trajectory is represented by the pair $S = (n, (x_0, x_1, \ldots, x_{k_n}))$, where $n \in \mathbb{N}$ represents the structure of the branching history, k_n is the number of the last node in the tree given by n, and the $x_i \in X$ are the phase coordinates of the particle immediately before the collision in the corresponding node of the tree.

We denote the probability density function that the original particle is born at x by $p_0(x)$, the probability density function that a collision at point x leads to m secondary particles at points $x_1, x_2, \ldots, x_m$ by $p(x \to x_1, x_2, \ldots, x_m)$, and the probability of absorption at point x by $g(x)$. We also need the probability $P_m(x)$ that particle x is split into m secondary particles after a collision. We denote this probability by

$$P_m(x) = \int p(x \to x_1, x_2, \ldots, x_m)\, dx_1 dx_2 \ldots dx_m.$$

In particular, we note that $P_0(x) = g(x)$.

Following [5], the probability density function of the branching trajectory S can be written as

$$p(S) = p_0(x_0) \prod_{(i, j_1, j_2, \ldots, j_m)} p\left(x_i \to x_{j_1}, x_{j_2}, \ldots, x_{j_m}\right) \prod_{x_i \in G} g(x_i). \tag{49.1}$$

The first product is computed over all of the nodes of the tree that have children, where i is the index of the parent node, m is the number of children of the parent node i, and $j_1, j_2, \ldots, j_m$ are the indices of the children nodes. The second product is computed over the subset G consisting of all the nodes that have no children.

We have to transform the probability density function to allow us to sample secondary particles successively. For this, conditional probabilities will be needed.

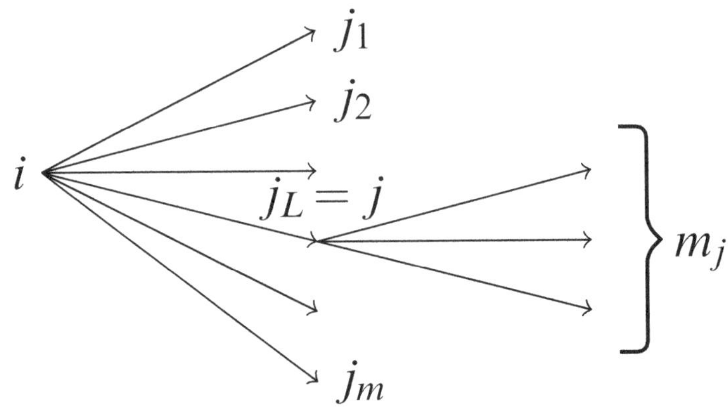

Fig. 49.1 Enumeration of the parent node and the children nodes

We introduce the conditional probability density functions as

$$p\left(x_k|x,x_1,\ldots,x_{k-1}\right)=\frac{\int p\left(x\to x_1,x_2,\ldots,x_m\right)dx_{k+1}dx_{k+2}\ldots dx_m}{\int p\left(x\to x_1,x_2,\ldots,x_m\right)dx_k dx_{k+1}\ldots dx_m},\quad k=1,2,\ldots,m-1, \tag{49.2}$$

$$p\left(x_m|x,x_1,\ldots,x_{m-1}\right)=\frac{p\left(x\to x_1,x_2,\ldots,x_m\right)}{\int p\left(x\to x_1,x_2,\ldots,x_m\right)dx_m}. \tag{49.3}$$

From (49.2)–(49.3), we obtain

$$p\left(x\to x_1,x_2,\ldots,x_m\right)=p\left(x_1|x\right)p\left(x_2|x,x_1\right)\ldots p\left(x_m|x,x_1,\ldots,x_{m-1}\right)P_m\left(x\right). \tag{49.4}$$

We define $p_n\left(x_j|x_0,x_1,\ldots,x_{j-1}\right)$ as follows. For a given n, we can find the numeric label of a parent node of the jth node, which we denote by i (see Fig. 49.1). We find all the children of this parent node, which we denote by $j_1, j_2, \ldots, j_m$ with $j_1 < j_2 < \cdots < j_m$. Because the node numbered j is a child of the node numbered i, we can state that there exists an l such that $j_l = j$, $1 \leqslant l \leqslant m$. Also we can count the number of children of the jth node, which we denote by m_j.

We define $p_n(x_j|x_0,x_1,\ldots,x_{j-1})$ by

$$p_n(x_j|x_0,x_1,\ldots,x_{j-1})=p(x_j|x_i,x_{j_1},x_{j_2},\ldots,x_{j_{l-1}})P_{m_j}(x_j).$$

We can rewrite the probability density function (49.1) as

$$p(S)=p(x_0)\prod_{i=1}^{k_n}p_n(x_i|x_0,x_1,\ldots,x_{i-1}). \tag{49.5}$$

Equation (49.5) is an algebraic transformation of (49.1).

In some cases, we represent $p_n(x_j|x_0, x_1, \ldots, x_{j-1})$, the probability to scatter to x_j, by

$$p_n(x_j|x_0, x_1, \ldots, x_{j-1}) = \Omega_n(x_j|x_0, x_1, \ldots, x_{j-1})e^{-l_n(x_j|x_0, x_1, \ldots, x_{j-1})}\Sigma(x_j). \tag{49.6}$$

This means that the probability of scattering to x_j is the probability of scattering in the direction of x_j multiplied by the probability of reaching x_j and colliding at x_j. In the above formula, $\Omega_n(x_j|x_0, x_1, \ldots, x_{j-1})$ represents the probability density function of scattering in the direction of x_j, $l_n(x_j|x_0, x_1, \ldots, x_{j-1})$ represents the optical path to x_j, and $\Sigma(x_j)$ represents the macroscopic total cross section at x_j. Also the multiplier r^{-2} is included in $\Omega_n(x_j|x_0, x_1, \ldots, x_{j-1})$ (one can write it separately but we don't do this to shorten the equations). So $p_n(x_j|x_0, x_1, \ldots, x_{j-1})$ is the probability that the particle scatters to a neighborhood of x_j.

49.3 Importance Sampling

Let P' be a measure on $\mathscr{H}$ that is induced by $p'(S)$. Let $p'(S) > 0$ everywhere that $p(S) > 0$. We can sample trajectories S_i with the biased probability P' and calculate the weighted sum as

$$Q^* = q(S)w(S), \tag{49.7}$$

where $w(S)$ represents the weight of an elementary event (i.e., a trajectory). To make Q^* an unbiased estimate of MQ, we choose a weight that is equal to the Radon–Nykodim derivative $w(S) = \partial P/\partial P'(S)$ [5].

In this case, $MQ^* = MQ$. This is the general form of importance sampling. Now we apply importance sampling to our probability space $(\mathbb{S}, \mathscr{H}, P)$. The Radon–Nykodim derivative is equal to the ratio of the probability density functions in (49.1) such that

$$w(S) = \frac{\partial P}{\partial P'}(S) = \frac{p(S)}{p'(S)} = \frac{p_0(x_0)}{p'_0(x_0)} \prod_{(i,j_1,j_2,\ldots,j_m)} \frac{p(x_i \to x_{j_1}, x_{j_2}, \ldots, x_{j_m})}{p'(x_i \to x_{j_1}, x_{j_2}, \ldots, x_{j_m})} \prod_{x_i \in G} \frac{g(x_i)}{g'(x_i)}.$$

We obtain the next rule that is formulated in [1]. The weight of a supertrack is computed as the product of the multipliers. The multiplier is a ratio of the unbiased and biased probabilities of the actual reaction channel.

We recall some known facts about importance sampling that are well stated in [5]. First, the probability density functions must satisfy the condition that $p'(S)$ can reach zero if $p(S) = 0$ for the same S. Second, to reduce the variance of Q^*

in comparison with the variance of q one can select the $p(S)$ to be proportional to $q(S)$. In our particular case, we should select $p'(S)$ such that particles are scattered mainly in the direction of the region of the detector.

49.4 Russian Roulette

If the expected contribution of a currently sampled trajectory is too small, then the fate of the trajectory can be determined from a game of Russian Roulette. If the trajectory survives Russian Roulette, the weight is increased; otherwise, the weight is multiplied by 0. We let $\nu(S) > 0$ denote the survival probability and we define a random variable L with a uniform distribution on the interval $[0, 1]$. When playing Russian Roulette, the weight of the trajectory is defined as

$$w(S) = \frac{1}{\nu(S)}\mathbb{I}(L < \nu(S)). \tag{49.8}$$

The estimator is given by (49.7). We take the average of Q^* to obtain $MQ^* = M_S\left(q\left(S\right)M_L\left(w\left(S\right)|S\right)\right)$. Because $M_L\left(\mathbb{I}\left(L < \nu\left(S\right)\right)|S\right) = \nu\left(S\right)$ we obtain $MQ^* = MQ$. So, if we choose weight in accordance with (49.8), the weighed estimate (49.7) is unbiased.

The variance of Q^* for Russian Roulette is larger than for analog Monte Carlo. But we have to take into account the computer time needed to reach the required level of statistical error. This time can be less because we save time by terminating the trajectories with small weights. The problem of the optimum choice of $\nu(S)$ for the supertrack approach is still open. We can state only that the optimum choice should depend on the programming realization, the computer architecture and so on.

49.5 Splitting

If the trajectory falls in the subset $\mathbb{T}$ of $\mathbb{S}$, then we split the trajectory into m new trajectories. These trajectories coincide before the split and they are sampled independently after the split. As a result, we obtain trajectories $S_1, S_2, \ldots, S_m$ with weights $w_1(S_1), w_2(S_2), \ldots, w_m(S_m)$. If we do not split the trajectory, then $m = 1$ and $w_1(S_1) = 1$.

In the case of splitting, the estimator takes the form

$$Q^* = \sum_{i=1}^{m} w_m(S_m)q(S_m). \tag{49.9}$$

If we take the average of this estimator, then we obtain

$$MQ^* = \int\limits_{\mathbb{S}\setminus\mathbb{T}} q(S)P(dS) + \sum_{i=1}^{m} \int\limits_{\mathbb{T}} q(S)w_i(S)P(dS).$$

If $w_1(S) + w_2(S) + \cdots + w_m(S) = 1$, then the estimator Q^* is unbiased, which means that $MQ^* = MQ$.

The variance of (49.9) is smaller than that of analog Monte Carlo because we increased the number of sampled trajectories while splitting. Also we saved the computer time because the new trajectories are sampled as one trajectory before splitting. The general rule of when to split a trajectory can be stated in the following form: when the particle enters the region where the detectors are concentrated.

49.6 Stratified Sampling

In stratified sampling, the area of the integration window is partitioned into subsets $\mathbb{T}_k, k = 0, 1, \ldots$. The Monte Carlo method is applied to each partition. The integral over the entire area is calculated as the weighted sum of the integrals over the partitions.

In the case of branching trajectories, it is convenient to split the sampled trajectory so that the new trajectories fall into different partitions. The result of the single sampling is the set of trajectories $S_1, S_2, \ldots, S_m$ from different partitions and their weights $w_1, w_2, \ldots, w_m$. The number of trajectories m is random. The estimator is

$$Q^* = w_1 q(S_1) + w_2 q(S_2) + \cdots + w_m q(S_m). \tag{49.10}$$

Our goal is to specify both the way in which to partition the set of trajectories and the rule by which to calculate the weight of each partition.

Two factors are important for understanding why stratified sampling is useful. First, the part of the new trajectories which they have in common is sampled only once, so this way of trajectory sampling allows us to save on computer time. Second, we can increase the probability of hitting the region where the detectors are concentrated, so the variance of Q^* can be reduced.

The sum (49.10) contains a random number of terms. We apply conventional technique and transform this random sum to an infinite sum over all $\mathbb{T}_k$. To do this, we have to choose exactly one trajectory from each $\mathbb{T}_k, k = 1, 2, \ldots$. The event A_k is that one of the trajectories falls in $\mathbb{T}_k$ when stratified sampling is used. If A_k has occurred, we take this trajectory and denote it by S_k. If not, we take any trajectory S_k from $\mathbb{T}_k$ and assign any weight $w_k(S_k)$ to it. Actually there exists a number such that all terms of (49.10) after this number are equal to zero.

We rewrite (49.10) as

$$Q^* = \sum_{k=1}^{\infty} q(S_k) w_k(S_k) \mathbb{I}(A_k), \tag{49.11}$$

where $\mathbb{I}(A_k)$ equals 1 if trajectory S_k was obtained during the trajectory sampling (i.e., if trajectory S_k was already in (49.10)), and equals 0 otherwise.

49.6.1 DXTRAN

DXTRAN is one of the most complex variance reduction techniques. This technique has been well described in [3] for Boltzmann tallies and in [1] using the supertrack approach.

Let us describe the DXTRAN game briefly. We assume that the region in which detectors are concentrated is small. We surround this region with a sphere, which is called the DXTRAN sphere. Each time that we sample the collision of a particle, we split the particle into two new particles. The first particle is called the DXTRAN particle. To sample the DXTRAN particle, we choose a random direction towards the sphere. The DXTRAN particle is scattered in this direction and transferred to the sphere without a collision. Once the DXTRAN particle is transferred to the sphere, the particle continues a normal (natural) run from the sphere. The DXTRAN particle is then excluded from the DXTRAN game.

The second particle that is created from the split is called the non-DXTRAN particle. This particle is sampled in the normal way. However, if the particle intersects the DXTRAN sphere before the next collision, then the weight of this particle is multiplied by 0.

According to the supertrack approach, in the DXTRAN game we should split a whole trajectory into two new trajectories. The first trajectory will contain the DXTRAN particle and the second trajectory will contain the non-DXTRAN particle. The DXTRAN trajectory is then excluded from the DXTRAN game.

From the DXTRAN game, we obtain a set of trajectories $S_1, S_2, \ldots, S_m$ and their weights $w_1, w_2, \ldots, w_m$. The number of trajectories m is random. The estimator Q^* for DXTRAN is given by (49.10) or (49.11).

In [3], DXTRAN is interpreted as a combination of splitting, Russian roulette, and importance sampling. However, we do not accept this interpretation as a proof that (49.10) is an unbiased estimate of MQ. A strict mathematical proof of DXTRAN for the Neumann–Ulam scheme is provided in [12]. In this paper, we investigate the case of non-Boltzmann tallies that cannot be calculated using Neumann–Ulam scheme.

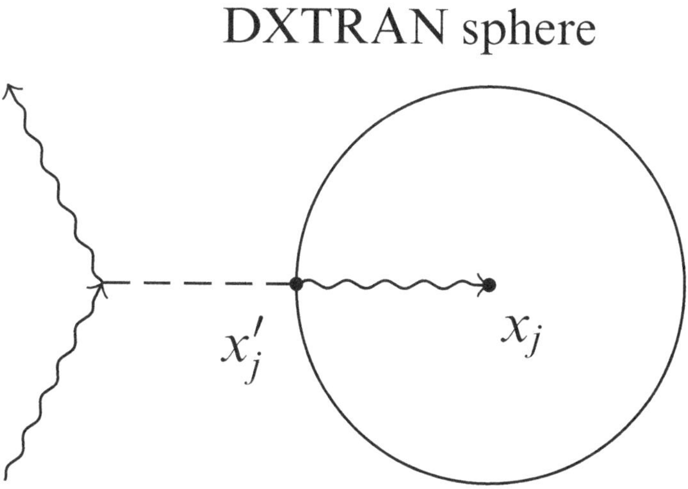

Fig. 49.2 Enumeration of the vertices in the DXTRAN technique

49.6.1.1 Partitioning the Set of Trajectories

We use the following subsets of S. Subset $\mathbb{T}_0$ contains all the trajectories that are not included in the DXTRAN game and all the trajectories that contain only the non-DXTRAN particles. Subset $\mathbb{T}_k$ is a subset of the trajectories that contain the DXTRAN particle after the kth application of the DXTRAN game to non-DXTRAN trajectory, $k \geqslant 1$.

Given a trajectory S, we can determine the subset $\mathbb{T}_k$ to which this trajectory belongs, and we can say at which node $D(S)$ the DXTRAN game has been played. We choose the point x_j' as the point where a particle enters the DXTRAN sphere (see Fig. 49.2). We define the function

$$p_n^{\mathrm{DXT}}(x_j|x_0, x_1, \ldots, x_{j-1})$$
$$= \Omega_n^{\mathrm{DXT}}(x_j|x_0, x_1, \ldots, x_{j-1}) e^{-l_n(x_j|x_0, x_1, \ldots, x_{j-1}) + l_n(x_j'|x_0, x_1, \ldots, x_{j-1})} \Sigma(x_j).$$

The function $\Omega_n^{\mathrm{DXT}}(x_j|x_0, x_1, \ldots, x_{j-1})$ represents the probability density function of scattering into a solid angle in the direction of the DXTRAN sphere, $\Omega_n^{\mathrm{DXT}}(x_j|x_0, x_1, \ldots, x_{j-1}) = 0$ if the direction of x_j does not point to the DXTRAN sphere, and the probability of scattering in any direction to the sphere equals 1. The function $e^{-l_n(x_j|x_0, x_1, \ldots, x_{j-1}) + l_n(x_j'|x_0, x_1, \ldots, x_{j-1})}$ represents the probability that a particle reaches x_j if the particle starts at x_j'. The probability of reaching x_j' equals 1 because the DXTRAN particle is transported deterministically to the sphere.

We rewrite the probability density function for DXTRAN trajectory in the form

$$p_k(S) = p(x_0) \prod_{\substack{j=1 \\ j \neq D(S)}} p_n(x_j | x_0, x_1, \ldots, x_{j-1}) \cdot p_n^{\mathrm{DXT}}(x_j | x_0, x_1, \ldots, x_{j-1}),$$

The difference between this equation and (49.5) is the multiplier, which corresponds to the deterministic transfer of a DXTRAN particle to the DXTRAN sphere. There is a probability measure on $\mathbb{T}_k$ induced by $p_k(S)$. Let us denote it by P_k, $k = 1, 2, \ldots$. It is important to notice that if the DXTRAN game is played the DXTRAN trajectories are sampled in accordance with the measures P_k.

49.6.1.2 Choosing Weights

Now we show how to choose weights $w_k(S)$ to make the DXTRAN game unbiased. If we take a term by term average of (49.11), then we obtain

$$MQ^* = \int_{\mathbb{S}} q(S) w_0(S) P(dS) + \sum_{k=1}^{\infty} \int_{\mathbb{T}_k} q(S) w_k(S) P_k(dS). \tag{49.12}$$

The first term corresponds to the non-DXTRAN trajectory. The probability measure that we use shows that the non-DXTRAN trajectory has been sampled in the natural way. The trajectories in the other terms have been sampled in a biased way according to $P_k(S)$. Our goal is to choose functions $w_k(S)$ such that we can transform (49.12) like

$$MQ^* = \int_{\mathbb{T}_0} q(S) P(dS) + \sum_{k=1}^{\infty} \int_{\mathbb{T}_k} q(S) P(dS) = \int_{\mathbb{S}} q(S) P(dS). \tag{49.13}$$

To set the first terms of (49.12) and (49.13) equal to each other, we choose $w_0(S)$ to be equal to 1 if $S \in \mathbb{T}_0$, and 0 otherwise. It means that the weight of a non-DXTRAN trajectory becomes 0 if the trajectory intersects the sphere. To set the other terms equal to each other, we choose the weight $w_k(S) = \partial P / \partial P_k(S)$. This results in a rule to calculate the weight multiplier of a trajectory

$$w_k(S) = \left. \frac{p_n(x_j | x_0, x_1, \ldots, x_{j-1})}{p_n^{\mathrm{DXT}}(x_j | x_0, x_1, \ldots, x_{j-1})} \right|_{j=D(S)}$$

$$= \left. \frac{\Omega_n(x_j | x_0, x_1, \ldots, x_{j-1}) e^{-l_n(x'_j | x_0, x_1, \ldots, x_{j-1})}}{\Omega_n^{\mathrm{DXT}}(x_j | x_0, x_1, \ldots, x_{j-1})} \right|_{j=D(S)}.$$

This rule is identical to the rules that were formulated in [1].

The question of the optimum choice of the DXTRAN sphere is still open. In accordance with common sense it seems to be close to optimal if one selects the sphere of minimal radius that contains the whole region of the detectors.

Conclusions
In this paper we showed that a probability theoretic approach is efficient for understanding the supertrack approach. In particular, the DXTRAN game can be treated as stratified sampling.

It is easy to understand that forced collisions [3] and forced detection techniques [7] can be treated as particular cases of stratified sampling too. It would be interesting to treat the implicit capture technique using a probability theoretic approach.

The classical way of investigating variance reduction techniques is to prove the unbiasedness and to estimate the variance. Due to the limited size of this paper, we did not precisely estimate the variances of the mentioned techniques.

The recommendations given above are based on the author's practice and coincide with the recommendations for the Neumann–Ulam scheme. The most comprehensive list of recommendations on how to use variance reduction techniques in the Neumann–Ulam scheme can be found in [3]. We hope that these recommendations are still valid when the supertrack approach is used. In any case, we recommend firstly to follow these recommendation when using supertracks. But the question about the optimum choice of the parameters of these techniques is still open.

References

1. Booth, T. E.: Monte Carlo variance reduction approaches for non-Boltzmann tallies. Nucl. Sci. Eng. **116**, 113–124 (1994)
2. Borisov, N.M., Panin, M.P.: Adjoint Monte Carlo calculations of pulse-height spectrum. Monte Carlo Methods Appl. **4**(3), 273–284 (1998)
3. Briesmeister, J.F. et al.: MCNP. A General Monte Carlo N-Particle Transport Code—Version 4C. Los Alamos National Laboratory Report, LA-13709-M (2000)
4. Dynkin, E.B.: Theory of Markov Processes. Pergamon Press, Oxford (1960)
5. Ermakov, S.M.: Method Monte Carlo and Adjoint Questions. Nauka, Moscow (1976) (in Russian)
6. Estes, G.P., Booth, T.E.: Supertrack Monte Carlo variance reduction experience for non-Boltzmann tallies. International conference on mathematics and computations, reactor physics, and environmental analyses, Portland, OR (United States) (1995). http://www.osti.gov/scitech/servlets/purl/10117121 of subordinate document. Cited 25 Feb 2014
7. Fredriksson, I., Larsson, M., Stromberg, T.: Forced detection Monte Carlo algorithms for accelerated blood vessel image simulation. J. Biophoton. **2**, 178–184 (2008)

8. Harris, T.E.: The Theory of Branching Processes. Die Grundlehren der Mathematischen Wissenschaften. Springer, Berlin (1963)
9. Lappa, A.V.: Weighted Monte Carlo estimate for calculation of high-order moments of additive characteristics of particle transport with multiplying. Comput. Math. Math. Phys. **30**, 122–134 (1990) (in Russian)
10. Lappa, A.V.: The Statistical Methods to Solve the Non-additive Problems of the Radiation Transport Theory. Doctoral dissertation. Saint Petersburg State University (1994) (in Russian)
11. Sevastyanov, B.A.: Branching Processes. Nauka, Moscow (1971) (in Russian)
12. Tsvetkov, E.A., Shahovsky, V.V.: The substantiation of DXTRAN modification of Monte Carlo method on basis of reciprocity relation for discriminating systems. Proc. Moscow Inst. Phys. Technol. **1**(2), 207–215 (2009) (in Russian)
13. Uchaikin, V.V.: Fluctuations in the cascade processes of the radiation transport through the substance. Part 1. Sov. Phys. J. (12), 30–34 (1978) (in Russian)
14. Uchaikin, V.V.: Fluctuations in the cascade processes of the radiation transport through the substance. Part 2. Sov. Phys. J. (2), 7–12 (1979) (in Russian)
15. Uchaikin, V.V.: Fluctuations in the cascade processes of the radiation transport through the substance. Part 3. Sov. Phys. J. (8), 14–21 (1979) (in Russian)

Chapter 50
The Dependence of the Ergodicity on the Time Effect in the Repeated Measures ANOVA with Missing Data Based on the Unbiasedness Recovery

Anna Ufliand and Nina Alexeyeva

50.1 Introduction

One way to solve the problem of missing data in the repeated measures analysis is an unbiased model with non-diagonal covariance matrix of errors [3]. The unbiased model can be obtained from the initial ANOVA model by subtraction of the displacement in individual means which is produced by repetition of the cross-averaging procedure [3]. The question arises, what characteristics of the obtained unbiased system make it more similar to the system with full data. The main problem was to investigate whether the ergodic property of the new system improves in comparison with the initial system. The ergodic property here should be understood in its physical meaning as the lack of the difference between time and space calculated means. As the result of this work, it was discovered that the ergodic property improvement depends on the time effect significance. The trend's increase(decrease) rate affects the degree of confidence with which we can claim about this fact.

50.2 The Repeated Measures Analysis of Variance and Missing Data

Consider the model of repeated measures analysis [1] of variance (ANOVA)

$$x_{ijt} = \mu + \alpha_i + \delta_{ij} + \beta_t + \gamma_{it} + \varepsilon_{ijt}, \tag{50.1}$$

A. Ufliand (✉) • N. Alexeyeva
Saint Petersburg State University, 7-9, Universitetskaya nab., St. Petersburg, 199034, Russia
e-mail: anna.uflyand@gmail.com; ninaalexeyeva@mail.ru

V.B. Melas et al. (eds.), *Topics in Statistical Simulation*, Springer Proceedings in Mathematics & Statistics 114, DOI 10.1007/978-1-4939-2104-1_50

where x_{ijt} is the data of the j-individual from the i-group at the time moment t, μ—general mean, α_i—group effect, β_t—time effect, γ_{it}—group and time interaction effect, $\delta_{ij} \sim N(0, \sigma_1^2)$—error, caused by the individuals variety and $\varepsilon_{ijt} \sim N(0, \sigma^2)$—the general model error (all errors are assumed to be independent). The amount of groups, individuals in group, and time points are equal to I, ν_i, and T respectively. Let M_{it} be the set of individuals from group i, who have an observation at the time point t; and denote m_{it} its cardinality, $\sum_t m_{it} = m_{i\cdot}$, $\sum_i m_{it} = m_{\cdot t}$, $\sum_t m_{t\cdot} = m_{\cdot\cdot}$. Let N_{ij} be the set of time points of the individual number j from the group i and denote n_{ij} its cardinality. In order to obtain the unique solutions of the systems of linear equations by means of LS Method for parameters estimation [5], the partial plan was considered:

$$\sum_{i=1}^{I} \frac{\alpha_i m_{i\cdot}}{m_{\cdot\cdot}} = 0, \quad \sum_{t=1}^{T} \frac{\beta_t m_{\cdot t}}{m_{\cdot\cdot}} = 0, \quad \sum_{i=1}^{I} \frac{\gamma_{it} m_{it}}{m_{\cdot\cdot}} = 0, \quad \sum_{t=1}^{T} \frac{\gamma_{it} m_{it}}{m_{\cdot\cdot}} = 0. \qquad (50.2)$$

In order to estimate parameters, the model (50.2) was divided into two parts: $x_{ijt} = z_{ij} + y_{ijt}$, where $\mathbb{E}z_{ij} = \mu + \alpha_i$, $\mathbb{E}y_{ijt} = \beta_t + \gamma_{it}$. When the data is complete z_{ij} is just the time mean $x_{ij.}$. In case of missing data the time mean becomes

$$x_{ij.} = \frac{1}{n_{ij}} \sum_{t \in N_{ij}} x_{ijt}, \qquad (50.3)$$

and $\mathbb{E}x_{ij.}$ is no more equal to $\mu + \alpha_i$.

Definition 50.1. The Cross Mean (CM) $A_{ij}(k), k = 1, 2, \ldots$ for the individual j from the group i is defined by the following recurrent equation:

$$A_{ij}(1) = \frac{1}{n_{ij}} \sum_{t \in N_{ij}} \frac{1}{m_{it}} \sum_{l \in M_{it}} (x_{ilt} - x_{il.}),$$

$$A_{ij}(k+1) = \frac{1}{n_{ij}} \sum_{t \in N_{ij}} \frac{1}{m_{it}} \sum_{l \in M_{it}} A_{il(k)}. \qquad (50.4)$$

Definition 50.2. The individual CM-displacement for individual j from the group i :

$$H_{ij} = \sum_{k=1}^{\infty} A_{ij}(k). \qquad (50.5)$$

The following theorem was proved in [3].

Theorem 50.1. *The models become unbiased after subtraction of the individual displacement* H_{ij} :

$$\mathbb{E}z_{ij} = \mathbb{E}(x_{ij.} - H_{ij}) = \mu + \alpha_i \tag{50.6}$$

$$\mathbb{E}y_{ijt} = \mathbb{E}(x_{ijt} - x_{ij.} + H_{ij}) = \beta_t + \gamma_{it}. \tag{50.7}$$

As a consequence, covariance matrices of errors stop being unity (the form of these matrices can be found in [2, 4]).

50.3 Balance Property of CM-Displacement

For the sake of notation simplicity, in this paper we are going to consider only one group of individual. We denote the number of individuals in it as N.

Consider the incidence matrix of missing data J with N rows and T columns, for every k construct the cross mean vector A with components $A_j(k)$, $j = 1, 2, \ldots, N$, denote the diagonal matrix Λ_N of dimension N with elements $\frac{1}{n_j}$, the diagonal matrix Λ_T of dimension T with elements $\frac{1}{m_t}$ and the stochastic matrix $P = \Lambda_N J \Lambda_T J^T$. In [3] it was shown that

$$A(k+1) = PA(k) = P^k A(1) \tag{50.8}$$

and the following theorem was proved.

Theorem 50.2. *Let* $m. = \sum_{t=1}^{T} m_t$. *The limit* $\lim_{k\to\infty} P^k$ *exists and is equal to the stationary matrix with identical rows:*

$$P^{\infty} := \lim_{k\to\infty} P^k = \begin{pmatrix} \frac{n_1}{m.} & \ldots & \frac{n_N}{m.} \\ & \ldots & \\ \frac{n_1}{m.} & \ldots & \frac{n_N}{m.} \end{pmatrix}. \tag{50.9}$$

In case, when $i = 1$ we have $H_j = \sum_{k=1}^{\infty} A_j(k)$, $H = (H_1, \ldots, H_N)^T$. Denote

$$x'_{jt} := x_{jt} - H_j\,. \tag{50.10}$$

Theorem 50.3. *The general mean does not change after subtraction of CM-displacement, i.e.*

$$x_{..} = \frac{1}{m_.} \sum_{j=1}^{N} \sum_{t\in N_j} x_{jt} = \frac{1}{m_.} \sum_{j=1}^{N} \sum_{t\in N_j} \left(x_{jt} - H_j\right) = x'_{..}\,. \tag{50.11}$$

To prove this fact the following balance property was introduced.

Lemma 50.1.

$$\sum_{j=1}^{N} n_j A_j(1) = \sum_{j=1}^{N} n_j H_j = 0. \tag{50.12}$$

Proof. By changing the order of summation in expression we obtain:

$$\sum_{j=1}^{N} n_j A_j(1) = \sum_{j=1}^{N} \sum_{t \in N_j} \frac{1}{m_t} \sum_{l \in M_t} (x_{lt} - x_{l.}) = \sum_{t=1}^{T} \sum_{j \in M_t} \frac{1}{m_t} \sum_{l \in M_t} (x_{lt} - x_{l.})$$

$$= \sum_{t=1}^{T} \sum_{l \in M_t} (x_{lt} - x_{l.}) = x_{..} - \sum_{l=1}^{N} \sum_{t \in N_l} x_{l.} = x_{..} - \sum_{l=1}^{N} n_l x_{l.} = 0\,.$$

Therefore $P^{\infty} A(1) = \mathbf{0}$, where $\mathbf{0}$ is zero vector. The second equality is obtained from

$$P^{\infty} H = P^{\infty} \sum_{k=1}^{\infty} A(k) = P^{\infty} \sum_{k=1}^{\infty} P^{k-1} A(1) = \sum_{k=1}^{\infty} P^{\infty} A(1) = \mathbf{0}.$$

The proof of Theorem 50.3. $x'_{..} = \frac{1}{m_.} \sum_{j=1}^{N} \sum_{t \in N_j} \left(x_{jt} - H_j\right) = x_{..} - \frac{1}{m_.} \sum_{j=1}^{N} \sum_{t \in N_j} H_j =$

$x_{..} - \frac{1}{m_.} \sum_{j=1}^{N} n_j H_j = x_{..}$,

since $\sum_{j=1}^{N} n_j H_j = 0$ (from Lemma 50.1).

50.4 Ergodic Property

For the ergodic systems the following fact is correct: mathematical expectation with respect to space is equal to mathematical expectation with respect to time. It is obvious that in case of full data means calculated with respect to space are equal to those calculated with respect to time. Also it is obvious, that in case of missing data this equality does not hold.

The objective was to investigate whether the subtraction of the displacement helps to improve the ergodic property.

Definition 50.3. Denote individual time-means $x_{j.} = \frac{1}{n_j} \sum_{t \in N_j} x_{jt}$ and space-means $x_{.t} = \frac{1}{m_t} \sum_{j \in M_t} x_{jt}$, respectively. The ergodic property is fulfilled if the arithmetic mean of individual time-means and space-means are equal, i.e.

$$x_* := \frac{1}{N}\sum_{j=1}^{N} x_{j.} = \frac{1}{T}\sum_{t=1}^{T} x_{.t} =: x^*, \tag{50.13}$$

where x_*—arithmetic mean of time-means and x^*—arithmetic mean of space-means.

Remark 50.1. If the ergodic property is fulfilled, then $x_* = x_{..} = x^*$, where $x_{..}$ is the general mean from (50.11).

Definition 50.4. Let $x'_{jt} = x_{jt} - H_j$ be the data with subtracted CM-displacement. The ergodic property improves if the following three inequalities are fulfilled:

$$|x'_* - x'_{..}| < |x_* - x_{..}|, \quad |x'^* - x'_{..}| < |x^* - x_{..}|, \quad |x'_* - x'^*| < |x_* - x^*|. \tag{50.14}$$

In other words, the ergodicity improves after subtraction of CM-displacement if the distance between time mean x_* and general mean $x_{..}$ decreases and the distance between space mean x^* and general mean $x_{..}$ decreases. It is clear that the third inequality is a consequence of the first two. In this paper we consider the first inequality in detail and provide a brief overview of the second inequality properties in the last section.

Lemma 50.2. *Let $\bar{H} = \frac{1}{N}\sum_{j=1}^{N} H_j$. The distance between space and time means:*

$$x'_* - x'_{..} = x_* - \bar{H} - x_{..}. \tag{50.15}$$

Proof. First,

$$x'_* = \frac{1}{N}\sum_{j=1}^{N}\frac{1}{n_j}\sum_{t\in N_j}\left(x_{jt} - H_j\right) = x_* - \frac{1}{N}\sum_{j=1}^{N} H_j = x_* - \bar{H}.$$

Second, Theorem 50.3 insures that $x'_{..} = x_{..}$ and the result follows.

Theorem 50.4. *Denote the matrix $Q = (I - P + P^\infty)^{-1}$ with elements $\{q_{jk}\}_{j=1,k=1}^{N}$, $\sum_{j=1}^{N} q_{jk} = Q_k$. Then $H = QA(1)$ and $\bar{H} = \bar{H}_1 + \bar{H}_2$, where*

$$\bar{H}_1 = \frac{1}{N}\sum_{j=1}^{N}\sum_{k=1}^{N} q_{jk}(\delta_{.}(k) - \delta_k + \varepsilon_{..}(k) - \varepsilon_{k.}), \tag{50.16}$$

$$\bar{H}_2 = \frac{1}{N}\sum_{j=1}^{N}\beta_{.}(j) + \delta_{.} + \varepsilon_{..} - \frac{1}{m_{.}}\sum_{j=1}^{N} n_j\delta_j - \frac{1}{m_{.}}\sum_{j=1}^{N} n_j\varepsilon_{j.}. \tag{50.17}$$

This theorem shows that it is possible to present the mean displacement as sum of two components, only one of which depends on the time effect.

Lemma 50.3. *Let* $x_{..}(j) = \frac{1}{n_j} \sum\limits_{t \in N_j} \frac{1}{m_t} \sum\limits_{l \in M_t} x_{lt}$ *and* $U = \{x_{..}(j)\}_{j=1}^{N}$, $V = \{x_{j.}\}_{j=1}^{N}$.

Then

1. $P^{\infty}(U - V) = \mathbf{0}$.
2. $\sum\limits_{k=0}^{\infty} P^k(U - V) = (I - P + P^{\infty})^{-1}(U - V)$.

Proof. 1) Similar to Lemma 50.1, $\sum\limits_{j=1}^{N} n_j(x_{..}(j) - x_{j\cdot}) = x_{..} - x_{..} = 0$.

2) Since $P^{\infty}(U - V) = 0$, the row $\sum\limits_{k=0}^{\infty} P^k(U - V)$ converges as k tends to infinity and its limit is:

$$\sum_{k=0}^{\infty} P^k(U - V) = \sum_{k=0}^{\infty}(P - P^{\infty})^k(U - V) = (I - P + P^{\infty})^{-1}(U - V).$$

The proof of the Theorem 50.4. According to its definition the displacement can be represented as
$H = U - PV + PU - P^2V + P^2U - \cdots =$
$= V + U - V + P(U - V) + \cdots + P^k(U - V) + \cdots =$
$= (I + P + P^2 + \cdots + P^k + \cdots)(U - V) + (I - P^{\infty})V$. Then $\bar{H} = \bar{H}_1 + \bar{H}_2$, $H_2 = H_{21} - H_{22}$, where $H_1 = (I + P + P^2 + \cdots + P^k + \cdots)(U - V)$, $H_{21} = V$, $H_{22} = P^{\infty}V$. By substituting the model $x_{jt} = \mu + \beta_t + \delta_j + \varepsilon_{jt}$ in $u_j = x_{..}(j)$ and in $v_j = x_{j.}$ $\sum\limits_{k=0}^{\infty} P^k(U - V) = (I - P + P^{\infty})^{-1}(U - V) = Q(U - V)$,
$\bar{H}_1 = \frac{1}{N} \sum\limits_{k=1}^{N} Q_k(\delta_{.}(k) - \delta_k + \varepsilon_{..}(k) - \varepsilon_{k.})$.

By analogy $\bar{H}_{21} = \frac{1}{N} \sum\limits_{j=1}^{N} \frac{1}{n_j} \sum\limits_{t \in N_j} \mu + \beta_t + \delta_j + \varepsilon_{jt} = \mu + \frac{1}{N} \sum\limits_{j=1}^{N} \beta_{.}(j) + \delta_{.} + \varepsilon_{..}$,

$\bar{H}_{22} = \frac{1}{m_{.}} \sum\limits_{j=1}^{N} \sum\limits_{t \in N_j} \mu + \beta_t + \delta_j + \varepsilon_{jt} = \mu + \frac{1}{m_{.}} \sum\limits_{j=1}^{N} \sum\limits_{t \in N_j} \beta_t + \frac{1}{m_{.}} \sum\limits_{j=1}^{N} n_j \delta_j + \frac{1}{m_{.}} \sum\limits_{j=1}^{N} n_j \varepsilon_{j.}$.

So $\bar{H}_2 = \frac{1}{N} \sum\limits_{j=1}^{N} \beta_{.}(j) + \delta_{.} - \frac{1}{m_{.}} \sum\limits_{j=1}^{N} n_j \delta_j + \varepsilon_{..} - \frac{1}{m_{.}} \sum\limits_{j=1}^{N} n_j \varepsilon_{j.}$, since $\sum\limits_{j=1}^{N} \sum\limits_{t \in N_j} \beta_t = \sum\limits_{t=1}^{T} \sum\limits_{j \in M_t} \beta_t = \sum\limits_{t=1}^{T} m_t \beta_t = 0$ because of the choice of differential effects.

Corollary 50.1. *The mathematical expectations of $\bar{H}_1$ and $\bar{H}_2$ are equal to*

$$\mathbb{E}\bar{H}_1 = 0, \ \ \mathbb{E}\bar{H}_2 = \bar{\beta} = \frac{1}{N}\sum_{j=1}^{N}\frac{1}{n_j}\sum_{t\in N_j}\beta_t. \tag{50.18}$$

respectively.

Theorem 50.5. *The theorem about the ergodicity improvement in terms of the first inequality from (50.14)*
Let $\xi = x_* - x_{..}$, $\eta = x'_* - x'_{..}$, and their variances $\mathbb{D}\xi = \tilde{\sigma}_1^2$, $\mathbb{D}\eta = \tilde{\sigma}_2^2$.
If $\sqrt{2} < k < \frac{|\bar{\beta}|}{(\tilde{\sigma}_1+\tilde{\sigma}_2)}$, then $\mathbb{P}(|\eta| > |\xi|) \leqslant \frac{2}{k^2}$.

In other words we obtained the upper estimate for the probability of the event of not improving the ergodicity after displacement subtraction.

Lemma 50.4. *For $\xi = x_* - x_{..}$ and $\eta = x'_* - x'_{..}$ the following is true: $\xi = \bar{H}_2$, $\eta = -\bar{H}_1$.*

Proof. $\xi = x_* - x_{..} = \frac{1}{N}\sum\limits_{j=1}^{N}\frac{1}{n_j}\sum\limits_{t\in N_j} x_{jt} - \frac{1}{m_.}\sum\limits_{j=1}^{N}\sum\limits_{t\in N_j} x_{jt} =$
$= \frac{1}{N}\sum\limits_{j=1}^{N}\frac{1}{n_j}\sum\limits_{t\in N_j}\mu + \beta_t + \delta_j + \varepsilon_{jt} - \frac{1}{m_.}\sum\limits_{j=1}^{N}\sum\limits_{t\in N_j}\mu + \beta_t + \delta_j + \varepsilon_{jt} =$
$= \frac{1}{N}\sum\limits_{j=1}^{N}\beta_.(j) + \delta_. + \varepsilon_{..} - \frac{1}{m_.}\sum\limits_{j=1}^{N} n_j\delta_j - \frac{1}{m_.}\sum\limits_{j=1}^{N} n_j\varepsilon_{j.} = \bar{H}_2$ in virtue of (50.17) of Theorem 50.4.
$\eta = x'_* - x'_{..} = x_* - \bar{H} - x_{..} = \bar{H}_2 - \bar{H} = -\bar{H}_1$ in virtue of Lemma 50.2.

Lemma 50.5. *Denote $G_k = \sum\limits_{t\in N_k}\sum\limits_{l\in M_t}\frac{1}{m_t n_l} - 1$. The variances of $\bar{H}_1$ and $\bar{H}_2$ are*

$$\mathbb{D}\bar{H}_2 = \sum_{j=1}^{N}\left(\frac{\sigma^2}{n_j} + \sigma_1^2\right)\left(\frac{1}{N} - \frac{n_j}{m_.}\right)^2, \tag{50.19}$$

$$\mathbb{D}\bar{H}_1 = \frac{\sigma_1^2}{N^2}\sum_{k=1}^{N} Q_k^2 G_k^2 + \frac{\sigma^2}{N^2}\sum_{k=1}^{N}\sum_{t\in N_k}\left(\frac{1}{m_t}\sum_{l\in M_t}\frac{Q_l}{n_l} - \frac{Q_k}{n_k}\right)^2. \tag{50.20}$$

Proof. First, by assumption, $\{\delta_j\}_{j=1}^{N}$ and $\{\varepsilon_{jt}\}_{j=1,t\in N_j}^{N}$ are independent and distributed identically inside the group with variances σ_1^2 and σ^2, respectively. Let us divide the stochastic component $\bar{H}_2$ into independent components that consist of δ_j and ε_{jt} : $\bar{H}_2 = \frac{1}{N}\sum\limits_{j=1}^{N}\frac{1}{n_j}\sum\limits_{t\in N_j}\beta_t + h_{21} + h_{22}$, where
$h_{21} = \frac{1}{N}\sum\limits_{j=1}^{N}\delta_j - \frac{1}{m_.}\sum\limits_{j=1}^{N} n_j\delta_j = \sum\limits_{j=1}^{N}\left(\frac{1}{N} - \frac{n_j}{m_.}\right)\delta_j$

$$h_{22} = \frac{1}{N}\sum_{j=1}^{N}\frac{1}{n_j}\sum_{t\in N_j}\varepsilon_{jt} - \frac{1}{m_{.}}\sum_{j=1}^{N}\sum_{t\in N_j}\varepsilon_{jt} = \sum_{j=1}^{N}\left(\frac{1}{Nn_j} - \frac{1}{m_{.}}\right)\sum_{t\in N_j}\varepsilon_{jt}$$

It is obvious that summands of both components are dependent, so the variance of the difference could not be calculated as the sum of variances of components.

The idea is to rearrange this difference into one sum, where each independent component has its own coefficient, that is to be found. As the result, all the summands are independent and the calculation of the final variance becomes easy.

$$\mathbb{D}h_{21} = \sigma_1^2\sum_{j=1}^{N}\left(\frac{1}{N} - \frac{n_j}{m_{.}}\right)^2, \mathbb{D}h_{22} = \sum_{j=1}^{N}\frac{\sigma^2}{n_j}\left(\frac{1}{N} - \frac{n_j}{m_{.}}\right)^2.$$

Second, similarly for $\bar{H}_1$ we obtain $\bar{H}_1 = h_{11} + h_{12}$, where

$$h_{11} = \frac{1}{N}\sum_{k=1}^{N}Q_k\left(\frac{1}{n_k}\sum_{t\in N_k}\frac{1}{m_t}\sum_{l\in M_t}\delta_l - \delta_k\right) = \frac{1}{N}\sum_{j=1}^{N}Q_k\left(\sum_{t\in N_k}\sum_{l\in M_t}\frac{1}{m_t n_l} - 1\right)\delta_k,$$

$$h_{12} = \frac{1}{N}\sum_{k=1}^{N}\frac{Q_k}{n_k}\sum_{t\in N_k}\left(\frac{1}{m_t}\sum_{l\in M_t}\varepsilon_{lt} - \varepsilon_{kt}\right) = \frac{1}{N}\sum_{k=1}^{N}\sum_{t\in N_k}\left(\frac{1}{m_t}\sum_{l\in M_t}\frac{Q_l}{n_l} - \frac{Q_k}{n_k}\right)\varepsilon_{kt},$$

$$\mathbb{D}h_{11} = \frac{\sigma_1^2}{N^2}\sum_{k=1}^{N}Q_k^2\left(\sum_{t\in N_k}\sum_{l\in M_t}\frac{1}{m_t n_l} - 1\right)^2,$$

$$\mathbb{D}h_{12} = \frac{\sigma^2}{N^2}\sum_{k=1}^{N}\sum_{t\in N_k}\left(\frac{1}{m_t}\sum_{l\in M_t}\frac{Q_l}{n_l} - \frac{Q_k}{n_k}\right)^2.$$

$$\mathbb{D}\bar{H}_1 = \frac{\sigma_1^2}{N^2}\sum_{k=1}^{N}Q_k^2\left(\sum_{t\in N_k}\sum_{l\in M_t}\frac{1}{m_t n_l} - 1\right)^2 + \frac{\sigma^2}{N^2}\sum_{k=1}^{N}\sum_{t\in N_k}\left(\frac{1}{m_t}\sum_{l\in M_t}\frac{Q_l}{n_l} - \frac{Q_k}{n_k}\right)^2.$$

The proof of the Theorem 50.5. From Corollary 50.1: $\mathbb{E}\xi = \bar{\beta}, \mathbb{E}\eta = 0$, variances for ξ and η were calculated in Lemma 50.5. The Chebyshev inequalities for ξ and η:

$$\mathbb{P}\left(|\xi - \bar{\beta}| \geqslant k\tilde{\sigma}_1\right) \leqslant \frac{1}{k^2}, \tag{50.21}$$

$$\mathbb{P}\left(|\eta| \geqslant k\tilde{\sigma}_2\right) \leqslant \frac{1}{k^2}. \tag{50.22}$$

The event of not improving the ergodicity is equivalent to the variable $|\eta|$ being greater than $|\xi|$ and $|\bar{\beta}|$ being the distance between mathematical expectations ξ and η. Let us consider the case, when $\bar{\beta} > 0$ (the other case can be considered similarly). If the distances between both variables and their mathematical expectations are less than k standard deviations, then $max|\eta| = k\tilde{\sigma}_2$, $min|\xi| = \bar{\beta} - k\tilde{\sigma}_1$. Then $\bar{\beta} > k(\tilde{\sigma}_1 + \tilde{\sigma}_2)$ results in $min|\xi| > max|\eta|$.

Therefore the ergodicity is not improved if at least one of the two following conditions is fulfilled: $|\xi - \bar{\beta}| > k\tilde{\sigma}_1$; $|\eta| > k\tilde{\sigma}_2$. The probability of each condition is less than or equal to $\frac{1}{k^2}$ from inequalities (50.21), (50.22). Therefore the probability of not improving of the ergodicity is less than or equal to $\frac{2}{k^2}$.

Remark 50.2. We have considered the part of the ergodicity improvement which refers to the first inequality from (50.14). Now we are going to provide a brief overview of the second inequality properties. Let vectors β, m, a, and $\mathbf{e}$ have length T, $\mathbf{e} = (1, 1, \ldots, 1)^T$, $a_k = \frac{1}{T} \sum\limits_{j \in M_k} \sum\limits_{\tau \in N_j} \frac{1}{n_j m_k}$ and $\sum\limits_{k=1}^{T} a_k = 1$. By analogy the following can be proved: $\mathbb{E}(x^* - x_{..}) = \frac{1}{T}\beta^{\mathbf{T}}\mathbf{e}$, after displacement subtraction $\mathbb{E}(x'^* - x'_{..}) = \frac{1}{T}\beta^{\mathbf{T}}\mathbf{e} - \beta^T a$.

It can be proved, that in most cases, when the time effect increases ($\beta_t < \beta_s$) and the amount of complete data decreases ($m_t > m_s$, $t < s$), the following inequalities are fulfilled: $\frac{1}{m_.}\beta^{\mathbf{T}} m = 0 < \beta^{\mathbf{T}} a < \frac{1}{T}\beta^{\mathbf{T}}\mathbf{e}$, which leads to the ergodicity improvement. Thus we have shown that in most practical cases we can observe the ergodicity improvement after subtraction of the displacement.

Conclusion

In order to investigate the conditions under which the initial repeated measures ANOVA system with missing data becomes more similar to the system with full data after its transformation into unbiased system by CM-displacement subtraction, we introduced the notion of the ergodicity improvement. We obtained the requirements which guarantee the ergodicity improvement with a certain probability.

References

1. Afifi, A.A., Azen, S.P.: Statistical Analysis: A Computer Oriented Approach. Academic Press, New York (1972)
2. Alexeyeva, N.P.: The ergodic method of missing data compensation for Split-Plot Designs with applications in longitude researches. In: Chirkov, R.M. (ed.) Mathematical Models. Theory and Applications, pp. 35–52. Publishing of the Saint-Petersburg State University, Saint-Petersburg (2010) (in Russian)
3. Alexeyeva, N.P.: Analysis of Biomedical Systems. Reciprocity. Ergodicity. Synonymy. Publishing of the Saint-Petersburg State University, Saint-Petersburg (2012) (in Russian)
4. Alexeyeva, N.P. et al.: Analysis of repeated cardiological incomplete data based on ergodic centralization of model. Bull. Almazov Center **3(8)**, 59–63 (2011) (in Russian)
5. Scheffe, H.: The Analysis of Variance. Wiley, Canada (1999)

Chapter 51
Mixture of Extended Linear Mixed-Effects Models for Clustering of Longitudinal Data

ChangJiang Xu, Celia M.T. Greenwood, Vicky Tagalakis, Martin G. Cole, Jane McCusker, and Antonio Ciampi

51.1 Introduction

Data from longitudinal studies include measurements of an outcome variable repeated over time on each study subject. These measurements are usually heteroscedastic and correlated within subjects; also, measurement times may be unequally spaced within subjects, may vary across subjects, and the number of measurements may vary from subject to subject. The Linear Mixed Effect (LME) model and its extension, the Extended Linear Mixed Effect (ELME) model [19], provide powerful tools for modeling longitudinal data from a homogeneous population. When data come from heterogeneous populations, the analyst may attempt to account for heterogeneity by modeling the outcome variable as a finite mixture of distributions.

C. Xu (✉)
Sainte-Justine University Hospital Center (CHU) Research Center, Montreal, QC, Canada H3T 1C5

C.M.T. Greenwood • V. Tagalakis
Lady Davis Institute for Medical Research Centre, Jewish General Hospital, Montreal, QC, Canada H3T 1E2

M.G. Cole
St. Mary's Research Centre, St. Mary's Hospital, Montreal, QC, Canada H3T 1M5

Department of Psychiatry, St. Mary's Hospital, Montreal, QC, Canada H3T 1M5

Department of Psychiatry, McGill University, Montreal, QC, Canada H3A 0G4

J. McCusker • A. Ciampi
St. Mary's Research Centre, St. Mary's Hospital, Montreal, QC, Canada H3T 1M5

Department of Epidemiology, Biostatistics, and Occupational Health, McGill University, Montreal, QC, Canada H3A 0G4

V.B. Melas et al. (eds.), *Topics in Statistical Simulation*, Springer Proceedings in Mathematics & Statistics 114, DOI 10.1007/978-1-4939-2104-1_51

To model a general multivariate outcome from a heterogeneous population, a rich literature on mixtures of distributions is available [8, 13, 14, 23]; these works are largely based on the EM algorithm [6]. A common concern is to reduce the number of parameters by simplifying the variance–covariance matrix while preserving flexibility. Specifically, Banfield and Raftery [2] proposed a general framework for the family of multivariate normal mixtures based on the spectral decomposition of the variance–covariance matrix. Also, McNicholas and Murphy [16] developed a parsimonious Gaussian mixture model (PGMM) family based on mixture of factor analyzers [15]. The PGMM family of models is well suited to the analysis of high-dimensional data because the number of covariance parameters is linear in the dimensionality of the data under consideration. These two families of models are implemented in the R [21] software packages mclust [7, 9] and pgmm [17], respectively.

However, these approaches are not adapted to the complexities of longitudinal data. Several recent papers deal with the specific nature of longitudinal data, but usually consider only some aspects of such data. Generalizing earlier work [3, 18], De la Cruz-Mesia et al. [5] developed a finite mixture model based on the LME model, which takes into account correlations explained by multi-level structures; however, they only explicitly consider uncorrelated and homoscedastic residual error matrices. In contrast, McNicholas and Murphy [17] proposed to model longitudinal data in the case of equally spaced fixed times as a Gaussian mixture model: using a modified Cholesky decomposition of the variance–covariance matrix as in Pourahmadi [20], these authors develop a family of parsimonious models that allow for both heteroscedasticity and correlation within subjects, but not for multi-level structures. Finally, Ciampi et al. [4] proposed to model longitudinal data from a heterogeneous population as a mixture of ELME models: in principle, such a mixture can accommodate most types of longitudinal data occurring in applications. In practice, however, the earlier version of the algorithm was marred by excessive computational cost and instability of results, especially when attempting to model random effects and correlated residual errors at the same time. Some of these concerns were addressed in Ji et al. [10], who developed some modifications to the EM approach of Ciampi et al. [4]; however, a reduction in computational time required parallel computing, and even with this reduction, computational costs remained high for the general case. To proceed further along this research direction, one has the option of developing faster algorithms for particular cases of ELME models, or to devise novel EM strategies.

In this paper we report some progress in both directions. In Sect. 51.2 we formulate the mixture of ELME Models (ELMEM) approach for modeling heterogeneity, while paying special attention to the case of AR(r) (autoregressive of order r) residual error structure. In Sect. 51.3, devoted to model estimation, we reformulate the standard EM algorithm; we propose a new variant of the EM algorithm for the particular case of equally spaced fixed times (EMAR); and further develop the most promising general approach proposed in Ji et al. [10], the EM with Monte

Carlo sampling (EMMC). In Sect. 51.4 we evaluate the statistical properties of our estimates by limited simulations. Two clinical examples are presented in Sect. 51.5. The discussion of Sect. 51.6 concludes the paper.

51.2 A Mixture of Extended Linear Mixed-Effects Models

Consider a data set containing M subjects from a heterogeneous population. The subpopulation of each subject is unknown. However, we assume that for a subject i in subpopulation k, the observed response is represented by the ELMEM:

$$y_i = X_i \beta_k + Z_i b_{ik} + \epsilon_{ik}, \tag{51.1}$$

where $y_i = (y_{i1}, \ldots, y_{i,n_i})'$, y_{ij} is the response variable of subject i at time t_{ij}, $j = 1, \ldots, n_i$, and

- X_i is a matrix of covariates of $n_i \times p$, and β_k is a vector representing fixed effects;
- Z_i is a design matrix of $n_i \times q$, $b_{ik} \sim N(0, \Psi_k)$ representing random effects;
- $\epsilon_{ik} \sim N(0, \Lambda_{ik})$ is a random vector of modeling errors, where $\Lambda_{ik} > 0$, positive definite.

The random effect vector, $\boldsymbol{b}$, and modeling error vector, ϵ, are assumed to be independent. The covariance matrix of y_i can therefore be written as:

$$\Sigma_{ik} = Z_i \Psi_k Z_i' + \Lambda_{ik}.$$

Let $\mathscr{C}_k$ denote the set of subjects in the cluster k, and $\alpha_k = \Pr\{\mathscr{C}_k\}$ be the mixture proportion for cluster k, satisfying $\sum_{k=1}^{K} \alpha_k = 1$. Let $\mu_{ik} = X_i \beta_k$. Then the set of observations for subject i of unknown subpopulation follows a multivariate Gaussian mixture distribution with probability density function:

$$f(y_i|\theta) = \sum_{k=1}^{K} \alpha_k \varphi(y|\mu_{ik}, \Sigma_{ik}), \tag{51.2}$$

where

$$\varphi(y_i|\mu_{ik}, \Sigma_{ik}) = \frac{1}{\sqrt{(2\pi)^{n_i}|\Sigma_{ik}|}} \exp\left\{-\frac{1}{2}(y_i - \mu_{ik})' \Sigma_{ik}^{-1} (y_i - \mu_{ik})\right\} \tag{51.3}$$

is the density of the multivariate normal distribution, and $\theta = \{\alpha, \beta, \Psi, \Lambda\}$ contain all unknown parameters. The covariance matrix of the modeling errors may be redefined as $\Lambda_{ik} = \sigma_k^2 V(\varphi_k, n_i)$, depending on some additional parameters. Then $\theta = \{\alpha, \beta, \Psi, \varphi, \sigma^2\}$. In this article, we use parameter notation without subscripts to represent the set of all corresponding parameters.

51.2.1 Autoregressive Errors

As in Pourahmadi [20], and in McNicholas and Murphy (2008) [16], we apply the Cholesky decomposition to the covariance matrix, Λ, of the residual error: we can therefore write

$$L\Lambda L' = D, \tag{51.4}$$

where L is a unique lower triangular matrix with diagonal elements equal to 1, and D is a unique diagonal matrix with strictly positive diagonal entries. L has the form:

$$L = L(\varphi, n) = \begin{bmatrix} 1 & 0 & 0 & \dots & 0 \\ \varphi_{2,1} & 1 & 0 & \dots & 0 \\ \varphi_{3,2} & \varphi_{3,1} & 1 & \dots & 0 \\ \vdots & \ddots & \ddots & \ddots & \vdots \\ \varphi_{n,n-1} & \dots & \varphi_{n,2} & \varphi_{n,1} & 1 \end{bmatrix}.$$

The modified Cholesky decomposition (51.4) may also be expressed in the form $\Lambda^{-1} = L'D^{-1}L$.

The values of L and D have interpretations as generalized autoregressive parameters and innovation variances [20], respectively. It can be shown that the residual error, ϵ_t, can be described by a generalized autoregressive model [20] for $t > 1$,

$$\epsilon_t + \varphi_{t,1}\epsilon_{t-1} + \dots + \varphi_{t,t-1}\epsilon_1 = \eta_t,$$

where η_t is called innovation and $\varphi_{t,j}$ are the corresponding elements of L. Equivalently, in matrix form: $L\epsilon = \eta$. If the innovations, η_t, are identically and independently distributed, then $D = \sigma^2 I$, where I is an identity matrix.

For an autoregressive process of order r, AR(r), a special case of the generalized autoregressive model, we have $D = \sigma^2 I$ and

$$L = L(\varphi, n) = \begin{bmatrix} 1 & & & & & \\ \varphi_1 & 1 & & \mathbf{0} & & \\ \vdots & \ddots & \ddots & & & \\ \varphi_r & \dots & \varphi_1 & 1 & & \\ & \ddots & & \ddots & \ddots & \\ \mathbf{0} & & \varphi_r & \dots & \varphi_1 & 1 \end{bmatrix},$$

where $\mathbf{0}$ represents zero entries. The lower triangular matrix may be rewritten as:

$$L(\varphi, n) = \sum_{u=0}^{r} \varphi_u J_n(u), \tag{51.5}$$

where $\varphi_0 = 1$, and

$$J_n(u) = \begin{bmatrix} \mathbf{0} & \mathbf{0} \\ I_{n-u} & \mathbf{0} \end{bmatrix},$$

is a matrix of $n \times n$, I_u is an identity matrix of $u \times u$.

51.3 Model Estimation

The likelihood function for model (51.1) is

$$l(\theta|y) = \sum_{i=1}^{M} \log f(y_i|\theta) = \sum_{i=1}^{M} \log \left(\sum_{k=1}^{K} \alpha_k \varphi(y_i|\mu_{ik}, \Sigma_{ik}) \right).$$

As in typical mixture problems, a direct maximization of the likelihood function is avoided. Instead, the EM algorithm [6] appears to be the method of choice. Let $\delta_{ik} = 1\{i \in \mathscr{C}_k\}$ be an indicator function for subject i being in cluster k, and let $\delta_i = (\delta_{i1}, \ldots, \delta_{iK})'$. Both the random effect vector, b, and cluster indicator vector, δ, are unobserved, and can be considered missing data or latent variables. However, if they were known, then the joint probability of Y_i, b_i, and δ_i could be written as:

$$\begin{aligned} f(y_i, b_i, \delta_i|\theta) &= \prod_{k=1}^{K} \{\alpha_k f(y_i, b_i|\theta, i \in \mathscr{C}_k)\}^{\delta_{ik}} \\ &= \prod_{k=1}^{K} \{\alpha_k \varphi(y_i|X_i\beta_k + Z_i b_{ik}, \Lambda_{ik})\varphi(b_{ik}|0, \Psi_k)\}^{\delta_{ik}}, \end{aligned}$$

where $\varphi(.)$, defined in (51.3), is the probability density function of the multivariate normal distribution. Taking the logarithm of both side, we obtain the following expression, known as the *complete data log-likelihood*:

$$\begin{aligned} l(\theta|y, b, \delta) &= \sum_{i=1}^{M} \log f(y_i, b_i, \delta_i|\theta) \\ &= \sum_{i=1}^{M} \sum_{k=1}^{K} \delta_{ik}\{\log \alpha_k + \log \varphi(y_i|X_i\beta_k + Z_i b_{ik}, \Lambda_{ik}) \\ &\quad + \log \varphi(b_{ik}|0, \Psi_k)\} \end{aligned} \quad (51.6)$$

$$= C - \frac{1}{2}\sum_{i=1}^{M}\sum_{k=1}^{K}\delta_{ik}\{-2\log\alpha_k + (\log|\Psi_k| + b'_{ik}\Psi_k^{-1}b_{ik})$$

$$+(\log|\Lambda_{ik}| + e'_{ik}\Lambda_{ik}^{-1}e_{ik})\},$$

where $C = \frac{1}{2}(qM + \sum_{i=1}^{M} n_i)\log(2\pi)$ is a constant, and $e_{ik} = y_i - X_i\beta_k - Z_i b_{ik}$.

We present here three variants of the expectation–maximization (EM) algorithm for estimating all the model parameters. The first one is a standard EM algorithm that works for any structure of the residual error covariance matrix. The second variant is a simplification of the first, for the particular case of autoregressive residual error and equally spaced fixed times: this variant leads to a substantial reduction in computing time. Lastly, the third variant is an alternative version of the EM algorithm, based on Monte Carlo sampling and valid for any structure of the residual error covariance matrix. We also discuss the methods for choosing initial values of the parameters and the number of clusters.

51.3.1 EM Algorithm

Instead of maximizing $l(\theta|y)$ directly, the EM algorithm aims to maximize the expectation of the complete data log-likelihood $E_{b,\delta}\{l(\theta|y,b,\delta)\}$. This is achieved by alternating between the E or expectation step and the M or maximization step, repeated over multiple iterations.

51.3.1.1 EM: The Standard EM Algorithm

E-Step

Let $\theta^{(s)}$ denote the value of the parameters after iteration s. Then the E-step at iteration $s+1$ involves the computation of a Q-function, $Q(\theta|\theta^{(s)}) = E_{b,\delta}\{l(\theta|y,b,\delta)|y,\theta^{(s)}\}$. Omitting the constant C in (51.6), we have:

$$Q(\theta|\theta^{(s)}) = -\frac{1}{2}\sum_{i=1}^{M}\sum_{k=1}^{K}\tau_{ik}\{-2\log\alpha_k + [\log|\Psi_k| + tr(\Psi_k^{-1}B_{ik})]$$

$$+ [\log|\Lambda_{ik}| + tr(\Lambda_{ik}^{-1}A_{ik})]\}, \quad (51.7)$$

where

$$\tau_{ik} = E\{\delta_{ik}|y,\theta^{(s)}\} = \Pr\{i \in \mathscr{C}_k|y_i,\theta^{(s)}\} = \frac{\alpha_k^{(s)}\varphi(y_i|X_i\beta_k^{(s)},\Sigma_{ik}^{(s)})}{\sum_{j=1}^{K}\alpha_j^{(s)}\varphi(y_i|X_i\beta_j^{(s)},\Sigma_{ij}^{(s)})},$$

$$A_{ik} = E\{e_{ik}e'_{ik}|y,\theta^{(s)}\} = (y_i - X_i\beta_k - Z_i\gamma_{ik})(y_i - X_i\beta_k - Z_i\gamma_{ik})'$$

$$+Z_i\Gamma_{ik}Z'_i \triangleq A_{ik}(\beta_k),$$

$$B_{ik} = E\{b_{ik}b'_{ik}|y,\theta^{(s)}\} = \gamma_{ik}\gamma'_{ik} + \Gamma_{ik},$$

and

$$\begin{aligned}
\Sigma_{ik}^{(s)} &= Z_i \Psi_k^{(s)} Z_i' + \Lambda_{ik}^{(s)}, \\
\gamma_{ik} &= E\{b_{ik}|y, \theta^{(s)}\} = \Psi_k^{(s)} Z_i' \Sigma_{ik}^{(s)^{-1}} (y_i - X_i \beta_k^{(s)}), \\
\Gamma_{ik} &= Var\{b_{ik}|y, \theta^{(s)}\} = (\Psi_k^{(s)} - \Psi_k^{(s)} Z_i' \Sigma_{ik}^{(s)^{-1}} Z_i \Psi_k^{(s)}).
\end{aligned}$$

All these quantities, and therefore the Q-function, $Q(\theta|\theta^{(s)})$, are straightforward to calculate using the above equations for given $\theta^{(s)}$.

M-Step

The M-step at iteration $s + 1$ aims to update the parameters θ by maximizing $Q(\theta|\theta^{(s)})$. Setting the first partial derivatives of the Q-function equal to zero and using the constraint $\sum_{k=1}^{K} \alpha_k = 1$ and $\Lambda_{ik} = \sigma_k^2 V(\varphi_k, n_i)$, we have:

$$\begin{aligned}
\hat{\alpha}_k &= \frac{1}{M} \sum_{i=1}^{M} \tau_{ik} \\
\hat{\beta}_k &= \left(\sum_{i=1}^{M} \tau_{ik} X_i' V_{ik}^{-1} X_i \right)^{-1} \sum_{i=1}^{M} \tau_{ik} X_i' V_{ik}^{-1} (y_i - Z_i \gamma_{ik}) \\
\hat{\Psi}_k &= \frac{\sum_{i=1}^{M} \tau_{ik} B_{ik}}{\sum_{i=1}^{M} \tau_{ik}} \\
\hat{\sigma}_k^2 &= \frac{\sum_{i=1}^{M} \tau_{ik} tr\{V_{ik}^{-1} A_{ik}(\beta_k)\}}{\sum_{i=1}^{M} \tau_{ik} n_i}
\end{aligned}$$

where $\sum_{i=1}^{M} \tau_{ik} \neq 0$ and $V_{ik} = V(\varphi_k, n_i)$. The parameters φ_k or equivalently the matrices V_{ik} can be estimated by minimizing the last term in the Q-function (51.7):

$$\xi(\varphi_k, \beta_k, \sigma_k^2) = \sum_{i=1}^{M} \tau_{ik} \{\log(\sigma_k^{2n_i} |V_{ik}|) + \sigma_k^{-2} tr(V_{ik}^{-1} A_{ik}(\beta_k))\}. \tag{51.8}$$

Substituting $\hat{\beta}_k$ and $\hat{\sigma}_k^2$ into (51.8), we have $\xi(\varphi_k, \hat{\beta}_k, \hat{\sigma}_k^2) = \sum_{i=1}^{M} \tau_{ik} \{n_i \log \hat{\sigma}_k^2 + \log |V_{ik}| + n_i\}$. Then the φ_k are estimated by

$$\hat{\varphi}_k = \arg\min\{\xi(\varphi_k) = \xi(\varphi_k, \hat{\beta}_k, \hat{\sigma}_k^2)\}.$$

The E-step and M-step are repeated until convergence. The individual or subject i is finally classified into the cluster $\hat{k}_i = \arg\max_k \{\tau_{ik}\}$.

51.3.1.2 EMAR: EM Algorithm for AR(r)

The standard EM algorithm usually converges slowly, due to the nonlinear optimization in the M-step. For an autoregressive residual error, however, the convergence can be accelerated. From (51.4) and (51.5), for autoregressive error and equally spaced fixed times, using the Cholesky decomposition, see Eq. (51.4), the covariance matrix for subject i in cluster k can be written as $V_{ik}^{-1} = L'(\varphi_k, n_i)L(\varphi_k, n_i)$. Then $\xi(\varphi_k, \beta_k, \sigma_k^2)$ of Eq. (51.8) is reduced to the following quadratic function of $\varphi_k = (\varphi_{k1}, \ldots, \varphi_{kr})$:

$$\xi(\varphi_k, \beta_k) = \sum_{i=1}^{M} \tau_{ik} tr\{V_{ik}^{-1} A_{ik}(\beta_k)\} = \sum_{u=0}^{r}\sum_{v=0}^{r} a_{uv}\varphi_{ku}\varphi_{kv},$$

where $\varphi_{k0} = 1$, and $a_{uv} = \sum_{i=1}^{M} \tau_{ik} tr\{J_{n_i}'(u)J_{n_i}(v)A_{ik}(\beta_k)\}$. Thus φ_k can be estimated as:

$$\hat{\varphi}_k = \arg\min \xi(\varphi_k, \beta_k),$$

a problem requiring only the minimization of a quadratic form, which can be performed by a standard highly efficient algorithm.

51.3.1.3 EMMC: EM Algorithm Using Monte Carlo Sampling

The indicator variables for cluster membership follow a multinomial distribution, and the probabilities of the distribution are estimated by $\tau_i = (\tau_{i1}, \ldots, \tau_{iK})$ in the E-step. Let $\delta_i^{(h)} = (\delta_{i1}^{(h)}, \ldots, \delta_{iK}^{(h)})$, $h = 1, \ldots, H$, be H samples of cluster membership indicators taken from the multinomial distribution, $Multinomial(1, \tau_i)$, where H is very a large number. Then we can use $\delta_i^{(h)}$ to replace τ_i in the Q-function (51.7), and have

$$Q_h(\theta|\theta^{(s)}) = -\frac{1}{2}\sum_{k=1}^{K} \sum_{\delta ik^{(h)}=1} \{-2\log\alpha_k + [\log|\Psi_k| + tr(\Psi_k^{-1}B_{ik})] + [\log|\Lambda_{ik}| + tr(\Lambda_{ik}^{-1}A_{ik})]\},$$

which consists of K extended linear mixed-effects (ELME) models. Let $\hat{\theta}_h = \arg\max_\theta Q_h(\theta|\theta^{(s)})$. Then $\hat{\theta}_h = (\hat{\theta}_{h,1}, \ldots, \hat{\theta}_{h,K})$ that can be estimated separately in each cluster using the efficient algorithm described in Pinheiro and Bates [19]:

$$\hat{\theta}_{h,k} = \arg\max_{\theta_k} \sum_{\delta ik^{(h)}=1} \{-2\log\alpha_k + [\log|\Psi_k| + tr(\Psi_k^{-1}B_{ik})] + [\log|\Lambda_{ik}| + tr(\Lambda_{ik}^{-1}A_{ik})]\}$$

The model parameters are then estimated by averaging over the H estimates:

$$\hat{\theta} = \frac{1}{H} \sum_{h=1}^{H} \hat{\theta}_h.$$

Since H needs to be large, the Monte Carlo based EM algorithm is computationally demanding.

51.3.2 Initial Values and Number of Clusters

The EM algorithm can be quite sensitive to the choice of starting values. A number of different strategies for choosing starting values have been proposed [14]. As Celeux et al. [3] and Ciampi et al. [4], we perform k-means clustering of regression parameters obtained from linear regressions on each individual or subject to obtain starting values.

The number of clusters in the finite mixture models may be estimated using Akaike information criterion (AIC) [1] or Bayesian information criterion (BIC) [22].

51.4 Simulations

We simulated data containing 200 individuals from four clusters that mimic four different patterns of time evolution (see Sect. 51.5): worsening, slowly worsening, slowly improving, and improving. The number of individuals in each cluster was chosen to be 30, 43, 57, and 70, respectively. The responses for individual i in cluster k were generated from the following model:

$$y_{ij} = \beta_{0k} + \beta_{1k} t_{ij} + b_{0,ij} + b_{1,ij} t_{ij} + \sigma_k^2 \epsilon_{ij},$$

where $j = 1, \ldots, n_i$, n_i is the number of measures for individual i, t_{ij} is the jth measured time point for individual i, $\beta_{s,k}$ are any fixed effects, $b_{s,ij} \sim N(0, \psi_{s,k})$ are random effects leading to cluster-specific and individual-specific patterns and correlation, and ϵ_{ij} are order 1 autoregressive AR(1). The true parameters are shown in Table 51.1.

We considered two simulation settings: (A) The measurement times are unequally spaced for each individual. The number, n_i, of observations was allowed to range from 15 to 25. (B) The measurement times are equally spaced and $n_i = 25$ for each individual. We generated 500 datasets for each simulation

Table 51.1 True parameters and average estimates of the parameters over 500 simulations assuming the true number of clusters are known, that is, $K = 4$

Method	Cluster	α	β_0	β_1	ψ_0	ψ_1	φ	σ^2
True parameters								
	1	0.150	8	−0.75	0.50	0.01	0.45	1.00
	2	0.215	6	−0.25	0.50	0.01	0.45	1.41
	3	0.285	4	0.25	0.50	0.01	0.45	1.73
	4	0.350	2	0.75	0.50	0.01	0.45	2.00
Estimated parameters								
EM	1	0.157 (0.028)	7.98 (0.231)	−0.734 (0.051)	0.45 (0.221)	0.013 (0.011)	0.447 (0.05)	1.014 (0.098)
	2	0.221 (0.024)	5.931 (0.306)	−0.223 (0.075)	0.372 (0.265)	0.015 (0.013)	0.451 (0.047)	1.438 (0.132)
	3	0.283 (0.031)	3.915 (0.267)	0.272 (0.062)	0.317 (0.267)	0.015 (0.012)	0.452 (0.041)	1.761 (0.136)
	4	0.339 (0.029)	1.957 (0.17)	0.754 (0.017)	0.398 (0.259)	0.01 (0.003)	0.451 (0.034)	2.013 (0.119)
EMAR	1	0.15 (0.024)	8.038 (0.231)	−0.739 (0.044)	0.431 (0.229)	0.013 (0.011)	0.395 (0.054)	0.847 (0.063)
	2	0.218 (0.035)	6.028 (0.3)	−0.237 (0.064)	0.335 (0.275)	0.019 (0.016)	0.4 (0.05)	1.19 (0.093)
	3	0.29 (0.039)	3.971 (0.29)	0.26 (0.063)	0.269 (0.266)	0.02 (0.016)	0.398 (0.04)	1.468 (0.081)
	4	0.343 (0.038)	1.937 (0.191)	0.748 (0.026)	0.39 (0.277)	0.012 (0.007)	0.401 (0.035)	1.691 (0.08)
EMMC	1	0.153 (0.024)	7.982 (0.219)	−0.745 (0.039)	0.49 (0.221)	0.011 (0.006)	0.448 (0.049)	1.01 (0.095)
	2	0.216 (0.016)	5.968 (0.279)	−0.243 (0.06)	0.501 (0.23)	0.01 (0.004)	0.447 (0.044)	1.417 (0.116)
	3	0.284 (0.021)	3.983 (0.234)	0.256 (0.054)	0.479 (0.208)	0.01 (0.004)	0.445 (0.038)	1.732 (0.118)
	4	0.347 (0.025)	1.994 (0.154)	0.752 (0.018)	0.485 (0.224)	0.01 (0.003)	0.449 (0.032)	1.999 (0.114)

The values in parentheses are sample standard deviations of the estimates

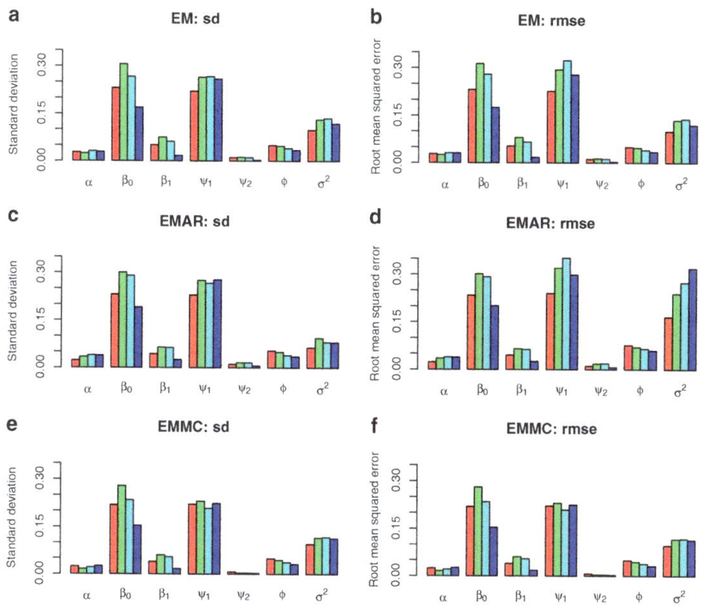

Fig. 51.1 Standard deviations and root mean squared errors of the estimated parameters over 500 simulations using EM, EMAR, and EMMC algorithms

setting, and estimate the model parameters and mixing coefficients for the simulated datasets. The number of clusters was chosen using both AIC and BIC model selection criteria.

For simulation setting (A), we estimate the parameters using algorithms: EM and EMMC. We cannot use EMAR because the times are unequally spaced. Table 51.1 shows mean and standard deviation of the estimated parameters over the simulations in which the BIC retrieves the true number of clusters. The standard deviations and mean squared errors are also shown in Fig. 51.1. It is seen that both EM and EMMC give similar results. Average run time is 6.72 h for EM and 98.26 h for EMMC with 200 samplings. The frequencies of the number of clusters selected by AIC and BIC are shown in Fig. 51.2.

For simulation setting (B), we estimated the parameters using algorithms: EMAR and EMMC. Average runtime over the 500 simulations is 0.76 h for EMAR and 5.49 h for EMMC. Table 51.1 shows mean and standard deviation of the estimated parameters over the simulations in which the BIC retrieves the true number of clusters. The standard deviations and mean squared errors are also shown in Fig. 51.1. The frequencies of the number of clusters selected by AIC and BIC are shown in Fig. 51.2.

Our simulation examples showed that BIC selects a more accurate number of clusters than AIC.

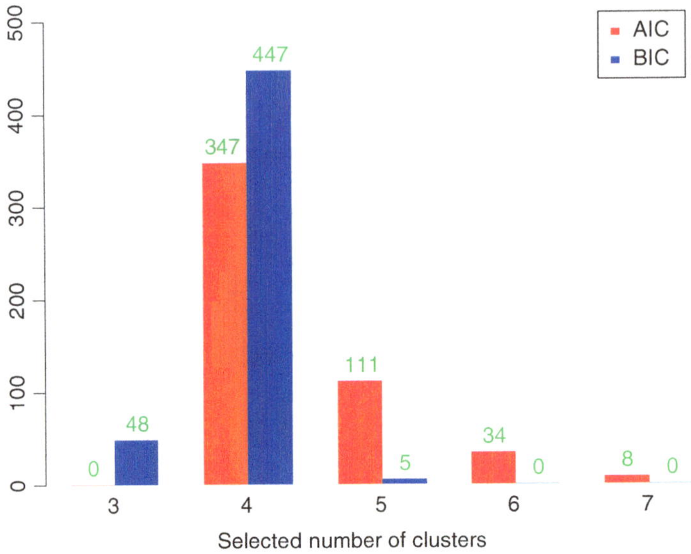

Fig. 51.2 Frequencies of cluster numbers selected by AIC and BIC for EM algorithm

51.5 Clinical Examples

Example 1. The data of this example were obtained from 99 patients with atrial fibrillation frequenting the anticoagulation clinic of a tertiary care centre between March 2001 and June 2012. The patients were treated with warfarin, an anticoagulant, in order to achieve a stable level of anticoagulation. The International Normalized Ratio (INR) was measured as a proxy for anticoagulation, and the purpose of INR monitoring was to ensure the achievement of a therapeutic INR level between 2 and 3. For each patient, the INR was measured several times, from the date of treatment initiation to a maximum of 240 days (8 months) thereafter, at regular intervals. We modeled the data as a sample from a mixture of K linear mixed models with autoregressive errors, where K was to be determined from the data. We fitted this model using the EMAR algorithm. The minimum BIC model consisted of two clusters containing 79 (80 %) and 20 (20 %) patients with parameters given in the upper panel of Table 51.2. Figure 51.3 shows typical patterns of INR trajectories. The larger cluster can be described as consisting of patients who are rapidly stabilized, with their INR remaining in a reasonable stable region of over the observation period. The smaller cluster contains patients who do not stabilize easily; some of them exhibit dangerous fluctuations in the last part of the observation period. The estimated parameters reflect the relative stability of the first cluster and the instability of the second one: error variance is lower in the first than in the second cluster, while autocorrelation, which suggests dynamic stability, is stronger in the first than in the second cluster.

Table 51.2 Estimates of model parameters and cluster number for INR and DI datasets

Dataset	Cluster	Number	α	β_0	β_1	ψ_0	φ	σ^2
INR	1	79	0.8004	−0.3166	3.2159	0.0553	0.1470	0.4914
	2	20	0.1996	1.7639	1.1251	0.0034	0.0235	1.5516
DI	1	43	0.3375	8.2806	−0.0529	6.2804	0.0910	6.5644
	2	33	0.2561	6.7320	−0.0475	10.1064	−0.0924	1.3856
	3	31	0.2579	9.9647	0.1114	10.7787	0.2549	4.3263
	4	20	0.1485	13.6804	0.0472	4.4929	−0.0718	1.1651

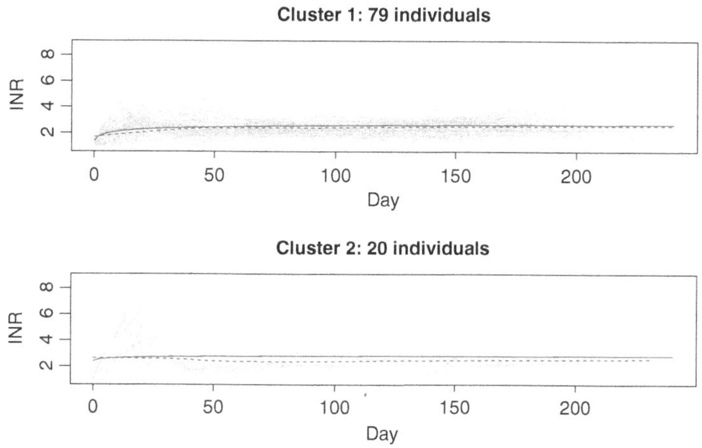

Fig. 51.3 Typical patterns of INR trajectories

Example 2. The data originated from a multi-centre clinical study in nursing homes including 127 elderly patients with delirium [12]. Delirium, a mental disorder relatively common in aging patients, was assessed using the Delirium Index, an instrument developed at St. Mary's Hospital [11]. We modeled the data as samples from a mixture of K linear mixed models with autoregressive errors. Again, times were equally spaced and fixed for all patients, so that the most convenient EM variant for parameter estimation was the EMAR algorithm. Results are summarized in the lower panel of Table 51.2 and in Fig. 51.4. The minimum BIC model consisted of four clusters. The shape of the average curves justifies naming the clusters as: *improving*, *slowly improving*, *worsening*, and *slowly worsening*. Note that we constructed our simulation models on the basis of these real data results. Both the curves and the parameter estimates reveal major difference between clusters, with the *worsening* cluster showing the largest autocorrelation and the *slowly worsening* cluster exhibiting the smallest overall variability.

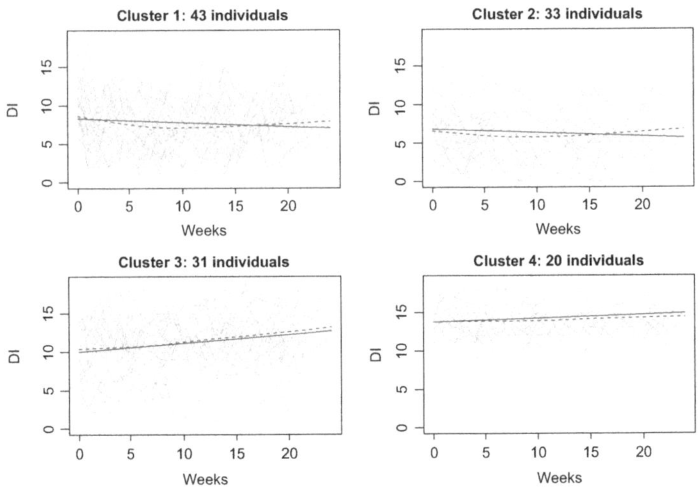

Fig. 51.4 Typical patterns of DI trajectories

51.6 Discussion

In this work we have presented a streamlined version of a general algorithm for modeling longitudinal data from a heterogeneous population. Returning to an EM algorithm proposed by Ciampi et al. [4], we have reformulated the main steps and developed two variants of EM estimation, one that aims at computational efficiency in the special case of equally spaced fixed times; the other that aims at greater generality.

The main progress reported here is the development of EMAR, a fast algorithm for fitting mixtures of ELMEMs for longitudinal data in the case of AR(r) residual errors and equally spaced fixed times. The acceleration of the EM algorithm for this special case owes to the work of McNicholas and Murphy [16], in particular, their use of the Cholesky decomposition of the residual error covariance matrix; however, this crucial step is integrated in a broader context which also includes random effects [5], so that we can simultaneously account for correlations arising from both multilevel and serial correlation features. The equally spaced fixed times case has traditionally attracted the attention of most researchers working with longitudinal data; while it is by no means the only interesting case in the applications, it will continue to occupy a central role, at least in the near future. Thus our progress, though modest in scope, does provide a new tool that may prove useful in contemporary data analysis.

The main merit of the reformulation of the standard EM algorithm rests on the generality of the mixtures of ELME models (Ciampi et al. [4]; Ji et al. [10]): it is indeed the broadest available framework for modeling longitudinal data from heterogeneous populations, as it allows for heterogeneities arising from several sources, including multilevel data structures, serial correlations, and heteroscedastic residual errors. However, the superior flexibility of the approach has yet to be fully exploited in practice.

Finally, the additional exploration of the EMMC approach shows promise for its generality, its stability, and the reasonably good statistical properties of the estimates it produces. Unfortunately, our experience continues to show that computational time cannot be reduced without resorting to parallel computing and/or superior computational facilities.

As we stated in the introduction, the theoretical advance presented in this work is modest. Our contribution here is to improve the feasibility of our general approach for fitting mixtures of ELMM to data. This has been achieved, on the one hand, by reducing the computing time for an important particular case by approximately 1/10, and, on the other, by further demonstrating the soundness of the EM–MC algorithm, a method that shows great promise for implementing EM algorithms with intractable M-step. In view of the increasing role played by longitudinal and multilevel data, we feel that it is important to produce ELMM-based reliable analytic tools that require realistic computing resources and that allow for population heterogeneity.

Future work will aim to make the flexibility of the mixture of ELMEM family more accessible in practice. We intend to proceed by developing, on the one hand, more computationally efficient algorithms for other special cases and particular submodels of the general family and, on the other, by devising novel computational tools to improve efficiency in the context of the EMMC algorithm and beyond.

Acknowledgements This work was supported by an operating grant from the Canadian Institutes of Health Research in partnership with CIHR—Institute of Aging; Public Health Agency of Canada (01228-000).

References

1. Akaike, H.: A new look at the statistical model identification. IEEE Trans. Automat. Control **19**(6), 716–723 (1974)
2. Banfield, J.D., Raftery, A.E.: Model-based Gaussian and non-Gaussian clustering. Biometrics **49**(3), 803–821 (1993)
3. Celeux, G., Martin, O., Lavergne, C.: Mixture of linear mixed models for clustering gene expression profiles from repeated microarray experiments. Stat. Model. **5**, 243–267 (2005)
4. Ciampi, A., Campbell, H., Dyachenko, A., Rich, B., McCusker, J., Cole, M.G. Model-based clustering of longitudinal data: application to modeling disease course and gene expression trajectories. Commun. Stat. Simulat. Comput. **41**(7), 992–1005 (2012)
5. De la Cruz-Mesia, R., Quintana, F.A., Marshall, G.: Model-based clustering for longitudinal data. Comput. Stat. Data Anal. **52**, 1441–1457 (2008)
6. Dempster, A.P., Laird, N.M., Rubin, D.B.: Maximum likelihood from incomplete data via the EM algorithm. J. Roy. Stat. Soc., Ser. B **39**(1), 1–38 (1977)
7. Fraley, C., Raftery, A.E.: Mclust: Software for model-based cluster analysis. J. Classif. **16**, 297–306 (1999)
8. Fraley, C., Raftery, A.E.: Model-based clustering, discriminant analysis and density estimation. J. Am. Stat. Assoc. **97**, 611–631 (2002)

9. Fraley, C., Raftery, A.E.: Model-based methods of classification: using the mclust software in chemometrics. J. Stat. Softw. **18**(6) (2007)
10. Ji, Y., Ciampi, A., Tagalakis, V.: Modified em algorithms for model-based clustering of longitudinal data. In: Golubi, A. (ed.), Proceedings COMPSTAT2012, pp. 367–378. Curran Associates, Inc. (2013)
11. McCusker, J., Cole, M., Dendukuri, N., Belzile, E.: The delirium index, a measure of the severity of delirium: new findings on reliability, validity, and responsiveness. Am. Geriatr. Soc. **52**(10), 1744–1749 (2004)
12. McCusker, J., Cole, M., Voyer, P., Monette, J., Champoux, N., Ciampi, A., Vu, M., Belzile, E.: Prevalence and incidence of delirium in longterm care. Int. J. Geriatr. Psychiat. 1152–1161 (2011)
13. McLachlan, G., Basford, K.: Mixture Models: Inference and Applications to Clustering. Marcel Dekker, New York (1988)
14. McLachlan, G., Peel, D.: Finite Mixture Models. Wiley, New York (2000)
15. McLachlan, G.J., Peel, D., Bean, R.W.: Modelling high-dimensional data by mixtures of factor analyzers. Comput. Stat. Data Anal. **41**, 379–388 (2003)
16. McNicholas, P.D., Murphy, T.B.: Parsimonious Gaussian mixture models. Stat. Comput. **18**(3), 285–296 (2008)
17. McNicholas, P.D., Murphy, T.B. Model-based clustering of longitudinal data. Can. J. Stat. **38**(1), 153–168 (2010)
18. Pauler, D.K., Laird, N.M.: A mixture model for longitudinal data with application to assessment of noncompliance. Biometrics **56**, 464–472 (2000)
19. Pinheiro JC, Bates DM (2000) Mixed-Effects Models in S and S-PLUS. Springer, New York
20. Pourahmadi, M.: Joint mean-covariance models with applications to longitudinal data: unconstrained parameterisation. Biometrika **86**(3), 677–690 (1999)
21. R Development Core Team R: A Language and Environment for Statistical Computing. R Foundation for Statistical Computing, Vienna, Austria (2013). URL http://www.R-project.org, ISBN:3-900051-07-0
22. Schwarz, G.: Estimating the dimension of a model. Ann. Stat. **6**(2), 461–464 (1978)
23. Titterington, D.M., Smith, A.F.M., Makov, U.E.: Statistical Analysis of Finite Mixture Distributions. Wiley, Chichester (1985)

Chapter 52
The Study of the Laplace Transform of Marshall–Olkin Multivariate Exponential Distribution

Igor V. Zolotukhin

52.1 Introduction

Requirement of the aftereffect absence for all coordinates of a random vector $Z = (Z_1, Z_2, \ldots, Z_k)$ implies that the vector is composed of independent exponentially distributed marginal components. Generalization of conditions for absence of aftereffect proposed by Marshall and Olkin [2] is as follows:

$$\mathsf{P}\left(Z_1 > z_1 + z,\ Z_2 > z_2 + z,\ \ldots,\ Z_k > z_k + z \Big/_{z_1 > z,\ z_2 > z,\ \ldots,\ z_k > z}\right)$$
$$= \mathsf{P}\left(Z_1 > z_1,\ Z_2 > z_2,\ \ldots,\ Z_k > z_k\right),$$
$$\forall z > 0,\ z_1 > 0,\ z_2 > 0,\ \ldots,\ z_k > 0 \qquad (52.1)$$

As shown by Marshall and Olkin, this holds if and only if

$$\bar{F}(z_1, z_2, \ldots, z_k) = \mathsf{P}(Z_1 > z_1,\ Z_2 > z_2,\ \ldots,\ Z_k > z_k)$$
$$= \exp\left[-\sum_{\varepsilon \in \mathcal{E}} \lambda_\varepsilon \max_{1 \leqslant i \leqslant k} \{\varepsilon_i z_i\}\right],\ z_i \geqslant 0, \qquad (52.2)$$

Here and below $\lambda_\varepsilon \geqslant 0$ are distribution parameters, $\mathcal{E} = \{\varepsilon\}$ is a set of k-dimensional indices $\varepsilon = (\varepsilon_1, \ldots, \varepsilon_k)$, each coordinate ε_i is equal to 0 or 1.

Following [1, ğ 2.2], we call $\bar{F}$ the reliability function. In particular, $\bar{F}(z) = \exp(-\lambda z)$ is the reliability function of the exponential distribution.

I.V. Zolotukhin (✉)
Russian Academy of Sciences, Institute of Oceanology, St.Petersburg, Russia
e-mail: igor.zolotukhin@gmail.com

V.B. Melas et al. (eds.), *Topics in Statistical Simulation*, Springer Proceedings in Mathematics & Statistics 114, DOI 10.1007/978-1-4939-2104-1_52

The distribution (52.2) is called the multivariate exponential Marshall–Olkin distribution (M–O). As in [2], we will denote MVE(λ_ε, $\varepsilon \in \mathcal{E}$) this class of distributions.

The case $k = 2$ only (BVE distribution in the notation of the authors) is discussed in detail in [2]. For this case, the distribution function is found. It is shown that it contains both an absolutely continuous and a singular parts. The moment generating function (Laplace transform) has been calculated.

In the multivariate case, the definition of MVE was given. It was noted that the distribution contains a singular component, and to find the Laplace transform of MVE is extremely difficult.

We present the following results:

- The reliability function of the projections of $\boldsymbol{Z}$ on any coordinate hyperplane is found. It is shown that all of these projections also have MVE distribution. The special operation helps to determine the parameters of such distributions using the parameters of the vector $\boldsymbol{Z}$ has been proposed.
- The explicit expression for the Laplace transform of the MVE distribution has been found.
- The formula to calculate the Laplace transform of the distribution of the projection of $\boldsymbol{Z}$ on an arbitrary coordinate hyperplane also has been found.

52.2 Main Results

We introduce the following notation.

Further the vector ε is used for the indication of coordinate hyperplane in the k-dimensional space.

We recall that ε is a k-dimensional vector $\varepsilon = (\varepsilon_1, \dots, \varepsilon_k)$, coordinates of which ε_i is equal to 0 or 1. Let $\mathbf{x} = (x_1, \dots, x_k)$ be the k-dimensional real vector.

Let $(\varepsilon, \mathbf{x})$ be the scalar product of vectors ε and $\mathbf{x}$. Let $\varepsilon\mathbf{x}$ be their coordinate-wise product. In other words, it is a vector whose coordinates are formed by multiplying the corresponding coordinates of the factors.

$$\bar{\delta} = \mathbf{1} - \delta, \; ||\delta|| = \sum_{i=1}^{k} \delta_i.$$

Examples. Let $\delta = (0, 1, 0)$ and $\mathbf{x} = (3, 4, 5)$.

We then have $(\delta, \mathbf{x}) = 4$; $||\delta|| = 1$; $\bar{\delta} = (1, 0, 1)$ and $\bar{\delta}\mathbf{x} = (3, 0, 5)$.

Then $\bar{F}(\varepsilon z)$ is the reliability function of projection of $\boldsymbol{Z}$ on coordinate hyperplane ε.

Now let's set the partial order relation in the set $\mathcal{E}$:

$$\forall \varepsilon, \delta \in \mathcal{E} \; \delta \leqslant \varepsilon, \text{ if for all } i \; \delta_i \leqslant \varepsilon_i;$$

$$\delta < \varepsilon \text{ if } \delta \leqslant \varepsilon \text{ and } \delta \neq \varepsilon.$$

At last, the sign "$\bullet$" would mean summation λ for the corresponding coordinate.

Examples. $\lambda_{1\bullet 0} = \lambda_{100} + \lambda_{110}$; $\lambda_{1\bullet\bullet} = \lambda_{100} + \lambda_{101} + \lambda_{110} + \lambda_{111}$; $\lambda_{\bullet \ldots \bullet} = \sum_{\varepsilon \in \mathcal{E}} \lambda_{\varepsilon}$.

And let us define the vector $\varepsilon \oplus \delta$, whose coordinates are calculated according to the following rule:

$$1 \oplus 0 = 1$$

$$0 \oplus 1 = 1$$

$$1 \oplus 1 = 0$$

$$0 \oplus 0 = \bullet$$

Let us take in consideration also the vectors ε, whose coordinates can take three values: 0, 1, $\bullet$. For these vectors we define the $\bar{\varepsilon}$ as follows:

$$\bar{\varepsilon}_j = 0 \text{ if } \varepsilon_j = 1; \ \bar{\varepsilon}_j = 1 \text{ if } \varepsilon_j = 0; \ \bar{\varepsilon}_j = \bullet \text{ if } \varepsilon_j = \bullet.$$

Example. Let $\delta = (0, 0, 1)$ and $\varepsilon = (1, 0, 1)$. Then $\overline{\varepsilon \oplus \delta} = (0, \bullet, 1)$.

Equation (52.2) can be written as

$$\bar{F}(z) = \bar{F}(\mathbf{1}\, z) = \exp\left(-\sum_{\delta \leqslant \mathbf{1}} \lambda_{\overline{\mathbf{1} \oplus \delta}} \max\{\delta z\}\right).$$

Theorem 2.

$$\forall\, \varepsilon \in \mathcal{E} \quad \bar{F}(\varepsilon\, z) = \exp\left(-\sum_{\delta \leqslant \varepsilon} \lambda_{\overline{\varepsilon \oplus \delta}} \max\{\delta z\}\right)$$

Remark 1. It follows that $\varepsilon \boldsymbol{Z} \in \text{MVE}(\lambda_{\overline{\varepsilon \oplus \delta}}, \ \delta \leqslant \varepsilon)$, so that the projection of random vector $\boldsymbol{Z}$ for any coordinate hyperplane also distributed according to Marshall–Olkin.

Particularly if $\varepsilon = (1, 0, \ldots, 0)$ $\bar{F}(\varepsilon z) = \bar{F}(z_1) = \exp(-\lambda_{1\bullet \ldots \bullet} z_1)$, and $\varepsilon \boldsymbol{Z}$ is distributed exponentially with parameter $\lambda_{1\bullet \ldots \bullet}$.

Let us denote

$$\psi(s) = \psi(s_1, \ldots, s_k) = \mathsf{E}\, e^{-s\boldsymbol{Z}} = \int_0^{\infty} \ldots \int_0^{\infty} \prod_{i=1}^{k} e^{-s_i z_i}\, dF(z_1, \ldots, z_k)$$

the Laplace transform of the distribution of MVE.

Theorem 3. *For any* $\boldsymbol{Z} \in MVE(\lambda_\varepsilon,\ \varepsilon \in \mathcal{E})$ *Laplace transform of its distribution is given by*

$$\psi(s) = \mathsf{E}\, e^{-s\boldsymbol{Z}} = \frac{1}{(\mathbf{1},\ s) + \lambda_{\bullet\ldots\bullet}} \sum_{\varepsilon\in\mathcal{E}} \lambda_{\bar{\varepsilon}}\, \psi(\varepsilon s). \tag{52.3}$$

Remark 2. Let the random variable X has an exponential distribution with parameter $\lambda_{\bullet\ldots\bullet}$, and vector $\boldsymbol{X} = (X,\ \ldots,\ X)$.

Laplace transform of such vector $\boldsymbol{X}$ is

$$\Psi(s) = \mathsf{E}\, e^{-s\boldsymbol{X}} = \frac{\lambda_{\bullet\ldots\bullet}}{(\mathbf{1},\ s) + \lambda_{\bullet\ldots\bullet}}.$$

Assuming that $p_\varepsilon = \dfrac{\lambda_\varepsilon}{\lambda_{\bullet\ldots\bullet}}$, this formula can be rewritten as

$$\psi(s) = \Psi(s) \sum_{\varepsilon\in\mathcal{E}} p_\varepsilon\, \psi(\bar{\varepsilon} s).$$

Hence an **MVE** distribution is the discrete mixture of the distribution $\boldsymbol{X}$ and its convolutions with projections of this **MVE** distribution on all its coordinate hyperplane.

Remark 3. In the bivariate case, the known expression for $\psi(s_1,\ s_2)$ found by Marshall–Olkin [2] coincides with (52.3) after some transformations.

Indeed, using the notation introduced above $\psi(s_1,\ s_2)$ can be written as:

$$\psi(s_1,\ s_2) = \frac{\lambda_{1\bullet}\,\lambda_{\bullet 1}}{(\lambda_{1\bullet} + s_1)(\lambda_{\bullet 1} + s_2)} + \frac{\lambda_{11}\, s_1\, s_2}{(\lambda_{\bullet\bullet} + s_1 + s_2)(\lambda_{1\bullet} + s_1)(\lambda_{\bullet 1} + s_2)},$$

this can lead directly to the form

$$\begin{aligned}\psi(s_1,\ s_2) &= \frac{1}{\lambda_{\bullet\bullet} + s_1 + s_2}\left[\lambda_{11} + \lambda_{01}\frac{\lambda_{1\bullet}}{(\lambda_{1\bullet} + s_1)} + \lambda_{10}\frac{\lambda_{\bullet 1}}{(\lambda_{\bullet 1} + s_2)}\right] \\ &= \frac{1}{\lambda_{\bullet\bullet} + s_1 + s_2}\Big[\lambda_{11} + \lambda_{01}\psi(s_1,\ 0) + \lambda_{10}\psi(0,\ s_2)\Big],\end{aligned}$$

where $\psi(s_1,\ 0) = \dfrac{\lambda_{1\bullet}}{(\lambda_{1\bullet} + s_1)}$, $\psi(0,\ s_2) = \dfrac{\lambda_{\bullet 1}}{(\lambda_{\bullet 1} + s_2)}$ are the Laplace transforms of (one-dimensional) exponential distributions for corresponding projections.

Theorem 4.

$$\forall\, \varepsilon \in \mathcal{E} \quad \psi(\varepsilon s) = \frac{1}{\sum_{\delta<\varepsilon} \lambda_{\delta\oplus\varepsilon} + (\varepsilon,\ s)} \sum_{\delta<\varepsilon} \lambda_{\delta\oplus\varepsilon}\psi(\delta s) \tag{52.4}$$

Remark 4. Since $\delta \oplus \mathbf{1} = \delta$, and $\lambda_{\mathbf{0}} = 0$, we can see that Theorem 3 is a special case of Theorem 4 if $\varepsilon = \mathbf{1}$.

Remark 5. It follows that the Laplace transform of the vector $\boldsymbol{Z}$ projection to coordinate hyperplane ε can be found by Theorem 3, but to do this we need to replace all the zeros on the "bullets" in the indices of all parameters λ.

Thus, when $\varepsilon = (1, 0, \ldots, 0)$

$$\psi(\varepsilon s) = \psi(s_1) = \frac{\lambda_{1\bullet\ldots\bullet}}{\lambda_{1\bullet\ldots\bullet} + s_1}.$$

52.3 Proofs

Proof of Theorem 2. We have $\bar{F}(\varepsilon z) = \exp\left(-\sum_{\delta\in\mathcal{E}} \lambda_\delta \max\{\delta\varepsilon z\}\right)$. Since $\forall\delta \in \mathcal{E}$ $\delta\varepsilon \leqslant \varepsilon$, we can summarize the first $\delta \leqslant \varepsilon$, and then by all γ such that $\gamma\varepsilon = \delta$. Note that if $\delta \leqslant \varepsilon$, then $\delta\varepsilon = \delta$.

$$\bar{F}(\varepsilon z) = \exp\left(-\sum_{\delta\leqslant\varepsilon} \max\{\delta z\}\left(\sum_{\gamma:\gamma\varepsilon=\delta} \lambda_\gamma\right)\right)$$

It is easy to see that for $\delta \leqslant \varepsilon$

$$\sum_{\gamma:\gamma\varepsilon=\delta} \lambda_\gamma = \lambda_{\overline{\varepsilon\oplus\delta}} \tag{52.5}$$

In fact,

if $\varepsilon_j = 1,\ \delta_j = 1$, then $\gamma = 1$,
if $\varepsilon_j = 1,\ \delta_j = 0$, then $\gamma = 0$,
if $\varepsilon_j = 0,\ \delta_j = 0$, then $\gamma = 0$ or $\gamma = 1$.

Such a rule exactly corresponds operations $\overline{\varepsilon \oplus \delta}$.

Proof of Theorem 3. In the proof we use the following representation of the random vector $\boldsymbol{Z} \in \text{MVE}(\lambda_\varepsilon,\ \varepsilon \in \mathcal{E})$. Let $\mathcal{E}_i$ $(i=1, \ldots, k)$—the set of indices ε, in which at ith position is 1; X_ε—independent exponentially distributed random variables with parameters $\lambda_\varepsilon \geqslant 0$. We assume $X_\varepsilon = +\infty$, if $\lambda_\varepsilon = 0$. The coordinates of the vector $\boldsymbol{Z} = (Z_1, \ldots, Z_k)$ let's define by the equality

$$Z_i = \min_{\varepsilon\in\mathcal{E}_i}\{X_\varepsilon\}. \tag{52.6}$$

Note that $Z_i \in E\Big(\sum_{\varepsilon\in\mathcal{E}_i}\lambda_\varepsilon\Big)$ (by the well-known property of the exponential law). Here, $E(\lambda)$ is the class of exponential laws. The reliability function for vector $\boldsymbol{Z}$ calculated by the formula (52.2), so that $\boldsymbol{Z} \in \mathrm{MVE}(\lambda_\varepsilon,\ \varepsilon \in \mathcal{E})$.

Let's denote W the minimum of random variables Z_i. Taking into account (52.6), we obtain $W = \min Z_i = \min\limits_i \left(\min\limits_{\delta\in\mathcal{E}_i}\{X_\delta\}\right) = \min\limits_{\delta\in\mathcal{E}} X_\delta$.

According to the addition formula

$$\mathsf{E}\, e^{-(s,\boldsymbol{Z})} = \sum_{\varepsilon\in\mathcal{E}} \mathsf{E}(e^{-(s,\boldsymbol{Z})};\ W = X_\varepsilon). \tag{52.7}$$

Here, following [1, ğ 4.2], we use the notation $\mathsf{E}(\xi;\ B) = \int\limits_B \xi(\omega)\mathsf{P}(d\omega)$.

By conditional expectation property

$$\begin{aligned}
&\mathsf{E}\left(e^{-(s,\boldsymbol{Z})};\ W = X_\varepsilon\right)\\
&= \mathsf{E}_{X_\varepsilon}\mathsf{E}\left(e^{-(s,\boldsymbol{Z})};\ W = X_\varepsilon,\ X_\delta \geqslant X_\varepsilon,\ \delta \in \mathcal{E}\backslash\{\varepsilon\}\right)\\
&= \int_0^\infty e^{-\lambda_\varepsilon x}dx \cdot \mathsf{E}\left(e^{-(s,\boldsymbol{Z})};\ X_\varepsilon = x,\ X_\delta \geqslant x,\ \delta \in \mathcal{E}\backslash\{\varepsilon\}\right)\\
&= \int_0^\infty e^{-\lambda_\varepsilon x}dx \cdot \mathsf{E}\left(e^{-(s,\varepsilon\boldsymbol{Z})}\, e^{-(s,\bar\varepsilon\boldsymbol{Z})};\ X_\varepsilon{=}x,\ X_\delta{\geqslant}x,\ \delta{\in}\mathcal{E}\backslash\{\varepsilon\}\right).
\end{aligned} \tag{52.8}$$

Let $\chi = (x,\dots,x)$, then, obviously, for $X_\varepsilon = x$ $\varepsilon\boldsymbol{Z} = \varepsilon\chi$,

$$\begin{aligned}
&\mathsf{E}(e^{-(s,\bar\varepsilon\boldsymbol{Z})};\ X_\varepsilon = x,\ X_\delta \geqslant x,\ \delta \in \mathcal{E}\backslash\{\varepsilon\})\\
&\qquad = e^{-x(s,\varepsilon)}\,\mathsf{E}\left(e^{-(s,\bar\varepsilon\boldsymbol{Z})};\ X_\delta \geqslant x, \delta \in \mathcal{E}\backslash\{\varepsilon\}\right).
\end{aligned} \tag{52.9}$$

Event $\{X_\delta \geqslant x,\ \delta \in \mathcal{E}\backslash\{\varepsilon\}\}$ is the intersection of independent events $\{X_\delta \geqslant x,\ \delta\bar\varepsilon > 0\}$ è $\{X_\delta \geqslant x,\ \delta \in \mathcal{E}\backslash\{\varepsilon\}\backslash\{\delta\bar\varepsilon > 0\}\}$. The vector $\bar\varepsilon\boldsymbol{Z}$ is independent of the second one.

Therefore

$$\begin{aligned}
&\mathsf{E}\left(e^{-(s,\bar\varepsilon\boldsymbol{Z})};\ X_\delta \geqslant x,\ \delta \in \mathcal{E}\backslash\{\varepsilon\}\right)\\
&\quad = \mathsf{P}(X_\delta{\geqslant}x,\ \delta \in \mathcal{E}\backslash\{\varepsilon\}\backslash\{\delta\bar\varepsilon{>}0\}) \cdot \mathsf{E}\left(e^{-(s,\bar\varepsilon\boldsymbol{Z})};\ X_\delta{\geqslant}x,\ \delta\bar\varepsilon{>}0\right)\\
&\quad = \exp\left(-x \cdot \sum_{\delta\in\mathcal{E}\backslash\{\varepsilon\}\backslash\{\delta\bar\varepsilon>0\}} \lambda_\delta\right)\mathsf{E}\left(e^{-(s,\bar\varepsilon\boldsymbol{Z})};\ \bar\varepsilon\boldsymbol{Z} \geqslant \bar\varepsilon\chi\right).
\end{aligned} \tag{52.10}$$

Let us show that

$$\mathsf{E}(e^{-(s,\bar{\varepsilon}\mathbf{Z})};\ \bar{\varepsilon}\mathbf{Z} \geqslant \bar{\varepsilon}\chi) = e^{-x\left[(\varepsilon,s)+\sum\limits_{\delta\bar{\varepsilon}>0}\lambda_\delta\right]} \cdot \mathsf{E}\, e^{-(s,\bar{\varepsilon}\mathbf{Z})} \tag{52.11}$$

Indeed, property (52.1) for distribution MVE follows

$$\bar{F}(\varepsilon(z+\chi)) = \bar{F}(\varepsilon z + \varepsilon\chi) = \bar{F}(\varepsilon z)\,\bar{F}(\varepsilon\chi), \tag{52.12}$$

$$\bar{F}(\varepsilon\chi) = \mathsf{P}(\varepsilon\mathbf{Z}>\varepsilon\chi) = \mathsf{P}\left(\bigcup_{\delta\varepsilon>0}\{X_\delta>x\}\right) = \exp\left(-x\sum_{\delta\varepsilon>0}\lambda_\delta\right), \tag{52.13}$$

$$E(e^{-(s,\bar{\varepsilon}\mathbf{Z})};\ \varepsilon\mathbf{Z} \geqslant \varepsilon\chi) = \int_0^\infty \cdots \int_0^\infty e^{-(s,\,\varepsilon(z+\chi))}\, dF\,(\varepsilon(z+\chi))\,. \tag{52.14}$$

As for any integrable function $g(z)$

$$\int_0^\infty \cdots \int_0^\infty g(z)dF(\varepsilon z) = (-1)^{\|\varepsilon\|} \int_0^\infty \cdots \int_0^\infty g(z)d\,\bar{F}(\varepsilon z),$$

then using (52.12)–(52.14)

$$\mathsf{E}(e^{-(s,\bar{\varepsilon}\mathbf{Z})};\ \varepsilon\mathbf{Z} \geqslant \varepsilon\chi) = e^{-x\left[(s,\varepsilon)+\sum\limits_{\delta\varepsilon>0}\lambda_\delta\right]}\, E\, e^{-(s,\varepsilon\mathbf{Z})}.$$

Now go substitute (52.11) into (52.10) and (52.10) into (52.9) and (52.9) into (52.8), and integrate. Substitute the result in (52.7) and obtain the desired result.

Proof of Theorem 4. The proof is along the same lines as Theorem 3. We set

$$W_\varepsilon = \min_{i:\varepsilon_i=1} \varepsilon\mathbf{Z} = \min_{\delta\varepsilon>0}\{X_\delta\}.$$

We have

$$\begin{aligned}
\mathsf{E}\, e^{-(s,\varepsilon\mathbf{Z})} &= \sum_{\delta\varepsilon>0} \mathsf{E}\left(e^{-(s,\varepsilon\mathbf{Z})};\ W_\varepsilon = X_\delta\right) \\
&= \sum_{\delta\varepsilon>0} \int_0^\infty \lambda_\delta e^{-x\lambda_\delta} dx \cdot \mathsf{E}\left(e^{-(s,\varepsilon\mathbf{Z})};\ X_\delta = x,\ X_\gamma \geqslant x,\ \gamma\varepsilon > 0,\ \gamma \neq \delta\right).
\end{aligned}$$

If $X_\delta = x, \quad \delta\varepsilon \boldsymbol{Z} = \delta\varepsilon\chi$ we get

$$\mathsf{E}\, e^{-(s,\varepsilon \boldsymbol{Z})} = \sum_{\delta\varepsilon>0} \int_0^\infty \lambda_\delta e^{-x\lambda_\delta} dx \cdot e^{-x(s,\varepsilon\delta)} \cdot$$
$$\cdot\, \mathsf{E}\left(e^{-(s,\bar{\delta}\varepsilon \boldsymbol{Z})};\ X_\gamma \geqslant x, \gamma\varepsilon > 0,\ \gamma \neq \delta\right). \tag{52.15}$$

But

$$\begin{aligned}
&\mathsf{E}\left(e^{-(s,\bar{\delta}\varepsilon \boldsymbol{Z})};\ X_\gamma \geqslant x,\ \gamma\varepsilon > 0,\ \gamma \neq \delta\right) \\
&= \mathsf{P}\left(X_\gamma \geqslant x;\ \gamma \in \{\gamma\varepsilon > 0\}\backslash\{\delta\}\backslash\{\gamma\varepsilon\bar{\delta} > 0\}\right) \cdot \\
&\quad \cdot\, \mathsf{E}\left(e^{-(s,\bar{\delta}\varepsilon \boldsymbol{Z})};\ X_\gamma \geqslant x,\ \gamma\varepsilon > 0,\ \gamma\varepsilon\bar{\delta} > 0\right) \\
&= \exp\left(-x \sum_{\gamma\in\{\gamma\varepsilon>0\}\backslash\{\delta\}\backslash\{\gamma\varepsilon\bar{\delta}>0\}} \lambda_\gamma\right) \cdot \mathsf{E}\left(e^{-(s,\bar{\delta}\varepsilon \boldsymbol{Z})};\ X_\gamma \geqslant x,\ \gamma\varepsilon\bar{\delta} > 0\right) \\
&= \exp\left(-x \sum_{\gamma\in\{\gamma\varepsilon>0\}\backslash\{\delta\}\backslash\{\gamma\varepsilon\bar{\delta}>0\}} \lambda_\gamma + x\left[(\varepsilon\bar{\delta},\, s) + \sum_{\gamma\varepsilon\bar{\delta}>0} \lambda_\gamma\right]\right) \mathsf{E}\, e^{-(s,\bar{\delta}\varepsilon \boldsymbol{Z})} \\
&= \exp\left(-x\left[(\varepsilon\bar{\delta},\, s) + \sum_{\gamma\in\{\gamma\varepsilon>0\}\backslash\{\delta\}} \lambda_\gamma\right]\right) \mathsf{E}\, e^{-(s,\bar{\delta}\varepsilon \boldsymbol{Z})}.
\end{aligned} \tag{52.16}$$

Substituting (52.16) into (52.15) and integrating, we obtain

$$\mathsf{E}\, e^{-(s,\varepsilon \boldsymbol{Z})} = \frac{1}{\sum_{\delta:\delta\varepsilon>0} \lambda_\delta + (s,\varepsilon)} \sum_{\delta\varepsilon>0} \lambda_\delta\, E e^{-(s,\bar{\delta}\varepsilon \boldsymbol{Z})}. \tag{52.17}$$

It remains to give (52.17) to (52.4). We'll do it in stages. First, making the substitution $\beta = \gamma\varepsilon$, such that $0 < \beta = \gamma\varepsilon \leqslant \varepsilon$, and taking into account (52.5), we write

$$\sum_{\delta\varepsilon>0} \lambda_\delta = \sum_{0<\beta\leqslant\varepsilon} \left(\sum_{\gamma\varepsilon=\beta} \lambda_\delta\right) = \sum_{0<\beta\leqslant\varepsilon} \lambda_{\overline{\beta\oplus\varepsilon}}. \tag{52.18}$$

When replacing in (52.18) $\beta' = \varepsilon - \beta$ (hereinafter signs subtraction and addition of the two-digit index means conventional subtraction and addition modulo 2)

$0 < \beta \leqslant \varepsilon \Leftrightarrow 0 \leqslant \beta' < \varepsilon$, $\beta \oplus \varepsilon = \beta' \oplus \varepsilon$ and

$$\sum_{\delta\varepsilon>0} \lambda_\delta = \sum_{0\leqslant\gamma<\varepsilon} \lambda_{\gamma\oplus\varepsilon}. \tag{52.19}$$

Similarly,

$$\sum_{\delta\varepsilon>0} \lambda_\delta \psi(\bar{\delta}\varepsilon s) = \sum_{0<\beta\leqslant\varepsilon} \left(\sum_{\delta\varepsilon=\beta} \lambda_\delta \psi(\bar{\delta}\varepsilon s) \right).$$

But $\delta\varepsilon = \beta \Leftrightarrow \bar{\delta}\varepsilon = \varepsilon - \beta$, as $\bar{\delta}\varepsilon + \delta\varepsilon = \mathbf{1}\varepsilon = \varepsilon$.

Hence

$$\sum_{\delta\varepsilon>0} \lambda_\delta \psi(\bar{\delta}\varepsilon s) = \sum_{0<\beta\leqslant\varepsilon} \psi((\varepsilon-\beta)s) \sum_{\varepsilon\delta=\beta} \lambda_\beta = \sum_{0<\beta\leqslant\varepsilon} \lambda_{\overline{\beta\oplus\varepsilon}} \psi((\varepsilon-\beta)s).$$

By replacing the $\beta' = \varepsilon - \beta$ we reduce the last equality to the form

$$\sum_{\delta\varepsilon>0} \lambda_\delta \psi(\bar{\delta}\varepsilon s) = \sum_{0\leqslant\beta'<\varepsilon} \lambda_{\varepsilon\oplus\beta'} \psi(\beta'\varepsilon). \tag{52.20}$$

Substituting (52.19) and (52.20) into (52.17) yields (52.4).

References

1. Borovkov, A.A.: Probability Theory (in Russian). Nauka, Moscow (1986)
2. Marshall, A.W., Olkin, I.A.: A multivariate exponential distribution. J. Am. Stat. Assoc. **62**(317), 30–44 (1967)

CPSIA information can be obtained at www.ICGtesting.com
Printed in the USA
LVOW01*1717091214

418018LV00001B/16/P

9 781493 921034